MW01599018

Information Sources in

Pharmaceuticals

Guides to Information Sources

A series under the General Editorship of
D. J. Foskett, MA, FLA
and
M. W. Hill, MA, BSc, MRIC

This series was known previously as 'Butterworths Guides to Information Sources'.

Other titles available are:

Information Sources in Grey Literature
  edited by C. P. Auger

Information sources in Metallic Materials
  edited by M. N. Patten

Information Sources in the Earth Sciences (Second edition)
  edited by David N. Wood, Joan E. Hardy and
  Anthony P. Harvey

Information Sources in Polymers and Plastics
  edited by R. T. Adkins

Information Sources in Energy Technology
  edited by L. J. Anthony

Information Sources in the Life Sciences
  edited by H. V. Wyatt

Information Sources in Physics (Second edition)
  edited by Dennis F. Shaw

Information Sources in Law
  edited by R. G. Logan

Information Sources in Management and Business
(Second edition)
  edited by K. D. C. Vernon

Information Sources in Politics and Political Science: a survey
worldwide
  edited by Dermot Englefield and Gavin Drewry

Information Sources in Engineering (Second edition)
  edited by L. J. Anthony

Information Sources in Economics (Second edition)
  edited by John Fletcher

Information Sources in the Medical Sciences (Third edition)
  edited by L. T. Morton and S. Godbolt

Information Sources in
Pharmaceuticals

Editor
W. R. Pickering

BOWKER·SAUR
London ● Melbourne ● Munich ● New York

British Library Cataloguing in Publication Data

Information sources in pharmaceuticals.
 1. Pharmacy. Information sources
 I. Pickering, W. Roy   II. Series
 615.4'07

ISBN 0-408-02518-2

Library of Congress Cataloging-in-Publication Data

Information sources in pharmaceuticals / editor, W. Roy Pickering.
 584p.   22 cm.—(Guides to information sources)
 Includes bibliographical references.
 ISBN 0-408-02518-2
 1. Pharmacy—Information services—Directories.
 2. Pharmaceutical industry—Information
services—Directories.   I. Pickering W. Roy.
 II. Series: Guides to information sources (London, England)
 [DNLM: 1. Drugs. 2. Information Services.   QV4 I42]
 RS56.I54   1989
 615'.19—dc20
 DNLM/DLC
 for Library of Congress   89-25128
                                                              CIP

Bowker-Saur is part of the Professional Publishing Division of Reed International Books, Borough Green, Sevenoaks, Kent TN15 8PH

Cover design by Calverts Press
Printed on acid-free paper
Printed and bound in Great Britain by
Biddles Ltd, Guildford and King's Lynn

Series Editors' Foreword

Daniel Bell has made it clear in his book *The Post-Industrial Society* that we now live in an age in which information has succeeded raw materials and energy as the primary commodity. We have also seen in recent years the growth of a new discipline, information science. This is in spite of the fact that skill in acquiring and using information has always been one of the distinguishing features of the educated person. As Dr Johnson observed, 'Knowledge is of two kinds. We know a subject ourselves, or we know where we can find information upon it.'

But a new problem faces the modern educated person. We now have an excess of information, and even an excess of sources of information. This is often called the 'information explosion', though it might be more accurately called the 'publication explosion'. Yet it is of a deeper nature than either. The totality of knowledge itself, let alone of theories and opinions about knowledge, seems to have increased to an unbelievable extent, so that the pieces one seeks in order to solve any problem appear to be but a relatively few small straws in a very large haystack. That analogy, however, implies that we are indeed seeking but a few straws. In fact, when information arrives on our desks, we often find those few straws are actually far too big and far too numerous for one person to grasp and use easily. In the jargon used in the information world, efficient retrieval of relevant information often results in information overkill.

Ever since writing was invented, it has been a common practice for men to record and store information; not only fact and figures, but also theories and opinions. The rate of recording accelerated after the invention of printing and moveable type, not because that in itself could increase the amount of recording but because, by making it easy to publish multiple copies of a document and sell them at a profit, recording and distributing information became very lucrative and hence attractive to more people. On the other hand, men and women in whose lives the discovery of the handling of information plays a large part usually devise ways of getting what they want from other people rather than from books in their efforts to avoid information overkill. Conferences, briefings, committee meetings are one means of this; personal contacts through the 'invisible college' and members of one's club are another. While such people do read, some of them voraciously, the reading of published literature, including in this category newspapers as well as books and journals and even watching television, may provide little more than 10% of the total information that they use.

Computers have increased the opportunities, not merely by acting as more efficient stores and providers of certain kinds of information than libraries, but also by manipulating the data they contain in order to synthesize new information. To give a simple illustration, a computer which holds data on commodity prices in the various trading capitals of the world, and also data on currency exchange rates, can be programmed to indicate comparative costs in different places in one single currency. Computerized data bases, i.e. stores of bibliographic information, are now well established and quite widely available for anyone to use. Also increasing are the number of data banks, i.e. stores of factual information, which are now generally accessible. Anyone who buys a suitable terminal may be able to arrange to draw information directly from these computer systems for their own purposes; the systems are normally linked to the subscriber by means of the telephone network. Equally, an alternative is now being provided by information supply services such as libraries, more and more of which are introducing terminals as part of their regular services.

The number of sources of information on any topic can therefore be very extensive indeed; publications (in the widest sense), people (experts), specialist organizations from research associations to chambers of commerce, and computer stores. The number of channels by which one can have access to these vast collections of information are also very numerous, ranging from professional literature searchers, via computer intermediaries, to Citizens' Advice Bureaux, information marketing services and information brokers.

The aim of the Guides to Information Sources (formerly Butterworths Guides to Information Sources) is to bring all these sources and channels together in a single convenient form and to present a picture of the international scene as it exists in each of the disciplines we plan to cover. Consideration is also being given to volumes that will cover major interdisciplinary areas of what are now sometimes called 'mission-oriented' fields of knowledge. The first stage of the whole project will give greater emphasis to publications and their exploitation, partly because they are so numerous, and partly because more detail is needed to guide them adequately. But it may be that in due course the balance will change, and certainly the balance in each volume will be that which is appropriate to its subject at the time.

The editor of each volume is a person of high standing, with substantial experience of the discipline and of the sources of information in it. With a team of authors of whom each one is a specialist in one aspect of the field, the total volume provides an integrated and highly expert account of the current sources, of all types, in its subject.

D. J. Foskett
Michael Hill

About the Contributors

Theodora Andrews

Is Professor of Library Science at Purdue University, US, and has been a visiting lecturer at several other universities in the United States over the years. She has played a leading role in various Libraries Association groups with special emphasis on pharmacy. As a writer, she has produced several books and articles but is perhaps best known as the author of *Guide to the Literature of Pharmacy and the Pharmaceutical Sciences* which was published in 1986.

Michael Archer

Graduated in Chemistry from Sheffield University, England and worked in the cosmetic industry before specializing in information work. Since 1982 he has headed the Information Service at Beecham's Development and Production site at Worthing.

Coyla Barry

Has held various posts in universities and industry since qualifying in Physiology and in Information Science. Currently she heads a literature section at Burroughs Wellcome in North Carolina and is responsible for the retrieval of both scientific and business information.

Simon Bell

Qualified both in Development Studies and in Information Science and now works in the School of Development Studies at the

University of East Anglia, England. He continues to work in Ethiopia, Nigeria and India and acts as a consultant on information handling and provision. One of his main interests is the problem of microtechnology usage in developing countries. He modestly disclaims expertise and prefers to be regarded as an academic with a deep interest in information issues in Africa.

Ed Collins

Is a pharmacist with qualifications from the Universities of Texas and Houston, US. He is currently Head of the Drug Information Service at Burroughs Wellcome in North Carolina. Before assuming this position, he practised pharmacy in the hospital environment for a number of years. He is very active in the drug information field and has written extensively both as a researcher and an information specialist.

Anita Crafts-Lighty

Is Managing Director of Biocommerce Data Ltd, a company she founded in 1986. After graduating from California Institute of Technology, she completed a PhD at Cambridge University and then filled senior information posts in industry before establishing her own company. Dr Crafts-Lighty is a well-known author and editor in the field of commercial biotechnology information and contributed the major work *Information Sources in Biotechnology* in 1986.

Tony Dayan

Studied Pharmacology and Medicine at the London Hospital before specializing in Neuropathology. After gaining experience in hospitals and industry, Professor Dayan took charge of the Department of Toxicology at St Bartholomew's Hospital. He has served on several British and European Community Advisory Committees concerned with pharmaceuticals, industrial products and agrochemicals.

Jonathon de Pass

Qualified in Medicine at Bristol University, England, in 1979 but has been a Healthcare Analyst since 1982. Dr de Pass is now Director of Barclays de Zoete Wedd Research Ltd.

Pauline Duckitt

Is Managing Director of Vital Information Ltd. She qualified from Cambridge University in Natural Sciences and City University,

England in Information Science. Subsequently, she worked in industry and a professional association before establishing her own business. She has remained active professionally and is currently a member of the Library and Information Sciences Advisory Council in the UK.

John Dunne

Is currently Programme Manager, Pharmaceuticals Unit, WHO in Geneva. He was educated at the London Hospital and subsequently at John Hopkins Hospital, Baltimore, and the University of California. Dr Dunne lectured at the London Hospital Medical College and held the post of Principal Medical Officer in the Department of Health, UK, before moving to Geneva. His publications include works on clinical trials design and the investigation of adverse drug reactions.

Tamara Eisenschitz

Originally qualified and did research in Theoretical Physics but turned to Information Science. Before being appointed by City University, England, she worked briefly as a patent searcher. Dr Eisenschitz has written extensively on patent issues and the problems of technology transfer and has published a book on intellectual property.

Yitai Gong

Is Director of the Shanghai Documentation and Information Centre in the Chinese Academy of Sciences, where he has worked for many years. His degrees include a BA in English Language and an M Phil in Information Science. He is active in Chinese library matters and has written extensively, mainly about bibliometrics.

Samantha Gurnah

Graduated in Biochemistry and has worked subsequently in research, university libraries and industry before moving into marketing and, in 1988, IMS International. She is currently Support and Development Manager for the international database, MIDAS.

Ivor Harrison

Is a qualified pharmacist and lecturer in Law and Practice of Pharmacy at the Welsh School of Pharmacy, University of Wales. He has contributed many articles in the field and is author of the three-volume reference work *Law on Medicines*, published in

1986. Dr Harrison is particularly interested in the problems of drug misuse and has acted as WHO consultant to several governments on supply and use of psychotropic drugs.

Nancy Hewison

Is Assistant Professor of Library Science at Purdue University, US, having qualified at Simmons College, Massachusetts, and then having held posts in university libraries at Oregon and Boston. She is active in Library Association matters and writes regularly about online searching.

Theresa Kot

Qualified as a pharmacist at the University of Sydney, Australia, and practised in the community and hospitals before turning to information work. Having previously worked in a poisons information centre, she was appointed Director of the New South Wales Drug Information Centre in 1979.

Sandy Mullen

Obtained his doctorate in Organic Chemistry at Glasgow University, Scotland, but has worked in West Germany at Bayer AG since 1982. He is responsible for information activites at the Pharmaceutical Research Centre and has developed a particular interest in trend analysis based on economics-related R&D information. Dr Mullen has written and spoken frequently on chemistry and information science and is currently Vice-President of the Pharma Dokumentationsring eV.

Ngaire Pettit-Young

Graduated from the University of New South Wales, Australia, in Librarianship and worked as a reference librarian for a few years before joining the NSW Drug Information Centre. She is actively involved in the medical libraries network, Gratis.

Roy Pickering

Qualified as a Biologist at Exeter University, England, before joining the Wellcome Foundation Ltd. He has been in charge of the company's central R&D Information function for some years and has served on various information management committees, including the Library and Information Services Advisory Council (UK) and the Board of the European organization Pharma Dokumentationsring eV. He is particularly interested in the problems of information management brought about by the application of technology.

Christopher Powell

Has a doctorate in Toxicology from the University of Surrey, England. Dr Powell now lectures at St Bartholomew's Hospital and continues research into toxic mechanisms.

Felix Rozanski

Is a lawyer educated at the Universities of Buenos Aires, Argentina, and Pittsburgh, US. Formerly a university professor he now co-ordinates the activities of the Study Centre for the Development of the Argentinian Pharmaceutical Industry (CEDIQUIFA) in Buenos Aires, a position which requires him to regularly contribute articles to journals and newspapers.

To Sasakawa

Studied at Keio University, Japan, qualifying in Library and Information Science. Since 1971 he has been employed by Yamanouchi Pharmaceutical Co. in Tokyo. He has written numerous papers, mainly about library matters and literature · databases, and has been chairman of the Japanese RINGDOC Users Group.

Rolly Simpson

Qualified at the Universities of Oglethorpe and Emory in the US. After working with medical units at the Georgia Institute of Technology and the University of Tennessee, he joined Burroughs Wellcome Co. in 1973 and is now head of the Product Literature Section.

Chanti Sundaram

Qualified in Medicine in Ireland and worked in general medicine before specializing in experimental toxicology at St Bartholomew's Hospital. Apart from lecturing, Dr Sundaram continues to work in the fields of carcinogensis and pathology.

Hugh Tilson

Is Director of the Division of Epidemology, Information and Surveillance at Burroughs Wellcome Co. in North Carolina. He obtained his MD at Washington School of Medicine and completed his training at several other American universities. He was an intern at Yale-New Haven Medical Centre before specializing in Public Health at Harvard. Dr Tilson held numerous senior posts in academic and public health before joining BW Co. in 1981. He is a prolific writer and a well-known public speaker.

Nico Van Putte

Is Head of the Scientific Information Department of Organon International, Holland. He studied Medicine at the University of Groningen before joining Organon. Dr Van Putte was President of the Pharma Dokumentationsring eV, an organization of European information managers, from 1982 to 1987, and remains active in European information matters generally.

Sandra Ward

Graduated in Chemistry at Durham University, England, and completed her PhD at Bath. She has spent most of her working life in the Pharmaceutical industry and is currently Director of Information Services for Glaxo Research Ltd. Dr Ward is a leading figure in the information profession and serves on numerous senior representative committees and boards. A Fellow of the Institute of Information Scientists, she received the 1987 IIS Award for services to information science.

Sue Wilder

Is Library Assistant in Purdue University's Pharmacy Library. She graduated at the University of Evansville, Indiana, but intends to qualify further in librarianship.

Contents

Introduction

W. R. PICKERING

The pronounced and often strange effects of various naturally occurring substances when administered to the human person have been known for a very long time. This book does not concern itself with the more bizarre drugtaking practices of primitive or civilized man but it is worth pausing to consider that the recognition of the remedial properties exhibited by some of these substances probably had its roots in this rather weird form of experimentation. Medicines evolved through a form of selection based on the observation of apparent benefit coupled with the absence of serious or lasting poisoning. This form of selection, unlike Darwinism, does not necessarily guarantee fitness for purpose; over the years, closer examination of remedial potions using more sophisticated methods has revealed hitherto unsuspected and worrying side-effects.

Much of what follows in this book is concerned with the careful measurement and comparison of the beneficial and harmful effects of candidate medicines. It is generally the case, although not infallibly, that potions are neither all good nor all bad but it is frequently pointed out that even seemingly innocuous substances can cause problems under certain circumstances and in sufficient quantity. The increasing awareness of possible side-effects inevitably caused concern amongst the more knowledgeable but, unhappily, it took a tragedy on the scale of the foetal abnormalities associated with thalidomide to bring matters to a head. There is now very stringent legislation administered through authoritative bodies and expert committees, which protects the consumer.

The high costs associated with drug development are due in part to the difficulty of discovering novel therapeutic activity but largely to the extended testing of candidate materials and the resulting high failure rate. There are no absolute rules governing whether a novel treatment should be given to patients: the decisions by regulatory and medical authorities are based upon the severity of the target disease, the alternative treatments available and the assessment of risk and likely benefit. Different criteria would be applied to, say, a therapy for AIDS or for cancer compared with one for the common cold or for pimples. The importance of soundly based information to these decision makers hardly needs stressing.

So far the use of the term 'pharmaceuticals' has been carefully avoided because it needs definition. The drugs referred to above would fall within its boundaries but it is more convenient for our purposes to extend the definition to embrace all remedial preparations produced by what has now come to be known as 'The Pharmaceutical Industry'. Thus items such as vaccines, antisera, diagnostics and other biologicals would be included but hygiene products and surgical equipment would not.

The biologicals sector of the industry arose later in time when the principles of the body's defence mechanisms were unravelled and exploited mainly during the last half-century. The burgeoning field of biotechnology now promises to reconcile the chemical and biological arms of pharmaceutical activity. Focusing on the industry in this way does not detract from the research and development undertaken elsewhere in universities and institutions but the fact remains that in view of the expenditure involved, industry provides the major outlet.

The pharmaceutical field is research-based and wide-ranging scientifically. Those engaged in this research are both consumers and generators of information and the contribution of industrial scientists to the general advancement of science is substantial, a fact which is sometimes overlooked. In view of the highly competitive market situation, however, pharmaceutical companies are careful to protect their intellectual property since this leads eventually to their profitability. Ironically, the principal means of accomplishing this, the patents system, was conceived originally as a way of transmitting information. The increasing collaboration between the academic and industrial sectors is managed through legal agreements and/or licensing arrangements on a mutual benefit basis and this practice can lead to further constraints on the free flow of information.

Most research work in pharmaceuticals is targetted, not only in a scientific sense but also commercially. As a consequence, there is a substantial and continuing demand for information concerning regional disease profiles, health care systems, market sizes, regulatory constraints, competitor activities, sales statistics and the like. A pharmaceutical company, in summary, tries to minimize wasted research and development effort and protect the knowledge it has acquired whilst, at the same time, ensuring that the conduct of its work conforms with international standards and is geared to major health problems. If the return on invested capital and labour is inadequate, the organization cannot survive in capitalist systems such as prevail in most Western countries. The principles are clearly different in socialist states but fewer pharmaceuticals arise from these sources.

The necessity of retrieving information from external sources has led to increased expenditure by the industry, particularly during the 'Information Explosion' of recent years. Coupled with the high costs of generating and managing information in-house, this rate of expenditure means that the industry spends very freely on information handling. The scale of the problem is such that there is no effective alternative to mechanization and it is perhaps one of the irritating aspects of the Information Age that the early perceptions for most people are that it is all about high technology and electronics. Such perceptions have adopted the wrong perspective since what the Information Age is (or should be) about is harnessing knowledge, improving the management of information and making it more generally available so that better informed decisions can be made for universal benefit.

The technology, although providing the pipework, pales into insignificance by comparison with the political and organizational problems which seem to dog international projects on this scale. Having said that, there is no doubt that information technology (IT) has had a profound impact in the developed world and will continue to extend its influence. The pharmaceutical industry was one of the first to benefit and is beginning to learn how to apply the techniques in-house with the prospect of further enhancing its performance. Companies are sufficiently close-knit organizationally to accommodate integrated information systems but there is still a long way to go.

Information flow within a pharmaceutical company can be visualized as a process of absorbing information from outside, adding to it by virtue of the company's own efforts and channelling the enriched information to protect and nurture developing products and, finally, releasing it outside once more to promote the safe and effective use of licensed products. The process continues long after marketing through post-marketing surveillance, a developing science concerned with monitoring the performance of pharmaceutical preparations after product launching. However carefully pharmaceuticals are tested prior to release, clinical trials are very limited in scope compared with the size and nature of the eventual consumer population. The regulatory authorities and all reputable companies are firmly committed to product monitoring in a controlled and co-ordinated manner.

It should be apparent by now even to those unfamiliar with the pharmaceuticals business that the industry generates information and disseminates it to all sectors of the community: researchers, advisers, prescribers, dispensers and the general public. To some extent this happens spontaneously but there is now in place a regulatory framework, anchored in the developed world but gradually extending elsewhere, which ensures that key information is obtained and provided. Regulatory authorities establish minimum acceptable standards and have to be satisfied: they are not all as demanding as, say, the Food and Drug Administration (FDA) of USA and standards do vary. Nevertheless, together with international bodies such as the World Health Organization (WHO), they afford a world-wide control over the marketing and use of pharmaceutical preparations. The WHO exists to try to establish and maintain global standards and guidelines but, inevitably, the role becomes increasingly concerned with the transfer of benefits to the less developed regions.

There is then an implicit responsibility on the part of the industry, as the prime generator of pharmaceutical information, to make this information generally available. There is, naturally, a reluctance to share secrets of potential commercial value with others but, once a product reaches the market, the need to safeguard intellectual property diminishes. In fact, there is at that time a desire to publish information and establish a sales platform by presenting the product in its most favourable light.

Most companies nowadays have a very clear policy about the maximum that can be claimed for a product and the minimum that can be withheld. Part II of the book considers these matters in more detail and addresses the issue of whether the industry does more to promote effective flow of information than is decreed by statute and commercial pressures.

Several of the conclusions of the Brandt Commission Report (1980) come to mind at this point, notably, the emphasis placed on the mutual dependence of developed and the less developed countries (LDCs) and the recommendation of large transfer of resources to LDCs. The prospect of increased trade alone should be sufficient to motivate industry, leaving aside the moral sense of obligation that many working in the industry feel. Nevertheless, it has to be recognized that pharmaceuticals are 'big business' and that products and the information to support them are mainly geared to major markets with an ability to pay. The sponsorship of orphan drugs in the US is a recognition that 'big diseases' in poor markets need attention and the recent moves of WHO towards manufacture might, in the future, provide outlets for preparations which companies find difficult to justify commercially.

The movement of information follows products in order to support them and, in general terms, if a region has advanced communications technology and can afford to buy in commercially available information, they can have access to a wide range of databases and supporting services such as libraries. There is, perhaps, a problem in less developed societies where, even if the technology dropped into place, the population would not be able to use it. Educational programmes take time and, in the meantime, information has to be delivered in a form which is in keeping with the needs and capabilities of the society.

Unhappily, disease does not respect these principles and problems and prospers relentlessly. The stark fact is that many parts of the world are inadequately served by advanced medicines and the necessary supporting information for their safe and effective use. But does the term 'safe' require redefinition; presumably different criteria apply if a population is under-nourished and disease-ridden? But who decides, and how? In simple terms, there is a prevailing imbalance brought about by two contradictory and powerful influences: on the one hand, the need for drugs and information in different parts of the world is determined by the geography of disease whilst, on the other, the generation of drugs and information is largely conditioned by socioeconomic factors. The net result is an emphasis on the diseases of the developed world. This is an effect which is well-recognized and much has been done in an attempt to counteract the widening of what has been termed the North-South divide. Realistically, one has to question whether much can be achieved in the face of such mighty forces. In a sense, this is a theme running through the book and there is an element of self-examination.

Two perspectives are presented: how those generating and disseminating information see their role and judge their success compared with the viewpoint of those who are primarily receivers and users of information. Hardly surprisingly, there are found to be differences of perception. No discussion on information can leave aside IT but the means of providing information in one society may be totally inappropriate in another and the high expectation for sophisticated data at a personal level in the most advanced societies is certainly not reflected world-wide.

The regions represented in this book have been selected to give a wide representation of requirements and achievements. They show a correspondingly broad range of political attitudes and solutions and residual problems. It must be stressed, however, that the views expressed are those of the authors. Even if strongly held personal convictions and interpretations are discounted, one is left with the feeling that there is a case for positive thought and action on a number of fronts. The demand for scientific and technical information depends upon the status of pharmaceutical science in the region but all regions without exception need information about the safe and effective use of medicines. In this latter respect, there is a responsibility on the part of the developed world as the main provider of medicines and, hence, of the pharmaceutical industry itself. WHO has a major role not only in improving information flow but also in sponsoring the discovery and manufacture of drugs needed in the poorer parts of the world.

The book is structured in three parts. The first deals in detail with the information that is necessary at all stages of pharmaceutical work – from the discovery of an active agent through development to its assessment as an effective remedial agent and its eventual release as a finished product conforming with all standards that society expects. This section finishes with an account of the work which continues after product release and the use of information thus gathered. The next section considers how the companies, national associations and the WHO (as the major relevant international body) perceive and fulfil their responsibilities. The third and final section examines the information scenarios in seven major regions of the world, as they influence the retrieval and use of pharmaceutical information. Because of the universal importance of drug information, the disease profiles and health care systems and relevant socioeconomic factors in the regions are outlined alongside descriptions of IT in place and the extent and nature of information available.

Readers familiar with the subject may find it surprising that Part I does not contain a chapter on medical information and, indeed, consideration was given to the merits of such a contribution. Medically qualified personnel are extensively involved both in the development and use of pharmaceutical agents and the pharmaceutical industry has recruited medical staff increasingly over the past decade or so. Since the main outlet for ethical products is also the medical practitioner, there can be no doubting the pivotal position of the medical profession in the field of pharmaceuticals and, as a consequence, the need for medical information. The reasons for not reviewing what is clearly a topic of fundamental importance is the existence of Morton & Godbolt's *Information Sources in the Medical Sciences*, a companion volume to this one, and interested readers are referred to this excellent work. Furthermore, much of what follows here is directly relevant to the information requirements of medical staff in their key roles of assessing the need for new products, investigating candidate materials in man and assuring the efficacy and safety of marketed agents. It was decided, therefore, that selective use of the two volumes should more than adequately meet the demands of most medical information departments and a separate chapter was considered unnecessary.

Those contributing to books of this nature have few pecuniary expectations, they do so because they are prepared to put considerable time and effort into making their knowledge and experience available to others. It has been a privilege and a pleasure to work with such a distinguished panel of contributors over the past few years. It only remains to express the hope that readers will enjoy and profit from the results of this collective labour.

Acknowledgements

My thanks are due to the contributors for their enthusiasm and willingness to adapt to the overall requirements of the book even when this involved them in extra effort. I should also like to express my gratitude to Douglas Foskett, Michael Hill, Ann Bell and Geraldine Turpie in their various editorial capacities for helpful comments and advice; to Susan Demuth who undertook the unenviable task of indexing without final text to hand until the last moment; and, finally but not least, to Sheila Blackburn for her expert secretarial assistance.

PART I

CHAPTER ONE

Chemical and physicochemical information

A MULLEN

Introduction

During the industrial revolution, Western society changed quickly from one based on an agrarian, handicraft economy to one dominated by industry and engineering. Today's innovations in communications, electronics and computers have brought about a further revolution which, coupled with the information explosion in science, has led us into the Information Age.

This chapter discusses the information requirements of the chemist in pharmaceutical research and development and presents the major sources available to him. The primary, secondary and tertiary literature will be explored while incorporating new developments and trends in each section. Chemical structure searching, reaction retrieval as well as non-structural chemistry also form focal points. Computerization is making constant inroads in all facets of the information scene, facilitating a better optimization of available information resources – these developments are reflected throughout.

The information needs of the chemist

Before looking more closely at the specific information requirements of the chemist, occupying a central role in pharmaceutical research and development, it would perhaps be advisable to address the question of what his actual objectives are and how efficient access to information can support his activities. The frequently quoted meagre success rate of one commercial drug resulting from ca. 10,000 compounds, synthesized and tested, from which arises the estimate of $ 125–140 million to develop a typical

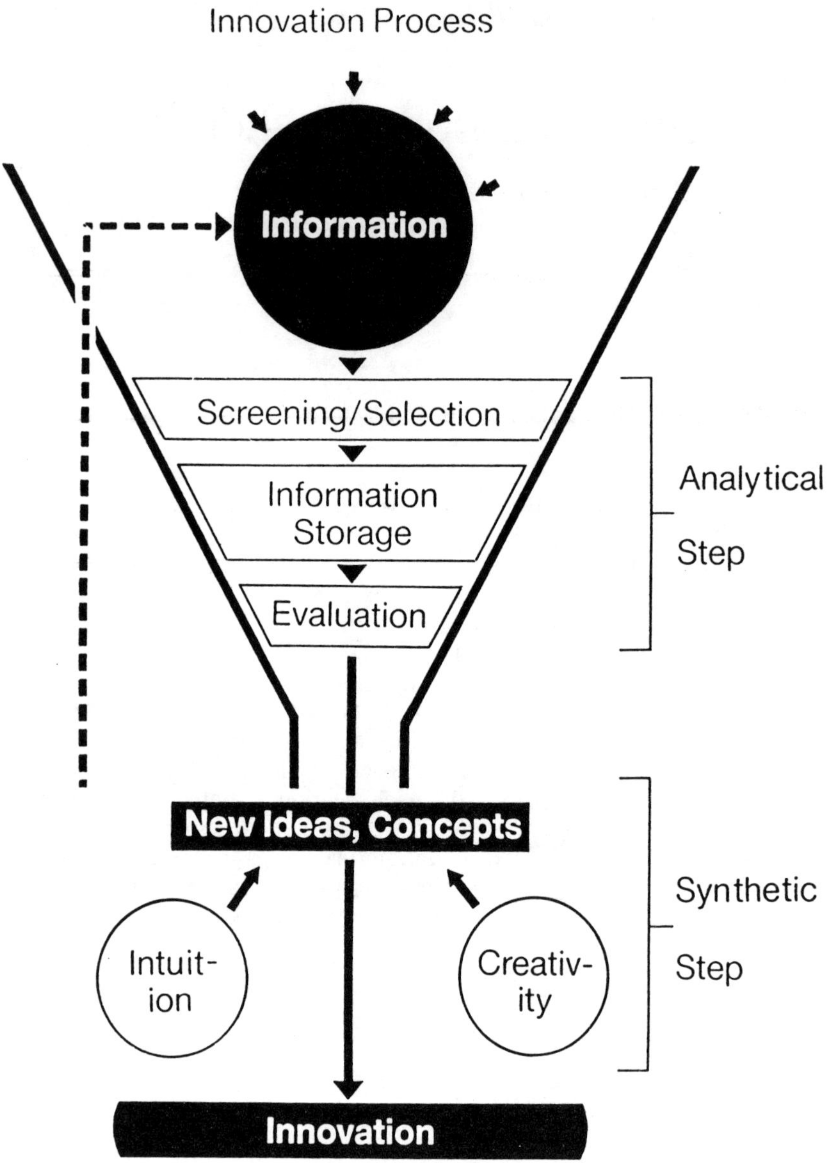

Innovation Process

Figure 1.1 The innovation process

product, are familiar figures to anyone acquainted with pharmaceutical research and development.

These statistics coupled with increasingly long development times, due in part to the stringent demands of registration authorities, often mean that 12 or more years can elapse between drug discovery and product launch (Eichin, 1988).

Speeding up this process and making more efficient use of resources must offer substantial benefits. Consequently, utilizing available information sources to make research and development work more effective and perhaps even more innovative are twin objectives for information technology.

However, the problem is not (thanks in part to the computerization of chemistry-based information) accessing sufficient information, but one of upgrading the material and extracting the relevant components. Information itself is useless unless it is applied to problem solving and to decision-support thereby catalysing the innovation process.

Available information makes a significant contribution to the innovation process as depicted in *Figure 1.1* (Möller, Mullen, Blunck, 1983). The analytical step involves the gathering of information, followed by selection and storage of relevant material and finally by its evaluation or upgrading. In *Figure 1.1*, the filter funnel represents this continual upgrading of information. New ideas and concepts can also lead to the need for more information to complete the cycle, which then, coupled with intuition/creativity, leads to the innovative step.

It cannot be claimed that information alone can increase creativity or generate new significant inventions. However, properly prepared and presented, information can facilitate and stimulate different approaches. The role of organizational framework in supporting creativity has recently been discussed (Bawden, 1986; Roberts 1988). Nevertheless, creativity and intuition – these remain the domain of man.

All the factors shown in *Figure 1.2* can contribute to the initiation of a new research product. Reliable, relevant, well packaged information is an essential ingredient (component) in all aspects of this process.

Once the need for a research programme has been identified, a state of the art review is normally commissioned. Any structure/activity relationships of compounds known to be active against the target disease as well as likely modes of action may generate new ideas/concepts. Molecular biology's contribution is increasing as a more rational approach to drug discovery gains credibility. If no biologically active substances or target receptor sites are known, the chemist traditionally resorts to selecting a wide range of

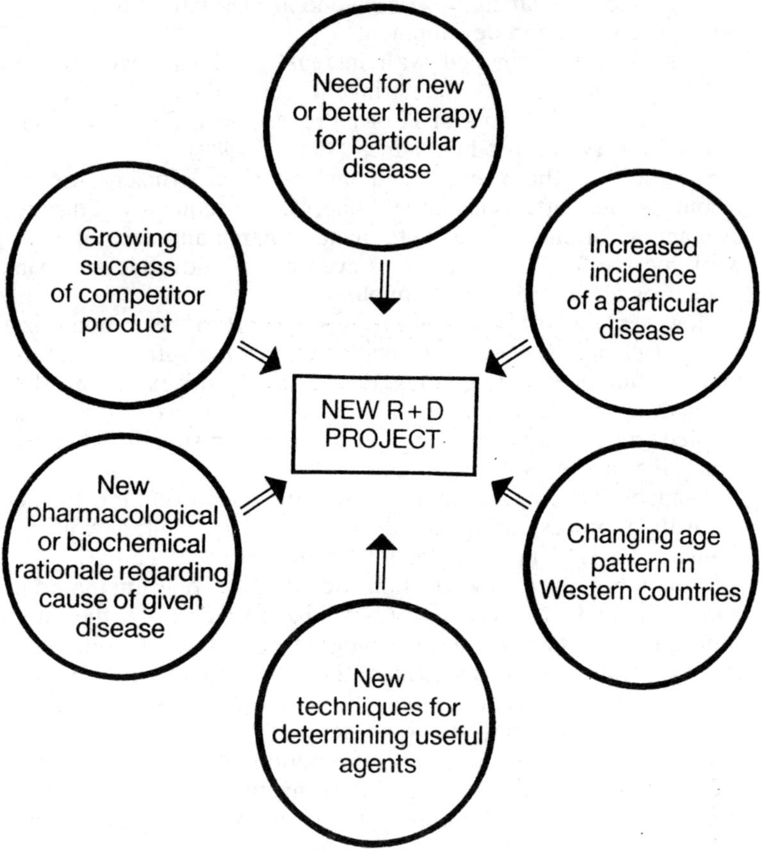

Figure 1.2 Driving forces behind development of a new drug

compounds which are then subjected to random biological screening.

In recent years, computerized tools such as molecular modelling have been used by the chemist to design and optimize his target molecules more efficiently (Roques, 1987; Unger, 1987).

Information technology will facilitate this process by reducing duplicated effort, by assisting in decision making through improved information storage and retrieval, preventing information overload and the integration of relevant information sources. All in all, the chances of success should be improved.

Growth in number of known chemical substances

Chemistry is in the very fortunate position of being the best documented branch of science, partly due to the fact that discrete, definable entities are involved.

It has been estimated that by 1830 around 2,000 individual chemical compounds had been identified. By 1860 (Kekule's benzene structure theory: 1865) this figure had increased to 3,000 and, by 1883 to 20,000. By 1919 the figure had expanded to 180,000. By 1935 ca. 350,000 and by 1964 ca. 1,280,000 substances are reported to have been known (Luckenbach, 1986; Zass, Rehm, 1985). As can be appreciated from *Figure 1.3*, there has been a dramatic increase in the number of substances registered since the inception of the Chemical Abstracts Registry Number in 1965.

At the end of 1988, the online CAS Registry file encompassed almost 9.5 million discrete compounds. This figure includes some substances registered as part of the pre-1965 registration project. Consequently, the total number of compounds currently known probably lies at around 11 million and is expanding at a rate of over 500,000 per annum. This expansion rate is equivalent to one new chemical compound being reported every minute of the year.

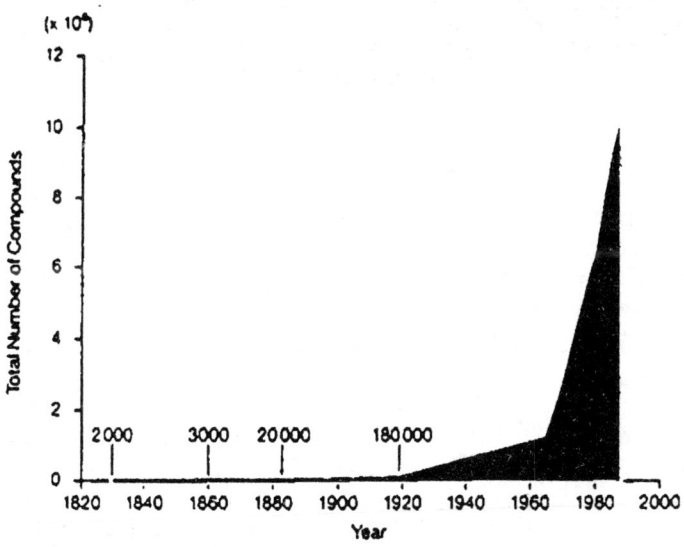

Figure 1.3 Growth in the number of known chemical substances, 1830–1987

The plethora of structures stemming from claimed Markush structures in patents are not included. To handle data on this scale defies manual methods and the situation can only be mastered using computer technology.

Expansion of chemical literature

The chemical literature relating to the above compounds is reported to be doubling in size every 8–10 years. In 1987, ca. 480,000 publications were reported in *Chemical Abstracts (CA)*, extracted from 12,000 different sources including periodicals, patents and research reports. By contrast, the number of scientific journals generally has been doubling only every 15 years (Economist, 1987).

However, the ca. 14,000 journals which report original chemical research are not equally significant. Based on citation studies it has been shown that the majority of the important contributions were distributed in the following manner (Wolman, 1983): 25% in 90 journals, 50% in 450 journals, 75% in 1,600 journals. Consequently, there are estimated to be between 1,500–2,000 important chemical journals.

DEVELOPMENTS IN CHEMICAL DOCUMENTATION

There has always been an appreciation by chemists of the value of the literature due to the fact that records of chemical experiments have a lasting value.

In 1817, the Gmelin Handbook (*Handbuch der theoretischen Chemie*) appeared in Frankfurt. This treatise represented the inception of chemical documentation, presenting as it did a medium for the selection, evaluation and accessible storage of chemical information. This basic work has continued in the inorganic area and now appears in English as the *Handbook of Inorganic Chemistry*. Gmelin's counterpart in organic chemistry, Beilstein's *Handbook of Organic Chemistry*, was first published in 1882.

In 1850 ca. 500 chemistry-based publications were appearing annually; this figure grew to over 10,000 by 1910 to reach the current level of over 500,000 per annum today. Due to the growth in the number of publications in chemistry it became necessary at an early stage to introduce an abstract journal. Therefore, in 1836, some 30 years before Kekule's benzene theory, the *Pharmaceutisches Centralblatt* was born, to be renamed in 1856 the *Chemisches Zentralblatt* which continued publication, producing good quality German-language abstracts of chemical publications, until 1969.

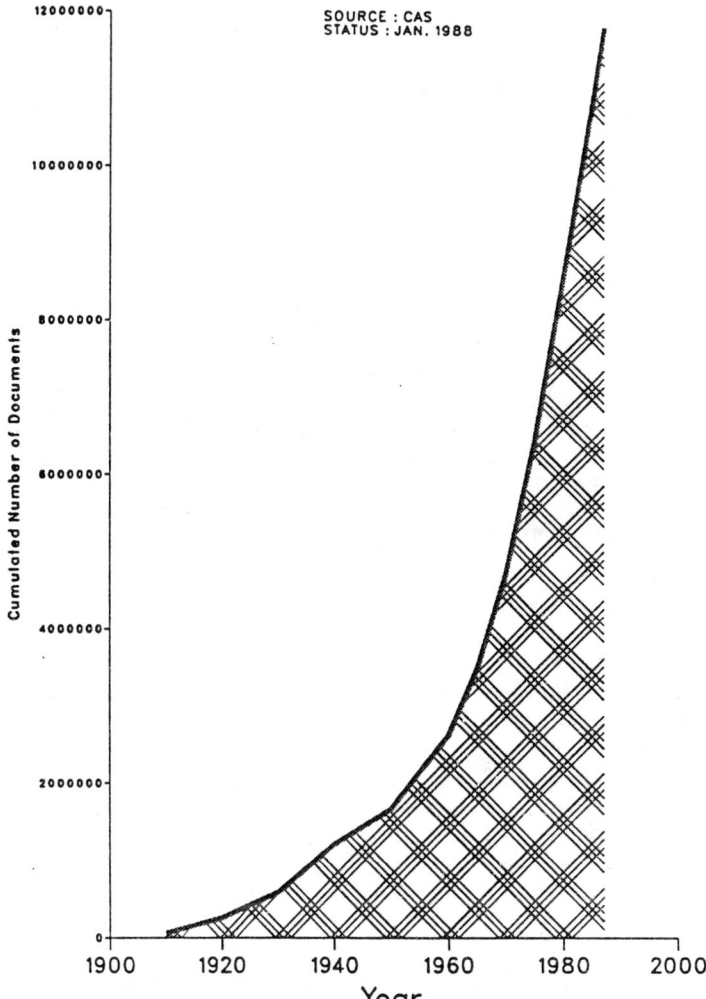

Figure 1.4 Total number of documents cited in *Chemical Abstracts* (1907–1987)

The American Chemical Society (ACS) started its Chemical Abstracts Service (CAS) in 1907. The enormous growth of this file covering literature of the period 1907–1987 is depicted in *Figure 1.4*. It is apparent that no chemist can keep abreast of the literature simply by reading primary or even abstract journals. The sheer scope and size of the *Chemical Abstracts* register caused this index to be listed in the *Guinness Book of Records* in 1986.

Need for computerization

The early batch systems used by information specialists, followed by online bibliographic files in the mid-1970s have been upgraded by graphic systems (*CIS, CAS-ONLINE, DARC, MACCS*) in the 1980s which facilitate chemical structure searching.

A description of the introduction of these online files can be found in the literature (Mullen 1980, 1981). These publicly available computerized databases are continuing to rapidly expand as is illustrated in *Figure 1.5*.

In mid-1989 there were already some 4,245 online files. Even if only a small proportion are purely chemistry related, this represents an important and powerful source of information. In line with their traditional activities in the chemical documentation field, the Chemical Abstracts Service now dominates the publicly available databases in the chemistry sector.

The acquisition and use of chemical information

The rest of this chapter will consider in more detail:

Primary classical (periodicals), secondary (abstract services) and tertiary (evaluated sources based on primary and secondary sources) information sources in chemistry including the impact of online files
Chemical structure and reaction retrieval
Specialized online files
New developments and trends.

Primary sources of information

Primary information sources are defined as those encompassing original published articles (journals, periodicals), theses, patents, congress reports etc. as well as unpublished documents such as laboratory journals/notebooks and other material of this nature (Neumüller, 1979–1988). It is beyond the scope, and not the aim, of this contribution to go into detail as far as individual titles are concerned.

The primary literature as an information source has been dealt with in various reviews (Anthony, 1979; Bottle, 1979; Neumüller, 1979–1988; Skolnik, 1982; Wolman, 1983; Ash, 1985).

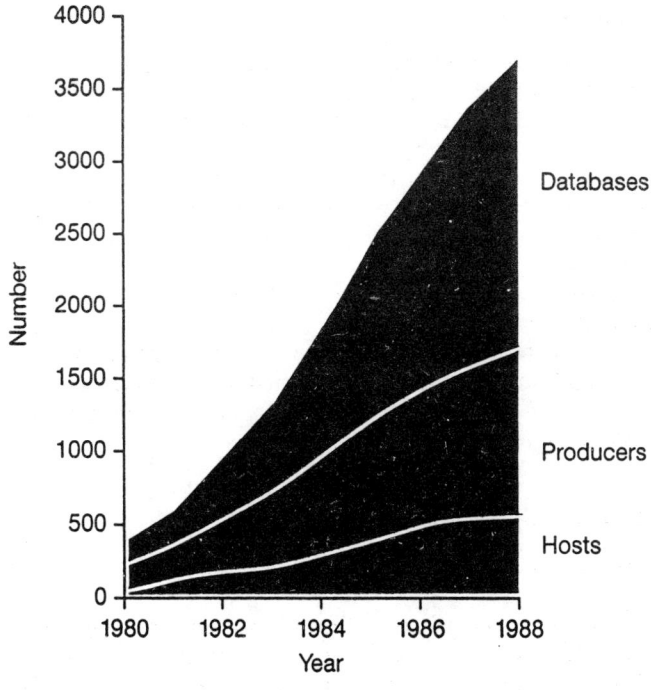

Source: Cuadra Assoc.

Figure 1.5 Development of online scenario 1980–1988

Journals and periodicals

For the purpose of communicating chemical knowledge from one generation to the next, the apprentice technique used by the alchemists soon gave way to books. The journal itself originated from the publication of texts of papers which had been presented at scientific meetings of learned societies, the objective being to disseminate this information to those members not present at meetings. The journal has remained in its present form since its inception with only minor changes taking place. More novel approaches over the last few decades have included the use of synopses, backed up by microfiche of the full text. The full text (excluding diagrams) of some 18 ACS journals going back to 1982 is now available online via STN International. The impact of

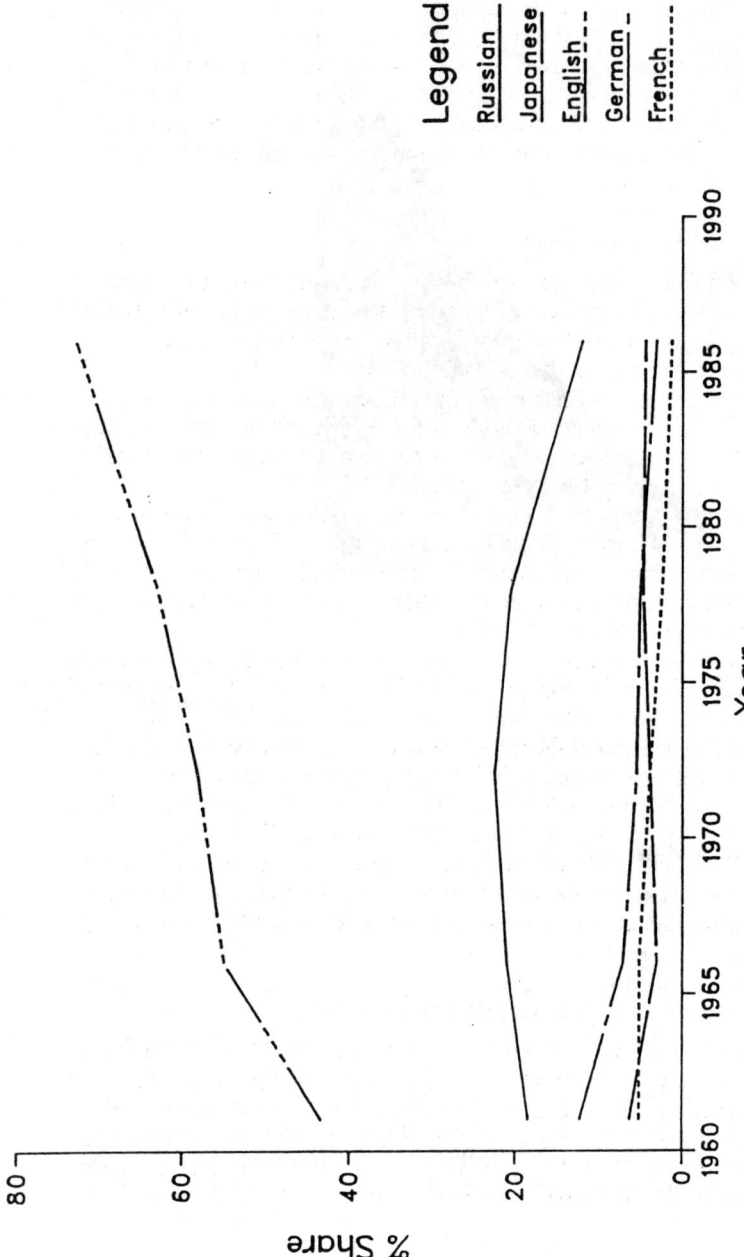

Figure 1.6 Relative share of different languages used in original publications (1960–1987) cited by *Chemical Abstracts*

electronic journal media will become greater as technology advances and their scope expands.

Currently the Chemical Abstract Service monitors some 12,000 current journals and over the last few years the number of papers abstracted has been fairly constant at 450,547 in 1981 to 476,162 in 1987. These figures do not signify a fall off in the expansion rate of primary publications but are primarily due to an increasingly stringent editorial policy.

PREDOMINANT LANGUAGES

The English language has become dominant in the field of chemistry as illustrated in *Figure 1.6* which shows the distribution of languages used in the original publications cited in the *Chemical Abstracts* file during the period 1960–1987.

While this distribution partly reflects the editorial policy of *Chemical Abstracts* (cf. Russian after 1978 in *Figure 1.6*), the trend towards widespread use of the English language for scientific publications cannot be overlooked (CAS, 1987).

Several studies have been directed at the decreasing ease of readability (Hartley, Trueman, Meadows, 1988) of the journal literature. The main factor is undoubtedly the use of more complex terminology and a higher density of information per journal page (Ash et al., 1985).

When the sources of the publications selected by Chemical Abstracts Services for processing are reviewed for the period 1909–1987 (CAS, 1987) their distribution is revealing (*Figure 1.7*).

The dominance of German papers in the early years of the 20th century reflects Germany's leading position in chemistry at that time. Nowadays, USA, USSR, Japan, W. Germany, UK, followed by France dominate the literature. The share of US-based publications has remained fairly constant at between 26 and 28% over the last 30 years (CAS, 1987). English, as the original language, has grown from slightly over 40% in 1960 to over 70% today.

ASSESSMENT AND COMPILATIONS OF JOURNAL TITLES

Citation studies have been carried out by ISI and others to ascertain which chemical journals appear to be the most influential (Skolnik, 1982). The annually issued *Chemical Abstract Service Source Index (CASSI)* provides a list of the most highly cited journals in the CA file. An extract from the current *Index* relating to medicinal chemistry is shown below.

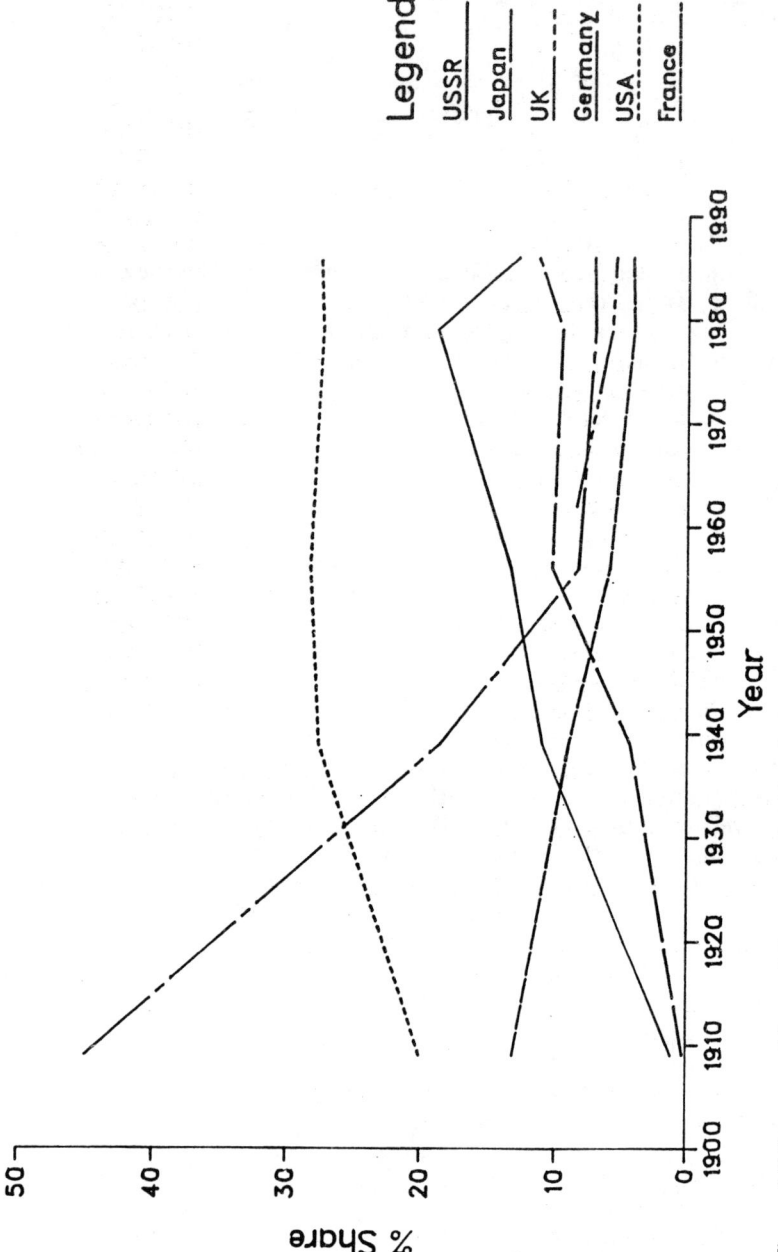

Figure 1.7 Sources of journal literature cited in *Chemical Abstracts*

Influential Journals in the Medicinal Chemistry Field

Position	Journal Title	CODEN
1	*J. Biol. Chem.*	JBCHA3
5	*J. Am. Chem. Soc.*	JACSAT
6	*Biochim. Biophys. Acta*	BBACAQ
10	*J. Phys. Chem.*	JPCHAX
11	*Tetrahedron Lett.*	TELEAY
13	*Biochem. Biophys. Res. Commun.*	BBRCA9
15	*J. Org. Chem.*	JOCEAH
17	*Am. J. Physiol.*	AJPHAP
18	*Biochemistry*	BICHAW
19	*Dokl. Akad Nauk SSSR*	DANKAS
21	*J. Chem. Soc., Chem. Commun.*	JCCCAT
23	*Biochem. J.*	BIJOAK
26	*J. Immunology*	JOIMA3
28	*Brain Res.*	BRREAP
33	*Nucleic Acids Res.*	NARHAD
35	*Methods Enzymol.*	MENZAU
36	*Cancer Res.*	CNREA8
37	*J. Bacteriol.*	JOBAAY
38	*Chem. Pharm. Bull.*	CPBTAL
39	*Nature*	NATUAS
40	*Biochem. Soc. Trans.*	BCSTB5
43	*Bull. Chem. Soc. Jpn.*	BCSJA8
46	*Endocrinology*	ENDOAO
47	*Eur. J. Biochem.*	EJBCAI

The above selection is based on the list of 1000 most frequently cited journals presented in the 1987 issue of *CASSI*. The CODEN is useful when searching for particular journals in online files (Neumüller, 1981).

The more important journal compilations include:

CASSI – Chemical Abstracts Service Source Index (annual)
Commonwealth Directory of Periodicals
Current Serials Received (annual) (British Library Lending Division)
Directory of Japanese Scientific Publications
Ulrichs International Periodicals Directory (annual) (Bowker Co.)
World List of Scientific Periodicals.

Most major scientific libraries will have one or several of the above works in their stocks.

An efficient functioning library which enables the user to gain rapid access to the original literature via conventional or electronic means remains indispensable to the research chemist.

Theses, congress reports and patents

THESES

A thesis or dissertation can generally be obtained by contacting the library of the institution where the thesis work was completed. As a rule, the thesis can either be borrowed, supplied as a photocopy, microfilm or microfiche.

Theses, congress and meeting reports are sometimes listed in *Chemical Abstracts* if the chemical content is sufficiently significant to warrant selection. Other sources for obtaining information about available theses are:

Dissertation Abstracts International (mainly USA and Canada)
Index to Theses Accepted for Higher Degrees in the Universities of Great Britain and Ireland (ASLIB)
Jahresverzeichnis der deutschen Hochschulschriften (W. German theses).

However, these important information sources tend to be underused as it is difficult to gain efficient access to their contents.

CONGRESS AND MEETING REPORTS

The proceedings of scientific meetings or conferences can be readily located if they have been cited in *Chemical Abstracts*. However, even then, the original publication is frequently in the form of an abstract. Nevertheless, these reports often represent a leading edge source.

A monthly compilation is issued by the British Library Lending Division entitled *Index of Conference Proceedings Received* which is fairly comprehensive and should be consulted if a *CA* search reveals nothing. Besides ISI's monthly publication *Index to Scientific and Technical Proceedings* (*ISTP*), there are online files such as *Conference Proceedings Index* and *Conference Papers Index* which provide access to over 1.25 million citations in this area. Additional information on sources can be obtained from a number of reviews (Bottle, 1979; Skolnik, 1982; Wolman, 1983).

PATENTS AND TECHNICAL PAPERS

For the chemist working in industrial research and development, patents provide an important source of information, 70% of which is believed not to be reported elsewhere. Werner von Siemens, one of the founding fathers of the German patent system, regarded (1886) the publication of new inventive ideas as being the essential value of a patent. This is really the basic idea behind a patent, in exchange for the exclusive protection given to an invention for a certain period of time (ca. 20 years) the information is made publicly available to all interested parties, thereby stimulating new ideas.

Each year, over one million patents are filed world wide, 10% of which are related to chemistry.

However, as Chapter 6 deals with patents, the main online sources of information will only be briefly considered here. To date, some 4,000 online files can be accessed and, of these, ca. 100 contain patent data, either as their focal point or as part of their general literature coverage. Some of the more important online patent files are shown in *Figure 1.8*.

Japanese technology is beginning to make its presence felt in the patents field; in *CA* some 55.5% of those documents stemmed from that country in 1987 compared to 11.8% in 1970 (*Figure 1.9*). The corresponding figures for the US are 8.0 and 23.7% respectively.

Derwent publications (*WPIL* file) is generally regarded as a major source and covers about 360,000 patents per annum of which approximately 160,000 are chemistry-orientated and some 14,000 drug-related. Patents from some 30 patent offices are documented.

It is a prime task of information departments to place only relevant information in the hands of research and development scientists i.e. in this case, to avoid information overload in the patents area. Interpreting patent information to gain an insight into research and development trends and competitor activities has been described in the literature (Mullen, Möller, Blunck, 1984; Mullen et al., 1988a, 1988b; Blunck et al., 1989).

RETRIEVAL OF DATA FROM PATENTS

Fragment codes for structure searching are available in the Derwent Publication files (*WPI/WPIL*) which facilitate the processing of queries involving Markush structures. New developments will involve a graphic search capability for Markush structures and an enhancement of the DARC software was released at the beginning of 1989, making the processing of Markush queries feasible for patents documented by Derwent Publications since 1987 together with those in INPI's new *PHARMSEARCH* file.

Graphic structure searching is possible in connection with patents documented in the *CA* and *PHARMPAT* files. However,

File	Hosts (sample)	Period from	No. of Records	Scope	Bibl. Data	Main Claim	All Claims	Examples	Patent Family	Codes	Abstract	cited Lit.	citing Lit.	Language	Structure searching graphic	Structure searching code	Update
BIOTECH ABSTRACTS	ORBIT	1982	60,000	intern.	+	-	-	-	-	-	+	-	-	Engl.	-	-	monthly
CA-File	STN	1967	7 Mio	intern.	+	-	-	-	-	IPC	+	-	-	Engl.	+	-	biweekly
CHINAPATS	ORBIT DIALOG	1985	2,500	CHINA	+	-	-	(+) structure	+ INPADOC	IPC	CA-Ref.-No Derwent-Ref.-No.	-	-	Engl.	-	-	monthly
CLAIMS/ CITATION	DIALOG	1947	3,396,011	US	-	+	-	-	-	IPC	-	+	+	Engl.	-	-	quarterly
CLAIMS/US PATENT ABSTRACTS	DIALOG ORBIT STN	1950	1,829,000	US	+	-	-	-	79: BE DE FR GB NE	IPC USPC	71	-	-	Engl	-	-	weekly
EPAT	TELESYST. QUESTEL	1978	205,000	EP	+	-	-	-	+ Europe, PCT	IPC	-	+	-	Engl. French German	-	-	weekly
FPAT	TELESYST. QUESTEL	1969	570,000	FR	+	-	-	-	+	IPC	-	+	-	French	-	-	weekly

File	Host	Year	No. records	Countries				Structure		Classification	CA-Ref.-No			orig. language			Update
INPADOC	ORBIT	1968	15 Mio	intern. 55 countries	+	–	–	–	–	IPC	–	–	–	orig. language	–	–	weekly
JAPIO	ORBIT	1976	2 Mio	JP	+	–	–	–	–	IPC JPC	+ Engl.	–	–	Engl.	–	–	monthly
PATDATA	BRS	1975	700,000	US	+	–	–	–	–	IPC USPC	+	+	–	Engl.	–	–	weekly
PATDPA	STN	1981	530,000	DE	+	–	–	–	–	IPC	+	(+)	–	German	–	–	weekly
PATOS	BERTELSMANN	1966	1 Mio	DE	+	+	–	–	–	IPC DEPC	+	–	–	German	–	–	weekly
PHARMPAT	STN	1987	5,000	intern	+	+	+	+ structure	+	IPC NCL	+	–	–	Engl.	+	–	weekly
WPI/WPIL	ORBIT DIALOG TELESYST. QUESTEL	1963	3.9 Mio	intern. 31 countries / Inv. from 1978	+	–	–	(+) structure	+	IPC MC PC	+ (81)	79 EP/PCT	–	Engl.	–	+	weekly

Figure 1.8 Leading patent files

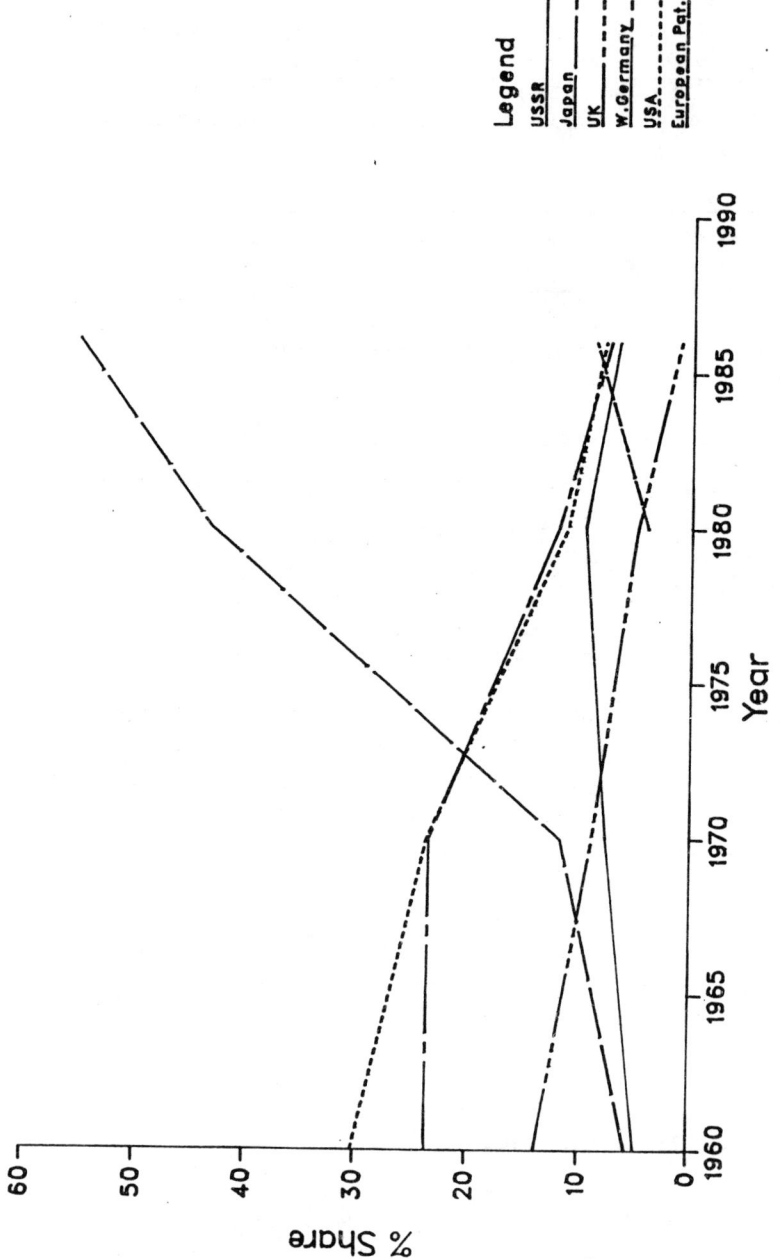

Legend

USSR
Japan
UK
W.Germany
USA
European Pat.

Figure 1.9 Country of issue of patents cited in *Chemical Abstracts*

in *CA*, only those substances actually synthesized and supported by physicochemical data are included.

Patents represent an extremely valuable source of information and a clear and ongoing view of the patent literature should form the basis of information support to research and development projects, in order to prevent effort being wasted. Comprehensive patent searching requires exhaustive scrutiny of several sources.

Secondary sources of information

Secondary sources (abstracting services) are based on the primary literature which is briefly reported in abstract form together with, typically, citation details, patent information and chemical structure and/or identifiers.

Chemical Abstracts Services

The Chemical Abstracts Registry Number (a unique machine-readable designation for a specific chemical entity) is used widely as a means of identifying chemical substances.

In *Chemical Abstracts*, the literature is subdivided into 80 sections. Sections 1–34 are published one week and 35–80 the next, two complete volumes being produced annually. The *CA* sections of principal interest to the medicinal chemist are:

Biochemistry Sections

1 Pharmacology
2 Mammalian Hormones
3 Biochemical Genetics
4 Toxicology
5 Agrochemical Bioregulators
6 General Biochemistry
7 Enzymes
8 Radiation Biochemistry
9 Biochemical Methods
10 Microbial Biochemistry
12 Nonmammalian Biochemistry
13 Mammalian Biochemistry
14 Mammalian Pathological Biochemistry
15 Immunochemistry
16 Fermentation and Bioindustrial Chemistry

Organic Chemistry Sections

21–30 Organic Chemistry
31 Alkaloids
32 Steroids
33 Carbohydrates
34 Amino Acids, Peptides, and Proteins

Applied Chemistry and Chemical Engineering Sections

47 Apparatus and Plant Equipment
48 Unit Operations and Processes
62 Essential Oils and Cosmetics
63 Pharmaceuticals
64 Pharmaceutical Analysis

Physical, Inorganic, and Analytical Chemistry Sections

65 General Physical Chemistry	75 Crystallography and Liquid Crystals
68 Phase Equilibriums, Chemical Equilibriums, and Solutions	78 Inorganic Chemicals and Reactions
72 Electrochemistry	80 Organic Analytical Chemistry
73 Optical, Electron, and Mass Spectroscopy and Other Related Properties	

However, as the contents of certain sections have been modified over the years, care should be exercised when using CA Section Numbers in an online search strategy by consulting *CA Index Guides*.

ACCESS TO CHEMICAL ABSTRACTS

PRINTED CA INDEXES

Conventional access to *Chemical Abstracts* is via the *Collective* or *Volume Indexes*. The dramatic growth in the *Collective Indexes* is shown in *Figure 1.10*.

The *Collective Indexes* have expanded markedly – from 4,823 pages for the period 1907–1916 to some 160,000 pages for the *11th Collective Index (CI)* (1982–1986).

The *11th CI* covers *CA Volumes* 96–105 and contains approximately 28 million index entries to more than 2.3 million *Abstract* references – all incorporated in 93 volumes.

An important tool, which should be used before consulting the printed *Indexes* of *CA* is the *Index Guide* which discusses the current CAS indexing philosophy, valid nomenclature, etc.

The *Indexes* themselves are organized in the following manner:

Authors' Index – lists names of authors, co-authors, inventors and patent assignees in alphabetical order and links them to *CA* abstract number

Chemical Substance Index – links *CA* index name, specific for a particular chemical substance, with the abstract number. The CAS Registry Number and index names are also included

Formula Index – provides *CA* index name, CAS Registry Number and abstract number for chemical substances via their molecular formula (Hill notation)

General Subject Index – links subject entries e.g. reactions, classes of substances, plant and animal species to the corresponding *CA* abstract number

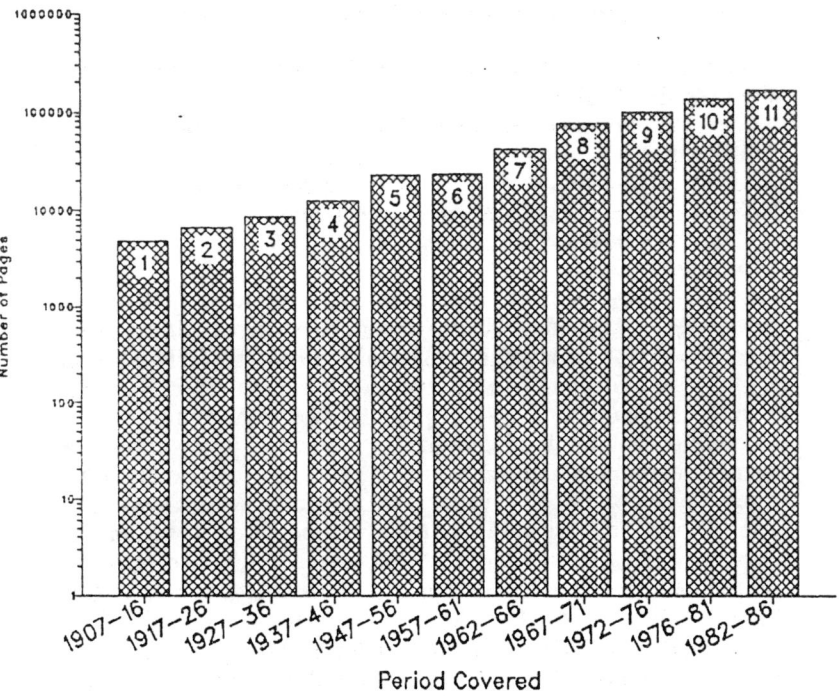

Figure 1.10 Growth experienced by *Collective Indexes of Chemical Abstracts* (1907–1986)

Patent Index – contains an alphabetical listing by the countries in which patents are listed within which the patents are arranged in ascending numerical order followed by the corresponding *CA* abstract number.

ONLINE ACCESS TO CA FILES

Online access to host computers spinning *Chemical Abstracts* tapes has been available for over a decade in various degrees of sophistication.

The *CA* files are offered by some seven different host computers via IPSS (International Packet Switching Service): BRS; Data-Star; DIALOG; ESA-IRS; ORBIT; STN International; Telesystemes-Questel.

The initial bibliographic *CA* file has been expanded to include comprehensive graphic (sub)structure searching capabilities on, at the moment, two host computers, Telesystemes-Questel (*DARC* software) and STN International (*Messenger* software).

Substructure searching can be conducted using either text structure input or graphic structure input to assemble the desired moiety. A whole series of terminals and personal computers (PCs) with appropriate software can be used to access the *Chemical Abstracts* files via the international packet switching networks (IPSS).

The most exhaustive range of *CA files* is provided by STN International:

CAOLD File – provides references to substances cited in *CA* before 1967. This file is being expanded stepwise ultimately to cover the period from the inception of *CA* in 1907. Only a limited amount of information is provided in this file: CA Accession Number, Document Type (for patents), Registry Numbers.

CA File – contains bibliographic data and *CA* index entries for over 8 million documents cited in *CA* since 1967. In addition, the complete abstract text is available for documents cited after 1970. The file may be searched using author names, Registry Numbers, keywords, a variety of index terms as well as the abstract text. Combinations of search terms may be employed to arrive at difficult-to-get information. One of the many advantages of this file is that immediate access to the information contained in four *Collective Indexes* is possible via the terminal keyboard. Moreover, the information is more up-to-date than the latest printed issue of *CA*.

CAS Registry File – the contents of the file include graphic structure diagram, CAS Registry Number, *CA* index names, molecular formula, up to 50 synonyms, number of available *CA* references. The file can be searched using the CAS Registry Number, molecular structure or substructure, name fragments or molecular formula.

In all the CAS files, search results in the form of CAS Registry Numbers can be transferred between different files to combine structure searching with concepts, making these files a very powerful facility in the hands of the chemist.

A whole series of documents describing CAS services can be obtained from *Chemical Abstracts* as well as other sources (Schulz, 1988).

Chemisches Zentralblatt

The *Chemisches Zentralblatt* was an important German language abstracting journal in the chemistry field which discontinued publication in 1969, after being in existence since 1830 as

Pharmaceutisches Centralblatt. It is often claimed that the pre-war coverage of *Chemisches Zentralblatt* is superior to that of *Chemical Abstracts*, particularly for literature from Eastern European countries. The present alerting service *ChemInform (Chemisches Informationsdienst)*, derived from *Chemisches Zentralblatt*, first appeared in 1970.

Current Abstracts of Chemistry and Index Chemicus (CAC&IC)

The *CAC&IC* publication, produced by the Institute for Scientific Information (ISI), is published on a weekly basis, giving chemists a current guide to chemical research and chemical technology. The *Current Abstracts of Chemistry* contains authors' abstracts of articles from over 110 selected journals reporting on the synthesis, isolation and identification of new compounds. Flow diagrams of reactions are extensively used to enable easy access to the data. The companion index to *Current Abstracts of Chemistry, Index Chemicus* contains the following information: molecular formula, author, subject as well as limited information on biological activity.

The *Index Chemicus* series made a brief appearance as an online file via Telesystemes-Questel, using the DARC software for graphic chemical structure searching. However, due to lack of commercial success the file was withdrawn in 1986.

Each year about 200,000 new compounds are added to *CAC&IC*, which currently contains over 3.5 million compounds. An interesting feature for the medicinal chemist is the broad assignment of biological activity, even if not fully standardized, alerting him to compounds of potential interest.

The *CAC&IC* file presents a challenge to ISI to demonstrate to users that it possesses its own unique features compared with *Chemical Abstracts*. In future, a way will probably be found to make at least part of the data accessible via graphic online retrieval, either in an in-house environment or via an external host.

Tertiary literature

The tertiary literature, based on primary and secondary sources, may represent an upgrading of information, when a subject matter is assessed in depth, yielding a treatise, handbook or monograph. This category also encompasses annual reports, dictionaries, encyclopaedias and other summary material.

Annual reports

Various learned societies and commercial publishers produce different series on an annual basis dealing with progress in a specific topic of interest to the medicinal chemist. Some of the better known review journals are:

Advances in Clinical Chemistry
Advances in Cyclic Nucleotide and Protein Phosphorylation
Advances in Drug Research
Advances in Heterocyclic Chemistry
Advances in Lipid Research
Advances in Magnetic Resonance
Advances in Neurochemistry
Advances in Organic Chemistry
Advances in Physical Organic Chemistry
Advances in Prostaglandin, Thromboxane and Leukotriene Research
Advances in Protein Chemistry
Annual Reports in Medicinal Chemistry
Annual Reports in Organic Synthesis
Annual Reports in NMR Spectroscopy

This compilation represents a small fraction of the available annual reviews. The standard literature on chemical information should be consulted for more detailed information on these types of sources (Bottle, 1979; Skolnik, 1982). Annual reviews afford the reader an excellent, reasonably up-to-date, overview of a particular topic together with reference to comprehensive bibliographic data for more detailed information. To date, these information sources are unavailable online.

Standard works

It would be inappropriate to review the standard reference material in the field of medicinal chemistry but some of the titles relevant to pharmaceutical research and development should be mentioned:

The Alkaloids
Beilstein
Chemistry of the Carbon Compounds (Rodd)
Chemistry of the Functional Groups (Patai)
Chemistry of Heterocyclic Compounds
Comprehensive Heterocyclic Chemistry (Katritzky and Rees)
Comprehensive Organic Chemistry

Gmelin – Handbook of Inorganic Chemistry
Houben-Weyl
Methodicum Chimicum
Organic Reactions
Organic Syntheses
Synthetic Methods (Journal of Synthetic Methods)

Only the Beilstein *Handbook* will be discussed here in any depth; information on other sources listed can be obtained from reviews by Mullen (1979) and Neumüller (1979–1988). Beilstein is selected as the major European source.

BEILSTEIN

Beilstein's *Handbook of Organic Chemistry*, first published in 1882, compiles data, after critical evaluation and verification, on the preparation and properties of carbon compounds (Lawson, 1987).

The following data are recorded (where available) for every organic compound, of which ca. 3 million are reported for the period 1830–1980 in Beilstein:

* natural occurrence and isolation
* preparation and purification on research and industrial scale
* physical properties
* chemical properties and reactions
* identification by derivatization and analytical procedures
* special handling techniques

The *Handbook* is produced at the Beilstein Institute in Frankfurt, W. Germany by a permanent staff of over 100 scientists plus external co-workers. More than 2,000 journals, monographs and patent series relating to organic chemistry are reviewed on a continuous basis.
The literature is covered in the following way:

	Period
	Period
Basic series	up to 1909
Supplementary series I	1910–1919
Supplementary series II	1920–1929
Supplementary series III	1930–1949
Supplementary series IV	1950–1959
Supplementary series V (English)	1960–1979

More than 70% of raw data from the source literature is rejected for inclusion in the *Handbook* either because the material is not novel or there is nothing original about the chemical reactions (Beilstein Institute, 1984). Cumulative indexes covering the six series facilitate access to the information in Beilstein.

Beilstein system numbers, which classify compounds into 4,720 classes, can be used to trace entries in the *Handbook*. The most convenient way to locate a compound in Beilstein is either to use the molecular formula indexes or Beilstein's Sandra PC software, which generates the Beilstein's system number from a graphically input structure. It is claimed that the programme is about 99% reliable in its selections.

BEILSTEIN ONLINE FILE

Over the last few years, considerable work has gone into producing a computerized version of Beilstein, thus overcoming the criticism of lack of currency whilst, at the same time, facilitating multi-dimensional access and graphic chemical sub-structure searching.

The file is to be or has already been made available on the Data-Star, DIALOG, ORBIT and STN (*Messenger* software) hosts.

Initially, Beilstein will be available for the period 1830–1959, so that some 1.6 million heterocyclic and over 100,000 acyclic substances will be in the online file. Up to 400 data fields will be assignable to each chemical structure. By the middle of the 1990s, it is hoped to have all the data listed in volumes of the basic and supplementary series online.

It will also be possible to obtain the file for in-house use. There are currently no plans to include CAS Registry Numbers in the file.

The database represents an important source of revised, critically evaluated information. Files of this type help counteract the effects of the information explosion because they select and package data in sympathy with user needs.

The availability of Beilstein online should expose it to a wider circle of chemists and ensure its place as one of the pillars of chemical documentation.

Chemical dictionaries and encylopaedias

As far as encyclopaedias are concerned the reader is referred to literature sources (Mullen, 1979; Neumüller, 1979–1988; Ash et al., 1985).

Ullmann's *Encyclopaedia of Industrial Chemistry* and Kirk-Othmer's *Encyclopaedia of Chemical Technology* are the main works in this area. Kirk-Othmer is searchable online via the BRS and DIALOG hosts, the full text of the file (without graphics) being available.

CHEMICAL DICTIONARIES

The outstanding dictionary in the chemical field is undoubtedly Neumüller's *Römpps Chemielexikon*, a German language compilation (with English translations of the entry headings) covering some six volumes and ca. 5,000 pages (Neumüller, 1979–1988). Each entry in the dictionary is critically reviewed, being supported by references to current literature. The scope and quality of this work by far exceeds that of its competitors.

The *Merck Index* is a reference book containing brief entries to ca. 2,000 biologically active substances together with supporting literature citations. This dictionary is available as a full text online file (without graphics) with the BRS, DIALOG and Telesystemes-Questel hosts. It is anticipated that a graphic version of the *Merck Index* file (with chemical structure diagrams) will appear in the not too distant future.

Heilbron's dictionaries cover some 190,000 chemical compounds with associated chemical and physical data. The following dictionaries are available as one united online file with a graphic display facility via the DIALOG host: *Amino Acids and Peptides, Carbohydrates, Dictionary of Organic Compounds* and *Dictionary of Organometallic Compounds*.

Other compilations such as the *USAN List* and Hawley's *Condensed Chemical Dictionary* are worth mentioning.

Chemical structure searching and reaction retrieval

The chemist has always preferred using structural diagrams in organic chemistry, nomenclature being less easily interpreted than the graphical representation of a chemical compound. Furthermore, as the IUPAC rules are not unambiguous, CAS extended them over time to eliminate choice of names. Unfortunately, however, there was already a situation in which a compound could well have a different name in each *Collective Index* (every 4–5 years) of *Chemical Abstracts*. Consequently, there are pitfalls for the unwary when searching for compounds via their nomenclature. The advent of linear codes [such as Wiswesser line notation (WLN), Dyson and Hayward] to represent structures did nothing to make these abstract systems more attractive to the chemist.

Software for chemical structure information based on such

methods was typified by CROSSBOW, which enabled batch processing during the late 1960s. Chemical Information System (CIS), produced by NIH/EPA was one of the first interactive online publicly available systems for teletype terminals accessing a chemical structure file. In this instance, the characters and symbols present on a standard keyboard were used to represent structural entries. The CIS/NIH file was launched in the mid-1970s. The 1980s saw the arrival of graphic-based chemical structure searching systems such as DARC and CAS-ONLINE. In parallel with these commercial systems, in-house developments were pursued by various chemical and pharmaceutical companies such as Upjohn (COUSIN), Pfizer (SOCRATES) and Bayer (RESY). For the most part, however, the use of commercial software products for manipulating chemical structural data has now become widespread notably *DARC* (Telesystemes-Questel) and *MACCS* (MDL).

Features of current online chemical structure retrieval systems

The main online software systems for chemical structure searching are *Messenger* (CAS-ONLINE) and *DARC* (Telesystemes-Questel), both of which can be used to manipulate vast quantities of data from, for instance, the Chemical Abstracts Registry File. Descriptions of these systems can be found elsewhere (Barnard, 1984; Schulz, 1988; Warr, 1988).

(1)

By means of either a teletype terminal for text structure input, a graphics terminal or PC and a suitable emulation package such as *EMU-TEK, Genesys, Infolog-G, PC-PLOT*, or *STN-Express*, fairly sophisticated substructure searching may now be undertaken in very large files of chemical compounds. For instance, searches for compounds conforming to the pattern (1) where G_1 = a set of defined groups, G_2 = another set of defined groups, and G_3 = open would now be regarded as routine. As of the end of 1988, Markush *DARC* has become available, which permits Markush-type queries in the *WPIL* (World Patent Index Latest) file and *PHARMSEARCH*, a recently constructed INPI patent file covering US, European and French patents. The new software enables substructure queries to be posed where, for example, the side chain is simply defined as being alkyl. This is in contrast to Generic *DARC*, where the generic group has to be exactly defined.

Hitherto, this type of enquiry could only be processed using fragment codes such as Ringcode or GREMAS (Ash, 1985).

PC SOFTWARE PACKAGES FOR SEARCHING IN EXTERNAL ONLINE CHEMICAL STRUCTURE FILES

Several PC software packages such as *DARC-Chemlink, MOLKICK* and STN-Express facilitate the offline input of chemical structures which can then be uploaded during the actual online session. After the search, the graphics and text may be captured, stored, redisplayed and/or printed. To date, there is no commercial software available for capturing the vectors from the online host structure display and converting them into connection table format, so that local chemical substructure searching could be carried out on the downloaded data. No doubt, this facility will become available in future.

An interesting feature of the *MOLKICK* software is that it is the first host independent interface for chemical structure queries. That is to say, once a structural query has been input in *MOLKICK*, there is a choice as to whether it is uploaded to STN or the Telesystemes-Questel host. When the Beilstein database becomes available on other hosts, more structure search options may well be developed.

IN-HOUSE SYSTEMS FOR STRUCTURE RETRIEVAL

The *MACCS* software produced by Molecular Design Limited (MDL) and, to a lesser extent the *DARC* package of Telesystemes-Questel, currently dominate the market arising from in-house structure retrieval needs associated with proprietary test compound files. The integration of such test compound files with biological results and other properties of the compounds can be tackled successfully using these products (Warr, 1988; Northup, 1987).

MACCS can be linked, via suitable interfaces (customization module), to various relational database management systems such as ORACLE, to facilitate the manipulation of and retrieval in research and development files.

Within the integrated research information system environment, which is now the ambition of many pharmaceutical companies, the bench chemist will have a wide range of infor-

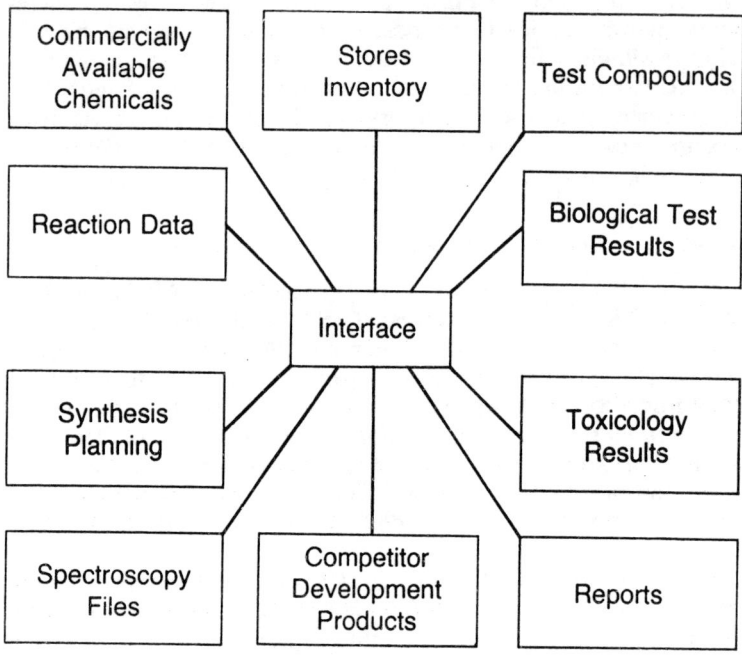

Figure 1.11 Components of a typical integrated research information system

mation resources at his disposal via his electronic workstation (*Figure 1.11*). There are many variations possible on this theme, but all envisage the controlled flow of information from multiple sources.

As the bench scientists, not necessarily skilled at information processing, will be searching these files, it is essential that the latter are assembled in such a manner that the infrequent user can successfully work with them.

Some end-user chemists will certainly be interested in searching external online files such as CAS ONLINE via the in-house network (Warr and Jackson, 1988) and for such activities both text and structure retrieval are essential. A development which will probably come, is a front end to the commercial host computers which will accept queries transmitted from graphic-based in-house systems such as those using *MACCS*. When this happens, the end-user will only have to learn a single set of commands. A natural consequence is that private literature files with structure searching possibilities will become increasingly used. Likewise, the increasing availability of 'off the shelf' files in formats such as *MACCS*, which provide information about commercial sources of chemicals, drugs under development and so on, will further accentuate the need to integrate the different components of research information systems and place them all at the disposal of scientists. The challenge and problems encountered in establishing in-house systems have been discussed by Warr (1988) and Ash et al. (1985).

Reaction retrieval systems and sources of chemicals

Chemical reaction retrieval is well documented in the literature (Ash, 1985; Willet, 1986; Deroulede, 1987; Warr, 1988). In the industrial environment particularly, the research chemist is concerned with synthesizing substances as efficiently and as rapidly as possible. The routes adopted should, therefore, be flexible (to enable analogues to be prepared) as well as short. Optimization of synthetic pathways is also of importance to the chemist involved in process development work. In this instance, inexpensive precursors linked to a short route with high yields are his ideal solution.

However, the concept of a chemical reaction is clearly more complex than that of a chemical compound as more factors must be taken into account:

multi-step reactions produce a whole series of precursors and products, $A \rightarrow B \rightarrow C \rightarrow D \rightarrow E \rightarrow F$
reaction retrieval must permit a whole series of questions to be addressed, $A \rightarrow B, A \rightarrow C, A \rightarrow D, A \rightarrow E, A \rightarrow F, B \rightarrow C, B \rightarrow D, B \rightarrow E, B \rightarrow F$, etc.
the reaction centres must be defined
yield, reaction conditions, solvent, temperature, pressure, etc. are also of interest
nature of altered bonds.

A number of reaction indexing systems are available for searching reaction files (Willett, 1986):

Fragmentation codes

Ringcode (Derwent) (used in online database)
GREMAS (IDC) *

Graphic-based programmes

SYNLIB *   (Synthesis Library)
REACCS *   (Reaction Access System)

ORAC *   (Organic Reactions accessed by Computer)
CONTRAST *   (Connection Table Reaction Analysis and Search
                     Technique – Pfizer in-house system)
CASREACT   (used in online file)
DARC-RMS *
* to date, mainly in-house

PUBLICLY AVAILABLE ONLINE REACTION FILES

The only widely used publicly available online systems for reaction retrieval are Derwent Publications' *Chemical Reactions Documentation File (CRDS)* and *CASREACT*.

CRDS employs Ringcode and gives access to Theilheimer's *Synthetic Methods of Organic Chemistry* series (1942–1974) and the *Journal of Synthetic Methods* (from 1975). The *CRDS* file contains over 63,000 reactions and is growing at a rate of 3,000 reactions per year. Access to the online file is, on account of the fragmentation code used, mainly restricted to experienced searchers (Finch, 1986).

The *CASREACT* file is based on a selection of reactions from ca. 100 journals from the *CA* sections 21–34. The file currently contains 250,000 single and multi-step reactions from 20,000 publications and is growing at the rate of 170,000 reactions per year. The period from 1985 is covered (Blower et al., 1988).

Searching is best undertaken either via CA Registry Number or an answer set from the CA Registry file. The *CASREACT* file must generally be used in conjunction with the CAS Registry file, where the (sub)structure searching ensues. The results of such a search are then transferred via a cross-file search from the CA Registry file to the *CASREACT* file where the corresponding reactions are retrieved.

However, this type of reaction documentation may result in, for example, one conversion of a functional group being documented several times per publication, i.e. for every change in the substitution pattern of the reactants even if they have no effect on the course of the reaction.

REACTION FILES FOR COMMERCIAL IN-HOUSE SYSTEMS

The main commercial software products used for in-house reaction retrieval are *REACCS, ORAC* and *SYNLIB*.

The following files, encompassing over 190,000 reactions, are available to the *REACCS* user:

	No. of reactions
CCR (Current Chemical Reactions)	75,000
CHIRAS (focal point: *Stereochemistry*)	10,000
CLF (Current Literature File)	32,000
JSM (Journal of Synthetic Methods)	25,000
Organic Syntheses	5,000
Theilheimer's *Synthetic Methods of Organic Chemistry*	46,000
	193,000

The *ORAC* user has currently over 70,000 reactions at his disposal including, for example, part of the Theilheimer file as well as the *ORAC* file from the University of Leeds.

The *SYNLIB* system, offers a file with ca. 40,000 reactions according to Borkent, Oukes, Noordik (1988). This paper also compares *ORAC, REACCS* and *SYNLIB* with one another with respect to searching capabilities and scope of available files.

Producers of reaction retrieval software are striving to extend the spectrum of files supplied to the user, while new databases will, in future also be made available by other producers, e.g. *ChemInform*.

Most companies aim to incorporate their reaction retrieval files into an integrated research information system environment (Hagadone, 1988; Mills et al., 1988).

SYNTHESIS PLANNING

Computer-assisted synthesis planning is a natural continuation of retrieval of reactions. Such programmes aim to help the synthetic or process chemist find an efficient route to synthesize a given target molecule, based on simple or inexpensive, commercially available precursors.

Generally, a retro-synthetic approach is made, the target molecule being input graphically and the software generates suggestions for precursors based on a library of 'transforms', as is the case with empirical programmes such as *CASP, LHASA, OCSS* and *SECS* (Braun, 1988; Trombini, 1987; Lee, 1987; Olsson, 1986). These library-based interactive retro-synthetic programmes such as *CASP* are not able to propose new chemical reactions but can suggest novel strategies for synthesizing a particular target molecule, based on established steps.

The other approach is based on logically orientated synthetic route planning where there is no adherence to information about individual reaction steps. Therefore, using this technique, also known as the non-empirical approach, a wide range of possibilities for synthetic route planning is generated, mainly based on mathematical combinations of bond exchanges (Herges, 1988). Some typical examples of such systems are *Cameo, Cyclops, EROS* and *IGOR* (Ash, 1985; Warr, 1988).

The *CASINO* software (Computer Aided Synthesis Inference for Organic Compounds) strives to combine the features of both approaches (Uchimaro et al., 1987).

Further development of the methods described above will eventually provide the chemist with a useful support for synthetic work, even if in some cases it only serves as a check list, identifying all possibilities.

Similar studies have been made regarding metabolic biotransformations of compounds e.g. *METAB, XENO* (Hrib, 1986), and *Metabolexpert* from Compudrug.

COMMERCIAL SOURCES OF CHEMICALS

While it is still possible to browse through numerous chemical catalogues to identify suppliers of particular chemicals, a more elegant and efficient method is to use one of the available databases either in-house or on an external online host computer:

Chemquest – this file, available on ORBIT as well as for in-house use in the MACCS and other formats, provides information on the sources of supply of over 80,000 chemicals including organic, biochemical and specialized compounds from over 50 suppliers. Each record contains nomenclature data, CAS Registry Number, compound description, price, synonyms and molecular formula. The file can be searched by structure and substructure.

CSCHEM – about 110,000 chemical compounds supplied by over 850 suppliers are listed in this STN file. Data such as CAS Registry Number, chemical or trade name and purity of product are included.

JANSSEN – this Telesystemes-Questel file contains entries to over 11,000 chemicals listed in the Janssen *Chimica Catalog*. Each record includes CAS Registry Number, nomenclature, price, molecular weight and molecular formula. Online ordering of chemicals is possible via the Telesystemes-Questel host.

Specialized online files

The number of chemistry-related online files has increased over the last few years (*Figure 1.12*). In this section, a brief overview will be given of specialist online files covering the following areas:

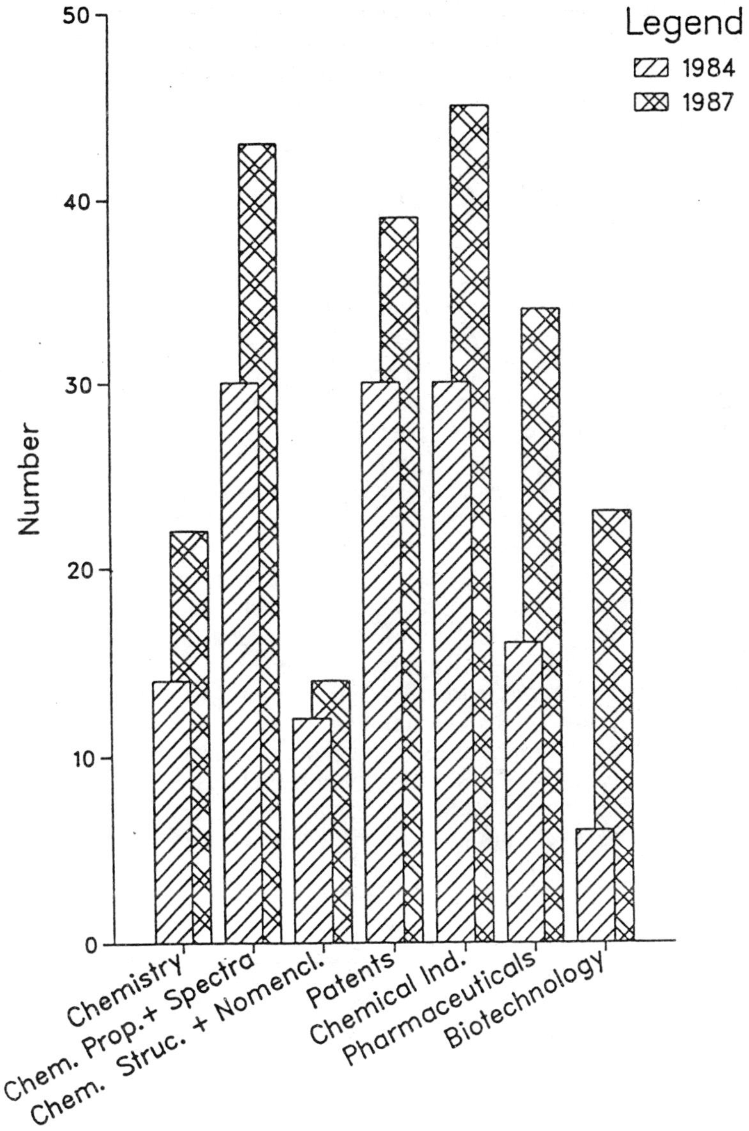

Figure 1.12 — chart

Legend
- 1984
- 1987

SOURCE : CUADRA ASSOCIATES

Figure 1.12 Growth of chemistry-related databases (1984-1987)

spectroscopy and crystallography, DNA/RNA and protein sequence files, as well as special property files.

Online spectroscopy and crystallography files

There are a number of spectroscopy files available which are of interest to the analytical chemist:

CNMR (C-13 Nuclear Magnetic Resonance Search System) – this database is available via CIS and contains 12,000 C-13 NMR spectra from the literature. Searching can be conducted in a variety of ways e.g. via CAS name, CAS Registry Number, molecular formula, individual shifts, intensities etc.

C-13 NMR Database/Infrared Databases – the file is produced by BASF and loaded on the STN host. The file contains data on ca. 83,000 C-13 NMR spectra. Chemical shifts are included as well as relaxation times, coupling constants, molecular and structural formulae of the compounds. Infrared spectral data on over 8,000 compounds are also included. Information on linewidth, intensity and experimental conditions is provided. A substructure searching capability is available

IRSS (Infrared Search System) – this IR spectra of ca. 6,500 compounds are compiled in this CIS file which can be searched via peak position, partial and complete molecular formula, chemical structure as well as CAS Registry Number

MSSS (Mass Spectral Search System) – this MS file, available via CIS and Telesystemes-Questel, currently contains the mass spectra of over 40,000 different substances. Each compound is identified by its CAS Registry Number, CAS nomenclature, molecular formula and molecular weight. Spectra can be searched, for example, via certain mass-to-charge ratios. Retrieved spectra may be listed or graphically output

WMSSS (Wiley Mass Spectral Search System) – about 130,000 mass spectra of chemical substances are present in this file on the CIS and INKADATA hosts. Each substance is identified by its CAS name, and CAS Registry Number, molecular weight and molecular formula. As before, it is also possible to search for spectra containing peaks of a given mass-to-charge ratio

MS ONLINE – this database, which is loaded on the CIS, INKADATA and the TDS hosts, contains over 80,000 MS spectra covering 68,000 different compounds. Interesting additional features are the PBM (Probability Based Matching) and STIRS (Self-Trained Interpretive and Retrieval Systems) programmes for matching unknown spectra against those in the file

Cambridge Structural Database – this file, no longer available through any of the major online hosts, tends to be used in in-house systems. The database contains published crystallographic data on organic and organometallic structures. Entries covering ca. 60,000 compounds are in the file. Each record contains CAS Registry Number, CAS name, molecular weight and formula, space group and cell parameters

NBS Crystal Structure Indentification File – this database contains over 120,000 records dealing with the crystallographic data of compounds. The data include reduced cell parameters, reduced cell volume, space group number and symbol, classification by chemical type, chemical formula and chemical name

Discussions of the use of NMR and MS files have recently appeared (van't Klooster et al., 1988; Tomellini, Hartwig, Woodruff, 1985). Such specialized files can only be searched effectively by experts from the analytical or other research departments.

DNA/RNA and protein sequence files

Medicinal chemists have become increasingly involved in bio-chemical processes. Within this context, scientists in the pharma-ceutical area have become primarily interested in DNA/RNA databases and protein or amino acid sequence files.

As these topics also receive attention in other chapters, only a brief overview of the major files (mainly for in-house use on a PC or mainframe) will be given here. Reviews of activities involving sequence data can be found in the literature (Bilofsky, Burks, 1988; Cameron, 1988; Sidman et al., 1988).

DATABASES

File	Number of sequences	Number of residues	Supplier
DNA/RNA			
EMBL	20,695	24,211,054	EMBL
GenBank	18,230	20,795,279	Intelligenetics
HIVN	38	231,597	G. Myers, Los Alamos
NBRF Nucleic	2,828	6,131,296	NBRF
PROMEGA	14	39,125	Promega Biotec
Protein			
HIVP	200	61,978	G. Myers, Los Alamos
NBRF Protein	9,138	2,377,999	JIPID+NBRF+MIPS
Swiss Prot	7,724	2,224,465	via EMBL

The analysis software permits users to retrieve sequence information in two ways – a direct query via database variables or a similarity search based on available key sequence data. Recently, Hitachi America has introduced a CD-ROM database for the GenBank, EMBL and NBRF files with appropriate searching facilities.

Special property files

Various aspects of the physical properties of chemical compounds are treated in a number of computerized files, a selection of which are reported below:

DETHERM-SDC – this database, available via INKADATA, contains data for calculating the thermophysical properties and phase equilibria of ca. 500 pure chemical substances and their mixtures in the fluid state e.g. density, enthalpy, entropy, heat of vapourization, specific heat and viscosity. Calculations can be carried out to determine properties under specific conditions

DIPPR – the *DIPPR* file, available via STN, contains textual information and numeric pure component physical property data for commercially important chemical substances. To date, information on 1,000 chemical compounds is located in the file e.g. enthalpies, heat capacities, vapour pressure, viscosity. Also included in the records are CAS Registry Numbers, IUPAC names and synonyms

Vapour Pressure Data File – the TDS host offers access to this database containing ca. 10,000 references on experimentally measured vapour pressures of organic and major inorganic compounds

Log P and Related Parameters Database – this online file, available via the TDS host, contains over 32,000 records, providing partition coefficients for over 15,000 organic compounds in ca. 300 solvents. Each record includes the CAS Registry Number, the compound name, log *P* value and p*K*a data. This information is useful for predicting the properties (absorption, solubility) and interactions of substances in chemical and biological procedures and processes. Included too, are measured values for steric and electrical effects as well as molar refractivity and other parameters of over 3,000 molecular fragments which can be used to predict log *P* values. Updating occurs twice per year when a total of about 1,250 records are added to the file.

The file is of particular interest to pharmacologists and chemists for correlations with such physicochemical phenomena as solubility, absorption and transport in biological systems as well as with many chemical and pharmacological properties of molecules.

Concluding remarks

Some examples of the vast array of information at the disposal of the chemist have been presented in this chapter. Future developments in this rapidly changing area should reflect the continual drop in the cost of data processing, the phased introduction of the PC and work station as well as increased computer sophistication, capacity and user-friendliness together with a more widespread interest in the application of computers to chemistry. The full impact of optical storage technology is currently difficult to assess. However, even at this stage, it is apparent that desk-top access to a very wide range of data and information resources will be a reality for many bench chemists in the near future.

The conventional role of the information specialist as an intermediary personally collecting and sorting information on behalf of users will change. The purpose will be the same but the methods will be different. Information systems need to be carefully constructed and maintained, whilst the introduction of integrated systems adds the further dimension of co-ordination. Presenting the desired information in the right format and at the right time remains the key objective whatever the methodology and demands dedicated information workers.

In future, manipulating retrieved data will become an integral part of an information system facilitating better data analyses, prediction and estimation. The development of chemical information networks will present major challenges to both the system developer and user as there is an urgent need for upgraded information to support decision-making processes within pharmaceutical research and development. Taking full advantage of technological advances particularly in research-based activities is a problem facing industry as it approaches the next decade.

References

Antony, A. (1979). *Guide to Basic Information Sources in Chemistry* (Jeffery Norton Publishing).

Ash, J., Chubb, P., Ward, S., Welford, S. and Willett, P. (1985). *Communication, Storage and Retrieval of Chemical Information*, (Ellis Horwood Series Chemical Science).

Barnard, J. M. (ed.) (1984). *Computer Handling of Generic Chemical Structures* (Gower Publishing Company).

Bawden, D. (1986). 'Information Systems and the Stimulation of Creativity.' *J. Information Science* **12**, 203–216.

Beilstein Institute (1984). *How to use Beilstein* (Springer-Verlag).

Bilofsky, H. S. and Burks, C. (1988). 'Genbank – Genetic Sequence Data Bank.' *Nucleic Acids Research*, **16**, 1861–1863.

Blower, P. E., Chapman, S. W., Dana, R. C., Erisman, H. J. and Hartzler, D. E. (1988). *Machine Generation of Multi-step Reactions in a Document from Single-step Input Reactions.* In *Chemical Structures* edited by W. Warr, pp. 399–408 (Springer-Verlag).

Blunck, M., Möller, E. and Mullen, A. (1985). *Statistical and Graphical Evaluation of Patent Analyses* (GDCh, 2. Vortragstagung der Fachgruppe Chemie-Information). Blunck, M., Busse, W. D., Meister, G., Möller, E., Mullen, A. and van Rooijen, L. A. A., (1989). 'Online Patent Searching – Rapid Answers to Critical Questions.' *Naturwissenschaften*, **76**, (3), 96–98.

Borkent, J. H., Oukes, F. and Noordik, J. H. (1988). 'Chemical Reaction Seaching Compared in REACCS, SYNLIB and ORAC,'. *J. Chem. Inf. Comput. Sci.*, **28**, 148–150.

Bottle, R. T. (ed.) (1981). *Use of Chemical Literature* (Butterworths).

Braun, H. W. (1988). 'Strategische Werkzeuge für den Chemiker'. *Chemische Industrie*, **40** (5), 43–53.

Cameron, G. N. (1988). 'The EMBL Data Library'. *Nucleic Acids Research*, **16**, 1865–1867.

CAS (1987). *Chemical Abstracts Service Statistical Summary, 1907–1986* (CAS).

Deroulede, A. (1987). 'Le point sur les systèmes d'information sur les réactions chimiques et synthèse assistée par ordinateur'. *Informations Chimie*, **289** (12), 143–146.

Economist (1987). 'Why Scientific Fact is Sometimes Fiction'. *Economist*, February 26, 103–104.

Eichin, K. H. (1988). *Internationale Perspektiven der Pharma-Industrie*, Pharm. Ind. **50**, 277–285, 415–423.

Finch, A. F. (1986). 'The Chemical Reactions Documentation Service: Ten Years On'. *J. Chem. Inf. Comput. Sci.*, **26**, 17–22.

Hagadone, T. R. (1988). *Current Approaches and New Directions in the Management of In-house Chemical Structure Databases.* In *Chemical Structures*, edited by W. Warr, pp. 23–42 (Springer-Verlag).

Hartley, J., Trueman, M. and Meadows, A. J. (1988). 'Readability and Prestige in Scientific Journals'. *J. Information Science*, **14**, 69–75.

Herges, R. (1988). *Reaction Planning.* In *Chemical Structures*, edited by W. A. Warr, pp. 385–398 (Springer-Verlag).

Hrib, N. J. (1986). 'Recent Developments in Computer Assisted Organic Synthesis'. *Annual Reports in Medicinal Chemistry*, **21**, 303–311 (Academic Press).

Lawson, A. J. (1987). 'Structure Graphics in: Pointers to Beilstein out'. *ACS Symp. Ser.*, **341**, 80–87

Lee, T. V. (1987). 'Expert Systems in Synthesis Planning: a User's View of the LHASA programme'. *Chemometrics and Intelligent Laboratory Systems*, **2**, 259–272.

Luckenbach, R. (1986). 'Chemie-Information: Quo vadis'. *Oesterr. Chem. Z.*. **87**, 284–289.

McWhirter, N. D. (ed.) (1986). *The Guinness Book of Records*, (32nd Edition, Guinness Books).

Mills, J. E., Maryanoff, C. A., Sorgi. K. L., Scott, L. and Stanzione, R. (1988). 'REACCS in the Chemical Development Environment'. *J. Chem. Inf. Comput. Sci.*, **28**, 153–155.

Möller, E., Mullen, A. and Blunck, M. (1983). *Support for Strategic Planning and Innovation in the Pharmaceutical/Chemical Industries via Information and Documentation Activities.* (1. Vortragstagung der GDCh Fachgruppe Chemie Information).

Mullen, A. (1979). 'Information Resources in Chemistry (Part 1)'. *Praxis der Naturwissenschaften, Chemie*, **28**, 243–247.

Mullen, A. (1980). 'Information Resources in Chemistry (Part 2). New Trends'. *Praxis der Naturwissenschaften, Chemie*, **29**, 185–194.

Mullen, A. (1981). 'Information Resources in Chemistry (Part 3). Online Systems'. *Praxis der Naturwissenschaften, Chemie*, **30**, 277–285.

Mullen, A., Blunck, M., Busch, T. and Möller, E. (1988a). *From Technological Current Awareness to Strategic Information*. (3e. Colloque sur l'information en Chimie [CNIC]).

Mullen, A., Blunck, M. Busch, T. and Möller, E. (1988b). *Trend and Competitor Analysis in the Pharmaceutical Area*. (196th National ACS Meeting Sept. 25–30, 1988).

Mullen, A., Möller, E. and Blunck, M. (1984). *Applications of PC 350 (DEC) for Online Searching Evaluation and Upgrading of Results from Patent and Literature Files*. (8th Int. Online Information Meeting).

Neumüller, O. A. (1979–1988). *Römpps Chemie-Lexikon* (Vols. 1–6, 8th Edition, Franckh'sche Verlagshandlung).

Northrup, A. (1987). 'Computer-aided Chemistry: Effective Management of Chemical Information'. *Biotechnology*, **5**, 455–457.

Olsson, T., (1986). 'LHASA – A Computer Programme for Synthesis Design and Selection of Protecting Groups'. *Acta Pharm. Suec.*, **23**, 386–402.

Rhymer, P. (1983). 'Zur Zukunft der chemischen Information und Dokumentation'. *Chimica*, **37**, 263–266

Roberts, E. B. (1988). 'What We've Learned Managing Invention and Innovation'. *Research Technology Management*, Jan.–Feb., 11–29

Roques, B. P. (1987). 'Molecular Recognition and Rational Design of Pharmacologically Active Molecules'. *C.R. Acad. Sci., Ser. Gen. Vie Sci.*, **4**, 193–209

Schulz, H. (1988). *From CA to CAS ONLINE* (VCH).

Sidman, K. E., George, D. G. Barker, W. C. and Hunt, L. T. (1988). 'The Protein Identification Resource (PIR)'. *Nucleic Acids Research*, **16**, 1869–1871

Skolnik, H. (1982). *The Literature Matrix of Chemistry* (John Wiley and Sons).

Tomellini, S. A., Hartwig, R. A. and Woodruff, H. B. (1985). 'Automatic Tracing and Presentation of Interpretation Rules and Use by PARIS: programme for the Analysis of IR Spectra'. *Appl. Spectrosc.*, **39**, 331–333.

Trombini, C. (1987). 'Planning Organic Synthesis I – Synthetic Plans and Processes'. *La Chimica e L'Industria*, **69**, 82–86.

Uchimaro, T., Tanabe, K., Hayashi, T. and Ouchi, A. (1987). 'A Novel Computer-aided design System for Organic Synthesis – The Overview of Casino'. *Chemistry Express*, **2** 417–420.

Unger, S. H. (1987). 'Computer-aided Drug Design in the Year 2000'. *Drug Inf. J.*, **21**, 267–275.

van't Klooster, H. A., Cleij, P., Luinge, H. J. and Kleywegt, G. J. (1988). *Computer-aided Spectroscopic Structure Analysis of Organic Molecules using Library Search and Artificial Intelligence*. In *Chemical Structures*, edited by W. Warr, pp. 219–234 (Springer-Verlag).

Warr, W. A. (ed.) (1988). *Chemical Structures* (Springer-Verlag).

Warr, W. A. and Jackson, A. R. H. (1988). 'End-user Searching of CAS ONLINE'. *J. Chem. Inf. Comp. Sci.*, **28**, 68–72.

Willet, P. (ed.) (1986). *Modern Approaches to Chemical Reaction Searching* (Gower Publishing Company).

Wolman, Y. (1983). *Chemical Information*, (John Wiley and Sons).

Zass, E. and Rehm, D. (1985). 'Chemieinformation in Lehre und Forschung'. *Frühjahrstagung der Online-Benutzergruppe der DGD 11–19 March 1985, DGD Schriftreihe (OLBG-6) 5/85*, pp. 84–101.

CHAPTER TWO

Biologicals and biotechnology information

A. CRAFTS-LIGHTY

Introduction

The pharmaceutical industry today is perceived by the general public as a manufacturer of synthetic (and therefore 'unnatural') chemical products, a view which is, of course, far from the truth. Medicine has always depended heavily on natural products, primarily plant-derived compounds, although animal products such as the anticoagulant enzymes of leeches have also earned a significant place in history. Today, extensive research is being conducted to identify, isolate and synthesize pharmacologically active substances from natural sources, mainly micro-organisms and plants. Increasingly, there is a trend to use biological rather than chemical systems to produce many of the drugs discovered, particularly where these are proteins. In addition, modern genetic manipulation techniques have opened up new opportunities to introduce genes for producing certain drugs into standard strains of 'production organisms' and more excitingly, to allow the design and manufacture of completely novel large molecules. Biological methods can offer energy and cost savings in some processes and therefore offer the industry both new products and improved efficiency.

A general product group termed biopharmaceutics now exists but many pharmaceutical manufacturing processes involve the use of biotechnology. The most widely quoted definition of biotechnology is 'the application of biological organisms, systems or processes to manufacturing and service industry'(1).

It is difficult to distinguish the boundaries of biotechnology but most people tend to associate the term with commercial or practical applications of molecular biology, genetics and microbiology. There is extensive overlap with basic biological and biomedical fields, particularly in the area of research literature, which is dealt with in Chapter 3.

Biotechnology is a term which became popular in the mid 1970s mainly to describe genetic engineering work. The pharmaceutical industry, however has been extensively involved with biotechnology since the 1940s when the large scale production of antibiotics by fermentation began. In the 1980s the main thrust has been on hormone replacement (insulin, growth hormone, etc.), antiviral drugs such as interferon and other cytokines, immuno-modulators to control the immune system and perhaps treat inflammatory and autoimmune diseases, and novel vaccines.

Biotechnological methods have also become important tools to elucidate basic biological processes, the understanding of which may lead to the development of new drugs which are organic chemicals to be synthesized conventionally. A good example of this is the cloning and expression of receptor proteins, performed to obtain purified material for X-ray crystallography. Once the three-dimensional structure is known, computerized molecular modelling methods can be used to design drugs which are inhibitors of receptor function.

Biotechnology is an interdisciplinary subject and as such presents some particular problems for information handling. The physical form of the information products and services available, however, is largely conventional, e.g. books, journals, market surveys, government regulations, databases. Relevant material is widely dispersed and can be drawn from the fields of medicine, biochemistry, genetics, microbiology, immunology, chemical engineering, chemistry and computing. Much of it is not easily identifiable as biotechnology to a non-scientist and many relevant items on techniques will not initially concern topics of direct relevance to pharmaceuticals.

International interest in biotechnology became widespread in the 1980s and most developed countries have government pro-grammes to encourage research and stimulate industrial use of biotechnological processes. The associated technology is seen as a significant opportunity for economic development although in some countries, political, environmentalist and religious opposi-tion to genetic manipulation is strong. One factor which has helped to limit public opposition is that the majority of early applications have been in health care. A 1988 survey by the US Pharmaceutical Manufacturers Association identified 81 biotech-nology-based drugs in advanced stages of development in the US, 40 of which were aimed at treating cancer. Only a few such products have actually been approved for sale; the majority being in various stages of clinical trials and preclinical research and development.

The commercial importance of biotechnology is widespread. The techniques are used in the development of pharmaceuticals and diagnostic tests but they also have applications in speciality chemicals, food processing, warfare, agriculture and mining. There are around 1000 companies developing and exploiting biotechnological techniques directly and thousands more supporting them with equipment and services. It is debatable whether a biotechnology industry as such exists, because the field is defined by a set of methods rather than a range of products, but it has certainly been seen as an expansion area by publishers.

The following sections will deal with the most important information resources which are relevant to biologicals or biotechnology for health care. A more detailed review may be found in *Information Sources in Biotechnology* (2).

Industrial microbiology and cell culture

Industrial microbiology is actually an ancient practice if one includes in it brewing and baking, but large scale pharmaceutical applications began with the manufacture of antibiotics. The main organisms involved are mainly bacteria and fungi which are grown in cultures and the desired antibiotic harvested from the fermentation broth. The active products are peptides or secondary metabolites which are secreted into the culture medium. Antibiotics may act in many different ways but generally inhibit some biochemical function essential to microbial survival or replication. The identification of new antibiotics traditionally involves screening thousands of strains of organisms for their ability to inhibit the growth of a pathogen and it is therefore important to maintain records and cultures of the organisms used in screens, of the infectious agents tested against and of those organisms/strains which produce antibiotics. In addition, in recent years the development of resistance to antibiotics has become an important clinical problem and it is thus useful to keep collections of resistant and non-resistant disease-causing strains to facilitate the study of the mechanisms of developing resistance. Companies isolating organisms producing potentially useful antibiotics or other drugs also need to store isolates of these for reference and therefore require in-house culture collections. The trend against using experimental animals for drug testing is also contributing to a requirement for maintaining cell cultures (usually mammalian) for in vitro toxicity and efficacy evaluation.

Organizing culture collections requires both a knowledge of

Table 2. 1. Possible text fields for culture collection catalogues

---

Accession number (unique identifier)
Type of organism (fungus, yeast, mammalian cell, etc.)
Organism name (genus, species, variant)
Isolate number
Genetic characteristics (plasmids, mating type, mutations, etc.)
Physiological characteristics (colony type, media required, etc.)
Products (hetorologous proteins expressed, antibiotics or enzymes
 produced)
Preservation (e.g. freeze dried spores, agar slope, etc.)
Date deposited
Source (e.g. tray number in liquid nitrogen banks)
Dates input/updated
Other (could include related research and development project number,
 comments on availability to outside researchers, number of isolates in
 stock, etc.)

---

microbiology and cell biology to select the correct storage
conditions, and attention to the design of associated records
databases. Table 2.1 lists features which a typical culture collection
catalogue record could contain. It is possible to go into greater
detail on most characteristics and it may be best to design separate
databases to handle different types of cells. The information
required on antibody producing mammalian cells could be quite
different from pathogenic fungi for example. A coding system has
been developed to analyse microbial characteristics (3) and this is
also often used. Relational techniques can help here by virtue of a
greater flexibility. Thus it would be possible to produce a data
model for an integrated system keeping track of all biological
entities but allowing discreet files for specific data collections. It
merely requires 'relating' data elements to link the files. Many of
the strains created or isolated by industry and maintained in
collections are never used commercially but the collection can
serve an important archival function.

For biologicals production, regulatory authorities such as the
US Food and Drug Administration (FDA) require the creation of
a Master Cell Bank to ensure that the product is always produced
from identical starting material which can be certified to be free
from contaminating organisms and unchanged by repeated passage in culture. Organisms which are critical to patented processes
or can themselves be patented must be deposited in an internationally recognized, publicly administered collection, although access
to them can be restricted. A list of such collections is given in
Table 2.2

Table 2.2 Important culture collections

Collection	Countries
American Type Culture Collection (ATCC)	USA
CAB International Mycological Institute	UK
Collection National de Cultures de Microorganisms	France
Culture Centre of Algae and Protozoa	UK
Culture Collection of the Institute for Fermentation	Japan
Deutsche Samlung von Microorganismen (DSM)	
(German Collections of Microorganisms)	F-R W. Germany
European Collection of Animal Cell Cultures	F-R W. Germany
National Collection of Pathogenic Fungi (NCPF)	UK
National Collection of Type Cultures (NCTC)	UK
National Collection of Yeast Cultures (NCYC)	UK
National Collections of Industrial and Marine Bacteria	
(NCIMB)	UK

Most of the culture collections listed also supply strains for research and testing purposes and publish printed catalogues of their stock at regular intervals. The American Type Culture Collection (ATCC) is probably the best known supplier but recently considerable attempts have been made to co-ordinate and improve access to European collections. Three important services have emerged; MiCIS, the Microbial Culture Information Service, a UK effort to create an on-line database containing the details of international services to aid in the location of strains of micro-organisms, MSDN, the Microbial Strain Data Network and MINE, the Microbial Information Network Europe, an inter-national operation to link European culture collections. MSDN and MINE are co-operative and complementary rather than competitive in their services. The World Federation of Culture Collections (WFCC), the Committee on Data for Science and Technology (CODATA) and the International Union of Micro-biological Societies (IUMS) are supporting the MSDN exercise.

CODATA is also sponsoring a separate service, the Hybridoma Databank, an international database to collect data on hybridoma lines. Most of these are hybrid mouse/human cells producing monoclonal antibodies. Such antibodies are increasingly incorpor-ated into in vitro diagnostic tests but also have in vivo diagnostic (tumour imaging) and even therapeutic potential (in treating cancer and septic shock and in drug delivery).

Competitive commercial culture databank services may be developed although the investment required could preclude any significant effort which is not subsidised. MIRDAB, the Micro-biological Resource Databank was a collaborative venture of Excerpta Medica (part of Elsevier), the European Federation of Cell and Virus Collections (EFCVC) and the WFCC. An ambitious database project covering cell lines, micro-organisms and genes, as yet, this has produced only one printed book (4), and does not look likely to be expanded or continued. Meanwhile, the only broad spectrum single guide to international culture collections is the printed *World Directory of Culture Collections*, a reference book published by the WFCC which lists holdings at a species level (5). A new series, *Living Resources for Biotechnology*, was announced by Cambridge University Press in 1988 and will include books on yeasts, filamentous fungi, bacteria and animal cells. These will review the major relevant culture collections and provide guides to preservation and identification procedures. Until quite recently, few culture collections have had computerized records owing to the enormity of the task of converting their manually maintained catalogues, but with the advent of services such as MINE, there is a rapidly increasing trend towards automation.

Equipment and reagents

Biotechnological techniques depend on the provision of special-ized equipment and reagents such as fermenters and monoclonal antibodies, although most of the materials used at a research level are found in all general microbiological and biochemical labora-tories. A few specialist companies provide specific materials for genetic engineering, including vectors, plasmids, DNA probes and synthetic genes. The supply of immunological reagents is a much broader field. Many polyclonal and monoclonal antibodies are used only in minute quantities for research purposes, often linked to specific derivatives for inclusion in immunoassays or as histological stains. Many of the companies supplying diagnostic tests for clinical use will also sell research reagents and many pharmaceutical companies now have divisions concerned with the manufacture of diagnostics. Diagnostic tests have likewise, be-come an important business area for many biotechnology com-panies, particularly those with a long term interest in developing therapeutic monoclonal antibodies. This is because the same antibodies can often be marketed as diagnostics much more quickly than a therapeutic product and generate the revenues needed to fund clinical trials of drugs. Biotechnology companies frequently develop their own monoclonal antibody-based assays for novel biologicals such as interferon and interleukin-2 in order to monitor their production processes.

Table 2.3. Equipment and reagent directories

Biological Substances. (1985). World Health Organisation
European Clinical Laboratory Buyer's Guide Edition. (1989).
   International Scientific Communications (Supplement to *European
   Clinical Laboratory*)
Guide to Biotechnology Products and Instruments. (1989). American
   Association for the Advancement of Science (Annual supplement to
   Science)
International Laboratory Buyer's Guide Edition. (1989). International
   Scientific Communications (Supplement to *International Laboratory*)
Laboratory Equipment Directory. (1989). Morgan Grampian.
   (Supplement to *Laboratory Equipment Digest*)
Linscott's Directory of Immunological and Biological Reagents.
   (1988–1989). Linscott's Directory.

The selection of specialized materials for a particular purpose is
generally done by the research or production scientists involved.
The main information resource used is a personalized collection of
suppliers' catalogues supplemented by directories. A selection of
relevant directories is given in Table 2. 3. The most notable of
these is Linscotts' *Directory of Immunological Reagents*, an annual
publication with quarterly supplements which lists over 30,000
antibody products. Additional information on new products may

Table 2.4. Laboratory supplies-oriented magazines

American Biotechnology laboratory (International Scientific
   Communications)
American Laboratory (International Scientific Communications)
Clinical Laboratory International (Pan European Publishing)
European Clinical Laboratory (International Scientific Communications)
International Biotechnology Laboratory (International Scientific
   Communications)
International Laboratory (International Scientific Communications)
Lab Products International (Pan European Publishing)
Laboratory Equipment Digest (Morgan Grampian)
Laboratory News (McLaren Publishers)
Laboratory Practice (United Trade Press)
Laboratory Products Technology (International Scientific
   Communications)
Life Science Lab Products (Corcoran Communications) *Medical
Laboratory World* (United Trade Press)
Medical Technologist and Scientist (AE Morgan Publications)

---

often be found in magazines aimed at laboratory personnel and Table 2.4 lists a number of publications which are primarily concerned with equipment and diagnostic tests. Some primary research journals, particularly *Nature* (Macmillan Journals) and *Science* (American Association for the Advancement of Science) carry extensive advertising concerned with laboratory supplies.

Genetic engineering and protein engineering

Genetic engineering is the main technique the public associates with biotechnology. This is the manipulation of an organism's genes to enable it to produce proteins and perform metabolic functions it would not naturally do. This can be accomplished by chemically-induced mutation but is distinct from screening for naturally occurring mutants or rare strains. More typically, genetic engineering involves using recombinant DNA techniques (cloning) to introduce a desired gene from a natural host into another, usually simpler, organism. The recipient then produces large amounts of the desired substance.

Genetic engineering methods have had a profound impact on the pharmaceutical industry over the last 5–10 years since they allow proteins such as insulin and growth hormone, previously extracted from animal or human cadavers, to be made in bacteria. This can lead to a simpler, more controlled and less expensive production process, allowing greater purity and yields with freedom from possible contaminating viruses.

The development of a recombinant DNA drug will usually first involve the isolation of the gene for a protein from human or animal tissue and its cloning into a microbial host (often initially using a plasmid vector and E. coli for simplicity). Once cloned the gene can be sequenced and by inclusion of appropriate promoter and enhancer sequences, expressed. Sometimes, a 'fusion protein' is produced by connecting the genes for two different proteins. This can then be cleaved after expression and may allow the product to be secreted extracellularly, making its subsequent purification easier. Once a functional product is obtained in large enough quantities, it can be tested preclinically and if promising, production and purification ('downstream processing') techniques developed prior to clinical trials.

Currently bacteria are prominent in their use as expression systems for recombinant proteins. However, the protein synthesis mechanism of bacteria differs significantly from higher organisms and hence it is not always easy to obtain expression of recombinant proteins in an active form. Other microbial expression systems (e.g. yeast and fungi) are under development, as are higher cells (e.g. insect, mammalian). Forms of hepatitis B vaccine and of the blood clot dissolving enzyme, tissue plasminogen activator (TPA) are now manufactured using yeast and mammalian cells respectively.

Genetic engineering is also moving into higher organisms as it is now possible to introduce genes into a mammal or plant. Recently such transgenic animals (sheep and mice) have been produced secreting pharmaceutically useful compounds such as Factor IX (a blood-clotting agent) and TPA in their milk. Such experiments suggest that 'molecular farming' could be a low cost method for producing some drugs in the future. Other researchers are evaluating genetically engineered plants as sources of pharmaceuticals although most current work on plants for drugs focuses on cell cultures to produce secondary metabolites.

Key resources to support such research and development are DNA and protein sequence databanks. Where novel vectors, genes or proteins are created (e.g. fusion proteins), these constructs are patentable but to ensure originality and to understand their structural relationship to similar genes or proteins, it is useful to compare one's results to others. In addition sequence data, per se, are tedious to read and almost impossible to analyse manually for interesting motifs (e.g. restriction enzyme binding sites). There is therefore a growing tendency not to publish complete sequence data in journal articles but merely to deposit them in a databank and discuss any important features in the paper. The databanks are all computer held and may also make available specialized software for sequence analysis. There is also a trend, led by the publication *Nucleic Acids Research* (IRL Press), to require authors to deposit a record of their sequence in a databank prior to publication of the paper. The compilers of protein (amino acid) sequence databanks and cell culture collections would like to see similar procedures followed for all protein sequences and articles describing novel cell lines or organisms. Such attitudes may eventually lead to the concept of databank publishing, whereby the creation of a database record is in itself regarded as a publication but as yet, this approach is resisted by most scientists who prefer to publish refereed papers.

Sequence database compilation is an enormous task and hence is co-ordinated internationally and subsidized by government agencies. The data however are regarded as 'public domain information' which should be 'freely' available to scientists although commercial operations can and do offer revised, added value versions of the data. Generally such data has been more easily accessible to academic scientists than industrial workers but this situation is improving with a growing realization of the importance of company research and development to the development of biotechnology.

The main DNA sequence collections are managed through GenBank, a US operation run by IntelliGenetics and the European Molecular Biology Laboratory (EMBL) nucleotide sequence library. These two operations co-operate very closely and although they still release data separately, aim to have identical databases by 1989. Up to 50% of the input is now obtained directly from authors. The remainder is derived from scanning about 70 molecular biology journals. It is hoped that data from patents will soon to be added to the EMBL library. The GenBank and EMBL databases have been distributed mainly on magnetic tapes but online access to GenBank is available on a commercial basis via Intelligenetics in the US. Daily updates to EMBL are accessible via European academic networks.

The amount of data involved is growing rapidly. The first EMBL release in 1982 covered only about 500,000 base pairs of sequences but by September 1988 the database had listings covering over 20 million pairs. It is estimated that around 150 sites now use these data, of which 15–20% are industrial. Certainly some form of access to this information is essential for any serious molecular biology effort.

Protein sequence data are also available through an international collaboration, PR International, which links the Max Planck Institute for Biochemistry in Martinsreid, West Germany with the Japan International Protein Information Database (JIPID) and the US National Biomedical Research Foundation (NBRF). Again data are currently collected mainly from journal articles. The compilers plan to process 3,000 new entries annually.

To support molecular biology research, a pharmaceutical manufacturer will require access to a wide range of biomedical journals, on-line databases and the databanks described above. It will also benefit from in-house software for sequence analyses and molecular modelling. Computer programmes running on microcomputers are available for some such tasks but sophisticated modelling of large protein molecules requires vast computing power and is often based at academic sites with the requisite supercomputers. As the underlying biology of many diseases becomes better understood, rational drug design is becoming an accepted part of pharmaceutical research and development and molecular design systems will more often be found in-house. They may be used both to design small organic compounds to inhibit proteins with enzymatic or receptor functions and to plan revisions to the structure of other natural proteins. Genetic engineering can then be used to create the modified protein for activity or functionality testing. Such 'protein engineering' methods open up vast opportunities for novel drugs and are already being used to prepare 'second generation' recombinant DNA products with improved characteristics such as increased in vivo half-lives. Although similar techniques have increased the thermal stability of some enzymes, it is proving difficult to modify the performance of proteins in a predictable and desirable way despite the very specific changes which can be made at precise sites using site-directed mutagenesis. The three-dimensional structure of many proteins is still unknown, often because they have not been crystallized, and hence it is often only the active sites of enzymes whose structure/function relationship is understood.

Enzymes

Enzymes are important to the pharmaceutical industry for several reasons. Some are themselves drugs, for example thrombolytic enzymes such as streptokinase and urokinase. Others are critically involved in disease processes and are the target of the drugs used to treat those conditions. Some enzymes are used to carry out complex reactions in the semi-synthetic manufacture of drugs such as the antibiotic penicillin. Many diagnostic tests incorporate enzymes. Restriction enzymes, which cut DNA at specific sequences are important reagents for molecular biology work.

Enzyme data are included in the protein sequence databases described earlier and a standardized nomenclature to describe the functionality of enzymes has been agreed by the International Union of Biochemistry (6). This is republished every few years with supplements published in the *European Journal of Biochemistry*. There are no major databases dealing specifically with information about enzymes of pharmaceutical interest but the UK company Biocatalysts has compiled Enzidex an extensive database of research and food-use enzymes which can be searched on a consultancy basis to source specific items.

Diagnostic methods

Bioassays, using experimental animals or isolated cells, can be time consuming and unreliable. They are increasingly being replaced, where possible, by more rapid, precise and reliable immunoassays which employ polyclonal, and now frequently monoclonal, antibodies. The labelling of antibodies used in immunoassays has been predominantly with radioisotopes; however, for safety and care of use, there is now a strong trend towards using nonisotopic labelling. Enzyme labelling is now common and enzyme immunoassays have significantly broadened the application of diagnostic tests.

Improvements in diagnostic tests also lead to new treatment opportunities. In particular with genetic diseases, 'gene therapy' to correct the hereditary problem may soon be possible. Since this involves human genetic engineering, ethical questions are raised, but some successes have already been reported in animal trials, and such techniques currently represent the best hope for treating certain fatal diseases. If developed, such remedies may operate much like vaccines with a viral vector (perhaps targeted to a particular group of cells) injected to give lifelong 'protection' by introducing the needed gene into some or all somatic cells. Germ cells would be unaffected. Approaches of this nature would be most suitable when a single protein such as an enzyme is missing and could be replaced. Genetic diseases involving chromosome translocation and consequent erroneous gene expression or the production of a defective and harmful protein which itself causes some of the symptoms, may be more difficult to treat.

Many genetic diseases are now being diagnosed by molecular biology techniques involving 'DNA probes' and restriction fragment length polymorphisms (RFLP). DNA probes are small labelled pieces of DNA with a complementary sequence to the target gene which detects the gene's presence by hybridization with it. They can also be used to detect infectious organisms and as with antibodies, can be labelled radioactively or with enzymes. RFLP methods are similar to 'DNA fingerprinting' and require the analysis of a pattern of bands produced in a gel after a sample of DNA is digested with specific enzymes. Certain patterns are indicative of particular genes (and diseases) and on a larger scale, each individual's DNA will produce a unique banding pattern overall which can be used for identity purposes. This 'human barcode' as it has been nicknamed, has obvious forensic applications but it is not yet possible to identify disease tendencies in the general population using such methods.

As yet there are few information resources concentrating specifically on modern diagnostic tests apart from the original papers and patents describing the methods involved and various papers assessing the clinical utility of tests in development or on the market. New tests available are listed in many of the equipment-oriented journals listed in Table 2.4 and in some of the directories given in Table 2. 3. The most relevant of these are *Linscott's Directory* and the *European Clinical Laboratory Buyers Guide*.

Clinical trials data and registration

The management of data about biological materials in preclinical and clinical testing is little different from handling information on synthetic chemicals. Regulatory authorities will require additional reassurance that preparations are free of contaminating viruses or mycoplasmas and further that the cell lines used in approved production processes are stable and do not vary from batch to batch. In practice, this usually requires the creation of a Master Cell Bank and documents describing standard operating procedures (SOPs) to cover growing up cells from this stock. The US Food and Drug Administration has published a number of *Points to Consider* documents on producing biological materials and is still to some extent formulating policy in this area as more and more biotechnology products reach clinical trials.

Company information

Chapter 7 deals with marketing and business information generally but it is appropriate to mention here that there are a number of information services dealing specifically with biotechnology companies and the markets for their products. Although biotechnology may not really be an industry itself, publishers and investors perceive it as an identifiable sector. Most of the information products available cover all the application areas of biotechnology but in practice, they are heavily oriented towards health care applications since this is where commercial activity is greatest.

In the 1980s a number of newsletters have been started which monitor the biotechnology industry. These contain news of clinical trials results, patents, research and development collaborations and advances, staff appointments and financial results. There are over 40 such specialist magazines available, mainly of US origin. A selected list is given in Table 2. 5. Many of these carry similar items, derived from press releases but others include original information, sometimes covering conferences. Most carry no advertising to defer production cost and hence are quite expensive. The notable exceptions are *Bio/Technology* (Nature Press) and *Genetic Engineering News* (Mary Ann Liebert), both of which are monthly glossies. *Bio/Technology* includes research papers and reviews as well as news articles. *Genetic Engineering News* has both short news items and background articles on particular companies, issues or markets.

The proliferation of newsletters, together with their research oriented counterparts (listed in Table 2.6), has inevitably led to a need for specialist indexing and abstracting services in both hard

Table 2.5. Biotechnology newsletters

Title	Publisher
Applied Genetics News	Business Communications
BioEngineering News	DJ Mysiecwicz Publishers
BioINVENTION	OMEC International
Bioprocessing Technology	Technical Insights
Biotech Business News	Bio/Medical Specialities
Biotech News	Springfield Information Services
Biotech Patent News	Biotech Patent News
Bio/Technology	Nature Press
Biotechnology Bulletin	IBC Technical Services
Biotechnology in Japan Newsservice	Japan Pacific Associates
Biotechnology Law Report	Mary Ann Liebert
Biotechnology News	CTB International Publishing
Biotechnology Newswatch	McGraw-Hill
BioVenture View	Io Publishing
DJM Enzyme Report	DJ Mysiewicz Publishers
European Biotechnology Newsletter	Editions Scientifiques Elsevier
Genetic Engineering Letter	Environews
Genetic Engineering News	Mary Ann Liebert
Genetic Technology News	Technical Insights
International Industrial Biotechnology	Cambridge University Press
New Biotech	Winter House Scientific Publications
New Biotech Business	Winter House Scientific Publications

Table 2.6. Biotechnology research journals

Bioprocess Engineering	Springer-Verlag
Bio/Technology	Nature Press
Biotechnology and Applied Biochemistry	Academic Press
Biotechnology Progress	American Institute of Chemical Engineering
Biochemical Genetics	Plenum Press
Cell	MIT Press
Cellular Immunology	Academic Press
EMBO Journal	IRL Press

Table 2.6. *Continued*

Gene	Elsevier
Hybridoma	Mary Ann Liebert
Immunology	Blackwell Scientific
International Journal of Peptide and Protein Research	Munksgaard
In Vitro Cellular and Developmental Biology	Tissue Culture Association
Journal of Antibiotics	Japan Antibiotic Research
Journal of Antimicrobial Chemotherapy	Academic Press
Journal of Applied Bacteriology	Blackwell Scientific
Journal of Bacteriology	American Society for Microbiology
Journal of Biological Response Modifiers	Williams and Wilkins
Journal of Biotechnology	Elsevier
Journal of Chemical Technology and Biotechnology	Blackwell Scientific
Journal of Clinical Immunology	Plenum Press
Journal of Fermentation Technology	Society of Fermentation Technology
Journal of Immunoassay	Marcel Dekker
Journal of Immunology Methods	Elsevier
Journal of Immunology	Williams and Wilkins
Journal of Industrial Microbiology	Elsevier
Journal of Interferon Research	Mary Ann Liebert
Journal of Molecular Applied Genetics	Raven Press
Journal of Molecular Biology	Academic Press
Journal of Tissue Culture	Tissue Culture Association
Lymphokine Research	Mary Ann Liebert
Molecular Biology and Medicine	Academic Press
Molecular and Cellular Probes	Academic Press
Molecular and General Genetics	Springer-Verlag
Molecular Immunology	Pergamon Press
Nature	Macmillan Journals
Nucleic Acids Research	IRL Press
Plasmid	Academic Press
Proceedings of the National Academy of Sciences	National Academy of Sciences
Protein Sequences and Data Analysis	Springer-Verlag
Protein Engineering	IRL Press
Science	American Association for the Advancement of Science
Yeast	John Wiley

Table 2.7. Biotechnology abstracting services

Title	Publisher	On-line hosts
Abstracts in BioCommerce	BioCommerce Data	Data-Star
		Dialog (1)
Biobusiness	BIOSIS	Data-Star
		Dialog
Biotechnology Abstracts	Derwent Publications	Dialog
		SDC-Orbit
Current Biotechnology	Royal Society of Chemistry	Data-Star
Abstracts		Dialog
		ESA-IRS

(1) Called *BioCommerce Abstracts and Directory.*

copy and database form. These are listed in Table 2. 7. The most up-to-date coverage is provided by BioCommerce Data's *Abstracts in BioCommerce (ABC)* and Derwent Publication's *Biotechnology Abstracts*. These two services overlap very little and taken together, probably provide the most cost-effective monitoring of developments in biotechnology. *Biotechnology Abstracts* covers research papers and patents while *ABC* indexes newsletters and other sources of business news. Both are published twice monthly, with *ABC* incorporating quarterly cumulated and indexed reference volumes. These also list additional citations as the *ABC* bulletin in printed form is designed primarily to offer current awareness. The *ABC* databases and reference volumes incorporate additional references where one abstract can summarize the content of several, similar articles, for example a popular story run by several newspapers. This unusual multicitation technique also means that the on-line version of *ABC* is very cost effective to search. It is available on both Dialog and Data-Star. Derwent's *Biotechnology Abstracts* is available on Dialog and Pergamon-Orbit-Infoline. Both are relatively small databases compared to *Chemical* or *Biological Abstracts* but contain many items not found in those conventional sources. With a fast moving competitive technology such as biotechnology, small pieces of information (many of which are never incorporated into published papers) can often be of great importance and may be hidden in an obscure conference presentation, patent application or newsletter report.

Specific biotechnology company information can be found using business-oriented abstracting services such as *Abstracts in*

Table 2.8. Biotechnology company directories

Bio1000	DJ Mysciewicz Publishers
BioScan	Oryx Press
BioText International Directory of Healthcare Biotechnology Company	PJB Publications
The 1988 GEN Guide to Biotechnology Companies	Mary Ann Liebert
The 1988/89 Biotech Industrial Directory	Biotechnology News
Genetic Engineering And Biotechnology Yearbook. North American and Japan Editions 1988/89	Elsevier Science Publishers
Genetic Engineering And Biotechnology Related Firms Worldwide Directory 1988/89	Sittig and Noyes
The Scrip Healthcare Biotechnology Financial Tables	PJB Publications
The International Biotechnology Directory 1989	Macmillan/Stockton Press
The U.K. Biotechnology Handbook '88	BioCommerce Data

BioCommerce but a number of specialist printed directories have also been published. Such books were very popular in the early 1980s but are difficult to keep up to date and many early editions have not been updated. A list of recent publications is given in Table 2. 8. The first online biotechnology directory became available in February 1989 via Dialog Information Services. It forms part of file 286, *BioCommerce Abstracts and Directory*, one of the online versions of *Abstracts in BioCommerce*. Detailed profiles of about 1,300 North American and European organizations (mainly companies) will be included initially but many more entries will be added.

Management of biotechnology information

Managing biotechnology and biologicals information as part of a larger pharmaceutical information programme is largely a matter of ensuring a broad outlook. It is desirable to employ information staff with a relevant scientific background such as biochemistry, genetics or microbiology instead of chemistry or pharmacology. Such personnel will be able to conduct better literature searches on-line and will have the subject knowledge to index books or articles held locally. With the exception of sequence databanks (which are usually searched directly by the end-user scientists), the types of documents and services available are broadly the same as for other types of pharmaceutical information and so will be easily accommodated by existing current awareness, enquiry, library and archival systems. Specialization of collections or services is likely to follow local management structure and hence could be arranged by therapeutic group, specific product or method. Some pharmaceutical companies will actually have biological products in development while others will be more interested in competitor intelligence or in the use of biotechnology as a research tool.

It is important to remember that there is a great deal of information generated on biological materials and their production, much of it outside the 'conventional' pharmaceutical literature. The fast pace of development in molecular biology has surprised many drug companies who have found themselves forced to license in new products from smaller specialist biotechnology companies. The 1990s however are likely to see more in-house health-care biotechnology research in large companies and the acquisition of more of the 'pure' biotech firms.

Biotechnology, particularly genetic engineering, does have one aspect which presents an additional challenge to the information professional as it becomes more widespread and that is one of public image. Information managers are becoming increasingly involved in informing their companies' employees, shareholders and customers about research and development activities and new products. Genetic engineering is widely misunderstood and often feared despite the fact that its health-care uses are mainly benign and aimed at saving life. Information personnel need to be aware of these attitudes since they may affect the thinking of company managers as well as the general public. Public opinion may also influence legislation, and restrictive regulations could necessitate relocating genetic work, a strategy already being used by some Danish and German companies in the face of increasingly constricting governmental attitudes. All in all, biotechnology presents a unique opportunity with rich promise for the future. Effective information management and presentation will materially assist scientists and business managers to realise this promise.

References

(1) Advisory Council for Applied Research and Development, Advisory Board for the Research Councils, The Royal Society. (1980). *Biotechnology. Report of a Joint Working Party* (HMSO).
(2) Crafts-Lighty, A. (1986). *Information Sources in Biotechnology* (Macmillan).
(3) Rogosa, M., Krichevsky, M. and Colwell, R. R. (1986). *Coding Microbial Data for Computers* (Springer Verlag).
(4) Elsevier. (1985). *Microbiological Resource Databank. MIRDAB catalogue. Volume 1.* (Elsevier Science Publishers).
(5) WFCC. (1988). *World Federation of Culture Collections Directory.* (World Federation of Culture Collections).
(6) IUB. (1978). *Enzyme Nomenclature.* (Academic Press).

CHAPTER THREE

Biological and biomedical information

P. DUCKITT

Introduction

The purpose of this chapter is to describe the sources and uses of biological and biomedical information in the context of their application within the area of pharmaceuticals. To cover such an extensive field within the confines of a single chapter requires a careful strategy and a few words of explanation. An attempt to enumerate a literature where the number of journals alone exceeds 10,000 would be futile. The alternative approach of describing the principal classes of information source important to the pharmaceutical industry and giving one or two examples of each type has been adopted. Wherever possible, sources have been identified which give the amount of detail necessary to identify individual items relevant to user requirements. Within the text specific important sources which are updated from time to time are mentioned. At the end of sections describing each of the classes of source there are references, where appropriate, to particular listings in a few major reference works on information sources. Although these will necessarily become dated, the research and experience captured in them remains relevant to all workers in the field. Finally, full bibliographic citations of these and other publications specifically mentioned are detailed at the end of the chapter.

Different names are often given to printed, online and optical disc-based formats of very similar publications. Wherever the name varies, the name commonly associated with the online version has been chosen because it is assumed that the online version will be the most familiar to information specialists within the pharmaceutical industry.

In association with source descriptions, the practical aspects of providing the corresponding information service within a pharmaceutical company are considered. What immediately follows describes the environment in which such information services must operate and what they are trying to achieve. Later in the chapter, the provision of services theme returns and the many options available to an information department regarding choice of sources and delivery format are explored together with service approaches which bring these complementary features together.

Biological and biomedical aspects of the drug development process

Biology is the science of the material life of organisms of all kinds. Biomedicine is generally used to denote that area of medicine concerned with the scientific study of the biological mechanisms underlying health and disease. Biotechnology is defined and considered elsewhere. The activities of gathering, interpreting and using biological and biomedical information are crucial to many stages of drug development.

Many noteworthy drugs, e.g. penicillin, have been discovered by a combination of observation and serendipity but that is not today's norm. Most drug discovery takes place within well defined research programmes. The establishment of these programmes (deciding in which areas of medicine to work) and the definition of projects (deciding which particular products will be sought) are fundamental to corporate planning. The systematic study of nucleic acid metabolism in micro-organisms and man which started in the 1940s serves as an example. From this basic research programme emerged drugs for the treatment of leukaemia, bacterial and protozoal infections, gout and other conditions of hyperuricaemia, and for protection in organ transplantation. Given a different development strategy other results may well have arisen.

The basic method of identifying therapeutic or prophylactic activity involves chemical compounds and appropriate biological systems or screens in which to test them. There are many sources of compounds. Most pharmaceutical companies have extensive files of registered research compounds that have been previously synthesized. Synthesis by design frequently takes place – either by modification of a known compound or construction of a molecule from scratch. Substances may also be recovered from natural sources such as plant extracts or microbial broths by processes of analysis and extraction or isolation. Alternatively, organisms of various kinds may be subjected to genetic manipulation in order to produce known or novel compounds. Researchers choose the compounds they think may have activity on the basis of their expert knowledge.

Ideal screening tests are those which can be performed quickly and in large quantities but which are also appropriate models for the pathological condition at which the drug is being aimed. In vitro tests are actively sought to minimize the use of experimental animals.

Before any of the compounds which show activity in the biological tests are selected as candidates for further development they must be assessed for possible relationship with known structure-activity and toxicity patterns and for their patentability.

The first step in biological characterization is to demonstrate the supposed utility of the compound in whole animals. The range of primary and secondary pharmacological effects in normal animals must be described and quantified. If an animal disease model is available (human arthritis can be mimicked in rats, for example) then the pharmacological effects can be tested in these animals as well. This early work provides the first evidence of dose-response relationships, pharmacokinetics (how the body acts upon the drug through absorption, distribution, metabolism and excretion) and likely toxicity.

Various tests are carried out on several animal species to quantify toxic effects. The 'ideal' drug has a wide gap between the dose levels at which it is pharmacologically or therapeutically effective and those at which it is toxic. The various forms of toxicity and the methods used to detect them are dealt with elsewhere in this book.

The selection and breeding of suitable strains of animals and their care before and during these studies is a necessary part of drug development. As with screening tests, methods for identifying toxic effects which will reduce the current huge investment in experimental animals are being sought.

Administering the compound, initially in minute quantities, to healthy volunteers establishes the human therapeutic dose which is safe to be given to patients in carefully controlled clinical trials. These findings have a major influence on whether the drug is to be progressed or not and provide valuable feedback into other development stages. Chemists and pharmacologists will receive guidance as to further molecular design once they know what pharmacological actions are produced in humans. Toxicologists, once they know the route of metabolism and excretion, can choose appropriate animal species for toxicity trials. The pharmacokinetic data are helpful in selecting appropriate formulations.

Information requirements and uses

Policy makers, clinical researchers, patent agents, marketing staff and those who provide product information to the outside world all use biological and biomedical information from time to time. However, the specific requirements of these groups are dealt with in later chapters. The main group of people requiring biological and biomedical information are research scientists involved in drug discovery, characterization and evaluation up to the point of clinical studies. The common label 'research scientist' does not mean that they form a homogeneous group with the same information requirements. Their functional specializations and the research programmes and development projects in which they are participating will determine their immediate needs. In addition, scientists will have their own individual exploratory research interests to pursue.

In general, information is needed to promote ideas and creativity; to solve problems; to manage day-to-day activities; to make decisions; to plan for the future; to keep in touch with peers. Bawden (1986) thoroughly reviews all the ways in which information can be used to stimulate ideas and creativity and to solve problems, specifically within the context of research and development in science and technology and from his own viewpoint within a pharmaceutical company. The most important information activities for these purposes are: providing an overall information-rich environment; including peripheral and speculative material; supplying interdisciplinary information; providing browsing facilities; encouraging informal channels of information transfer; and giving information geared to individual preferences and requirements.

The need for information for day-to-day management is most obvious where project management is concerned. Activities cut across functional specializations and involve various feedback loops which must be exploited in order to achieve effective use of resources. Internally generated information is naturally more valuable in meeting this requirement.

Information is required for many disciplines, including molecular biology, genetics, biochemistry, physiology, pathology, pharmacology and toxicology. Microbiology and zoology must be added to the list if the company produces antimicrobials and endoparasiticides. Researchers make their decisions based on a fund of knowledge about the state-of-the-art in each discipline. Any slight movement away from their existing specialities will necessitate new background information to create or extend this vital underpinning. Researchers also need specific data which have already been established. They can compare these with the unknown aspects of what they are investigating in order to reduce their uncertainty and make more effective decisions. External information in the form of intelligence and opinion on conditions in the outside world which will affect the company's activities (epidemiology, health requirements, activities of other companies and so on) will also be useful.

Finally amongst the general requirements, there is a need for information exchange within the researchers' peer groups, an essential part of keeping knowledge up to date. Some of this exchange will come through scanning the published literature but this only supplements the informal exchange which takes place within the framework of professional meetings. Research scientists need to be constantly alerted to such meetings.

The more specific information requirements of researchers are determined by the particular functions they undertake. Those who are responsible for obtaining compounds from biological sources, for instance, need to be aware of the families, genera and culture conditions that have produced useful compounds previously and, indeed, those that have been non-productive. Compound choice and design are also dependent on what is known about the physiology of body systems, biochemistry of particular processes, pathology of target diseases and pharmacology of particular compound groups. Information about screens and disease models used elsewhere can influence the design of new ones. Information on all aspects of toxic effects can usefully be taken into account before animal testing in determining initial dosage levels. All target organisms and experimental models will have different characteristics which must be taken into account in experimental design, in extrapolating results to other species, and in ensuring their care and welfare. Information resources to cover these aspects will be requested.

Information sources appropriate to the use of biological systems for the production of drugs and research compounds, and to the application of biotechnological methods to elucidate basic biological processes, are discussed in Chapter 2. Information about drug metabolism and toxicology for regulatory purposes is covered in Chapter 4.

Information characteristics

The externally generated biological and biomedical information that is commonly drawn upon in the drug development process has the same characteristics as any other original information. It exists in five states: ideas, observations, experience, intelligence and opinion. Ideas may come from intuition or from the ordering and reviewing of existing information. Observations are the products of experimentation. Intelligence tells us what is going on outside ourselves. Opinion represents people's judgments and beliefs.

The ways in which this information is recorded and transferred follow the characteristics of scientific information in general. Much of it is not made public in any formal way but is transferred directly between people. Formal transfer via publication in the scientific literature ensures that information in its various states is recorded in an impersonal way, open to anyone who cares to look. As well as fulfilling these archival functions, the published literature also has important social and economic consequences, as personal recognition and other professional rewards such as research funding are achieved largely as a result of published work. Primarily, however, formal publication assures information quality through the practice of refereeing.

The essential nature of scientific information transfer is summarized by de Solla Price (1981) and Kronick (1985). The basis of informal communication is the invisible college of people working in a subfield of knowledge. As such a group gets organized, its communication system moves from a membership list for offprint or preprint exchange via an informal research newsletter to a subsection of a professional society with its own meetings and eventually to the more formal research journal. Once it becomes established as a speciality, the number of members increases beyond that which can support informal communication and the process continues to produce a hierarchy of fields and subfields of knowledge, each containing units more closely related to each other than to those outside. The formal scientific information transfer system is based on scientific papers. Each gives rise to new opportunities to ask questions and to solve problems and so the number of scientific papers grows at an exponential rate. A growing number of journals alone is not sufficient to compensate for this explosion of information and there occurs progressive compression of the contents of scientific papers into abstracting and indexing journals, reviews, monographs and data compilations. Only by this means does the volume of information become manageable. Such a process provides the distinction between primary publications and secondary publications. According to the Council on Biological Sciences Information (1970), primary publications are citable works which give sufficient details to enable peer review for the validation of the results and conclusions recorded. Secondary publications in the strict sense provide means by which scientists are made aware of and are assisted in obtaining access to primary publications, but more broadly can be thought of as those which give access to recorded knowledge.

The amount of information thus presented would be completely unmanageable were it not for the fact, noted by Bradford, that the relevant sources for any particular subject are concentrated in a relatively small part of the literature. Kronick (1985) describes this phenomenon and de Solla Price (1981) rationalizes it in terms of the principle of cumulative advantage, a constant process of positive feedback resulting in success building on success whether of a particular author or a journal publishing in a particular subject area. Whatever the theoretical basis, the practical result is the clustering of information sources into core and peripheral collections which, if carefully selected, can ease the process of tracking down information for particular purposes.

Biology and biomedicine share these same characteristics with the rest of scientific information but with major additional factors. The Council on Biological Sciences Information (1970) summed them up thus: 'The special problems in the handling of biological information arise from the diversity of biological subject matter and the complexity of biological approaches towards phenomena of the living world . . . This diversity is reflected in the twenty major categories that the National Science Foundation finds necessary for biology, as opposed to four for chemistry and ten for physics; in the existence of hundreds of societies and associations in biology; and . . . in the large number of journals that biologists have found it necessary to establish in order to satisfy real or imagined specialized needs . . . Although the problems of handling biological information are no different in kind from those in the physical sciences, factors of diversity and complexity make them vastly different in magnitude.'

The diversity and complexity have resulted in some particular trends in the characteristics of biological and biomedical information which are worth noting briefly. Fragmentation has, paradoxically perhaps, resulted in the need to bring together different disciplines and fields to tackle particular problems. There has therefore been an increasing number of mission-oriented information sources as opposed to traditional discipline-oriented sources. In an atmosphere of heightened competition for scarce research resources, various publication strategies have been developed which increase the profile of authors and their work. Notable amongst these are the reduced size of the unit of publication, giving a larger number of publications in the form of short communications of various kinds, and the use of English as the language of publication, irrespective of the author's native tongue. Publishers are also trying to reduce the time from receipt of manuscript to publication by methods such as publishing from camera-ready copy, accepting the manuscript in machine-readable form and experimenting with electronic publication. The publishers of secondary information sources are also intent on speeding up the processing of primary literature. Whilst most people work in specialist fields, there remains a general need to know what is happening in other areas. This shows itself in the escalating number of review-type publications which seek to collate and comment on primary publications in selected areas. The net result is more publications about information sources, and about information sources in ever smaller areas, such as this chapter!

The final publication trend within biology and biomedicine is one which it is to be hoped will not become widespread. This is the deliberate withholding of research results for commercial reasons, beyond the usual restraint which precedes a patent application. A case was recently reported in the popular scientific press (Anon, 1989) where an American biotechnology company announced that it had isolated and sequenced the genome of the virus that causes non-A, non-B hepatitis and had developed a prototype test to identify carriers. The company was taken to task by virologists, notably within the public sector, for not publishing its findings. This nicely demonstrates the point that all scientists are ultimately dependent on the prompt and free dissemination of new information.

The sources

The sources are described by class, with references to additional material being added at the end of each section where appropriate. There are some useful items which discuss many source classes for one subject area and give extensive lists of individual sources. They are grouped here for ease of reference.

See: Turner (1987) for biochemistry; Rowland (1984) for biochemistry, biophysics and molecular biology; Wyatt (1987b) for microbiology; Whyte (1984) for medical microbiology; Calam (1984) for pharmacology and therapeutics; Hamilton (1984) for immunology and transplantation. Turner (1987), Rowland (1984) and Wyatt (1987b) contain discussion of sources on methods and techniques. Sources of information on experimental animals are covered in Wyatt (1987b) and Roberts (1984).

Non-published sources

There are several possible titles for this section, all of which are imperfect in some way. Essentially, it embraces informal, people-based sources that fall outside the formal methods of scientific publication. Two principal aspects are dealt with here – the identification of individuals and organizations possessing expertise in particular areas, and the identification of conferences and the like at which people with common interests meet to discuss and exchange information. These sources are included because they are accessible to outsiders. There are other methods of communication between members of an invisible college which might be useful sources, were they accessible. Research newsletters, for instance, are especially relevant to the life sciences and are described by Wyatt (1987a) as 'an invisible literature – the untraceable publishing the uncitable'. Electronic bulletin boards, which merit the same description, provide a further example. As the circulation of both is limited to the invisible college which generates them, they are more relevant to the scientists concerned than to the information professional.

EXPERTS AND CENTRES OF EXCELLENCE

Skilled and knowledgeable practitioners often provide the quickest route to information, especially when practical details or opinions about current research or emerging areas are required. The information obtained in this way is already selected and evaluated.

As well as identifying such individuals, it is also useful for the information specialist to select organizations with corporate expertise in particular areas. These centres of excellence – perhaps research institutes, government-funded bodies, campaigning organizations, trade associations, learned societies or professional organizations – often create and organize their own information files. They frequently provide, for example, collections of published and unpublished literature, together with data and statistics which will be indexed in depth for retrieval by their own information department. Some of this information may already be available in the form of publications or databases but the centre itself will often supply a superior service in terms of selection, evaluation and speed.

Indexes and directories of expertise, research and services, which point to individuals and organizations, are now readily available. It is always advisable to establish how the entries have been selected, i.e. whether the inclusion of a person or organization merely acknowledges their existence or constitutes a positive recommendation. The sources tend to cover particular geographical areas and/or particular subject specialities.

Researchers, scientists and engineers in the UK's universities, polytechnics and government laboratories can be traced through

British Expertise in Science and Technology (Longman Cartermill) or *Current Research in Britain* (British Library). Other examples of the growing number of online directories are *ExpertNet* and *PDQ*. *ExpertNet* contains biographical profiles of US medical experts, especially those available for medico-legal consultation. *PDQ* (National Cancer Institute) is a database on cancer treatment with a directory section containing information on US physicians, organizations and institutions specializing in cancer care and research.

Individuals actively researching (and publishing) in particular fields can be identified via indexing and abstracting services (see below). It is necessary for this purpose to use those services which routinely include the author's organization and address, such as *BIOSIS Previews, Embase* and *Scisearch*.

Membership lists of learned societies and professional organizations provide names of individuals and sometimes indicate special areas of expertise. The *World of Learning* (Europa) lists learned societies by country. Details of medical societies in the UK are given in the *Medical Directory* (British Medical Association).

Once individuals have been identified, biographical sources can be checked for further details. Examples of these are the *Medical Sciences International Who's Who* (Longman), *Who's Who in Science in Europe* (Longman), *Who's Who of British Scientists* (Simon Books) and *American Men and Women of Science* (R.R. Bowker).

Most directories of research, however, concentrate on organizations, thus providing one route to identifying centres of excellence. Examples are *European Research Centres, Pacific Research Centres*, both directories of organizations in science, technology, agriculture and medicine, and *Medical Research Centres*, all published by Longman. Medicine in the UK is covered by Wiley's *Medical Research Directory*. The *Research Centers Directory* (Gale Research) is a comprehensive guide to non-profit and university research in the US and Canada but does not include government research.

Other directories do not specifically cover research but identify associations working in discrete subject areas or on behalf of particular groups of people. The *Directory of British Associations* and *Associations in Ireland*, the *Directory of European Associations* in two volumes and *Pan-European Associations* (all CBD Research) and the *Encyclopedia of Associations* (Gale Research) include many of these.

The various departments of national governments provide a vast repository of expertise and knowledge. The US government is willing to share theirs but it is necessary to find the appropriate contact. *Information USA* (Viking Penguin) provides a suitable guide to help locate such persons.

Information centres or libraries may be identified using helpful sources such as the *Aslib Directory of Information Sources in the United Kingdom* in two volumes (Aslib) and *European Sources of Scientific and Technical Information* (Longman). Specialist directories refer to slightly smaller subject areas within the UK, e.g. the *Directory of Medical and Health Care Libraries in the United Kingdom and Republic of Ireland* (Library Association), *Life Sciences Libraries in and around London* (Imperial College) and the *Guide to Government Libraries* (British Library), which also lists many other 'official' libraries. The *Directory of Health Sciences Libraries in the United States* (American Medical Association) covers nearly 3000 libraries in the US.

Other useful directories can be traced through the *Directory of Scientific Directories* (Longman) and *International Directories in Print* (Gale Research).

See also: Morton and Godbolt (1984) for details of medical libraries in the UK, Ireland, US and Canada; Chen (1981i) for a list of directories; Chen (1981ii) for a list of biographical sources; Chen (1981iii) for a list of professional society directories.

CONFERENCES AND MEETINGS

In his review of information systems and the stimulation of creativity, Bawden (1986) emphasizes the importance of conferences and similar meetings. He points out that 'the contribution of the information professional here will be to provide the sorts of tools already available (indexes of expertise, referral services, meetings calendars, etc.), and indeed to expand their scope where possible, and try to maximize receptiveness within their organizations to the importance of these informal channels.'

Information on forthcoming conferences and meetings tends to be scattered. Most information departments find that no single source meets the diverse needs of a pharmaceutical company and spend considerable time and effort on scanning a variety of sources for details.

Amongst the sources devoted entirely to future meetings are: *Forthcoming International Scientific and Technical Conferences* (Aslib), the *Annual International Congress Calendar* (Union of International Associations, Brussels), *World Meetings: Medicine, United States and Canada* and *Outside United States and Canada* (Macmillan), the *Calendar of Congresses of Medical Sciences* (Council for International Organizations of Medical Sciences,

Geneva), *Fairbase* (Fairbase Datenbank GmbH) and *Forthcoming Events* (BIOSIS).

Many publications whose principal information content lies in other areas include notices of forthcoming meetings. Amongst abstracting and indexing services examples are *Current Biotechnology Abstracts* (Royal Society of Chemistry) and *British Medicine: A Monthly Guide to the Literature* (Pergamon Press). Many primary journals contain meetings calendars, especially when they are organs of professional organizations. The *British Journal of Pharmacology* and the *Journal of the American Medical Association* are just two examples. Newsletters are also a useful source of announcements.

Published sources

Although the distinction between primary and secondary published literature can readily be made in terms of the way they are used, as discussed earlier, the physical formats in which primary and secondary publications appear are not so easily disentangled. Primary published sources are discussed under the headings journals, publications from meetings, reports, theses and patents, whilst headings for secondary published sources are indexing and abstracting services, reviews, books and data compilations. It is recognized that this categorization is imperfect and that there are many exceptions. Journals may contain reviews as well as scientific articles. Meetings papers may be published in journals. Reviews may contain information which satisfies the definition of primary literature, as may certain books. Nevertheless, published sources have been divided in this way because each category tends to be used in a different manner for particular information requirements and also because there are different sources and routes for tracking down members of each category.

Primary published sources

JOURNALS

Journals (serials or periodicals) contain many different kinds of information, including didactic information, opinion and analysis. However, the main importance of journals lies in the information derived from experiments and observation that are contained within scientific research papers.

The information requirements of biological and biomedical researchers within the pharmaceutical industry will largely be satisfied through the provision of relevant journal articles. Scientific and clinical researchers typically want information in a raw form, in contrast to some other categories of information users in the pharmaceutical industry. They want to assess the quality and validity of the research for themselves.

It is impossible to find a reliable figure for the number of journals in the fields that would encompass the full range of biological and biomedical interests within the pharmaceutical industry. Based on the numbers of journals scanned by the major indexing and abstracting services and on the extent of their overlap, a reasonable estimate must be around 10,000 titles. Needless to say, journals are not all trying to do the same thing nor are they of equal quality. When selecting journals as sources of information, it is necessary to be aware of the variations that are possible in subject orientation, format of articles, intended readership, geographical emphasis, sponsorship, editorial policies, speed of publication and circulation policy. Some of these factors may affect the general quality of the journal and others will affect the value of a source journal when trying to answer a particular kind of enquiry or satisfy a special information need.

The Council on Biological Sciences Information (1970) stated 'It is possible to identify – although probably no one would dare do so in print – a list of about a thousand journals in which upwards of 90 per cent of the significant original work in biology appears.' This relates to the concept of core journals described above. Knowledge of the core journals in a field can be used to establish a collection for reference purposes, to choose sources for current awareness scanning, or to select titles in order to answer a specific question. Although it may be difficult or impossible to define the core journals by scientific standard alone, there is a more objective means of evaluation based on methods of citation analysis. By counting the number of citations to a particular journal in the journal articles indexed for *Scisearch* in a given period, and by relating these to the number of articles published in that journal, lists of core journals can be derived for any subject area (Garfield 1976). The results of such studies are regularly published in *Current Contents* and are useful in providing quantitative data to supplement qualitative assessments and value judgements. Another quantitative approach to producing a set of core journals in a specialized field is to carry out a computerized search of an appropriate indexing and abstracting service and to subject the retrieved items to a frequency analysis of the journal titles from which they are drawn.

Ulrich's International Periodicals Directory (R.R. Bowker) can be searched for journals covering particular subject areas. Specialist libraries often supply lists of their journal holdings and the larger libraries produce computerized catalogues which can be used to identify journals by subject, language, and country of publication. Examples of note are *Serline* (National Library of Medicine), the *British Library Catalogue: Science Reference and Information Service* and the *Swiss Libraries Biomedical Journals*. All of the major producers of indexing and abstracting services publish lists of the journals they cover. In printed form, their value generally lies in providing accurate bibliographic information but when they are available in computerized format or in classified form they can be used as an additional searchable source of subject listings.

Information about new journals appears in announcements, advertisements and reviews in major journals such as *Nature, Science*, the *Lancet* and the *New England Journal of Medicine*. A systematic check can be carried out via the various national bibliographic services, for example *BNBMARC* (British Library) and *LCMARC* (Library of Congress) which contain a record for each new journal title.

These are means of searching for and identifying journal titles. Searching the contents of journals by means other than scanning the issues themselves requires indexing and abstracting services which are described next. The principal services for journals relevant to biology and biomedicine (although most cover other literature types in addition) are *Chemical Abstracts* (American Chemical Society), *BIOSIS Previews* (BIOSIS), *Medline* (National Library of Medicine), *Embase* (Excerpta Medica), *Scisearch* (ISI) and *CAB Abstracts* (CAB International).

See also: Dannatt (1984) for a list of guides to periodicals and a list of significant biomedical journals (selected on the basis of three different criteria); Chen (1981iv) for a list of significant medical journals.

PUBLICATIONS FROM MEETINGS

Despite the continuing success of scientific conferences and meetings, the value of their published output is disputed (Whimster, 1988). The main point of contention is that the papers and abstracts relating to conference presentations are often published before the conference takes place and may not be subjected to the procedure of peer review that accompanies other methods of publication. There are, however, things to be said in favour of meetings papers. Not all papers or abstracts are pre-prints, which attract most of the criticism. Depending on the publisher, many will be reviewed. Some presentations end up being published in journals as discrete papers, some will appear in a single volume along with other contributions from the same conference producing essentially a monograph, and other proceedings are published regularly as a series, being virtually indistinguishable from a journal. Even in their less defensible forms, there is still a role for meetings papers as an information source of value to the pharmaceutical company information department seeking biological and biomedical information. If published soon after the event, even if only in abstract form, they can be used in a number of ways. They provide: current awareness of who is working on what and where; reviews of the current state of the art; insights into approaches that have not been successful; and sometimes records of discussion that add to the basic value of the papers themselves.

The problem with the use of meetings papers as a source of information is that they are often difficult to trace if they are not published as part of a journal. Of the major indexing and abstracting services in the field, *BIOSIS Previews* contains signifcant coverage of biological and biomedical meetings. There are several sources which are devoted to meetings but these cover all areas of science and technology, including biology and biomedicine. The main ones are *Index to Scientific and Technical Proceedings* (ISI) and *Conference Papers Index* (Cambridge Scientific Abstracts). Other services index the proceedings as a whole rather than individual papers, for example the *Directory of Published Proceedings* (InterDok Corp.) and *Conference Proceedings Index* (British Library).

See also: Chen (1981v) for a list of the main published conference series.

REPORTS

Reports originating from outside the information department's own organization may be a minor source of primary published information for the biological and biomedical research worker but should not be ignored. They may document fully, for instance, research work being carried out by or on behalf of a government department or international organization under research grants or they may bring together expert evidence, analysis and data associated with the investigation of a particular problem. Annual reports of research institutes frequently list current work that is not yet published and provide useful lists of publications. The level of detail is often high in reports and results may appear quickly.

Unfortunately, reports are often hard to find. Many are distributed selectively and might be described as being only partially published. They are a major component of the so-called 'grey literature' and are more difficult to identify and obtain than other kinds of literature. *BIOSIS Previews* (BIOSIS) covers research reports in biology and biomedicine. *NTIS* (National Technical Information Service) indexes technical reports arising from US government funding in all subject areas. SIGLE (European Association for Grey Literature Exploitation) attempts to cover this type of literature. Official publications produced by the government publishing offices of the US and UK respectively are listed and indexed in the GPO and HMSO catalogues. For the UK there is an additional source, *British Official Publications (Non HMSO)* (Chadwyck-Healey).

THESES

Theses or dissertations produced as full or partial requirements for research degrees are another category of 'partially published' primary information sources. Most significant findings recorded in theses are subsequently published elsewhere, but the value of theses lies in the full bibliographies they contain, the wealth of technical detail recorded in them and the information they provide about the research being carried out in particular institutions.

Dissertation Abstracts Online and *BioTheses* (University Microfilms International) both cover several countries but principally the US and Canada. The *Index to Theses* (Aslib and Expert Information Ltd) covers the UK and Ireland. SIGLE also contains some theses.

PATENTS

Although patents are dealt with in detail in Chapter 6, it is worth mentioning them briefly here because of their value to biological and biomedical researchers as technical literature. Patents grant a limited monopoly in exchange for divulging as much technical detail as would be necessary to allow someone skilled in the art to reproduce the invention. As the inventions relate to products, processes and now even genetically manipulated organisms, their practical value to researchers is considerable. They also serve as indicators of research activity and hence have a current awareness role.

Patents are covered by various life sciences indexing and abstracting services as technical literature as well as by specialist patent databases. They are worth searching if a cross-section of patents is desired along with other types of literature in a subject search. Examples of such indexing and abstracting services are *CAS* (Chemical Abstracts Service), *BIOSIS Previews* (BIOSIS), *Life Sciences Collection* (Cambridge Scientific Abstracts), *Biotechnology Abstracts* (Derwent Publications), *Current Biotechnology Abstracts* (Royal Society of Chemistry) and *PASCAL*

(CNRS). *Scisearch* (ISI) includes patents as cited references and thus links them to the mainstream scientific literature.

See also: Simmons (1988) for a detailed discussion of indexing and abstracting services covering patents.

Secondary published sources

INDEXING AND ABSTRACTING SERVICES

Indexing and abstracting services provide references to publications and some indication of content. Their purpose is to allow retrieval of relevant literature in a defined field of interest, to provide a substitute for the original document through an outline of the salient points in an abstract, and to provide an alerting service to current publications.

These services vary considerably in form according to the amount of information derived from the original document and the amount and type of information added by the supplier. At one end of the spectrum are contents lists of journals and books, such as the *Current Contents* series (ISI) which includes *Life Sciences* and *Clinical Practice* editions. The contents for each journal issue are displayed more or less as they appear in the original publication and indexes are added to allow searching by author and by significant words in the titles. Such services are designed for scanning by the end-user. They provide a more convenient alternative to visiting a library and consulting each journal and extend the range of sources that can be examined. Their success depends on the scanner being able to identify relevant articles from a combination of the title, the journal in which it appears and the author. The identified articles have then to be obtained.

In the middle of the range are services which group references together according to broad subject area, rather than source, and add subject indexing to represent the content of the entire article. The index terms may be uncontrolled or controlled. Uncontrolled terms usually supplement the words in the title and are based on the terminology used by the author or the indexer's subject knowledge. Controlled indexing terms are taken from a thesaurus and are chosen to represent the important concepts in the original article, independent of the title words or other terminology used. The author's abstract, perhaps with slight editing, is often added. Examples of such services are *Medline* (National Library of Medicine) and the *Life Sciences Collection* (Cambridge Scientific Abstracts).

Other services extend this basic pattern in various ways. They often provide more extensively edited or even specially written abstracts. The indexing systems used may be extremely sophisticated to allow, for example, precise representation of chemical substances, retrieval of groups of related topics or concepts, and linkage of terms to prevent false combinations during retrieval. Major examples are *Ringdoc* (Derwent Publications) and *Chemical Abstracts* (American Chemical Society).

One major indexing service which differs from the others is *Scisearch* (ISI). It uses the technique of citation indexing. This allows a search for documents which have cited a known publication by exploiting the scholarly practice of referring to related older work. In this way, a subject search can be extended forwards in time. Extrapolating this concept to documents which have cited references in common allows the definition of coherent, active areas of research which are termed 'research fronts'. These can be used as additional subject retrieval keys. Citation indexing has the advantage of being, by its nature, multidisciplinary and thus a useful addition to the range of retrieval tools for the biomedical searcher in the pharmaceutical field.

All these indexing and abstracting services allow, to a greater or lesser extent, identification of the latest developments in a topic, retrieval of a corpus of literature on any subject, and identification of individuals and institutions working in particular subject areas. Depending on the sophistication of the indexing system, the searcher can identify concepts, chemical substances, diseases, organisms, techniques and the precise relationships between them. Where the services are available in computerized form, it is possible to perform searches with both high recall and high precision and to restrict easily the results by additional criteria such as language and date of publication.

Services vary in their coverage of types of publication. The most important indexing and abstracting services for coverage of journals, reports, theses, meetings papers, reviews and books have been mentioned in the appropriate sections. Although the information specialist will want to use tens or scores of different services in providing for the full range of information needs in biology and biomedicine, basic requirements can normally be met from the literature coverage provided by the main services in the field: *Chemical Abstracts* (American Chemical Society), *BIOSIS Previews* (BIOSIS), *Medline* (National Library of Medicine), *Embase* (Excerpta Medica), *Scisearch* (ISI) and *Commonwealth Agricultural Bureaux Abstracts* (CAB International). For other services, the most up-to-date sources are now the directories of online databases. Of these, the two most comprehensive are the *Cuadra Directory of Online Databases* (R. R. Bowker) and the *Database of Databases* (M. E. Williams, Inc.). Directories of CD-ROM products, e.g. the *International Directory of Information*

Products on CD-ROM (TFPL), are becoming more useful as the number of indexing and abstracting services available in this form increases.

See also: Hassanaly and Dou (1988) for discussion of indexing and abstracting services in chemistry; Snow (1988) for discussion of indexing and abstracting services in biological sciences; Van Camp and Seeley (1988) for discussion of indexing and abstracting services in health sciences.

REVIEWS

Reviews involve critical selection and condensation of the primary literature in a particular area. At best, they can lead to new interpretations of work and provide insights into a subject, such that they should really be considered a primary contribution to research. At worst, they are little more than aggregations of abstracts. Whatever their individual merits or deficiencies, reviews are becoming important information sources in biology and biomedicine because of the aforementioned growth in volume of primary publications accompanied by fragmentation of subject fields.

Reviews are also valuable in providing a foundation for moving into a new area or aspect of research. They can save much time and effort by encapsulating the state of the art and identifying key issues, methods and individuals. They also provide a useful tool for the information department setting up current awareness profiles, in that their bibliographies can be used to identify rapidly the most productive primary sources in a particular subject area.

Reviews appear in many different places and in many guises: as articles in journals, as conference papers, and as contributions to treatises, handbooks and encyclopaedias in addition to review journals and review series. Theses and journal articles reporting primary research frequently include substantial and important reviews. Journals and series devoted to reviews are usually identified by suitable titles such as *News in Physiological Sciences, Progress in Allergy, Current Topics in Nutrition and Disease, Methods in Enzymology, Advances in Immunology, Recent Advances in Neuropathology, Trends in Biochemical Sciences, Annual Review of Genetics* and so on, seemingly without limit.

Identifying review articles in journals is a straightforward procedure if they are indexed as such by the secondary services. *Medline, BIOSIS Previews* and *Chemical Abstracts* (amongst others) provide this information. Reviews appearing in series are less commonly covered by the major services. *BIOSIS Previews* is a good source and the *Index to Scientific Reviews* (ISI), as its name suggests, is devoted entirely to reviews from both multi-authored works and journals. Review series can be found through the *Irregular Series and Annuals* part of *Ulrich's International Periodicals Directory* (R. R. Bowker).

See also: Chen (1981vi) for a list of important review series.

BOOKS

Books contain many different types of information for many different purposes. Occasionally they will be the vehicle for primary publication of information but more often share the characteristics of reviews. The purpose of a book may be educational (a textbook or treatise), for reference (a handbook), for instruction (a manual), or for the advanced discussion of highly specialized areas (monographs). All of these will have their uses from time to time in providing information to biological and biomedical staff working in pharmaceuticals.

New books will be identified through book reviews (potentially valuable if critical) and listings in relevant journals or newsletters but although the more significant may be recorded there, the number that can be considered is relatively small. Publishers' catalogues and lists from major specialist booksellers give greater coverage of titles and synopses of new and forthcoming books, although they cannot be expected to be evaluative. Bibliographic services such as *BNBMARC* (British Library), *LCMARC* (Library of Congress) and *Whitaker's* (J. Whitaker and Sons Ltd.) and computerized catalogues of major libraries such as *CATLINE* (National Library of Medicine), *British Library Catalogue: Science Reference and Information Service* and *British Library Catalogue: Document Supply Centre (Monographs)* could in theory be used for current awareness and retrospective identification. However, the representation of the subject content in these sources is often too superficial to be useful for searching at the level required to support research. Unfortunately, many of the major subject-specific indexing and abstracting services do not cover books or only cover them patchily. *BIOSIS Previews* and the *Index to Scientific Book Contents* (ISI) are arguably the best in this respect, providing access to individual chapters as well as to the whole work.

DATA COMPILATIONS

Data compilations are to be found in many different forms fulfilling a great variety of functions. They may contain numeric values; define and standardize terminology; or give concise summaries of properties. Numeric data, text and bibliographic references are frequently mixed. They masquerade under a number of titles: handbooks, dictionaries, encyclopaedias, compendia, databanks, and so on. Their value lies in the quick access they provide to data extracted from a number of primary sources. The best of them also evaluate the data before publication so that the information is more reliable. Data compilations are essentially sources for reference and looking up particular items of information, often in support of laboratory work or other information searching.

A few titles will have to suffice as examples as the scope of this kind of information source is so diverse. The *CRC Handbook of Biochemistry and Molecular Biology* (CRC Press) gives data on proteins, nucleic acids, lipids and carbohydrates. The *Hazardous Substances Data Bank* (National Library of Medicine) extracts toxicological values from the literature and also provides the bibliographic references. *Martindale* (Royal Pharmaceutical Society of Great Britain) summarizes the actions and uses of drugs and lists the available forms.

Unfortunately, there is no obvious single source covering this kind of information. Some of the sources listed for reviews and books will identify a few. Where they exist as computerized databanks they may be identified through the directories which also list online and CD-ROM indexing and abstracting services. Scanning the reference sections of appropriate libraries is probably the best way of getting to know them.

See also: Thornton (1984) for a list of standard medical reference sources including sections on dictionaries, nomenclature and terminology, databooks, and statistical sources; Chen (1981vii) for a list of encyclopaedias; Chen (1981viii) for a list of dictionaries; Chen (1981ix) for a list of handbooks; Chen (1981x) for a list of tables, databooks and statistical sources.

Retrieving the information

Choice of sources

It is important to choose the classes of information source which are appropriate to solving particular information problems. Choice will be influenced by the structure and characteristics of each category. Many biological and biomedical information problems arising in practice benefit from the use of several complementary sources. Bawden and Brock (1982) illustrate this very well in their report of a study of different approaches to chemical toxicology searching, which was carried out largely within the pharmaceutical industry. Although toxicology produces extreme conditions for information requirements (it is usually essential to ensure comprehensive retrieval) the principles established in the study can usefully be applied to other areas of biological and biomedical information work. The study showed the need to use a variety of sources for comprehensive retrieval, with judicious combination allowing deficiencies of particular sources to be overcome. To provide an effective service requires combining printed sources, computerized indexing and computerized data compilations. Printed sources should not be dismissed as 'old hat'. Where older material is important they will be the only sources. It may often be important to follow up citations in the more recent literature in order to get back to material which would otherwise be virtually irretrievable. The value of good library and in-house document collections, giving access to books, reports and indexing and abstracting publications which might otherwise be considered outdated, was shown in the study. On the other hand, computerized sources, both indexing and abstracting services and data compilations, were shown to be necessary for completeness. Whenever searching was restricted to a single source, results were poor. Needless to say, the resourceful information department will use experts and centres of excellence as necessary to make good any deficiencies in its own resources.

Selection of appropriate sources is often a matter of familiarity and experience. For indexing and abstracting sources several criteria need to be considered before a final selection is made. Coverage does not relate just to subject matter although this is the major determining factor in selection. It is also necessary to appreciate the size of the service – how many records and how many sources. The types of literature covered, the criteria for inclusion of items from any particular source, and geographical coverage are all important. Other factors which may be significant are the time span covered by the service, the currency of the material it contains, the aspects that are searchable, indexing quality, availability and quality of abstracts, and availability for searching.

For computerized sources – whether indexing and abstracting services, full texts of journals, reports or books, or data compilations – there are computer-based aids to selection in the form of cross-file searching facilities. These allow limited searches in a wide range of databases and assessment of results in terms of numbers of hits obtained. Though crude, this often reveals as potentially useful sources that might otherwise have been overlooked.

Information delivery options

In providing an information service in biology and biomedicine, it is not sufficient to consider sources in isolation. The methods of making those sources available must also be considered. The service provided is thus the result of the interactions between the perceived needs of the users (both enquirers and searchers) and the available technology. Medawar (1985) has written that 'the problems raised by the – as it sometimes seems – oceanic volume of scientific knowledge are essentially technological problems, for which adequate technological solutions are rapidly being found; computers like those used in the great medicobiological extracting services such as Excerpta Medica of Amsterdam endow a medical scientist with an almost infinitely capacious exosomatic memory with prompt and trustworthy powers of information retrieval.' This simplifies the situation perhaps but is essentially true.

The main delivery options available are hard copy and computer-based. Computer-based options can be further divided, depending on communication possibilities. Many information sources (indexing and abstracting services, data compilations and full texts of journals, books and reports) are available online via remote computers. It is also possible to buy complete indexing and abstracting services or particular subsets for mounting on in-house mini or mainframe computers (via tapes) or microcomputers (via diskettes). Increasingly, sources are becoming available in optical disk (usually CD-ROM) form for use in-house.

Although it is usual nowadays for hard copy to be generated from computer-based technology, it still maintains the advantage of ease of use with no need for special knowledge or equipment. For older material, of course, there is often no alternative to print on paper. The advantages of online searching over manual searching of a printed equivalent are well known to the information professional. Briefly, the process is much quicker and easier, more complex searches can be undertaken using more access points, and the information is more current. Manual searching does, however, allow a degree of browsing which is not usually possible or, to be more accurate, is too expensive with online sources to be practicable. CD-ROM searching systems have the same advantages as online systems (although the sources are not usually so current) and, because usage costs are not incurred once a particular source has been purchased, browsing is a practical possibility.

Buying in services on tape or diskette or downloading from online sources allows the information department to select particular items, add extra details, and repackage according to the requirements of the users. Current awareness bulletins, in hard copy or computerized form, are easily produced and the cumulative files are available for subsequent retrospective searching. The same technology is also used by commercial information service suppliers and repackaging of information, especially for multidisciplinary, mission-oriented areas, is increasingly common. Chemical Abstracts Service and BIOSIS combine their data to produce a series of printed current awareness services called *BIOSIS/CAS Selects*. CD-ROM 'libraries' – selections from general indexing and abstracting services plus entire specialized services – are now available for areas such as cancer and AIDS. *ADONIS* (Stern and Campbell, 1988) is another ambitious repackaging project using CD-ROM. It is a trial document delivery service supplying over 200 biomedical journals, from four different publishers, to major document centres. If successful, it is possible that texts of journal articles in this form will become more generally available and that alerting services to the contents of the journals will be available before the appearance of the journals on paper.

Service options

The ways and means by which biological and biomedical information is provided within the pharmaceutical operation are varied. Some organizations have a separate part of an information department devoted to this area. Alternatively, information provision may be organized through individuals or small teams servicing the requirements of programmes, projects or departments, perhaps drawing on central information services and support. The basic task is the same in all cases: to match the information needs of the users with the content of the information sources in the form of a manageable service. The whole range of information sources – informal and formal, people-based and literature-based, primary and secondary – are needed to meet the varied information requirements for the drug development process. Selectivity is vital because of the volume of information available.

With so many source and delivery options, the trend in the recent past has been for professional intermediaries to undertake as much as possible of the searching on behalf of the research staff. Developing technology and an increased awareness of information sources are now leading to a partial reversal of this process. User-friendly (in relative terms) systems are available for online searching. These can be most effective when linked with just one database. *Medline*, for example, is searched by many end-users using the *Paperchase, Grateful Med* and French Minitel-based systems. CD-ROM systems are extremely popular with end-users and probably have the biggest potential for development.

Such developments have led in some instances to a reassessment of the role of the information professional in providing biological and biomedical information services. The information professional is responsible for identifying and supplying in suitable forms relevant information sources, for supporting their use by research staff, and for taking on the more complex searching tasks while the research staff carry out their own routine searching in a limited number of sources. A parallel development has been the contracting out of specific aspects of information provision to commercial companies. This option is normally used in order to help the information department cope with peaks in demand or to provide access to sources or services for which there is only an occasional demand. Using external contractors can work satisfactorily as long as they are at least as well qualified to do the job as internal staff and as long as security is not jeopardized.

The exact mix of services offered will very much depend on the size of the company, the needs in the area of biological and biomedical information and the resources available. Ward (1987) has written a detailed account of how one successful pharmaceutical company has achieved this mix. As ever, the art lies in making a judicious selection from the many possibilities outlined here and in constant re-appraisal.

References

Anon. (1989). 'Withheld data blocks hepatitis research'. *New Scientist*, 14 January, 29.

Bawden, D. (1986). 'Information systems and the stimulation of creativity'. *Journal of Information Science*, **12**, 203–216.

Bawden, D. and Brock, A. M. (1982). 'Clinical toxicology searching: a collaborative evaluation, comparing information resources and searching techniques'. *Journal of Information Science*, **5**, 3–18.

Calam, D. H. (1984). *Pharmacology and therapeutics*. In *Information sources in the medical sciences*. eds. L. Morton and S. Godbolt, 3rd edn, pp. 188–207. London: Butterworths.

Chen, C-C. (1981). *Health sciences information sources*. (i) pp. 281–308; (ii) pp. 315–321; (iii) pp. 601–603; (iv) pp. 526–535; (v) pp. 547–553; (vi) pp. 344–373; (vii) pp. 52–62; (viii) pp. 63–93; (ix) pp. 94–139; (x) pp. 140–153. Cambridge, Massachusetts: MIT Press.

Council on Biological Sciences Information (1970). *Information handling in the life sciences*. Washington, D. C.: Division of Biology and Agriculture, National Research Council.

Dannatt, R. J. (1984). *Primary sources of information.* In *Information sources in the medical sciences.* eds. L. Morton and S. Godbolt, 3rd edn, pp. 17–43. London: Butterworths.

Garfield, E. (1976). 'Significant journals of science'. *Nature,* **264**, no 5587, 609–615.

Hamilton, D. N. H. (1984). *Immunology and transplantation.* In *Information sources in the medical sciences.* eds. L. Morton and S. Godbolt, 3rd edn pp. 281–289. London: Butterworths.

Hassanaly, P. and Dou, H. (1988). *Chemistry.* In *Manual of online searching strategies.* eds. C. J. Armstrong and J. A. Large. pp. 157–236. Aldershot: Gower.

Kronick, D. A. (1985). *The literature of the life sciences,* pp. 57–70. Philadelphia: ISI Press.

Medawar, P. (1986). *The limits of science.* Oxford: Oxford University Press.

Morton, L. and Godbolt, S. (1984). *Libraries and their uses.* In *Information sources in the medical sciences.* eds. L. Morton and S. Godbolt, 3rd edn, pp. 1–16. London: Butterworths.

de Solla Price, D. (1981). *The development and structure of the biomedical literature.* In *Coping with the biomedical literature.* ed. K. S. Warren. New York: Praeger.

Roberts, D. C. (1984). *Morbid anatomy: clinical and experimental.* In *Information sources in the medical sciences.* eds. L. Morton and S. Godbolt, 3rd edn, pp. 224–249. London: Butterworths.

Rowland, J. F. B. (1984). *Biochemistry, biophysics and molecular biology.* In *Information sources in the medical sciences.* eds. L. Morton and S. Godbolt, 3rd edn, pp. 142–173. London: Butterworths.

Simmons, E. S. (1988). *Patents.* In *Manual of online searching strategies,* eds. C. J. Armstrong and J. A. Large. pp. 84–156. Aldershot: Gower.

Snow, B. (1988). *Biological sciences.* In *Manual of online searching strategies.* eds. C. J. Armstrong and J. A. Large. pp. 237–278. Aldershot: Gower.

Stern, B. T. and Campbell, R. (1988). *ADONIS: the story so far.* In *CD-ROM: fundamentals to applications.* ed. C. Oppenheim, pp. 181–219. London: Butterworths.

Thornton, J. L., revised by H. R. Hague (1984). *Standard reference sources.* In *Information sources in the medical sciences.* eds. L. Morton and S. Godbolt, 3rd edn, pp. 70–89. London: Butterworths.

Turner, J. M. (1987). *Biochemical sciences.* In *Information sources in the life sciences.* ed. H. V. Wyatt, 3rd edn, pp. 57–79. London: Butterworths.

Van Camp, A. J. and Seeley, C. (1988). *Health sciences.* In *Manual of online searching strategies.* eds. C. J. Armstrong and J. A. Large. pp. 279–322. Aldershot: Gower.

Ward, S. E. (1987). *Sources, services and technology: what opportunities do they provide?* In *The future of industrial information services.* eds. L. Wood and R. Haigh, pp. 36–58. London: Taylor Graham.

Whimster, W. F. (1988). 'Conference previews'. *British Medical Journal,* **297**, 1000.

Whyte, B. H. (1984). *Medical microbiology*. In *Information sources in the medical sciences*. eds. L. Morton and S. Godbolt, 3rd edn, pp. 250–280. London: Butterworths.

Wyatt, H. V. (1987a). *Reading for profit: current awareness*. In *Information sources in the life sciences*. ed. H. V. Wyatt, 3rd edn, pp. 11–18. London: Butterworths.

Wyatt, H. V. (1987b). *Microbiology including mycology, virology, tissue culture, immunology and use of animals*. In *Information sources in the life sciences*. ed. H. V. Wyatt, 3rd edn, pp. 80–100. London: Butterworths.

CHAPTER FOUR

Toxicology and drug metabolism information

A. D. DAYAN, C. J. POWELL AND S. SUNDARAM

Purposes and structure of toxicity testing

In the development of a new drug, toxicity testing has three principles, which apply equally to the main active ingredient, probably a new chemical entity (NCE), and to the other materials used in the chosen formulation. They are almost all concerned with the prediction of hazards in man, or other target species, by extrapolation from laboratory experiments to the circumstances of clinical use.

The objectives are:

(a) Demonstration of toxic effects – nature, intensity or frequency of occurrence, dose and duration of exposure required to produce them, species affected (if found in vivo), and any likely suggestion about mechanism – to aid extrapolation.

(b) Within the limits of the types of experimentation done, broadly to show what types of toxic effect are not produced.

(c) To note any novel actions that may be of serendipitous value in suggesting a new use for the NCE.

As general classes, these three headings cover the active ingredient, the formulated product, impurities and breakdown products and any modifying effect of the circumstances of use in particular patients, e.g. youth or old age, nutritional state, sex, pregnancy, interactions with other drugs and the consequences of disease in altering the desired response or the body's handling of the preparation.

In practice, because different types of study are required to reveal and investigate major classes of adverse action, several distinct types of experimental investigation are performed. These have been empirically devised to reveal most forms of toxicity and to generate the associated data required for their interpretation.

They are all based on demonstrating harmful effects in animals under conditions which experience has shown to permit reasonably confident extrapolation to man. Typically some or all of the following studies will be required:

(a) Acute toxicity – effects of a single high dose in animals followed for up to fourteen days, formerly quantified in the LD50 test, but now more often measured in a semi-quantitative limit or acute toxicity test. The experiments are usually done in the mouse or rat and, as a very restricted pharmacological and functional exploration, in a non-rodent species.

(b) Subacute-chronic toxicity – results of repeated dosing over several days to about one year with very wide clinical, haematological, biochemical, autopsy and histopathological assessment. Such studies are usually done in the rat and a non-rodent species, commonly the dog, but sometimes a primate.

(c) Carcinogenicity – examination of the potential of an NCE to produce tumours on very prolonged administration, normally to the mouse for 18 months and the rat for 2 years.

(d) Reproduction toxicity – investigation of effects on the formation of mature gametes, on the ability of mature animals to mate, on the capacity for fertilization and on intra-uterine and postnatal development of the fetus and neonate to sexual maturity.

(e) Genetic toxicology – whether the substance can affect DNA at the level of the genome, the gene or the chromosome, considered independently as a source of heritable abnormality (mutation), and as a probable predictor of at least one major class of carcinogen – 'genotoxic" compounds as opposed to 'non-genotoxic" (epigenetic) chemicals which appear to cause tumours without directly affecting DNA.

(f) Irritancy – a topically applied substance requires testing to show that it does not cause local tissue damage at the site of application, whether external or internal.

(g) Sensitization and allergenicity – the property of certain substances to cause either a specific immune response against themselves or other antigens, including a more general (auto) immune response against normal body constituents.

(h) Functional or pharmacological toxicity – the dynamic, functional effects that drugs may have because of their intended or related pharmacological properties, which may have adverse consequences if carried to excess, e.g. bradycardia due to beta-blockade, or the anticholinergic actions of tricyclic anti-depressants.

Last, the presence or absence of toxicity must always be considered in relation to the pharmacokinetics of the formulation administered, i.e. the rate and extent of absorption, the disposition and persistence in the body of the active agent, the nature, site and rate of formation of metabolites and their disposition, and the rate and routes of excretion of the parent compound and its metabolites.

As experience has shown that the types of effects being sought and the circumstances required to produce them differ so much, very different types of experimentation are necessary to investigate their occurence or absence.

The detailed pattern of experiments appropriate to a given substance will depend on its properties, likely usage and on specific regulatory as well as scientific requirements. Experimentation is driven by the scientific and ethical need to know about toxic hazards before patients are put at risk and the clinical and equally important need to balance therapeutic benefit against toxic cost. There is also the economic and regulatory need to comply with national rules to achieve speedy and efficient drug development.

The general nature of drug development and the place of toxicity testing are reviewed in Burley and Binns (1985; see especially Chapter 2 by Brimblecombe and Dayan).

There is extensive discussion of the general nature and evaluation of toxicity testing and pharmacokinetics in Clayson, Krewski and Monro (1985), and of the analogous area of food constituents in *Environmental Health Criteria Nos. 6* (1974), *30* (1982) and *57* (1986). A valuable older survey is given by Paget (1970). Pharmacokinetic models are generally assessed by Thorne, Jackson and Smith (1986).

Ad hoc surveys of broad classes of methods and their predictive value appear from time to time in a number of journals, especially in *Regulatory Toxicology and Pharmacology* and *Risk Analysis*, as well as in the standard toxicological journals which include *Archives of Toxicology, Fundamental and Applied Toxicology, Human Toxicology*, and *Toxicology and Applied Pharmacology*.

Almost all formal toxicity testing is done in animals, because it would be unethical and illegal to experiment directly on man with substances of unknown properties. Carefully evaluated results from animals represent the best available surrogate for man (Paton, 1984; Office of Technology, 1986), and even chronic effects can be produced in animal experiments within a sufficiently brief time to be of practical value. The extrapolation between species that is required in predicting toxic risks in man must be done with caution and with critical awareness of the difficulties sometimes posed by individual species, strain and sex-related differences (Calabrese, 1983; Walker and Dayan, 1986; Tardiff and Rodricks, 1987).

Functional toxicity and pharmacology

Most of these effects can only be discovered pragmatically by considering the known pharmacodynamic actions of the drug and making appropriate observations during toxicity tests, provided that the procedures are not so intrusive that they affect the other principal observations to be made (if they do, then additional groups of animals will be required) e.g. measurement of blood glucose level if a hypoglycaemic sulphonylurea was being studied, or plasma electrolytes and acid-base balance if a diuretic was under investigation.

A further component is the mandatory need to check every compound for major but unrealized fundamental or pharmacodynamic effects, especially if it is to be given parenterally. The need for such 'safety pharmacology" studies and screening methods for effects on the central and peripheral nervous systems, the cardiovascular and respiratory systems, and preferably for effects on smooth muscle (gastrointestinal tract, uterus), renal function, and, as appropriate, on other organs is reviewed at length in Zbinden and Gross (1979). Together, these aspects merge into the non-routine pharmacological exploration of possible functional actions, which may be detected or suspected because of findings in standard toxicity tests, but which were not appropriate for the study of functional disorders (e.g. as suggested by alterations in behaviour, or a particular pattern of lesions in several organs). Such phenomena or suspicions should be explored empirically, using techniques focused on the action of interest (consult standard textbooks and journals for physiological and pharmacological techniques).

Genetic toxicity testing

The purpose of genetic toxicity testing (also known as mutagenicity testing) is to discover whether a substance can affect DNA under standardized experimental conditions.

Since introduction of the first convenient method, the Ames test, in 1973 (Ames et al., 1973) there has been a considerable explosion of techniques and means to standardize genetic toxicity testing, and of assessments of their value in predicting genetic and carcinogenic hazards to man. New procedures are still being proposed and modifications suggested to improve the sensitivity or reliability of established methods.

At the regulatory level, there is general agreement about the value of a main set of tests, possibly with additional or modified studies for specific types of chemicals or actions: in vitro, a point mutation assay in several strains of S. typhimurium (Ames test), a study of chromosomal damage in dividing lymphocytes in vitro (usually human), possibly a point mutation test in lymphocytes, in vivo chromosome studies or the comparable micronucleus test in bone marrow cells in a rodent, and perhaps a dominant lethal test in a rodent.

All the in vitro experiments are done with and without hepatic S-9 microsomes as a partial surrogate for in vivo xenobiotic metabolism, and in them and in all in vivo experiments, a wide range of concentrations or doses of the test and standard comparator compounds is examined. These and many other reasonably well understood methods are fully discussed and critically analyzed in the publications of the UK Environmental Mutagen Society (UKEMS, 1983, 1984) together with recommendations for the statistical analysis and preliminary interpretation of results.

General surveys and inter-laboratory trials of the reproducibility and experimental validity of the principal test methods have been published by de Serres and Ashby (1981), de Serres (1983), Parry and Arlett (1985), Montesano et al. (1986), Tennant et al. (1987) and Ashby et al. (1988).

Attention has recently been paid to the value of selected genetic toxicity tests in predicting carcinogenicity in rodents and in man by using data from standardized animal carcinogenicity tests and a limited number of substances proven to be carcinogenic in humans. This assessment, which is crucial in determining the utility of genetic toxicity tests in general drug development, has been well reviewed by Montesano et al. (1986), Zeiger (1987), Shelby (1988) and Ashby (1988).

The leading journals in the field include *Environmental Mutagenesis, Mutagenesis* and *Mutation Research*. Valuable articles and reviews often appear in journals concerned with cancer, and in such review publications as *Environmental Health Perspectives*.

Immunotoxicity

Drugs may be designed to affect the immune system, for instance as general immunosuppressants, such as azathioprine and cyclosporin, or as immunostimulants, such as thymopentin; or, they may have a general immunosuppressant action whilst being administered for other purposes, e.g. cytotoxic agents and corticosteroids.

It is becoming increasingly recognized, too, that some compounds are capable of exciting a specific immune response to themselves, or to an impurity or metabolite (notably penicillins). Skin rashes and the other reactions of cutaneous hypersensitivity are amongst the commonest signs of drug allergy. Effects on the skin are sometimes loosely termed 'contact dermatitis', although that is really a more specific disorder, as its name implies.

Other compounds may cause autoimmune reactions against body tissue antigens, such as alpha-methyldopa producing haemolytic anaemia, and the lupus erythematous-like disorders occasionally associated with hydralizine and hydantoins.

The entire subject of immunotoxicity is reviewed by Descotes (1986), Berlin et al. (1987) and van Loveren and Vos (1989).

Reactions involving the skin are discussed in Ritz and Buehler (1980), Marzulli and Maibach (1986), and especially in the journal *Contact Dermatitis*.

The area is rapidly changing and its general clinical and pharmaceutical importance is uncertain. It has been easy to blame any unexpected response on 'allergy' or 'idiosyncracy', which has led to much confusion. Dewdney (1983) has given a mordant analysis of requirements to be met before blaming immunity for unwanted actions.

Pathology

Pathology is the study of disease processes and is traditionally divided into general pathology, concerned with the basic reactions of cells and tissues underlying all disease processes, and systematic pathology which deals with the specific responses of specialized tissues to injury. As a science, it encompasses investigation of the aetiology of disease, the mode of its origin and development (pathogenesis), and study of the consequent morphological and functional changes in tissues. Its importance in toxicology lies in the extensive use of the concepts and techniques of pathology to detect and analyze the harmful effects of chemicals.

The branches of particular relevance to experimental pathology include morbid anatomy and histopathology, chemical pathology, haematology and comparative pathology. Morbid anatomy is the study of the structural changes that can be seen in diseased tissues and organs. Chemical pathology involves interpretation of the results of quantitative and qualitative analysis of body fluids in association with the disease state. Haematology is the study of blood and blood-forming organs and their disorders, and is of especial interest as the haematological system is frequently the target of drug toxicity. Comparative pathology treats the diseases of animals and man comparatively and a good working knowledge of background biological variation in the response of laboratory animals to NCEs and of the differences between changes in them and in man, is important for valid interpretation of experimental results. Investigation of the toxicity of a drug should always include general histopathological examination of relevant tissues and organs as well as measurement of appropriate haematological and biochemical parameters.

Many relevant databases including *Index Medicus, Excerpta Medica* and *Current Contents, Life Sciences,* as well as other general information sources on morbid anatomy and histopathology are described in *Information Sources in the Medical Sciences* (Morton and Godbolt, 1984), which also mentions the standard textbooks and leading journals.

The main journals providing coverage of useful articles on haematology include *Acta haematologica,* the *American Journal of Haematology, Blood,* the *British Journal of Haematology* and *Folia Haematologica.* Articles on laboratory animal haematology may also be found in the standard medicine, pathology, veterinary and toxicology journals such as the *British Medical Journal, American Journal of Pathology, Journal of Comparative Pathology, British Journal of Experimental Pathology, Laboratory Investigation, Lancet, Veterinary Record, Journal of the American Veterinary Medical Association, Veterinary Journal, Archives of Toxicology, Journal of Toxicological Sciences* and others.

The various textbooks providing information about blood and blood-forming organs and their diseases range from introductory texts such as *Short Textbook of Haematology* (Thompson and Proctor, 1984) and *Essential Haematology* (Hoffbrand and Pettit, 1984) to weighty monographs such as *Disorders of the Blood* (Thompson, 1977) and *Blood and its Disorders* (Hardisty and Weatherall, 1982). *Practical Haematology* (Dacie and Lewis, 1984) is a useful guide to haematological techniques such as staining, counting and enzyme methods, blood-cell cytochemistry and laboratory diagnosis. Publications such as *Schalm's Veterinary Haematology* (Jain, 1975), *Comparative Mammalian Haematology* (Hawkey, 1975), *Comparative Clinical Haematology* (Archer and Jeffcott, 1977) and *Atlas of Laboratory Animal Haematology* (Sanderson and Phillips, 1981) are essential guides to comparative haematology.

Articles on chemical pathology can also be found in the major medicine, pathology, toxicology and veterinary journals mentioned, and in *Clinical Chemistry, Biochemical Journal, Journal of Laboratory and Clinical Medicine, Clinical Science, Biochemistry, Biochimica et Biophysica Acta, Research in Veterinary Science,*

Veterinary Clinical Pathology, Laboratory Animals, Comparative Biochemistry and Physiology, Advances in Veterinary Science and Comparative Medicine and others. *Veterinary Clinical Pathology 1888–1988* (McSherry and Valli; *Journal of Comparative Pathology*, 1988), a brief history of the development of the science over the past 100 years, is particularly interesting to read.

Standard texts such as *Fundamentals of Clinical Chemistry* (Tietz, 1987) and *Clinical Chemistry in Diagnosis and Treatment* (Zilva, Pannall and Mayne, 1988) provide basic general information on laboratory principles, analytical procedures and the structure, function, and quantification of the constituents of body fluids. Comparative chemical pathology is dealt with by such publications as *The Rat in Laboratory Investigation* (Farris and Griffith, 1949), *Textbook of Veterinary Clinical Pathology* (Medway, Prior and Wilkinson, 1969), and *Veterinary Clinical Pathology* (Doxey, 1971).

Among the various academic organizations that may provide further useful sources of information on all aspects of pathology are the Royal College of Pathologists, the British Society of Toxicological Pathologists, the Royal Society of Chemistry, the Royal Society of Medicine and the British Medical Association. Reference to publications such as the *Directory of British Associations and Associations in Ireland*, the *Directory of European Associations* and the *Encyclopaedia of Associations* (US) will provide information on a considerable number of medical and scientific societies.

Reproductive toxicology

Reproductive toxicology is concerned with the adverse effects of various exogenous agents, such as chemicals (e.g. cortisone and diethylstilboestrol), biological agents (e.g. rubella virus) and physical agents (e.g. radiation) on the reproductive system. They may cause structural and functional abnormalities at any stage of the reproductive cycle, which can be conveniently divided into gametogenesis, the gestational period and postnatal life. Adverse effects on the process of gametogenesis including mutation may result in reduced fertility, sterility, and abnormality or early death of the developing embryo. Toxicity during the gestational period results in spontaneous abortion and structural and functional abnormalities of the embryo or foetus. Postnatal toxicity may be due to exposure via breast milk, to direct exposure of the offspring, or to maternal factors, such as hormonal imbalance and impaired nutrition, impinging on the wellbeing of the offspring. Postnatal defects may also follow prenatal exposure with lesions subsequently affecting later development.

The relevant databases and important textbooks, monographs and journals are discussed in the appropriate sections of *Information Resources in Toxicology* (Wexler, 1988). Briefly, basic information on reproductive toxicology can be obtained from general textbooks of toxicology such as *Casarett and Doull's Toxicology* (Klaassen, Amdur and Doull, 1986) and *Principles and Methods of Toxicology* (Hayes, 1989). Detailed information can be had from more specialized texts such as *Reproductive Toxicology* (Mattison, 1983), *Handbook of Teratology* (Wilson, 1977), *Reproductive Hazards of Industrial Chemicals* (Barlow and Sullivan, 1982), *Developmental Toxicology* (Snell, 1982) and *Physiology and Toxicology of Male Reproduction* (Lamb and Foster, 1988). Practical aspects of the subject are covered by such books as *Practical Teratology* (Taylor, 1986) and *Experimental Embryology and Teratology I* (Woollam and Morris, 1974). Safety evaluations of chemicals potentially hazardous to reproduction and guidelines for monitoring exposed populations are outlined in the publications of various organizations such as the OECD, ECETOC, WHO, FDA (US), EPA (US), MAFF (UK) and MHW (Japan), and in such monographs as *Guidelines for Studies of Human Populations Exposed to Mutagenic and Reproductive Hazards* (Bloom, 1981).

Articles on reproductive toxicology appear regularly in the leading general toxicology, reproductive toxicology, developmental biology, cancer and review journals such as *Archives of Toxicology, Fundamental and Applied Toxicology, Teratology, Teratogenesis, Carcinogenesis and Mutagenesis, Journal of Reproduction and Fertility, Progress in Clinical and Biological Research, Cancer Research, Carcinogenesis, Environmental Health Perspectives* and the *Annual Review of Pharmacology and Toxicology*.

Carcinogenicity assessments

Life time bioassays for carcinogenicity are conducted on rodent species for reasons of cost, utility, brevity and experience. In practice various strains of rat, mouse or hamster are used. Histological diagnosis of tumours which develop spontaneously or are chemically induced is, with one or two notable exceptions, uncontroversial. There are few diagnostic compendia of rodent neoplasms. The IARC/WHO *Pathology of Tumours in Laboratory Animals* covers in separate volumes *Tumours of the Mouse* (1979), *Tumours of the Rat* (1987) and *Tumours of the Hamster*

(1982). These multi-author books contain black and white reference photographs of the histological appearance of most commonly occuring tumours. *The Pathology of Laboratory Animals: Volumes I and II* (1978), is an alternative reference source. This multi-author compendia covers most common laboratory animal species and deals with all aspects of their diseases, including infective agents, degenerative conditions and neoplasms. The International Life Sciences Institute (ILSI) monograph series focuses on the neoplastic and non-neoplastic pathology of the most frequently used laboratory species in safety evaluation studies. Volumes covering the endocrine system (1983), the urinary system (1986), the digestive and respiratory systems (1985), the nervous system (1988), and the genital system (1987) have been published.

The diagnosis of mouse liver tumours is a frequent problem and has proved a fertile ground for divergent opinions. *Mouse Liver Neoplasia Current Perspectives* contains 11 multi-author chapters on the biological behaviour of these tumours, but little on their implication in safety terms. The induction of these tumours and their functional characteristics is also addressed in *Mouse Liver Tumours*, a supplement to *Archives Toxicology* in 1987.

The relevance to humans of increased incidences of rodent neoplasms is without doubt one of the most controversial and disputed issues in toxicology and has a substantial literature. Overviews include *Human Risk Assessment* (1987) which covers the role of animal selection and the extrapolation of animal data for human risk assessment. The *International Agency for Research on Cancer Monograph Series* are invaluable reference resources. Forty-two volumes have been published since 1972 and new titles are frequently added. They provide reviews of the literature and give an assessment of the status of each compound in one of four categories as either: a known human carcinogen, a probable or possible human carcinogen. Volumes 6 and 21 on sex hormones; 13, 24 and 26 on miscellaneous pharmaceuticals; 10 on carcinogenesis assays; 12 on cancer screening tests and 15 on cancer incidences are particularly useful.

The *Surveys of Compounds Tested for Carcinogenic Activity* has been published in 14 editions since 1947 and is available in the UK, from Microinfo Ltd, Alton, Hants. It consists of indexed lists of bioassay test results. The data is presented compound by compound, in a stylized manner, but does not give an assessment of the data.

In product safety evaluation it is commonplace to regard the short term in vitro and in vivo tests for DNA damage as distinct from the conventional two year rodent bioassays. In practice these assessments are carried out separately, often with the short-term tests being performed as a pre-screen. The literature tends to be divided into either short-term tests or bioassays, with surprisingly, or perhaps dishearteningly, little overlap. Welcome exceptions include *The Handbook of Carcinogen Testing* (1985) with sections covering the philosophy, conduct, problems and interpretation of conventional two year bioassays, as well as short term and in vitro methods.

The *Evaluation of Short-term Tests for Carcinogens* reports an International Programme on Chemical Safety collaborative study. This is a mammoth assessment, in two volumes and 125 chapters, of alternatives to conventional life-time carcinogenicity tests. The results of this rigorous testing programmeme are likely to be a useful reference source for those conducting or interpreting short-term tests, especially when unexpected results arise.

Further information is available in *Carcinogenesis and Mutagenesis Testing* (1984), containing 24 multi-authored chapters on the theory and practice of in vitro and in vivo assays for cancer inducing potential. *Long-term and Short-term Assays for Carcinogens* (1986) also makes a critical appraisal of the scientific strengths and limitations of these tests.

Original publications on the genotoxicity and mutagenicity of specific compounds can be found in the following journals: *Mutation Research, Applied Methods in Oncology, Canadian J. Genetic Cytology, Genetics, Proceedings American Society Cancer Research, British J. Cancer* and *Toxicology in vitro*.

On-line searching can be accomplished via *Cancer lit/CAZZ* which is hosted by Data-Star/Blaise/Dialogue. The database is compiled by the National Library of Medicine in the US, covering carcinogenesis abstracts and cancer therapy abstracts from 1963 to date. Searches can be conducted by approved keyword using *MedLine* vocabulary.

Experimental models of carcinogenicity and factors which alter the incidence of chemically induced tumours have a vast literature. A working overview from the chemical safety perspective can be gained from chapter 5 *of Casarett and Doull's Toxicology* (1986, pp 99–173). For a more detailed approach to specific topics in carcinogenesis *Advances in Cancer Research*, published annually, contains 8–10 authoritative reviews on specific areas of current research. Each volume is indexed and extensively referenced. While the articles tend to be specialized, they are good starting points from which to delve into the literature.

Inhalation toxicology

Inhalation studies are usually performed with rodents, dogs or primates when human exposure to the compound is likely to be via the inhalation route. Thus anaesthetic gases and bronchodilators would be tested in this way. There appear to be comparatively few resources devoted exclusively to this aspect of toxicology. Articles on the toxicology of specific compounds appear in the following journals: *Archives Pathology, Environmental Research, Environmental Health Perspectives, Laboratory Investigation, American J. Pathology, Chemical Biological Interaction, Toxicology Applied Pharmacology, Fundamental Applied Toxicology, American Reviews Respiratory Disease, CRC Critical Reviews Toxicology, Archives Toxicology, Human Toxicology, The Toxicologist,* and *Toxicology Pathology.* A useful step into the inhalation toxicology literature (and equally to other areas of toxicology) can be made using *Toxicology Abstracts.* Entries are divided into subject heading according to chemical type and effects. An annual index lists literature reviews in alphabetical order and abstracts by keyword and author. It is published in monthly parts and includes legislation and standard, but not clinical toxicology. *Assessment of Inhalation Hazards* published in conjunction with the International Life Sciences Institute is a multi-authored volume containing the proceedings of a meeting held in February 1989 which was devoted to the integration and extrapolation of animal and human data.

Drug metabolism and kinetics

Most drug metabolism studies are performed in industrial laboratories and, sadly, are often not published. The literature therefore is a poor reflection of the level of commercially available information. Review articles can be of great value and occur in appropriate journals such as *Pharmacological Reviews, Reviews in Biochemical Toxicology* and *Drug Development Research.* Articles on specific aspects of drug metabolism are frequently published in the following journals: *Agent Actions, Advances in Pharmacology, Analytical Biochemistry, Archives Biochemistry Biophysics, Biochemistry, Biochemical Pharmacology, Biochemistry Biophysics Research Communications, Biochimica et Biophysica Acta, Drug Metabolism Disposition, European J. Biochemistry, Federation Proceedings, J. Biochemistry, J. Biological Chemistry, J. Pharmacology and Experimental Therapeutics, Methods in Enzymology, Molecular Pharmacology, Xenobiotica.*

Because the literature is dispersed, secondary information sources such as *Chemical Abstracts* or *Chem Line* from the US are the best starting points for compounds which have not entered clinical use. The specific or generic chemical name or *CAS* number can be combined with appropriate terms such as Biotransformation/metabolites/pharmacokinetics/distribution. Worthwhile introductions to the methods used in drug metabolism can be found in Gibson and Skett (1986) and La Du et al. (1972), which also both contain directions for further reading. Once a drug enters clinical development metabolism studies are vital to enable the relevance of the animal safety studies to be assessed. The most productive sources of information on clinical metabolism and kinetics are likely to be *Med Line* and *Excerpta Medica* and for methods see Gorrod and Beckett (1978).

In addition, the proceedings of meetings, such as those from European Drug Metabolism Workshops (published at two yearly intervals – the 11th was in 1988) serve as valuable sources of reference. Review articles on specific aspects of drug biotransformation can be found in *Progress in Drug Metabolism*. Each volume contains 6–10 articles, is indexed and contains a contents list of earlier volumes.

References

Ames, B. N., Durston, W. E., Yamasaki, E. and Lee, F. D. (1973). 'Carcinogens are mutagens. A simple test system combining liver homogenates for activation and bacteria for detection.' *Proc. Natl. Acad. Sci USA*, **70**, 2281–2285.

Archer, R. K. and Jeffcott, L. B., eds. (1977). *Comparative clinical haematology* (Blackwell Scientific Publications).

Ashby, J., de Serres, F. J., Shelby, M. D., Margolin, B. H., Ishidate, M. and Becking, G. C. (1985). *Evaluation of short-term tests for carcinogens*. Vols I and II. (Cambridge University Press).

Barlow, S. M. and Sullivan, F. M. (1982). *Reproductive hazards of industrial chemicals: An evaluation of animal and human data* (Academic Press).

Benirschke, K., Garner, F. M. and Jones, T. C., (eds). (1976) *The pathology of laboratory animals*. Vols I and II (Springer-Verlag).

Berlin, A., Dean, J., Draper, M. H., Smith, E. M. B. and Spreafico, F. (eds). (1987). *Immunotoxicology* (Martinus Nijhof).

Bloom, A. D. (1981). *Guidelines for studies of human populations exposed to mutagenic and reproductive hazards* (March of Dimes Birth Defects Foundation).

Bridges J. W., Chasseaud, L. F. and Gibson, G. G., (eds.) (1988). *Progress in drug metabolism* Vol. 10 (Wiley).

Burley, D. M. and Binns, T. (1985). *Pharmaceutical Medicine*.

Clayson, D. B., Krewski, D. and Monro, I. C. (1985). *Toxicological risk assessment* Vols I and II. (CRC Press).

Calabrese, E. (1983). *Principles of animal extrapolation*, (Wiley).

Chambers, P. L., Henschler, D. and Oesch, F., (eds.) (1987) 'Mouse liver tumours.' Symposium of the European Society of Toxicology, Rome, 1986.

Archives Toxicology Supplement 10 (Springer-Verlag).

Dacie, J. V. and Lewis, S. M. (1984). *Practical haematology*. (Churchill Livingstone).

Dayan, A. D. and Walker, S. R. (1986). *Long-term animal studies. Their predictive value for man*. (MTP Press).

Descotes, J. (1986). *Immunotoxicology and drugs and chemicals*. (Elsevier).

Dewdney, J. D. (1983). 'Clinical diagnosis of drug hypersensitivity,' in *Immunotoxicology*, eds G. G. Gibson, R. Hubbard, and D. V. Parke (Academic Press).

Douglas, J. F., (ed.) (1984). *Carcinogenesis and mutagenesis testing*. (Humana Press).

Doxey, D. L. (1971). *Veterinary clinical pathology*. (Balliere, Tindall and Cassell).

ECETOC. (1983). *Identification and assessment of the effects of chemicals on reproduction and development*. Monograph No. 5 (European Chemical Industry Ecology and Toxicology Centre).

EPA. (1978). *Proposed guidelines for registering pesticides in the United States; hazard evaluation: humans and domestic animals*. Federal Register, 44: (No. 91): 37336–403. (US Environmental Protection Agency).

Farris, E. J. and Griffith, J. Q., (eds) (1949). *The rat in laboratory investigation*. (J.B. Lippincott and Co.)

FDA (US). (1966). *Guidelines for reproduction studies for safety evaluation of drugs for human use*. (US Food and Drug Administration, Bureau of Foods).

FDA (US). Advisory Committee on Protocols for Safety Evaluations, Panel on Reproductions (1970) 'Report on reproduction studies in the safety evaluation of food additives and pesticides.' *Toxicol. Appl. Pharmacol.*, **16**, 264–96.

FDA (US) (1982). *Toxicological principles for the safety assessment of direct food additives and colour additives used in food*. (US Food and Drug Administration, Bureau of Foods).

Gibson, G. G. and Skett, P. (1986). *An introduction to drug metabolism*. (Chapman Hall).

Gorrod, J. W. and Beckett, A. H. (1978). *Drug metabolism in man*. (Taylor and Francis).

Hardisty, R. M. and Weatherall, D. J. (1982). *Blood and its disorders* (Blackwell Scientific Publications).

Hawkey, C. N. (1975). *Comparative mammalian haematology*. (Heinemann Medical Books Limited).

Hayes, A. W., ed. (1989). *Principles and methods of toxicology*. 2nd edn (Raven Press).

Hoffbrand, A. V. and Pettit, J. E. (1984). *Essential haematology*. (Blackwell Scientific Publications).

IARC/WHO Monographs on the *Evaluation of carcinogenic risks to humans*: Vol. 6, *Sex hormones* (1974); Vol. 10, *Some naturally occurring substances* (1976); Vol. 13, *Some miscellaneous pharmaceutical substances* (1977); Vol. 21, *Sex hormones II* (1979); Vol. 24, *Some pharmaceutical drugs* (1980); Vol. 26, *Some antineoplastic and immunosuppressive agents* (1981). (WHO Book Sales).

Jain, N. C. (1986). *Schalm's veterinary haematology* (Lea and Febiger).

Jones, T. C., Mohr, U. and Hunt, R. D., (eds.) *Monographs on pathology of laboratory animals: The endocrine system* (1983); *The respiratory system* (1985); *The urinary system* (1986); *The genital system* (1987); *The nervous system* (1988) (Springer-Verlag).

King, L. J., (ed.) (1989). *Toxicology Abstracts*.

Kitchen, I. (1984). *Textbook of invitro practical pharmacology*. (Blackwell Scientific Publications).

Klaassen, C. D., Amdur, M. O. and Doull, J., (eds.) (1986). *Casarett and Doull's toxicology: the basic science of poisons*. 3rd edn (Macmillan Publishing Company).

Klein, G. and Weinhouse, S., (eds.) (1988). *Advances in Cancer Research*, **51**.

La Du, B., Mandel, H. and Way, E. (1972). *Fundamentals of drug metabolism and drug disposition*. (Williams and Watkins).

Lamb, J. C. and Foster, (eds.) (1988). *Physiology and toxicology of male reproduction*. (Academic Press).

MAFF (UK). (1982). *Pesticides safety precautions scheme. Working document B.8, testing for reproductive toxicology*. (Pesticides Registration and Surveillance Department, UK Ministry of Agriculture, Fisheries and Food).

Marzulli, F. and Maibach, H. (1986). *Dermatotoxicology*. 3rd edn (McGraw Hill).

Mattison, D. R., (ed.) (1983). 'Reproductive toxicology.' *Progress in Clinical and Biological Research*, **117**.

McSherry, B. J. and Valli, V. E. O. (1988). 'Veterinary clinical pathology 1888–1988.' *Journal of Comparative Pathology*, **99**(1).

Medway W., Prior, J. E. and Wilkinson, J. S. (1969). *Textbook of veterinary clinical pathology* (Williams and Wilkins Co.)

Milman, H. A. and Weisburger, E. K., (eds.) (1985). *The handbook of carcinogen testing* (Noyes Publications), p 624.

Mohr, U., (ed.) (1989). *Assessment of inhalation hazard*. (Springer-Verlag, ILSI Monograph).

Montesano, R., Bartsch. H. and Tomatis, L., (eds.) (1976). *Screening tests in chemical carcinogenesis*. (IARC/WHO Scientific Publications No. 12).

Montesano, R., Bartsch, H., Vaino, H., Wibourn, J. and Yamasaki, H., (eds.) (1986). *Long-term and short-term assays for carcinogens: a critical appraisal*. (IARC/WHO Scientific Publications, No. 83).

Morton and Godbolt. (1984). *Information sources in the medical sciences*. (Butterworths).

MHW (Japan). (1982). *Draft for new guidelines for reproduction studies for safety evaluation of drugs*. (Ministry of Health and Welfare, Japan).

OECD. *Guidlines for testing of chemicals, section 4: health effects*: 414 *Teratogenicity* (1981); 417 *One generation reproduction toxicity study* (1983); 418 *Two generation reproduction toxicity study* (1983). (Organisation for Economic Cooperation and Development).

Office of Technology Assessment. (1986). *Alternatives to animal use in research, testing and education*. (US Government Printing Office).

Paget, G. E. (1970). *Methods in toxicology* (Blackwell).

Paton, W. (1984). *Man and mouse. Animals in medical research*. (Oxford University Press).

Parry, J. M. and Arlett, C. F., (eds.) (1985). *Comparative genetic toxicology*. (Macmillan).

Popp, J. A., (ed.) *Mouse liver neoplasia. Current perspectives*.

Ritz, H. L. and Buehler, J. V. (1980). 'Planning, conduct and interpretation of guinea pig sensitisation patch tests' in *Current Concepts in Cutaneous Toxicity*, eds. Drill, V. A. and Lazar, P., (Academic Press).

Roloff, M. V. (1987). *Human risk assessment*. (Taylor and Francis). Sanderson, J. H. and Phillips, C. E. (1981). *Atlas of laboratory animal haematology* (Oxford University Press).

de Serres, F. J. and Ashby, J., (eds.) (1981). *Evaluation of short-term tests for carcinogens: report of the international collaborative study*. (Elsevier/North Holland) New York.

Snell, K., (ed.) (1982). *Developmental toxicology* (Croom-Helm).

Snell, K., (ed.) (1985). 'Biochemical disturbances of fetal development.' *Biochem. Soc. Trans.*, **13**, 73–91.

Tardiff, R. G. and Rodricks, J. V. (1987). *Toxic substances and human risk. Principles of data extrapolation*. (Plenum).

Taylor, P. (1986). *Practical teratology* (Academic Press).

Tennant, R. W., Margolin, B. H., Shelby, M. D., Zeiger, E., Haseman, J.K., Spalding, J., Caspary, W., Resnick, M., Statiewicz, S., Anderson, B. and Minor, R. (1987). 'Prediction of chemical carcinogenicity in rodents from in vitro genetic toxicity assays. *Science*, **236**, 933–941.

Thompson, R. B. (1977). *Disorders of the blood: a textbook of clinical haematology* (Churchill Livingstone).

Thompson, R. B. and Proctor, S. J. (1984). *A short textbook of haematology*. (Pitman Publishing Ltd).

Thorne, M. C., Jackson, D. and Smith A. D. (1986). *Pharmacodynamic models of selected toxic chemicals in man. Vol 1: Review of metabolic data; Vol 2: Routes of intake and implementation of pharmacodynamic models.* (MTP Press) Lancaster.

Tietz, N. W., ed. (1987). *Fundamentals of clinical chemistry.* (Churchill Livingstone).

Turusov, V. S., (ed.) *Pathology of tumours in laboratory animals: Tumours of the rat* (1987); *Tumours of the mouse* (1979); *Tumours of the hamster* (1982) (IARC/WHO Scientific Publications).

van Loveren, H. and Vos, J. G. (1989). 'Immunotoxicological considerations' in *Advances in applied toxicology.* eds. A.D. Dayan and A. J. Paine (Taylor and Francis).

WHO. (1974). *Principles and methods for evaluating the toxicity of chemicals, part 1. Environmental health criterion 6.* (WHO).

US National Cancer Inst. (1987). *Survey of compounds which have been tested for carcinogenicity.* (1985–86) Vol. 10. (US National Cancer Inst.)

UKEMS. (1983). *Report of the UKEMS sub-committee on guidelines for mutagenicity testing, part I: basic test battery; part II: supplementary tests* (UKEMS).

WHO. (1984). *Principles for evaluating health risks to progeny associated with exposure to chemicals during pregnancy. Environmental health criteria 30.* (International Programme on Chemical Safety (IPCS), World Health Organisation).

WHO. (1986). *Principles for evaluating health risks to progeny associated with exposure to chemicals during pregnancy. Principles of toxicokinetic studies.* (World Health Organisation).

Wilson, J. G. and Clarke Fraser, F., (eds.) (1977). *Handbook of teratology* (4 vols, Plenum Press).

Wexler, P. (1988). *Information resources in toxicology* (Elsevier).

Woollam, D. H. M. and Morris, G. M., (eds.) (1974) *Experimental embryology and teratology I.* (Elek Science).

Zbinden, G. and Gross, F., (eds.) (1979). *Pharmacological methods in toxicology.* Pergamon, Oxford.

CHAPTER FIVE

Technical development and production information

M. ARCHER

Introduction

Following the synthesis of a new chemical entity, its successful screening for useful pharmacological activity and testing to ensure its toxicity is suitably low compared to its active level, it may then be considered for clinical assessment and development. Some of these potential drugs will successfully pass all the development phases and enter production. This chapter describes the sources of information that are of potential use in the development and production areas of the pharmaceutical industry. Development within the pharmaceutical industry covers several inter-related avenues of work during which the new medicine progresses from the research to the production stages as indicated in *Figure 5.1*. In the process development phase the preparation of the active ingredient must be developed from the small-scale research synthesis to give a safe, commercially viable process. In order to confirm that the preparation is successful, methods of analysing the product and identifying and quantifying the impurities and decomposition products must be developed. These analytical methods are essential in the pharmaceutical development phase when the active chemical entity is incorporated into a useful dosage form acceptable to the patient, the physician, the pharmaceutical company and the regulatory authorities. Quality assurance and good laboratory practice are involved in most areas discussed to ensure that the work is carried out to the highest scientific standards and that it is appropriate and comprehensive. The work culminates in the production process, which will also be subject to regulatory compliance, quality control, packaging, engineering, safety and environmental control.

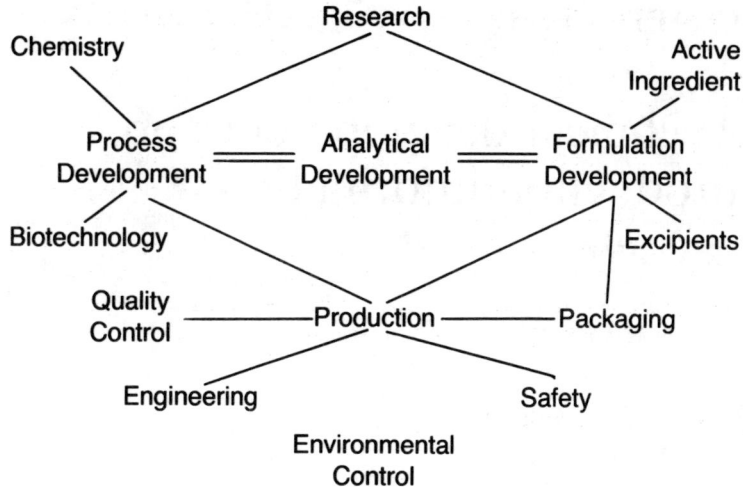

Figure 5.1 The work associated with of pharmaceutical development and production.

Much of the development work on a pharmaceutical product is carried out in parallel. However, this chapter will cover the wide range of information requirements involved in the order: process development, analytical development, pharmaceutical development and production. It will start by discussing the importance of internal information and end by outlining the general requirements of an Information Centre, the focal point for background as well as specific information provision.

Internal information

'The primary information source for a product is the detailed data gathered during the research phase – the chemical, pharmacological, toxicological, pharmaceutical and clinical reports produced before marketing,' (Williams et al., 1984).

'Information retrieval is a major problem facing the chemical industry. During the time from when a reaction is first attempted to when it is scaled-up for production, many chemists in various departments may have worked on the project. To prevent duplication of expensive experimental work, an efficient method for the storage and retrieval of data is critical,' (Mills et al., 1988).

Pharmaceutical companies have special needs for the storage and indexing of such data not least because of regulatory requirements but also for general business efficiency. In this respect it is particularly important to avoid duplication of effort when projects are being passed from the research phase to the development phase, then on to the production phase. Whereas research scientists will probably transfer to new projects, development and production scientists are likely to be working on earlier projects for long periods, perhaps up to 15 years. Frequently, also, research takes place at different locations from development and production activities which makes it even more important to have efficient transfer of information. Most pharmaceutical companies invest heavily in appropriate information storage and retrieval methods to optimize information transfer.

Process development

Process development is an essential stage leading to the manufacture of the active ingredients of pharmaceuticals. The active ingredient can be manufactured by chemical synthesis or by biotechnology and the information needs of workers are determined accordingly. Promising candidate drugs are passed to the process scientists who devise processes for the manufacture of the drug which must be economical, have a high yield, be safe, use starting materials that are reasonably readily available, preferably use equipment that is already installed, be patentable and not already patented and produce material consistently of the required quality. Most of the criteria stem from the major consideration that an economical process is required in order that the company may profit from the sale of the product. The high yield requirement follows, not only because the amount and cost of the starting materials can then be reduced, but also because the work involved in extraction and purification can be reduced together with the quantity of waste material for disposal.

The process must be safe and if hazardous materials or reactions have to be used, there is the additional expense of providing suitable equipment and enclosure. The starting materials should be available in the required quantities, or processes for producing them must be devised and due consideration has to be given to possible future requirements. It is preferable not to have to install new manufacturing plant at the beginning of the lifetime of a new product of uncertain market penetration, so the process ought if possible to utilize existing plant. If any stages in the production process are already patented by other companies then licence agreements will probably have to be negotiated.

The process development scientist, therefore, needs, initially, information on alternative methods of preparation of chemical entities, whether by chemical or biotechnological methods. If information is not available on the specific material the question may be extended to compounds of similar structure. A chemical entity may be the new drug substance or precursor in its synthetic pathway. It is important to know whether the projected starting materials are likely to be available in the required quantities and whether they have been registered under hazardous substance legislation.

Information on types of reactions, chemical and fermentation, and about the conditions for optimum yields is needed as well as all available data on the hazards associated with the reaction. The potential hazards of the starting materials, solvents and by-products increase in importance as the quantities handled are larger and are, therefore, even more relevant at the production stage. There are sources of information available to assist in answering all these enquiries as outlined below.

Books

There have been many hundreds of books published in the field of organic chemistry, many of which are useful in development work. Mentioned here are just a few titles which have particular value in this area.

Beilstein (The Beilstein Institute) covers the organic chemistry literature up to 1979 (5th supplement) and is unique in that all the data it contains has been verified by the Beilstein Institute. As its literature coverage dates from before the beginning of the century it is useful for preparation methods of relatively simple precursors or reagents which have not been investigated recently and are not immediately available commercially.

The Dictionary of Organic Compounds (Chapman and Hall Ltd.) lists the most important chemical compounds of the present day and gives key physical and chemical data (including synthesis) as well as some hazard data. It is well indexed under systematic and common names as well as molecular formula and Chemical Abstracts Registry Number. A separate volume covering anti-biotic substances has been published.

Rodd's Chemistry of Carbon Compounds is a series covering general chemistry classified by type of compound. It contains much useful background chemistry and reaction information.

The Compendium of Organic Synthetic Methods (Harrison and Harrison, Wiley) is a multivolume work designed for use by bench chemists planning syntheses. It is a systematic listing based on the functional groups of the starting materials and products; however, one disadvantage in this area of work might be seen as its coverage of reactions giving low yields or using exotic reagents.

Synthetic Reagents (Pizey, Ellis Horwood) is a set of books covering the literature on the use of specific synthetic reagents in detail, on average four reagents in each of five volumes, and indicates selectivity and optimum conditions for the reagents.

On the other hand Fieser and Fieser's major work *Reagents for Organic Synthesis* (Wiley) is more encyclopaedic; volume 1 published in 1967 reviews the literature on reagents up to 1966, and the series extends to volume 13 (1988) which is concerned with 1986 and 1987.

Two annual publications from Wiley which are standard references in chemical development work are *Organic Syntheses* and *Organic Reactions*. The former contains methods for the preparation of organic compounds which have been verified as satisfactory by an independent laboratory, the latter dedicates chapters to specific reactions discussing limitations, influences and optimum conditions for each.

One book which is invaluable in the chemical development area is the *Aldrich Chemical Company Catalogue* which lists 20,000 chemical substances available for purchase from that company. It thus gives an initial indication of the commercial availability of substances. The catalogue, however, also affords other inform-ation such as references to the substances covered by *Beilstein, Fieser and Fieser, Merck Index* and the *Registry of Toxic Effects of Chemical Substances* (mentioned later). It is also a very convenient source of Chemical Abstracts Registry Numbers below.

Hazardous chemical reactions comprise an area which until fairly recently has not been well covered in the literature. The first book to deal specifically with this aspect was Bretherick's *Handbook of Reactive Chemical Hazards* (Butterworths). This book is a compilation of reactions which are known to be potentially violent and hazardous and provides information on 9000 elements and compounds. It is written for the chemist and is indexed by molecular formula thus circumventing problems of differing nomenclature. As well as indicating hazardous reactions between chemicals, it gives stability data on single specific compounds and general data on classes of compounds, which could indicate whether a new compound is likely to cause problems.

A similar but smaller compilation, listing 3550 hazardous reactions, is the *Manual of Hazardous Chemical Reactions* NFPA 491M published by the National Fire Protection Association. As this latter publication uses chemical names for indexing, more care is needed to ensure the correct one is used.

The Association of The British Pharmaceutical Industry is revising its *Guidance Notes on Chemical Reaction Hazard Analysis* originally published in 1981 and revised and re-titled *Guidelines*, in late 1988. This publication gives basic information on methods of evaluating hazards likely to arise during the manufacture of chemicals.

There are many books covering sources of chemicals usually produced by national chemical manufacturers associations. For instance, in the UK the Chemical Industries Association produce *Chemicals '86* listing chemicals and allied products marketed by member firms, whilst in France the Union Des Industries Chimiques produce a *Directory* covering 700 French manufacturers and over 6000 products (which is also available on floppy disc). *The Directory of World Chemical Producers* published by Chemical Information Services Ltd. lists over 5000 producers worldwide and 50 000 entries. At over £200 this annual publication is still of considerable use if bought only every 3–4 years. The products of the American chemical industry are catalogued in *Chem Sources – USA* (Directories Publishing Co. Inc.) over 104 000 chemicals appearing here from 855 companies.

A consideration when devising new production methods is whether the starting materials or intermediates have been registered under the new hazardous substances regulations. In Europe all chemicals which were on the market between 1971 and 1981 were registered by their suppliers in the *European Inventory of Existing Chemical Substances (EINECS)*. Should a substance not be included in this list, then, if quantities of over one tonne per annum are to be produced, full toxicological studies could well be required to establish its safety. The interpretation of this legislation is under discussion at the present time. Medicinal products and their ingredients are exempt and it was originally believed that so also were pharmaceutical intermediates. However, it now appears that this may not be the case. There is also discussion as to whether the one tonne per annum qualification level will be maintained or whether all new compounds will be subject to the notification procedures. This area of legislation will become more important in the future and information will be an essential ingredient. In the USA the Toxic Substances Control Act has an equivalent directory issued by the Environmental Protection Agency.

Biotechnology has been a significant growth area in recent years as indicated by the number of books published on the subject; Table 5.1 gives a short bibliography of relevant texts. However a comprehensive coverage of information sources on this subject is available from the Biotechnology Information Service of the

Table 5.1. Biotechnology bibliography

Atkinson, B. and Marituna, F. (1983). *Biochemical Engineering and Biotechnology Handbook* (Macmillan)

Callam, C. T. (1987). *Process Development in Antibiotic Fermentations* in *Cambridge Studies in Biotechnology 4* (Cambridge University Press).

Demain, A.L. and Solomon, N. A. (eds.) (1986). *Manual of Industrial Microbiology and Biotechnology* (American Society for Microbiology).

Duffy, J. I. (ed.) (1980). *Chemicals by Enzymatic and Microbial Processes: Recent Advances*, Chemical Technology Review No. 161, (Noyes Data Corporation).

Godfrey, T. and Reichelt, J. (1983). *Industrial Enzymology: The Application of Enzymes in Industry* (Macmillan).

Higgins, L. J., Besty, D. J. and Jones, J. (eds.) (1985). *Biotechnology: Principals and Applications* (Blackwell).

Ho, C. S. and Oldshue, J. Y. (eds.) (1987). *Biotechnology Processes Scale-up and Mixing* (American Institute of Chemical Engineers).

McGregor, W. C. (ed.) (1986). *Membrane Separations in Biotechnology* in *Bioprocess Technology Volume 1* (Marcel Dekker).

Moo-Young, M. (1985). *Comprehensive Biotechnology* (Pergamon Press).

Prentis, S. (1984). *Biotechnology – A New Industrial Revolution* (Orbis).

Rehm, H. J. and Reed, G. (eds.) (1981). *Biotechnology: A Comprehensive Treatise, Volume 1* (continuing) (Verlag-Chemie).

Stanbury, P. F. and Whitaker, A. (1984). *Principals of Fermentation Technology* (Pergamon Press).

British Library, which lists books, journals, databases and directories in areas such as patents, health and safety, research information and biological control.

Journals

The journals referred to in this area are numerous, including most of the general organic chemistry and biotechnology titles. *Chemical Abstracts* covers all the major areas of interest in pharmaceutical development, including much in the biotechnology field. However, the full publication is now so large that it is best used as a database, as outlined below, and as *CA Selects* where abstracts relevant to specific subjects are published in fortnightly updates. A list of potentially important titles in this series for process development appears in Table 5.3(a). In order to cover a large number of primary journals, *Current Contents* can be acquired from the Institute of Scientific Information who reproduce the contents pages of several hundred journals in weekly publications classified by wide subject areas. The *Chemistry, Physics and Earth Sciences* and *Life Sciences* titles are useful in the chemistry and biotechnology areas respectively. Probably of more direct use to chemists, but covering fewer journals, is *Methods in Organic Synthesis* from the Royal Society of Chemistry. This publication gives abstracts and schematic diagrams of reactions from papers on organic synthesis which have appeared in the major journals in the field. Other important journals in the area are the *Journal of Synthetic Methods* from Derwent Publications, which updates Theilheimer's standard work entitled *Synthetic Methods* and goes to make up the CRDS database mentioned below, and *Current Chemical Reactions* from ISI.

Chemical supplier information can be found in *Manufacturing Chemist* (UK), *European Chemical News, Chemical Engineering News* and *Chemical Marketing Reporter* (US).

A list of the journals of use in the application of biotechnology is given in Table 5.2. In the biotechnology field *Biotechnology Abstracts* (Derwent Publications) covers the major journal literature and patents with good abstracts. The competitive publication from the Royal Society of Chemistry, *Current Biotechnology*

Table 5.2. Biotechnology journals useful in development

Biocatalysis – industrial exploitation of biological catalysts including relation to process design and subsequent downstream processing (Harwood Academic Publishers).

Bio/technology – international monthly for industrial biotechnology (Nature).

Biotechnology and Bioengineering – 18 issues per year (Wiley).

Biotechnology Letters – applied microbiology and biochemistry in combination with chemistry and process engineering.

Developments in Industrial Microbiology – microbiology applied to industrial materials, processes, products and their associated problems (Society for Industrial Microbiology).

Enzyme and Microbial Technology – international monthly journal on biotechnology research and reviews (Butterworth Scientific).

Journal of Chemical Technology and Biotechnology – studies related to biotechnology and chemical technology research conversion into products and processes (for the Society of Chemical Industry by Elsevier).

Journal of Microbiological Methods – bimonthly journal with special emphasis on biotechnology, environmental and applied microbiology, industrial microbiology and others (Elsevier).

Process Biochemistry – industrial biochemistry in food, pharmaceutical, brewing, effluent and water treatment (Turret Business Publications).

Table 5.3. *CA Selects* of use in development and production

(a) In development
Asymmetric synthesis and induction
Biological information transfer
Catalysis (organic reactions)
Chemical instrumentation
Computers in chemistry
Enzyme applications
Fermentation chemicals
Isomerization and catalysts
Optimization of organic reactions
Organic reaction mechanisms
Organic stereochemistry
Organometallics in organic synthesis
Phase transfer catalysis

(b) In analysis
Automated chemical analysis
Gas chromatography
High performance liquid chromatography
Ion chromatography
Mass spectrometry
Organic analytical chemistry
Paper and thin-layer chromatography
Pharmaceutical analysis
Thermal analysis

(c) In formulation
Controlled release technology
Cosmetic chemicals
Drug and cosmetic toxicity
Formulation chemistry
Pharmaceutical chemistry (journals)
Pharmaceutical chemistry (patents)
Candidate topic for 1989 – Drug delivery systems and dosage forms

(d) In production
Chemical hazards, health and safety
Chemical processing apparatus
Distillation technology
Liquid waste treatment
Pollution monitoring
Recovery and recycling of wastes
Solvent extraction
Ultrafiltration
Candidate topic for 1989 – Chemical engineering operations

Abstracts, includes 'technocommercial' business news and legal issues, but originally did not cover patents, although this has now been rectified.

Databases

Chemical Abstracts is of great utility in answering many of the questions raised by development scientists. The preparation methods for a specific chemical is a common requirement and use of the Chemical Abstracts Registry Number (RN), a unique number given the Chemical Abstracts Service to each chemical indexed by them, suffixed by the letter 'p' (for preparation) readily retrieves all relevant references. Use of the *CAS Online* structure searching facility, available on the STN host can extend the search to preparations of similar compounds if needed. If information of this type is available it is most likely to be found in *Chemical Abstracts*, but another source worth investigating is *Index Chemicus* published by the Institute of Scientific Information (ISI). *Index Chemicus* covers the synthesis, identification and isolation of new compounds. Over 4,000,000 compounds are covered with a growth rate of 180,000 per year. About 115 journals are abstracted and more than 90% coverage of new organic compounds is claimed. It does not cover patents. Structures and synthetic routes are given in the abstracts. It is based on Wiswesser Line Notation, a method of coding chemical structures into computer readable and sortable text strings, and limited substructure searching is therefore possible. The database is no longer available online and access is at present by use of annual cumulated Wiswesser indexes which are to be withdrawn. Another method of access to this very useful database is therefore required.

Another frequent information requirement is the patent situation of preparative routes. This is well covered by the *World Patents Index* from Derwent Publications and available online from several vendors. Structure searching is possible using the Derwent Chemical Coding, a system of code numbers applied to parts of the structure, which has been simplified by a graphical input computer programme called *TOPFRAG* which automates the coding. It is often the case that scientists can anticipate the companies likely to hold patents in their areas of interest and the *World Patents Index* is easily searched for patent assignee. Databases specifically relating to chemical reaction information are now becoming more numerous and readily available.

The *Chemical Reactions Data Service* from Derwent Publications which is available online through the Orbit system gives access to more than 55000 reactions and new synthetic methods which have been reported since 1942. It also covers known processes using new reagents or methods and interesting applications or extensions of known reactions. It is based on Theilheimer's *Synthetic Methods of Organic Chemistry* and current awareness is provided by the *Journal of Synthetic Methods*. However searching is not easy for the infrequent user since it is based on a combination of keywords and fragmentation codes.

Chemical Abstracts Service have recently instigated a reaction database called *Casreact*, available through the STN online system. This is searched using the ubiquitous Registry Number specified to relate to the reactants, reagents, solvents, catalysts or products. The cost of use of the database includes a charge for the number of registry numbers used in the search. However, a set of numbers can be transferred from a substructure search and this is charged as a single unit. The database dates from 1985 and contains over 20,000 records with more than 200,000 single-step reactions. It is growing at a rate of 15,000 records per year extracted from 100 journals.

Other reaction retrieval systems are available for use on externally supplied databases and for compilation of inhouse reaction records. The best known are *REACCS* (REaction ACCess System) from Molecular Design Ltd., San Leandro, California, *SYNLIB* (SYNthesis LIBrary) from Distributed Chemical Graphics Inc, Philadelphia, Pennsylvania and *ORAC* (Organic Reactions Accessed by Computer) from the Wolfson Unit for Computer Aided Design, Leeds, UK. The databases available with these systems prepared by academic chemists abstracting the primary literature and on Theilheimer's *Synthetic Methods of Organic Chemistry*, now contain tens of thousands of reactions. A comparison of searching these three systems is given by Borkent et al. (1988). Reaction databases are a growth area; however their usefulness is determined by the size of the database that is available and the construction of the database is the time-consuming step. All the systems mentioned above will prove more valuable, the larger they become.

Computer programmes and expert systems are being developed to assist in the design of chemical syntheses and to predict products and yields from reactions. *LHASA* (Logic and Heuristics Applied to Synthetic Analysis) is a programme developed at Harvard and Leeds Universities which will, starting from the product required, perform a retrosynthetic analysis and suggest synthetic pathways, intermediates and starting materials. The programme is highly interactive with the user, using graphical input and output and suggesting pathways by use of its database of types of reactions, called reaction transforms. *Cameo* (Computer Assisted Mechanistic Evaluation of Organic reactions) on the other hand, is used to predict the outcome of chemical reactions. It does not use a database, but performs a detailed mechanistic evaluation of reactions to estimate what they will yield.

Biotechnology Abstracts and *Current Biotechnology Abstracts* are both available online, the former on Dialog and Orbit, the latter on Data-Star. The Life Sciences Collection from Cambridge Scientific Abstracts is searchable through Dialog and is available on compact disc for in-house use. It consists of 18 subject-oriented subfiles, amongst which are *Biotechnology Research Abstracts* and *Microbiology Abstracts: Industrial* and *Applied Microbiology* titles. The collection is based on material from over 5,000 journals and also includes selections from books, serials, proceedings and patents. Nearly 100,000 abstracts are added each year. The compact disc version holds items from 1982 onwards (each disc holding two years of the database) and it is updated semi-annually. The *MICIS* database from The Laboratory of the Government Chemist contains information on the properties, growth, preservation and storage and use of organisms listed in various culture collection catalogues. This is a new databank and will become increasingly useful as the quantity of information it contains increases. A problem with *MICIS* is the lack of consistency between the entries from different culture collections, there appearing to be discrepancies in field definitions.

In the area of chemical source databases there is a need for a comprehensive database listing the suppliers of chemicals in bulk. The databases that are available for this purpose are dominated by the suppliers of small quantities of materials for laboratory use rather than production. Information on the possible bulk availability of chemicals can still be obtained from these databases by considering the quantities offered and prices whilst it may be possible to trace a bulk supplier through a contact listed. *Fine Chemicals Directory* from Fraser Williams was originated by a group of pharmaceutical companies and structure searching capabilities were incorporated through the use of Wiswesser Line Notation. The database is now available online as *Chemquest* on Orbit and on microfiche or on tape for loading onto in-house computers where it can be interrogated using more sophisticated structure searching software. The STN service offer an online version of *Chem Sources – USA*, listing over 110,000 chemicals available in the US. The associated database *CSCORP* gives the names and addresses of the suppliers. The Chemical Exchange Directory SA is a company which collects and maintains an online library of commercial/technical data sheets on chemicals available throughout the entire world. This system acts as an advertising service for suppliers of chemicals who pay a fee to have their datasheets included. Once again this is a potentially valuable databank which needs to expand. Soris (Specialist Organics Information Service) set up by a consortium of UK chemical companies is an enquiry service which will direct customers to UK suppliers of chemicals in bulk. The UK emphasis of this service limits its usefulness and a European equivalent is required.

As well as specific supplier databanks some of the hazardous chemicals databanks mention large-scale manufacturers of chemicals. The *Ecdin* (Environmental Chemicals Data and Information Network) on Datacentralen and *Hazardline* on BRS have fields in their records for bulk suppliers in Europe and the US respectively.

As mentioned earlier bulk hazardous chemicals are controlled under European and American legislation based on inventories of existing chemicals. Both inventories are available as databases, the European *EINECS* directory on compact disc, the American *TOSCA* directory online via Dialog.

Analytical development

Analytical procedures must be developed for the active ingredient, the starting materials in its synthesis and for other ingredients in the formulations. Specifications for all significant ingredients must include confirmation of their identity, quality and significant impurities with criteria for acceptance of batches. There must be quality control at each stage in the manufacturing process and evidence that the finished product is of the correct chemical and physical structure and purity. Other tests which may be required by some authorities concern solubility, crystal form, polymorphism, particle size, solvation, oil/water coefficients and pKa/pH values where applicable. Methods of analysing the active ingredient, its impurities and degradation products in the final formulation can also be used in stability and shelf-life tests carried out at various temperatures, humidities and in various light conditions. Test methods are necessary to indicate any reactions between the ingredients or between the ingredients and the packaging and also to monitor any sorption of the formulation ingredients to the container or leaching of any pack component into the medicine. Uniformity of dosage and release rates of controlled release formulations are other criteria of importance. The methods used in these analyses include high performance liquid chromatography (HPLC), thin layer chromatography

(TLC), gas liquid chromatography, infrared and ultraviolet spectroscopy, and thermal analysis.

Books

The analytical area of development work is well covered in Beckett and Stenlake's *Practical Pharmaceutical Chemistry*. Volume 1 has a chapter on medicaments in formulations including quality control of formulated products and sterility testing whilst volume 2 deals with the physical techniques of analysis such as chromatography, thermal analysis and particle size analysis. Clarke's *Isolation and Identification of Drugs* (Pharmaceutical Press) deals in four parts with the analysis of drugs in pharmaceuticals, body fluids and post-mortem material. Part 1 covers analytical techniques in general, part 2 gives analytical and toxicological monographs on 1300 drugs, part 3 has 66 indices of analytical data, e.g. melting points, retention indices and infrared peaks and, finally, part 4 describes the reagents used in analytical procedures.

The annual publication *Analytical Profiles of Drug Substances* is seen by its editor Florey as a supplement to the official compendial standards giving more physical and chemical data along with methods of synthesis, physical and biological degradation and metabolism.

Journals

Some of the major journals in this area are listed in Table 5.4. *Chemical Abstracts* is a major information source having one section dedicated to pharmaceutical analysis and relevant *CA*

Table 5.4. Journals relevant to analytical development

Analyst
Analytical Chemistry
Analytical Proceedings
European Spectroscopy News
Journal of the Association of Official Analytical Chemists
Journal of Chemometrics
Journal of Chromatographic Science
Journal of Chromatography
Journal of High Resolution Chromatography Communications
Journal of Liquid Chromatography
Laboratory Practice
LC Liquid Chromatography HPLC Magazine
Pharmacopoeial Forum

Selects are given in Table 5.3(b). *Analytical Abstracts* from the Royal Society of Chemistry is a more specific source and its coverage of analysis is less comprehensive than *Chemical Abstracts*. The *Ringdoc* profiles booklets (Derwent Publications), a weekly coverage of the pharmaceutical literature with informative abstracts, have a title on *Drug Analysis and Methodology*.

Databases

Chemical Abstracts, Analytical Abstracts and *International Pharmaceutical Abstracts* are the main databases used in analytical development, all having sections dedicated to the subject. The main advantage of *Analytical Abstracts* online is the ability to specify whether the term being searched is the analyte (the substance being analysed), the matrix (the medium being analysed) or the concept (technique, reagent or field of study). It covers 300 core journals and is only available online from 1980 onwards. Other databases that might be interrogated are described elsewhere in this chapter.

Formulation development

In 1937 Samuel E. Massengill Co. distributed a new liquid dosage form of sulfanilimide, one of the first of the modern generation of drugs, in the US. Diethylene glycol was the vehicle used by the company's chemist as it was able to dissolve the required quantity of drug. The chemist did not investigate the toxicity of the solvent and over 100 people died from taking the 'medicine'. As a result, legislation was passed ensuring that manufacturers do not market a new drug until they have persuaded the Food and Drug Administration that it is safe (Sherman, 1988).

This disastrous occurrence illustrates that the safe formulation of any drug is a vital part of the development of a medicine. Drug substances are almost always given to patients in a formulated medicine, only rarely in their pure form. The formulation of a medicine is designed to make available to the patient the correct quantity of drug at the correct rate in a stable, safe and acceptable form. For the manufacturer, the formulation must be readily produced in bulk at the right price and be stable during storage under various or specified conditions. For the doctor, the formulation available must be relevant to the case – oral formulations are useless if the patient is unconscious. Ideally, therefore, the formulation must be stable, safe, reproducible, amenable to large-scale manufacture, aesthetically acceptable, free from contamination (microbiological or foreign body) and, above all, in satisfying these requirements the bioavailability of the active ingredient must not be compromised. It is usual for new drugs to appear in formulations suitable for administration by several different routes and Table 5.5 gives an idea of the range of dosage forms available. These various criteria shape the infor-

Table 5.5. Range of dosage forms available for different administration routes

Administered via	Dosage forms
Ear (Aural, Otic)	Ear drops – solutions or suspensions, ointments
Eye (Ophthalmic)	Ointments, solutions – sterile, pH near that of tears
Mouth (Oral)	Capsules – medication enclosed in gelatin shell (hard or soft) Lozenges and pastilles for dissolution in the mouth Tablets – compressed Coated tablets – protective coating for controlled release or taste masking Layered tablets – for controlled therapeutic effect or incompatible ingredients Pills – now most made by tabletting in pill shape Powders – mixed with water – can be effervescent Granules Syrups – sweetened solution Elixir – sweetened alcoholic solution Other solutions and suspensions Emulsions (oil-in-water or water-in-oil) Mouthwashes, gargles
Parenteral	Injections, infusions in the form of solutions, suspensions or emulsions Implants
Rectum/vagina	Suppositories (vaginal – pessaries) melt or dissolve Tablets, capsules, ointments, enemas, powders, solutions
Skin (Topical)	Ointments – greasy, semisolid preparations Creams – semisolid emulsions, water-in-oil or oil-in-water Tinctures (natural drug extract), liniments (oily or alcoholic solutions or suspensions for rubbing), lotions, solutions suspensions and gels Pastes and cerates (ointments with high dispersed solids or waxes respectively) Aerosols, powders, dressings, soaps and shampoos

mation needs of the formulation scientist. The preformulation stage includes collecting information on the fundamental properties of the drug substance such as stability, hygroscopicity, solubility, dissolution rate, partition coefficient, dissociation constant, particle size and shape, melting point, polymorphism, density, flow characteristics, and compressibility which might affect decisions on the formulation of the product.

Formulations consist of the active therapeutic agent and various non-therapeutic agents usually called excipients (other terms used are additives, adjuvants, bases, necessities or vehicles). Amongst these are listed antioxidants, binders, coating agents, colours, diluents, dispersing agents, emulsifiers, fillers, flavours, lubricants, perfumes, preservatives, propellants, solubilizers, stabilizers, surfactants, sweeteners, tablet disintegrants, thickeners and wetting agents. Collecting basic information on the properties of the excipients is also part of preformulation.

Guidance on the choice of excipients can be obtained by comparison with previous in-house formulations, those of competitor products or by considering the properties of new materials. If new excipients are to be used they will be considered in the same way as new drug substances by the regulatory authorities and will require the same lengthy and expensive testing unless they are already permitted in the food industry. It is, therefore, advantageous for formulation scientists to be able to obtain information on chemicals used in the food industry.

Once candidate excipients have been selected they must be shown not to react with the active agent or with the other excipients. The parameters which might affect the bioavailability of the active ingredient are particle size, crystal form, solubility, the excipients present, tablet shape and hardness, coatings and salt form of active agent. Much work is undertaken on using the dosage form to control or sustain the release of drugs or to carry a drug to where it is needed before releasing it. In a recent market survey it was reported that nearly 120 companies in the United States are investigating novel drug delivery systems using transdermal, nasal, liposomal, infusion, controlled release implants, oral time-release, microsponge technology and erythrocytic targeting systems.

Other tasks which are the responsibility of the Formulation Development Department are the assessment and testing of packaging, the conduct of scale-up and pre-production studies and, from the experience thus gained, the issuing of manufacturing instructions to the Production Departments and contributing to trouble shooting. New formulations must be supplied for clinical trials, specifications must be agreed with the Quality Assurance Department and data supplied for product registration purposes.

Books

Of the books available in this area of work the most relevant is the *Handbook of Pharmaceutical Excipients* published in 1986 jointly by the Pharmaceutical Society of Great Britain and the American Pharmaceutical Association. This is an updated, English language version of the *Katalog Pharmazeutischer Hilfsstoffe* which was prepared by a working party from Ciba-Geigy, Hoffmann-La Roche and Sandoz and published in 1974. In the more recent version 134 monographs are included which describe materials commonly used as pharmaceutical excipients. Information given includes name and synonyms, uses, suppliers, descriptions (most containing electron photomicrographs), the major pharmacopoeial specifications, typical physical properties, stability, recommended storage conditions, safety and handling data, incompatibilities and references to the sources of this data. As the concept of the book came from industrial pharmacists, it is not surprising that it covers very precisely the information which is needed.

The US FDA *Product Excipient Listing* is a 2900 page computer-generated report listing all excipients used in pharmacological products [both prescription and over-the-counter (OTC)] registered with the FDA for sale in the US. The excipients are listed alphabetically with the products that contain them, the manufacturers of the products and the dosages. Although not giving properties of these materials it does give ideas on what type of excipient can be considered for new formulations. This compilation is possible because of the need in the US to declare inactive ingredients on labelling or data sheets, as is the case in other countries such as Australia, France, Italy, the Scandinavian countries and W. Germany. The DHSS in the UK have proposed that manufacturers should give details of all ingredients of medicinal preparations whether active or not. If this becomes law, a DHSS compilation of ingredients may soon be available and since most pharmaceuticals sold in the UK are also sold in the US and are listed in this FDA publication, any argument against such legislation appears futile.

Florence's *Materials Used in Pharmaceutical Formulation* discusses the problems faced by those who formulate medicines in four related areas of work, film coating oral dosage forms, tablet lubricants, polymeric materials for controlled release and fatty alcohol mixed emulsifiers and emulsifying waxes. The contributors are all expert in these fields of formulation and this text is invaluable to formulation and quality control scientists.

Pharmaceutics: The Science of Dosage Form Design edited by Aulton is aimed primarily at pharmacy undergraduate students and has been written to replace Cooper and Gunn's *Tutorial Pharmacy*. Its emphasis on design of dosage forms and all aspects of drug delivery make it a valuable source of information for all working in the formulation area of the pharmaceutical industry. The subject is dealt with in five sections covering physicochemical properties and principles, bioavailability and biopharmaceutics, drug delivery systems, microbiology relevant to dosage forms and, finally, large scale production of medicines.

The fourth edition of Ansel's book *Introduction to Pharmaceutical Dosage Forms* is also intended for pharmacy students, to 'introduce them to medicinal and pharmaceutic substances, their formulation into discrete dosage forms and drug delivery systems and their utilization in patient care'. General chapters cover routes of administration, biopharmaceutic considerations and dosage form design including ingredients, formulation and current good manufacturing practice. These are followed by chapters on specific routes of administration and related dosage forms, e.g. oral suspensions, aerosols and inhalations and transdermal systems, making it another basic general text for the formulation scientist in industry and elsewhere.

Pharmaceutical Dosage Forms is also the title of a series of practical books published by Marcel Dekker. The two volumes on *Parenteral Medications* edited by Avis, Lachman and Lieberman cover the formulation and manufacture of small and large volume parenterals, environmental control, sterilization and clean room operations as well as regulatory, quality assurance and GMP considerations. The three volumes on *Tablets* edited by Lieberman and Lachman cover aspects from pre-formulation through to production and quality assurance of compressed tablets, layer tablets, effervescent, chewable and medicated lozenges. Another relevant series of books is published under the general title of *Drugs and the Pharmaceutical Sciences*. Particularly noteworthy are *Sustained and Controlled Release Drug Delivery Systems* edited by Robinson (volume 6) and *Novel Drug Delivery Systems: Fundamentals, Developmental Concepts, Biomedical Assessments* edited by Chien (volume 14).

Marcel Dekker have recently announced a major running series entitled *Encyclopedia of Pharmaceutical Technology* encompassing, according to their literature, everything from the properties of the drug to the delivery system and production and providing 'the most up-to-date and easily accessible reference work available on pharmaceutical technology'.

Trissel's *Handbook of Injectable Drugs* summarizes the primary research on the stability and compatibility of injectables. Particularly useful are stability, pH and concentration data together with the details of the formulation where appropriate. A list of colorants that are allowed in solid oral drugs is available from Capsugel. The list comprises 34 colorants with special regard to hard gelatin capsules. In general, colours used in the food industry are permitted in oral drugs and information on these can be obtained from The British Food Manufacturing Industries Research Association who provide lists of permitted natural and synthetic colours. In the area of general food additives that may be allowable in drugs, the publications of the joint Food and Agriculture Organization of the United Nations and the World Health Organization Expert Committee on Food Additives is the authoritative source. Their *Food Additives Data System* (*FAO Food and Nutrition Paper 30*) gives names, synonyms, functional class evaluation status, and permitted uses in food of additives.

Journals

Table 5.6 lists the major journals, mostly English language, that have been found particularly valuable in pharmaceutical development.

ABSTRACTING JOURNALS AND DATABASES

Chemical Abstracts has the widest coverage of any comparable abstracting service or database. The pharmaceuticals section (63) and the pharmaceutical analysis section (64) are of particular interest to formulation scientists. Moreover, papers appearing in other sections are cross-referenced to these sections if related to pharmaceuticals. Because of its wide subject coverage, *Chemical Abstracts* allows general subject searching, e.g. stability of emulsions, for which references might be found in the food and paint subject areas. Chemical substances are clearly indexed in *Chemical Abstracts*, each unique compound being given a Registry Number. However, as the numbers do not indicate chemical structure or relationships, classes of compounds, e.g. salts of a particular drug, can only be searched by finding and entering each individual derivative or by constructing family searches with online structure searching systems. It is possible in *Chemical Abstracts* to use controlled indexing terms but these are not as precise for pharmaceutical searches as in the more specific databases, a disadvantage of wide database coverage. It is also possible by

Table 5.6. Journals relevant to formulation development

Acta Pharmaceutica Suecica – Research in pharmaceutics journal from the Swedish Academy of Pharmaceutical Science (in English). First published 1964, 6 issues a year.

Advanced Drug Delivery Reviews – New Elsevier journal giving critical review articles on current and emerging aspects of research into design and development of advanced drug delivery systems and their application to experimental and clinical therapeutics – publication erratic at present.

American Pharmacy – American pharmacy practitioners journal.

BIBRA Bulletin – British Industrial Biological Research Association Bulletin giving information on toxicity and legislation on drugs and additives, especially food additives.

Chemist and Druggist – British retail pharmacy journal.

Drug and Cosmetic Industry – manufacturing and packaging information in drugs and cosmetics.

Drug Development and Industrial Pharmacy – Concerned with all aspects of the development and production of drugs and pharmaceutical products, this journal examines both technological and organizational aspects of industrial pharmacy.

Drug Intelligence and Clinical Pharmacy – journal for drug therapy decision makers – clinical.

Drug and Therapeutics Bulletin – up-to-date reviews on drugs and their uses for doctors from the publishers of *Which*.

Drugs Made in Germany – International views of politics and science with contributions on development, manufacture, marketing, distribution and therapeutic use of drugs.

Food Drug Cosmetics Law Reports – legislation etc from FDA.

Gold Sheet – quality control newsletter.

International Journal of Pharmaceutical Technology and Product Manufacture – includes dosage form design and process development excipients and medicinal dosage forms and product manufacture.

International Journal of Pharmaceutics – physical, chemical, analytical, biological and engineering studies related to drug delivery.

Journal of Controlled Release – the science and technology of controlled release of active agents.

Journal of Parenteral Science and Technology – original research reports or review articles in the parenteral area.

Journal of Pharmaceutical Sciences – monthly research journal of the American Pharmaceutical Association.

Journal of Pharmacokinetics and Biopharmaceutics – includes studies in drug release from dosage forms, dosage form developments and evaluations.

Journal of Pharmacy and Pharmacology – development and formulation of drugs and medicines.

Manufacturing Chemist – UK chemical industry journal.

PAGB Bulletin – bulletin of the Pharmaceutical Association of Great Britain.

Pharma International – multilingual practical pharmaceutics journal.
Pharmaceutica Acta Helvetiae – Swiss pharmaceutics journal.
Pharmaceutical Journal – journal of the Royal Pharmaceutical Society of
    Great Britain.
Pharmaceutical Technology – pharmaceutical research and applications.
Pharmacopoeial Forum – drug standards development and official
    compendia revision (*United States Pharmacopoeia*).
Pharmazie – journal of the East German pharmaceutical industry.
Pharmeuropa – drug standards development and official compendia
    revision (*European Pharmacopoeia*).
Powder Technology – science and technology of wet and dry particulate
    systems.
Scrip – news items from the pharmaceutical industry worldwide.

---

searching online through STN, *Chemical Abstracts* own system, to search the abstract. Thus physical property data can be located, e.g. pH or density, which might not appear in the main index. The *CA Selects* titles of potential interest in this area are listed in Table 5.3(c).

International Pharmaceutical Abstracts produced by the American Society of Hospital Pharmacists covers the pharmacy and drug related international literature pertinent to all phases of drug development and use. It contains about 200,000 references from 600 journals, but does not cover patents. Indexing is based on controlled terms, Chemical Abstracts Registry Numbers and free text, the abstract being searchable in online systems. It has in the past only been available online on the Dialog host, but other hosts including Data-Star are now offering *IPA* and a compact disc version from Cambridge Scientific Abstracts which will also contain *Drug Information Fulltext* is due late in 1988. Compared with *Chemical Abstracts*, its coverage is restricted but, being a specific database for drug information, output from searches tends to be more relevant.

Index Medicus or its online equivalent *Medline* indexes the medical literature from over 70 countries, covering 3200 journals and growing at about 250,000 articles per year, about 50% of which since 1975 have abstracts in English. The detailed indexing of this database makes it possible to obtain very specific results from searches. The controlled indexing terms (MESH) are hierarchical in structure, each term being associated with a MESH number. For example, Drug Dosage Forms are encompassed by D26.394.225 and truncated searching (i.e. items with anything following and including 225) will retrieve papers on all drug dosage forms whereas Delayed Action Dosage Forms are incorporated in D26.394.225.276 and this number truncated will retrieve all types of delayed action dosage forms. More specifically, if only Enteric-coated tablets are of interest then D26.394.225.276.695 is searched. Thus both general topics and specific applications can be investigated. Another feature of the *Medline* indexing is the ability to link terms with descriptor codes in order to ensure that only relevant papers are retrieved. If, for instance, papers on toxicity problems relating to a pharmaceutical ingredient are required, linking the ingredient with the descriptor code for adverse effects or toxicity or poisoning will avoid retrieval of papers that mention the ingredient and toxicity etc. but with these indexing terms being unrelated. Medline is available on several online hosts and in various compact disc versions.

The *Excepta Medica* database (*Embase*) indexes articles from over 3500 primary journals on human medicine and related disciplines and dates back to 1974. The main indexing system of the database has recently been changed to a system called Emtree Classification which is very similar to the *Medline* system. For example, Emtree classification D3 refers to pharmaceutical vehicles and additives, D3.30 to drug additive, D3.30.310 to flavouring agents whilst D3.30.310.850 is sweetening agents. Linking with descriptor codes is also now possible as in the *Medline* file. *Embase* is available on most major online hosts but the compact disc version anticipated in 1987 has not been released although a subfile of papers relating to cancer, *Cancer-CD* is available in this form. The full compact disc version would be most welcome for frequent users as it would reduce costs.

Information on patents is available in the *World Patents Index* from Derwent Publications, with the pharmaceutical coverage of the file, *Farmdoc*, dating from 1963. Searches relating to formulations can be constructed using Manual Codes in the B12-M series which covers formulation types (e.g. B12-M02 refers to a cream or an ointment). A new database from Chemical Abstracts Service called *Pharmpat* contains chemical and biochemical patents on drugs and pharmaceutical chemistry with in-depth substance and subject indexing. Coverage dates only from January 1987 but the database already contains over 5000 patents.

An online, full-text version of Trissel's *Handbook of Injectable Drugs* is available from Dialog in a file called *Drug Information Fulltext*. Included in the same file is the fulltext version of the *American Hospital Formulary Service – Drug Information*, which records chemistry, stability, absorption, pharmacokinetics, uses, precautions and dosages of selected drugs. It has just been announced that these two publications along with *International Pharmaceutical Abstracts* will be available on CD-ROM later in 1988. *Martindale* is also loaded online and on CD-ROM (as part of the Micromedex system) so that all three major reference works in this area are now available in a variety of forms with increased timeliness and versatility of searching.

The *Ringdoc* database from Derwent Publications addresses a substantial part of the pharmaceutical literature and has the major benefit of informative abstracts. However, the service is one of the few online databases that requires an up-front subscription, a fact which discourages some potential users.

A database of interest to scientists working on the formulation of topical products is the recently announced *Kosmet* produced by the International Federation of Societies of Cosmetic Chemists and loaded on Data-Star; it is the first cosmetics database. Other databases of occasional value are the *Pharmaceutical and Healthcare Industry News Database* (Data-Star), the *Pharmaceutical News Index* (Dialog and others) and *Pharmaprojects* (Data-Star). These may all be used to follow up announcements of new formulation types that have been launched, patented or are available for licence.

Production

The information needs of the production area are extremely varied and because of this they will be discussed by areas of work and types of information rather than consideration of the nature of the materials involved. It might be assumed that production staff have less need for information but the truth is that they have less time to obtain information and therefore need it packaged differently from research and development staff. Facts are rather more important than literature references.

Information needs

The new pharmaceutical is by now a marketable product and economic manufacturing processes will have been devised to enable its production. The internal information generated earlier is of major importance at this stage. Documentation describing how the active ingredient and its formulation must be manufactured are basic not only because the process has been registered with the regulatory authorities and must be closely adhered to, but also because, by this time, the procedure is being passed on to non-technical staff for whom operating procedures must be closely defined. The manufacturing procedure has been agreed and a product licence awarded on evidence of safety, efficacy and quality has been granted. However, manufacture can be carried out only at production sites which have manufacturing licences. Once the product goes into production the scale of work is increased such that safety, environmental control and pollution control considerations become paramount. The environmental impact must be controlled both within the factory to preserve the health of the workforce and also in the immediate vicinity of the factory to preserve the health and goodwill of the local inhabitants.

The production is usually carried out on a large factory site, which generates information needs in its own right. The engineering department will deal with building requirements, mechanical engineering, electrical engineering and computer technology and chemical engineering. Downstream processing in production areas to retrieve expensive reagents and the use of advanced separation technology, whether filtration, membrane separation or distillation, to recover extra product from waste streams can make a significant difference to the economics of production. Furthermore, the quality of the product and the requirements for further processing can be affected together with the amount and quality of the waste, all of which are factors which influence disposal decisions.

Packaging the finished product is the final act and requires information on topics ranging from types of container and closure required to mechanical packaging lines.

Sources of information

In the UK, it is essential to refer to the Medicines Act and all the regulations, statutes and guidelines resulting therefrom which are published by the HMSO. HMSO issues a *Daily List* of items published by the office available on subscription and either sent daily or batched and dispatched weekly. These are cumulated into monthly and annual lists of Government Publications and the complete list of publications is available as a monthly microfiche subscription under the title *HMSO in Print*. Information on current legislation arising from a particular Act is recorded in the HMSO's *Statutes in Force*. Another way of keeping up-to-date in this area is through a subscription to Butterworth's *Law of Food and Drugs*, a multivolume loose-leaf publication and updating service encompassing the UK's relevant legislation.

All members of the Association of British Pharmaceutical Industry receive a *Weekly Service Sheet* which discusses and reviews new legislation in the pharmaceutical sector.

American legislation on drugs can be followed through the *Federal Register*, the daily record of the proceedings of the US Government, available in hard copy or microfiche form from the US Government Printing Office. The *Food Drug Cosmetic Law*

Reports from the Commerce Clearing House Inc., Chicago, summarizes the legislation in these areas, and the suppliers will provide the fulltext where required; the service is a convenient current awareness tool. All the regulations actually in force in the United States are published each year in the *Code of Federal Regulations* (*CFR*). This publication refers to all US regulations and the Food and Drug Administration legislation is covered under title 21, available separately from the US Government Printing Office. Background documentation pertaining to FDA regulations including unpublished documents obtained under the Freedom of Information Act, can be accessed in the online database *Diogenes* available on Dialog and Data-Star.

In order to obtain a manufacturer's licence the manufacturer has to show compliance with good manufacturing practice (GMP). 30 countries issue guides to what they consider is good manufacturing practice, amongst which is the UK's *Guide to Good Pharmaceutical Manufacturing Practice* (third edition, 1983) issued by the DHSS. In the United States, the FDA issues the *FDA's Current Drug GMP Final Rule* and similar guides/guidelines on sterile drug products produced by aseptic processing, inspection of bulk pharmaceutical chemical manufacturing, inspection of computerized systems in drug processing and the general principles of process validation.

International guidelines are prepared by the World Health Organisation with *Good Practices in the Manufacture and Quality Control of Drugs* and the Pharmaceutical Inspection Convention, European Free Trade Association, including *Basic Standards of GMP for Pharmaceutical Products* and the recently revised *Guidelines for the Manufacture of Active Pharmaceutical Ingredients*.

For the engineering professional, the standard reference works are Perry's *Chemical Engineers Handbook* published by McGraw-Hill and Kempe's *Engineers Year-book* published by Morgan-Grampian. Perry contains sections on process control, process economics, waste management, size reduction and enlargement, distillation, biochemical engineering and 21 others providing information concerning most chemical engineering activities. Kempe provides information in many diverse areas of engineering in over 90 sections of a two volume work. For a detailed list of information sources in the engineering sector, the book by Anthony (1986) in this series is recommended and, for the chemical engineering discipline specifically, *Chemical Engineering Data Sources* by Jankowski (1986) provides relevant data.

Standards are of particular importance to engineers in general and membership of the British Standards Institute ensures that one of the largest collections of national and international standards is obtainable by loan or purchase at discounted rates. Membership also secures a subscription to *British Standards News* which exposes details of proposed changes to standards for comment. Non-members of the Institute may borrow standards through the local public library service, designated libraries within each region being responsible for holding a complete set of British Standards, and all libraries are able to arrange loans of foreign standards through the BSI. The annual *British Standards Catalogue* is an essential library tool and an index is available online in the BSI database of Infoline. Several commercial organizations offer standards services, Technical Indexes of Bracknell provide an updating service on microfilm tailored to individual requirements and review 90% of the world's most commonly used industrial standards. Technical Indexes also produces a suppliers catalogue service on microfilm covering areas such as process engineering, laboratory equipment and manufacturing and materials handling equipment. Online databases affording access to equipment suppliers information are *Dequip* and *Deteq* on STN International. Both are produced by Dechema Deutsche Gesellschaft fur chemische Apparatewesen, the former service listing equipment in the fields of chemical engineering and biotechnology, the latter in environmental engineering. There are several bibliographic databases covering the various engineering disciplines. *Compendex* and *EI Engineering Meetings* produced by Engineering Information Inc. span all the disciplines; the former scans 4500 journals, technical reports and monographs and the latter individual papers from engineering meetings worldwide. *Compendex* is the largest engineering database available, containing over 2.2 million records at the time of writing and is well indexed using a controlled vocabulary [*Subject Headings for* Engineers (SHE)]. The remaining relevant databases are discipline specific. *Chemical Abstracts* covers the subject of chemical engineering well and several titles in the *CA Selects* series are relevant – see Table 5.3(d). *Chemical Engineering Abstracts* produced by the Royal Society of Chemistry is available both as an online database and as a current awareness journal. It is the major English language chemical engineering database on the Data-Star, ESA-IRS and Infoline online hosts. Other discipline specific databases include *ISMEC* (*Information Service in Mechanical Engineering*) available on Dialog and ESA-IRS and covering the literature of mechanical engineering since 1973 and *INSPEC* (*Information Services for the Physics and Engineering Communities*) available on Dialog, Orbit and STN and incorporating electrical, electronic and computer abstracts since 1969.

A frequent information requirement of chemical engineers is physical property data on chemicals. What promises to be a useful reference source has just been published by the Science Reference and Information Service: *A Guide to Sources of Physical Property Data* lists 594 sources of physical property data comprising the largest collection of such reference material in Western Europe. *DIPPR (Physical Property Data on Pure Chemicals)* is a database recently added to the STN armoury and contains physical property data on 766 commercially important chemicals. The data were compiled through a project of the Design Institute for Physical Property Data organized by the American Institute of Chemical Engineers.

The Research Association of the Paper and Board, Printing and Packaging Industries (PIRA) can help satisfy the information requirements of staff responsible for packaging. Not only does PIRA provide an online database in this subject area but the Association also offers research, consultancy and training. The PIRA database on Infoline is not unique since Dialog hosts the *International Food Information Service* database – *Packaging Science and Technology Abstracts* – and information may also be found in the more generalized pharmaceutical and engineering databases depending on the nature of the enquiry.

The safety information requirements of production personnel extend far beyond the basic hazard information on specific chemicals. The *Barbour Index* is an updating service on microfiche which contains the full text of legislation, standards, guidance notes and booklets relating to all aspects of industrial safety. The microfiche are updated each quarter so that all current information is readily available. The databases collected by the Health and Safety Executive in the UK (*HSELINE*), the International Labour Organisation (*CISDOC*) and the National Institute for Occupational Health and Safety (*NIOSHTIC*) are all searchable online and all three sources are available together on one compact disc from Silver Platter Inc. This CD-ROM version means that bibliographic information on industrial safety from all over the world is now immediately available in a convenient form. Systems which give safety data on chemicals in the form of datasheets are accessible both online and on compact disc systems. The *Environmental Chemicals Data Information Network* (*ECDIN*), online from Datacentralen in Denmark, is the European version giving general hazard data, first aid instructions, waste disposal precautions and environmental impact data together with the major European producers alongside and relevant EEC Directive information such as labelling requirements. *Hazardline* on BRS gives similar information but considers in great detail the US legislation with which users of the material must comply. The major periodical in this area is the *American Industrial Hygiene Association Journal* which usually publishes very practical papers about the factory environment. Medical contingencies are covered by the *British Journal of Industrial Medicine* and more general hazard information appears in the *Dangerous Properties of Industrial Materials Report* (the updating journal to Sax's book of the same name).

Pollution and environmental control are matters of prime concern on any large production site. Garner's *Control of Pollution Encyclopaedia* published by Butterworths is designed as a guide to the rules and regulations prevailing and consists of two looseleaf volumes and a supplementary updating service. The *Environmental Data Service (ENDS)* reviews environmental and pollution issues. There are in addition several bibliographic databases including *Pollution Abstracts* produced by Data Courier, *APTIC (Air Pollution Technical Information Center)* and *Enviroline* from the Environment Information Center all of which are available on Dialog and may prove useful.

General requirements for an information centre serving development and production functions

As well as providing a focus for accessing external information sources in specific subject areas, an Information Centre on a pharmaceutical development and production site will hold the standard reference texts needed from time to time by all personnel. Perhaps the most important requirement is an up-to-date set of pharmacopoeias. Pharmacopoeias comprise monographs specifying the assay, impurity levels, ranges of significant physical parameters and other properties of drugs available in the country of publication. If the drug meets the standards laid down it is then considered to be of suitable quality to be marketed in that country. The *British Pharmacopoeia (BP)* is a two-volume work with a companion volume of infrared reference spectra. The latest edition was released in 1980 and annual addenda have been published since that date; at the time of writing a new edition is imminent (1988). Volume 1 of the 1980 edition refers to drug substances whilst volume 2 deals with preparations, reagents and relevant general information. Veterinary products are covered in their own publication the *British Pharmacopoeia (Veterinary)*, the latest edition of which was published in 1977 and again a new edition is due in 1988.

The *United States Pharmacopoeia (USP)* is produced every 5

years and annual updates are available on a subscription basis. The *National Formulary*, which deals with pharmaceutical ingredients, is now part of this publication. A companion journal *Pharmacopoeial Forum* (published six times a year) introduces in-process revisions of the monographs in the *USP* with discussion of reasons and policy and requests for comments.

The *European Pharmacopoeia* (*EP*) relates to most of the countries of Western Europe. The second edition (1980) was published in a loose leaf form so that it could be readily updated. A companion journal similar to the US *Pharmacopoeial Forum* has just (1988) been announced and should prove beneficial.

WHO has compiled a list of current pharmacopoeias from all over the world, over 30 countries having their own publications in addition to the several regional and international versions. This index can be obtained from the WHO in Geneva.

Remington's *Pharmaceutical Sciences*, first published in 1886 but now in its 17th edition (1985), consists of almost 2000 pages and by virtue of its size alone is 'a comprehensive treatise covering the scientific foundations of pharmaceutical research, development, control and manufacture'. Part 8 of this publication discusses pharmaceutical preparations and their manufacture, ranging from pre-formulation through methods of manufacture to the coating of tablets.

A text which is essential to any information centre dealing with pharmaceutics is *Martindale: The Extra Pharmacopoeia*. The aim of this book is to provide a concise summary of the properties, actions and uses of drugs and medicines for the practising pharmacist and medical practitioner. In fulfilling this objective, it coincidentally provides an invaluable source of information for those concerned with drug development and manufacture. *Martindale* is also available as a fulltext online version on Dialog and Data-Star thus increasing the versatility of searching and cross referencing.

The *Merck Index* started life in 1889 as a catalogue of the products sold by the Merck Company. It has since developed into a primary information source for chemists and pharmacists and is now in its 10th edition (1983). Data on over 10,000 chemicals and drugs are recorded including identification, physical and toxicity data, therapeutic and commercial uses and bibliographic citations to the chemical, biomedical and patent literature. It also is available as a fulltext online version on Dialog.

An information centre serving the wide subject interests involved in pharmaceutical manufacture and production benefits from a library of general encyclopaedic texts. The *Encyclopaedia of Chemical Technology*, usually referred to by the names of its original editors, *Kirk-Othmer*, is a 25-volume set published by Wiley. It ranges over most areas of industrial chemistry and technology and nearly all the contributions are very thorough with extensive bibliographies. It is recommended as a primary source for enquirers seeking general background information. It too is available online through vendors such as Dialog. A compact disc version can be purchased and for libraries with the appropriate hardware, is an attractive proposition since searching is rendered far easier and more versatile.

Finally, the *Handbook of Chemistry and Physics* (often known by the nickname *The Rubber Book* as it is published by the Chemical Rubber Company) is published annually, and contains a multitude of tabulated properties of chemicals and general physical data.

References

Borkent, J. H., Oukes, F. and Noordik, J. H. (1988). 'Chemical Reaction Searching Compared in REACCS, SYNLIB and ORAC'. *Journal of Chemical Information and Computer Science*, **28**, 148–150.

Jankowski, D. A. and Selover, T. B., (eds.) (1986). *Chemical Engineering Data Sources* (AIChE Symposium Series 247 volume 82, American Institute of Chemical Engineers).

Mills, J. E., Maryanoff, C. A., Sorgi, K. L., Scott, L. and Stanzione, R. (1988). 'REACCS in the Chemical Development Environment 1.' *Journal of Chemical Information and Computer Science*, **28**, 153–155.

Sherman, M. (1988). 'Infamous drugs: A quick history.' *Drug and Cosmetic industry*, February, 37–44, 78.

Williams, G. A., Beaumont, C., Roach, M., Ward, S. E. and Wallis, D.A. (1984). 'Information services in the development and use of medicines Part 1: Resources and services'. *The Pharmaceutical Journal*, September 29th, 381–383.

CHAPTER SIX

Intellectual property

T. EISENSCHITZ

Introduction

Manufacturers of pharmaceuticals are interested in protecting their processes, products and presentation. Intermediate results and discoveries also require protection. Development and testing processes are costly to design and products need to be on the market for many years and in many countries in order to be profitable. One can understand, therefore, the high value that the industry places on protective measures. The instruments of protection are collectively called intellectual property (IP) and since protection and access to information are closely linked, the various forms of protection are of great interest to information scientists.

A product under development and in the early post-launch years can be covered by a patent which protects the new compound and/ or the method of making it while research and testing are completed prior to licensing. When the product finally reaches the market, this protection extends for a few more years while costs are recouped and profits are made. In most countries the patent lifetime is 20 years from the date of application.

The name of the product can also be protected. This happens in two ways: the active chemical is given a generic name over which no individual has rights but the brands of the product as sold are given specific names which are usually registered trade marks. These brands consist of the active ingredient plus other agents or excipients.

Finally the 'get up' and packaging of a product can be protected. This protection applies to all the labelling, packaging, mounting, colouring and other characteristics by which a manufacturer's product may be identified. A number of rights are relevant to this process, namely some aspects of registered trade marks, copyright and an informal right to overall appearance protected by an action called passing off. A registered trade mark is important in this respect because it can involve distinctive colours and designs as well as a brand name. Copyright protects all original texts and drawings. Passing off protects any distinctive feature of a product from being copied. Not all countries deal with unofficial marks in this manner.

These are the forms of intellectual property and they will be considered in more detail later in the text. However, prior to this, attention is drawn to the increasing role of European legislation in the application of intellectual property rights.

Intellectual property and the EEC

The European single market to come to pass by 1992 is of great significance to industry. It will abolish all barriers to trade within the EEC and this will encourage use of Community-wide brand names and even greater use of Community-wide patents.

Membership of the EEC has distorted the application of intellectual property rights among member states. The main function of the EEC is to provide conditions for free movement of goods and services whereas intellectual property rights are all national. They allow proprietors to prevent others from dealing in their products within national boundaries. This tends to re-establish the boundaries which the EEC is trying to abolish and is therefore in conflict with the objectives of the Community. The EEC has resolved this problem by asserting that whilst IP rights can be maintained, they may only be exercised in certain ways; their effect is thus limited so as not to over-ride the principles of the common market (Korah, 1986).

In practice this means that goods may be placed on the market for the first time as chosen by their owner but, once released into the EEC, competitors are permitted to deal in them. For instance, competitors can buy in one country and sell in another thereby undercutting the local producer. In these ways, some competition is enforced even within monopoly enterprises. Licensing agreements therefore, have to be framed to take this into account. It is worth noting that EEC law takes precedence over national law where there is conflict. EEC law is enshrined in the Treaties, cases are reported in special series of law reports such as *Common Market Law Reports* and new EEC regulations and directives are notified in the *Official Journal of the EEC*. This information is available online as the database *Celex* which is most easily available in Britain within the host Justis on the Context retrieval system. It is also available directly from the EEC Commission.

Patents

The nature of a patent

A patent is a bargain between an inventor and the State. The inventor receives a limited-time monopoly to exploit his invention. This is the most public and visible aspect of the patent grant. However, the main purpose of the system is to facilitate the spread of knowledge of technological advances and for this reason the quid pro quo of the monopoly is for the inventor to deposit for the public a written detailed description of the idea. The monopoly lasts for around 16–20 years depending on national law, but the information continues to be available and useful indefinitely. In most countries, specifications are published and are available for sale or consultation through specific libraries. Increasingly, patents are being searched online in bibliographic or full text databases for both technical and commercial/marketing purposes.

Patents are instruments of national or regional law. Every patent granted establishes the bargain in one country or a small group of countries so that patents may appear in families of similar documents relating to protection of the same invention in a number of territories. The key characteristics of such patents are the date, country and number relating to the very first application of the series and are called collectively the priority data.

The essence of patentable inventions is that they should embody knowledge which is new and not obviously deducible from that already known. These criteria are called novelty and inventive step or non-obviousness, respectively. Novelty means that searchers looking in the prior art (all material published before the priority date) should not be able to find a published record of the same work. Inventive step requires perusal not only of the same subject area but also of related technologies to locate any material from which the invention under discussion might have been deduced. This is a rather subjective criterion (Eisenschitz, 1987).

The details of any one nation's law are best clarified as and when they are required. There are a number of compilations in existence (e.g. *Patents Throughout the World*) which assist in this process.

How to apply for a patent

There are three major systems of patent administration. These are deferred examination, traditional patents and the Comecon system. Before these are described, there are some general considerations which apply universally. (For a survey of patent problems see Phillips, 1985). First, who can apply for a patent? Any inventor can apply for his own invention. In most countries including the UK, where the inventor is an employee, the employing organization can apply in its own right. The major exception to this rule is the US where the applicant must always be the inventor. An employee will then assign his rights to the company he works for if this is appropriate. Corporate patentees are then called 'assignees' and this usage is quite common internationally.

What happens when there is more than one claim to an invention? When an application is lodged with a Patent Office, the date and time are recorded. In most countries s/he who submits first is awarded the patent in the event of dispute about priority. Exceptions are the US and Canada where to establish the right to the patent, an inventor must show when s/he first had the idea and then that s/he worked on it steadily to bring it to fruition. However, the US authorities are considering a change to a first-to-file system in the interests of international harmonization (EPO, 1986b). After the initial application the International Paris Convention signed in 1883 allows up to one year for additional filings in other countries.

Patent grant procedure usually involves an examination of the contents of a written specification in comparison with the laws and practice of the country concerned in order to check that the invention satisfies all requirements of patentability. These are especially (but not exclusively) demonstrations of novelty and inventive step. Because these criteria are defined in terms of the prior art, a search of what seems to be the most relevant prior art is part of the procedure.

DEFERRED EXAMINATION

The essence of deferred examination is that the prior art search is carried out in advance of the examination. The applicant is given time to consider the results of the search and s/he then requests substantive examination which is examination of the substance or contents of the specification. The specification is published twice; once after the prior art search and again simultaneously with the grant. In this way details of all inventions are published whether or not a patent grant ensues.

The deferred examination system is now in widespread use. Most European countries use it, as do Australia, Japan and many others. It began in Holland in the mid-1960s as a way of cutting the backlog of applications by allowing withdrawal of a hopeless case without the need for a full examination. There are many slight variations in procedure. From the information point of view a particularly important achievement is the standardization of early publication. This is the publication, set at 18 months after the priority date, of the application and the results of the prior art search. The standardized date ensures that specifications relating to members of the same family all appear at around the same time.

The British national system is one of deferred examination (Patent Office, 1986a,b). It is also possible to obtain a European Patent in which up to 13 states may be designated, including Britain (EPO, 1986). It is also possible to have a centralized search carried out under the Patent Co-operation Treaty (PCT) (WIPO, 1985,a). Member states are from all over the world and the patents enter each national system for grant. The US is a member of the PCT. These are the three ways by which a British patent can be obtained, as well as the patents of many other important countries and they are all based on deferred examination. For the European and PCT routes, protection in more than three countries is required for them to be economical. From the searcher's point of view, this consolidation of applications reduces the number of documents to be checked and also the range of languages required to be read. European patents are written in English, German or French; PCT applications are published in one of these three languages or in Russian or Japanese.

THE TRADITIONAL PATENTING SYSTEM

This was the principal means of granting patents until superseded by deferred examination. When the examiner receives a patent application he embarks on the full examination and carries out the prior art search as part of this process. Only one document is published and this is after the specification has been accepted by the examiner. Publication may be simultaneous with the grant but there is more likely to be a waiting period as this is the first opportunity for public comment. The grant would then follow after any objections have been dealt with or at the end of a fixed period.

The UK Patents Act of 1949 provided for the traditional type of patenting system. Quite a number of ex-colonial countries inherited systems based on these principles upon independence, and a number still use them. The US system is of this type. US patents last for 17 years from the date of grant, subject to renewal fees being paid. However, a new Act has recently been passed allowing the term to be extended explicitly for pharmaceutical patents in circumstances where safety testing has lasted so long that there has not been enough time to earn a reasonable return on investment from sales, (US Patent Term Extensions Act, 1984), when a five-year extension is permitted. At present this is a unique piece of legislation but is something the industry has been pressing for in many other countries.

THE COMECON SYSTEM

This system is used by the USSR and its CMEA partners. It is a dual arrangement in which foreign applicants receive an ordinary patent of the traditional type but national inventors are given an inventors' certificate which confers recognition and an appropriate reward. Exploitation is left in the hands of the State. The two types of document are arranged in a single numerical sequence when published and are equally available through patent libraries and information services. Within the current negotiations for revision of the Paris Convention there is pressure to include inventors' certificates in its scope.

Contents of the specification

Patent documents are highly standardized which is a great help to readers as the basic contents and layout are very similar in all cases. The three main components are the front page, the body of the specification and the claims.

A modern front page of a British patent (shown in *Figure 6.1*) contains the title, abstract and drawing together with bibliographic details mainly added by the Patent Office. Each item is labelled with a number of the left hand side (e.g. the application number is 21). These are INID numbers allocated by WIPO (the World Intellectual Property Organization) which concerns itself with international standardization. They are the same in all countries and intended as field labels for computer processing. More generally, they indicate to a searcher which names and numbers are which if the specification is in an unfamiliar language. The title alone is often too brief to be helpful but it is informative if combined with the abstract.

Two classifications are available, the domestic and the inter-national. The fields in which the examiner carries out his searches are shown in the domestic classification. Symbols are also given for the international classification administered by WIPO and are included to aid multi-country searching (WIPO, 1986). Many countries use only the international classification and have no separate domestic system. Two exceptions are the UK and the US both of which apply major national classification schemes.

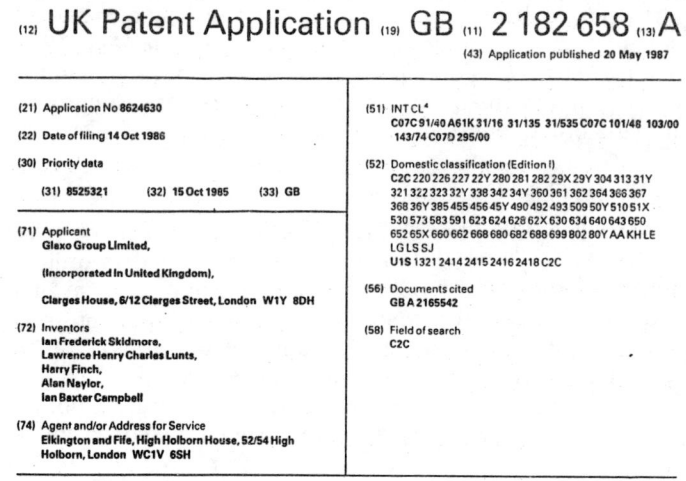

(12) UK Patent Application (19) GB (11) 2 182 658 (13) A

(43) Application published 20 May 1987

(21) Application No 8624630

(22) Date of filing 14 Oct 1986

(30) Priority data

(31) 8525321   (32) 15 Oct 1985   (33) GB

(71) Applicant
Glaxo Group Limited,

(Incorporated in United Kingdom),

Clarges House, 6/12 Clarges Street, London W1Y 8DH

(72) Inventors
Ian Frederick Skidmore,
Lawrence Henry Charles Lunts,
Harry Finch,
Alan Naylor,
Ian Baxter Campbell

(74) Agent and/or Address for Service
Elkington and Fife, High Holborn House, 52/54 High
Holborn, London WC1V 6SH

(51) INT CL⁴
C07C 91/40 A61K 31/16 31/135 31/535 C07C 101/48 103/00
143/74 C07D 295/00

(52) Domestic classification (Edition I)
C2C 220 226 227 22Y 280 281 282 29X 29Y 304 313 31Y
321 322 323 32Y 338 342 34Y 360 361 362 364 366 367
368 36Y 385 455 456 45Y 490 492 493 509 50Y 510 51X
530 573 583 591 623 624 628 62X 630 634 640 643 650
652 65X 660 662 668 680 682 688 699 802 80Y AA KH LE
LG LS SJ
U1S 1321 2414 2415 2416 2418 C2C

(56) Documents cited
GB A 2165542

(58) Field of search
C2C

Figure 6.1 Front page of a modern British patent

Documents found in the search are listed usually on the front page but under the European and PCT systems they are at the back of the specification and give details of the degree of relevance of each reference and which claims, if any, would be affected.

The body of the specification contains the disclosure to the public commencing with an introduction setting out the background and prior art of the inventions. It is usually quite brief in case further prior art is subsequently discovered which changes the perspective of the invention. A few countries notably the US, require complete disclosure of the prior art as known to the applicant. As a result, US patents usually include a comprehensive description of the inventor's startpoint although the description may not be exhaustive since some of the disclosure can be made privately to the Patent Office. There follows the specific description of the invention. This will be in general terms elaborated by examples. In chemical patents there may be many examples as reactions can be carried out in different ways and many compounds have analogous behaviour.

The claims set out the inventive aspects which are to be protected and stake the legal claims. The applicant normally claims widely so that an infringer cannot make an insignificant change as a means of circumventing the patent. The examiner is concerned to ensure that no part of the prior art is included in the claims and that the inventive aspects are supported by the descriptions and examples given in the main text. A great many claims are submitted starting with the most general form of the invention; specific details may be essential to the operation but are over-specific as far as the concept is concerned. As a consequence details are added one by one in a sequence of claims.

Specifically pharmaceutical inventions

What kinds of inventions can be protected in the pharmaceutical area? These are mainly in chemistry and biotechnology and the major categories are as follows:

- a new compound will be protected however made or used;
- a known compound will be protected when used in a novel process or made in a new way;
- a combination will be protected when a compound needs to be combined with others for storage or for delivery to its site of action;
- a compound produced using a purified naturally occurring micro-organism or a genetically engineered organism;
- genetically engineered organisms and purified naturally occurring organisms and the techniques required to produce and handle them.

Patent systems are particularly well used by the pharmaceutical and other chemically-based industries; this is because the patent protects the research and development rather than just the exploitation of final products. All patents are required to describe novel products or processes, even though they need not have been constructed or carried out in practice. There are in fact many paper inventions but all of these should have been thoroughly assessed. In chemistry this requirement can be modified by virtue of the Periodic Table. It is possible (or at least plausible) to claim an invention for a whole host of compounds formed by homologous substitutions on the basis of just one or two compounds made and tested in the laboratory. These compound representations are called Markush formulae after the American chemist who was the first to file a patent application involving such generic formulae in his claims. Their use is now widespread and has led to the evolution of a special kind of patent, the selection patent, to deal with unexpected discoveries made concerning speculative compounds.

At the turn of the century while working with dyestuffs, Markush realized that activity stemmed from a small part of a molecule. Changes in a regular fashion could be achieved by substitution of homologues and Markush had to defend and justify the practice of taking this seriality for granted in patent claims. Once accepted, this approach became common practice and was then extended to the newly emergent pharmaceutical industry and to pesticides, veterinary products and similar fields. Because of the interaction with living organisms, however, outcomes are less predictable and justification much less clear. Related compounds do not have the same smoothly varying activity. Nevertheless, by the force of tradition, patents in these fields are submitted and accepted with Markush claims. The assumptions can be defended by arguing that the main claim is for a new compound in its own right and that activity in an organism is a secondary issue. However, since the patent would not have been applied for without significant activity, this reasoning is unsound. An example of a Markush formula is given in *Figure 6.2*. It involves the following symbols: X, Y, Ar, R^1 and R^2. Each of these symbols embodies a number of choices so that their combination gives rise to a very large number of options indeed, of which 18 examples are given. They serve to illustrate the invention and do not exclude

(54) Dichloroaniline derivatives

(57) Compounds of the formula (I)

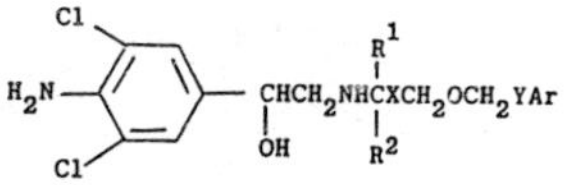

wherein

X represents a bond, $C_{1\text{-}8}$ alkylene, $C_{2\text{-}6}$ alkenylene or $C_{2\text{-}6}$ alkynylene and
Y represents a bond, $C_{1\text{-}4}$ alkylene, $C_{2\text{-}4}$ alkenylene or $C_{2\text{-}4}$ alkynylene, the sum total of carbon atoms in X and Y being not more than 8;

Ar represents a phenyl group substituted by one or more substituents selected from nitro, $-(CH_2)_q R$ [where R is $C_{1\text{-}3}$ alkoxy, $-NR^3R^4$ (where R^3 and R^4 each represent a hydrogen atom, or a $C_{1\text{-}4}$ alkyl group, or $-NR^3R^4$ forms a saturated heterocyclic amino group which has 5-7 ring members), $-NR^5COR^6$ (where R^5 represents a hydrogen atom or a $C_{1\text{-}4}$ alkyl group, and R^6 represents a hydrogen atom or a $C_{1\text{-}4}$ alkyl, $C_{1\text{-}4}$ alkoxy or $-NR^3R^4$ group), and q is 1 to 3], $-(CH_2)_r R^7$ [where R^7 represents $-NR^8SO_2R^8$ (where R^8 represents a $C_{1\text{-}4}$ alkyl, phenyl or $-NR^3R^4$ group), $-NR^5COCH_2N(R^5)_2$ (where each of the groups R^5 represents a hydrogen atom or a $C_{1\text{-}4}$ alkyl group), $-COR^9$ (where R^9 represents hydroxy, $C_{1\text{-}4}$ alkoxy or NR^3R^4), $-SR^{10}$ (where R^{10} is a hydrogen atom, or a $C_{1\text{-}4}$ alkyl group optionally substituted by hydroxy, $C_{1\text{-}4}$ alkoxy or NR^3R^4), $-SOR^{10}$, $-SO_2R^{10}$, $-CN$, or $-NR^{11}R^{12}$ (where R^{11} and R^{12} represent a hydrogen atom or a $C_{1\text{-}4}$ alkyl group, at least one of which is $C_{2\text{-}4}$ alkyl substituted by a hydroxy, $C_{1\text{-}4}$ alkoxy or NR^3R^4 group), and r is 0 to 3], $-O(CH_2)_q COR^9$, or $-O(CH_2)_r R^{13}$ [where R^{13} represents hydroxy, NR^3R^4, $NR^{11}R^{12}$ or a $C_{1\text{-}4}$ alkoxy group optionally substituted by hydroxy, $C_{1\text{-}4}$ alkoxy or NR^3R^4, and t is 2 or 3].

R^1 and R^2 each represents a hydrogen atom or a $C_{1\text{-}3}$ alkyl group, the sum total of carbon atoms in R^1 and R^2 being not more than 4;

and physiologically acceptable salts and solvates (e.g. hydrates) thereof, have a *stimulant action* at β_2-*adreno-receptors* and may be used in the treatment of diseases associated with reversible airways obstruction such as asthma and chronic bronchitis.

GB 2 182 658 A

Formulae in the printed specification were reproduced from drawings submitted after the date of filing, in accordance with Rule 20(14) of the Patents Rules 1982.

Figure 6.2 Typical Markush formula

other forms; they will, however, be those that have been synthesized as all known possibilities must be disclosed. Applicants are keen to disclose as many variations as possible because, in the event there is any challenge to the patent, only the examples given will be accepted as being truly invented and worthy of protection. It must be emphasized that the number of individual compounds in such cases may run into millions because the symbols do not necessarily represent specific sets of compounds but rather whole groups such as 'any alkyl'.

Most chemical patents adopt the selection patent scheme due to their complexity. Selection patents are issued for a patented group or for component sub-groups. Each member must possess an unexpected property not found in the rest of the group and this property must involve some inventiveness in its own right whereupon a second patent is granted because the compounds were never truly invented in the first place. An example of a selection patent is given by the amoxycillin case. Amoxycillin is a synthetic penicillin from Beechams which has a particularly high absorbance rate in the stomach. The group of penicillins to which amoxycillin was applied for. Once the right questions on absorbance properties have been established, the outcome might be viewed as obvious. However, Beechams were just exploring, they were not searching for particular properties and it was decided, therefore, by the Court of Appeal that there was sufficient inventiveness for a patent to be granted. In this instance, it happened that the company which owned the wider patent also sufficient inventiveness for a patent to be granted. In this instance, it happened that the company which owned the wider patent also applied for the selection patent but it could just as well have been a different company. Outside the chemical area, choosing among alternatives is much less complex and it is usually assumed that technical reasoning will cover all eventualities. As a consequence, selection patents are much harder to obtain in non-chemical fields. It could be, however, that protection of new higher organisms is a potential area for selection patents. Descriptions of the organisms are vague and as they become more widely used it is possible that additional protection will be required (Goodier, 1988).

This discussion illustrates a major disadvantage for patents as sources of technical information, that is, facts are often obscured in lists of possibilities. In chemistry, particular reagents and reaction conditions are presented as ranges of possibilities. Similarly, the organisms and parts of organisms handled by genetic engineers are presented vaguely with scope for argument as to how much is really known. Even the apparatus used in experiments is described generally even though in biochemistry and some chemical applications, experimental conditions need to satisfy stringent requirements. In general, the workable alternatives and the associated constraints have often not been elucidated when the patent application is filed and much routine work remains designated as 'mere workshop adjustments'. The resulting confusion and extra work leads many researchers to feel that their time is better spent working from first principles and ignoring the information revealed by patents altogether (Jalloq, 1982).

Most of the patents described above deal with new compounds. Much work needs to be done on the reactions which can bring a compound into being in sufficient quantity and economy of effort for it to be a commercial possibility. An interesting side-issue is the fact that computerized systems for predicting reaction pathways are being developed and refined. If it ever became a routine procedure to check the computer files and follow their recommendations, then the resulting compounds would no longer be produced only by means of human inventiveness. Their synthesis could become the standard work of technicians and, at that time, patents would no longer be available as the results of such syntheses could be judged to be obvious (Blick, 1979). In the same way, much work in biochemistry is regarded as potentially worthwhile because patentable but reliable techniques are being devised which might in time make patentability much harder to justify.

If a patent involves micro-organisms, the mere reading of the specification is insufficient. In order to repeat the invention one needs access to the organism itself and for this purpose they are usually deposited in recognized culture collections. Access to the collection is allowed after early publication in most deferred examination countries, but only after grant in the US. The European Patent Organization allows access only by an accredited expert within the period between early publication and grant but most countries permit full access by anyone. Details of the depository are given in the associated specification. This is a controversial issue because possession of living material is a much greater step towards performing an invention than reading a written description.

Internationally, the mutual recognition of depositories and the level of description required is regulated by the Budapest Convention of 1977 which came into force in 1980. These procedures are administered by WIPO which has an up-to-date list of member states and of depositories.

As yet there has been no consideration of disclosure for patents involving higher organisms.

Use of the information

Patent information can be useful to the technical or commercial departments of a firm in various ways.

TECHNICAL INFORMATION

For a chemist, the valuable information could include:

structure and properties of new compounds
how they are made
systems of storage on the shelf or of delivery to their site of operation (i.e. pharmaceutical formulations)
means of preservation of a substance.

In a similar vein, knowledge of biological techniques, micro-organisms and other related topics recorded in patents can prove to be useful.

If precise details are needed, then a patent may give just what is wanted or fall far short of the mark. Often what is useful in the patent may not, in fact, be the precise details as problems do not usually recur in identical form. What is usually better is an idea of the types of problems and solutions encountered throughout a range of patents so that the enquirer can adapt these to his own circumstances. This statement, of course, does not apply under circumstances where the aim of the work is to reproduce an invention for legal and competitive purposes; the purely informational aspects being discussed are of much greater and more general importance as they are not limited by time or nationality.

TECHNO-COMMERCIAL INFORMATION

Such information is derived by statistical analysis. Companies and individuals active in a particular subject field can be identified and work similar to that of the searcher may be particularly interesting. The most active countries, research centres and individual inventors will be revealed and the most productive industries in a country or worldwide identified. Clearly many deductions arising from such analyses have commercial value in addition to technical interest and benefit.

Currently, much research is being undertaken into the use of patent statistics for commercial forecasting and, as a consequence, a whole new class of patent users has been generated. Recognition of this developing interest in patent statistics is evidenced by the recent release by the Battelle Institute of a package called *Patents-PC* which appears to be similar to *Patstat* (on a brief assessment) and which analyses the output of an online search of any patent database (Battelle, 1987).

LICENSING

British patents licences are recorded in the Register of Patents kept at the Patent Office. A particularly interesting legal provision is that of licences of right which enables any applicant to obtain a licence to use the patent in question. The terms of use will be by agreement or, in the event of disagreement, will be settled by a patent examiner after appropriate representations. Endorsed licences of right are notified in the *Official Journal of Patents*, the weekly publication from the Patent Office and subsequently in unofficial registers as well as the official record. Depending upon the nature of the searchers' interests information of patent licences and licences of right is another aspect of commercial awareness.

The availability of patent literature

Patents tend to be held separately from other forms of literature in special libraries. Because of this they are often overlooked by researchers who have normally been conditioned by their training to think almost entirely in terms of the journal literature. Work on the accessibility of patent literature and its overlap with the contents of journal articles has tended to be either very broad, e.g. chemical, electrical and mechanical or very narrow, e.g. sewage processing (Eisenschitz, 1984).

Chemists tend to make better use of patent information compared with other subject specialists. This is due largely to the policy of *Chemical Abstracts* which integrates patents with other literature. Those covered are principally those patents describing newly synthesized chemicals whereas known chemicals applied in new ways are not usually included. The policy is to cover material of interest to chemists in this sense, and most process and other patents are deemed to be of interest to industrial applications rather than to chemists per se. Thus there is a grey area in which some material is covered and others not. In the central domain of new chemicals, the coverage is almost perfect at least for countries where the documents are readily accessible and the language is not too obscure (Oppenheim, 1974).

The abstracts of *Chemical Abstracts* are very informative. Details of the compound and its method of synthesis are given as is information about basic physiochemical properties. In a study of citations to patents found in the *Science Citation Index*, Tewnion found that the majority of chemical patents cited the relevant abstract number from *Chemical Abstracts*. Information referred to was available in the abstract and it is likely (although not proven) that the patent was not consulted at all in many cases (Tewnion, 1982).

Apart from *Chemical Abstracts*, patents are abstracted in the journals of many smaller specialist organizations. Such organizations tend to concentrate on small specialized fields but may cover a wider subject area and a number of countries within their remit. There is always a problem with borderline subject coverage; some items will be included and others not. In general, an important criterion is the quality of indexing together within priority details and patent number (Mann and Hellyer, 1980). Examples of high quality specialist patent services in restricted subject areas are afforded by the American Petroleum Institute's file *APIPAT* and the Institute of Paper Chemistry's *PaperChem*. In common with *Chemical Abstracts*, these publications embrace wide geographical coverage in their chosen subject areas.

Ultimately the best service is offered by those organizations that specialize in patent information and give complete coverage in one or many countries. Derwent Publications' *WPI* and *CPI* printed journals are prime examples of multi-country coverage. From the collection of material in the database, a variety of indexing and abstracting bulletins are printed which meet the needs of a variety of users. These several publications are highly regarded in industrial circles and are widely circulated. In many instances, patent information is being used directly as input to further research thus fulfilling the original intention of the patent system. Evidence of this use is manifested in the increasing number of printed bulletins being produced for single or very small groups of customers (Shenton, 1987).

Inpadoc is similarly a multi-country database. Its coverage is much wider than Derwent's (55 as against 33 countries) and it is particularly useful for equivalents searching. For single countries, US patents are available from a number of sources, in particular through Orbit for Derwent databases or using *Claims* on Dialog. Other major single country databases are *JAPIO* from the Japanese patent information organization and *Chinapats* for patents from the Peoples' Republic of China, both available on Orbit. The French patent office provides *EPAT* and *FPAT* covering European and French patents, respectively searchable through Questel.

One argument against using patent literature is that any worthwhile information will be reproduced in more accessible form in journal articles and it is worth considering this assumption further. Overlap is highest in chemistry at around 20%, a low level of duplication when the importance of chemical inventions is taken into account (OTAF, 1977) but in the situation where entire research projects are closely delineated by groups of patents, the overlap with articles then becomes much higher (Eisenschitz,

Lazard and Willey, 1986). It should be borne in mind that articles can cover any aspect of the project and the overlap with patents is never complete since articles and patents are written for different purposes and at different times throughout the project lifecycle. A recent survey of a group of British pharmaceutical patents showed an overlap of 60%, much higher than any previous studies. However, within that total it was found that a subset of patents of high scientific interest yielded a 100% overlap whereas another subset of more commercially oriented variations produced a much lower correlation. There could have been further material in the trade literature but this was not investigated. Whilst the above relates particularly to chemical compounds, some areas of genetic engineering would probably mirror these high overlaps. On the other hand, some subject areas could be expected to generate lower overlaps since the scientific interest of patent contents is very variable. For instance in a food processing study the total overlap was only 25% suggesting that patent literature in this area is of little scientific interest. Another main reason for lack of journal publication could be commercial value, irrespective of scientific concerns. One drawback in the use of the general literature is the fact that patents are rarely cited in journal references which makes it difficult for a researcher to locate them. Finally, it remains true that many of the operational details are given in patents and nowhere else.

Searching

It is apparent then that one can search for patents by both using general literature services or using those specific to patents. The former are useful in that the patent search is integrated with the searches for all other forms of information but for completeness of subject coverage and width of country coverage, specialist services are required. Nowadays, both general and specific services are available online so they can be accessed as and when required. Furthermore, the range of searchable fields in each online record means that one is not restricted to those indexes provided by the database compiler.

SPECIALIST SERVICES

All national patent offices maintain their own integral collection of national patents and appropriate indexes. Most countries collect, in addition, the patents of a selection of other countries and hold these as single country fields i.e. each national collection is distinct and must be searched separately. The Dutch national patent collection (which has provided the basis of what is now the searching branch of the European Patent Office) is different and all the patents are classified in the same scheme and are searchable as one field. This collection, therefore, is an important exception.

The commercially available online systems are much easier to search as a single integrated collection; two are available, Derwent and *Inpadoc*. A number of single country services are also available online which are easlier to use than the corresponding printed versions and complement the multi-country services. Online services, in general, provide bibliographic information and sometimes an abstract. However, the hard copy has to be consulted for drawings and the claims and body of the specification.

SINGLE COUNTRY SERVICES

For many years only US patents were available in database form, but recently, French, Japanese, Chinese, European and PCT databases have become accessible.

US patents have been produced in hard copy using computer technology since 1971 and, as a consequence, it has been possible to lease tapes of the full texts of US patents since that time. However, the records are only complete from 1975 onwards as there were still some unsolved problems relating to diagrams. The Orbit services of Pergamon Orbit Infoline in the US offers a separate US database of front-page information and claims in a file set up by Derwent Publications; this is the same information as seen on the British front page and the missing documents of 1971– 1975 have been added. Another US producer IFI Plenum has US patent bibliographic details mounted in a file called *Claims* which extends back to 1950 for chemical patents and back to 1963 for non-chemical patents; it is valuable because of this extent of backlog. The unique chemical indexing system used has its origins in the DuPont Company. It is mounted on Dialog, Orbit and STN International and is accessible from most parts of the world. The only US full text field available at present is *LEXPAT* dating back to 1975 which is mounted by Mead Data Central of Ohio, a host experienced in dealing with full text.

The French Patent Office has created a set of databases of French patents awarded since 1969 and of all European patents since their onset in 1979. These files are called *FPAT* and *EPAT* respectively and are mounted on the French host Telesystemes Questel.

British patents will soon be available as a separate database from the Patent Office and this too will include European patents.

Chinapats and *JAPIO* covering Chinese and Japanese patents respectively are both available via the Orbit host. To ascertain the latest state of patent databases, since new ones are likely to start up at any time, it is advisable to consult the available directories and lists which are regularly published. Any general online directory will cover them but, in particular, the journal *World Patent Information* has an updating feature on online databases.

MULTI-COUNTRY SERVICES

There are two notable services: *Inpadoc* covers 55 sources of patents (Piltch and Wratschko, 1978) and Derwent Publications covers 33 (Dixon and Oppenheim, 1982). *Inpadoc*'s strength is its very wide coverage but only bibliographic information is provided and there is a duplication of the various national patent offices' files. Derwent covers fewer countries but nearly all the material is produced in-house which enables the product to be closely tailored to its users' needs. Industry is the main market so the information provided by Derwent stresses the applications and the industries in which an invention is used whereas the patent granting authorities tend to emphasize the technical advance made by the invention. For instance, an improvement in a vessel for culturing micro-organisms can be related either to the organisms it is intended for or their metabolic products (which is the reason for making the effort). The two aspects are complementary and satisfy different market niches.

Inpadoc – The International Patent Document Centre

The Centre is a private company set up by the Austrian Government in association with WIPO. Any patent-issuing authority can become a member. Each member sends weekly or monthly records to the Centre in Vienna where they are entered without modification into the computer. The result is a computer database from which a series of indexes is printed directly onto COM microfiche. Patent offices of member states have access to all the output and are obliged to provide free access to most of the indexes in fulfilment of WIPO's aim of encouraging technological development through access to information and products. Most records go back to 1968 but one must check the starting date for each country if in doubt.

   The database is available by direct line from Vienna but is also hosted by Pergamon Orbit Infoline in Europe and America and by Dialog and STN. All words, numbers and dates are searchable and, on Orbit, Derwent accession numbers have been added to all records to enable cross-file searching. This feature is particularly useful in that it allows an abstract to be consulted for all patents and claims for US patents.

The COM indexes arrange the patents in order by IPC, applicant/patentee, inventor and country and give all previous family members. The weekly collection is published together as a *Gazette* and quarterly, annual and five-yearly cumulations of each index are published separately. Because the file has become so large, Orbit now issues another file called *Inpanew* which contains the most recent 15 weeks of the *Gazette*.

For comprehensive patent family searching *Inpadoc* has no rival. Families are sought by priority data and multiple priorities are followed through by repetition of the sequence to give the extended family. However, for subject matter searching *Inpadoc* is only moderately useful as there is only the International Patent Classification available as a guide. If classification searching is required, a file on microfiche only, called *Capri*, is available and this holds older patents of many countries reclassified by IPC. The aim is really to provide an instantly searchable patent index for developing countries and the service is not readily available elsewhere.

For statistical processing and commercial intelligence, a valuable facility on Orbit is the GET command. Once a set of references has been created GET enables time series and ranked lists to be produced (Questel has a similar command called Memsort).

Derwent Publications Ltd.

Derwent covers 33 sources of inventive information comprising the publications of 29 national patent offices, those of the EPO and WIPO-PCT and also two defensive publication journals, *Research Disclosures* and *International Technological Disclosures*. These last mentioned journals publish brief accounts of inventions for the purpose of establishing priority so that no one else can apply for a patent later.

All patents are entered onto the Derwent computer; even if a new document is an equivalent it is noted and the patent number added to the database. Additional information is added if appropriate and includes designated states, examiners' search reports and additional assignees. A new abstract is written for some of the more important countries. If a new document is created, it is given an informative title, abstract and indexing terms by Derwent staff with the aim of bringing out the information of value to industrial searchers as well as the inventive information useful to patent offices.

The product thus consists of bibliographic details and abstracts and encompasses complete database and various printed outputs containing the information repackaged in a variety of ways for different user groups. Once the database has been prepared it costs little to print in different formats for small groups of users. The database on pharmaceutical patents goes back to 1963 and other chemical areas started at various dates up to 1970. All other subjects started in 1974.

The chemical file called *Chemical Patents Index* (CPI) embodies deep chemical coding designed for substructure searching. Most other matter is classified in a relatively simple manner and the resultant file is called, somewhat confusingly, *World Patent Index*. Now that the services are online the entire searchable collection, including the original *CPI*, is called *WPI*. Anyone can access the online service but use of the deep indexing facility is constrained to full subscribers, i.e. those who take the hard copy service as well. The online service is available on the Orbit, Questel and Dialog hosts. National classifications are not displayed but there are the general Derwent Classes supplemented by the International Patent Classification terms as applied by the national offices. Equivalents searching can be undertaken using the Derwent accession number which means that only members of the immediate family can be traced. A way around this difficulty is to use related priorities which are listed and search priorities directly to obtain the extended family. Another disadvantage is that far fewer countries are included than in the *Inpadoc* service.

The *WPI* file was closed in 1980 because it was getting so large and a second file *WPIL* was created for subsequent input. The original *WPI* remains open to receive members of existing families as these will be given the same accession number assigned to the original invention and will be added to that record. A further problem arises when multiple priorities are involved as more than one Derwent record may be within the same family. The system is again more cumbersome to use than *Inpadoc* for family searching. The one area of family searching in which Derwent does excel is that of non-convention priorities. This may seem to be a contradiction in terms, but these are specific circumstances in which an application is made outside the 12 month limit both before there are any publications of application or patents in the family. In many countries, as long as there is no publication the application would still be patentable despite the fact that it clearly belongs to the same family. Such patents are difficult to trace as their priority details are different. Derwent makes an effort to include them in families by actually reading the contents and checking back to a known family. Note that in the larger, more active patenting authorities such as the US and European countries prior art is defined to cover pending applications for novelty purposes so that non-convention priorities do not arise. Non-convention applications are particularly relevant to the pharmaceutical area where developments can take place over a long enough time scale to open up possible markets after the time limit has expired.

Statistical analyses

Patent information nowadays is used as much if not more in strategic business planning and marketing as in research and development and other technical departments.

Derwent has organized its statistical functions into a package, *Patstat*, to be run on a micro-computer (Oppenheim, 1983). The system is used for commercial trends analysis, and is known as *Patstat-plus* in its present version which has been developed and refined beyond that described by Oppenheim. It is a very powerful facility, allowing for a number of cross-correlations as well as basic rankings and time series.

In Orbit a dataset can be created suitable for statistical operations using the command GET. The operation ties up a lot of CPU time and has to be done immediately following the selective search (Terragno, 1984), using *Patstat-plus*, the set created is downloaded into the micro so that statistical operations can be undertaken at leisure. Although only selected information needs to be downloaded, downloading can still take a considerable time and must be considered a disadvantage of the system. Questel has its own online statistical software called *Memsort* and, by using this, the Derwent package can be bypassed for quick analyses on both Questel and Orbit. It is unclear at present which approach is preferred by users but, as with most of Derwent's services, their statistical package seems to be most appropriate for large scale, frequent users. Occasional and small-scale users may be better off using general online software involving no additional outlay, i.e. one pays only for what one uses.

Patstat was developed for use with Derwent's *WPI* files but has recently been extended to accept data from its US Patent Office files as well. The results of analyses from both sources can be merged and edited.

Online searching of chemical structures

So far this chapter has concentrated on searching for patent bibliographic information but because chemical patents act as a register of newly synthesized chemicals and also because of the prevalence of Markush type claims, patent searchers need to be able to carry out complex structural searches. There are two problems here (Barnard, 1984):

- to represent three-dimensional chemical structures in a form suitable to enter onto a computer;
- to represent Markush groups in one formula rather than as all the individual members.

Chemical structure representation has been solved by use of line notation and connection tables. Current research concentrates on allowing the chemist to input structures by drawing with a light pen. The computer then converts these lines to appropriate notation.

The second problem is more recondite. Markush groups can encompass many thousands of compounds and it is impractical and meaningless to try and enumerate them all. What is required is a compressed code which allows one formula to be put in and the extent of variation indicated just as in the patents.

Until recently, the only commercially available system which allows both input and retrieval of Markush formulae has been Derwent Publications fragmentation code (Jackson, 1984). Very simply, fragments are coded separately and then the bits are linked together. A series of compounds is handled by means of a range (from . . . to) on the computer. A front-end system called *TOPFRAG* has just been introduced for use with these codes; the user inputs a diagram and the programme works out all the codes. At present, only one structure at a time can be input but the system will be enhanced to include at least limited generic searching. However, the climate being right for progress, various other systems are being developed which are based on new ideas about topological representation and which allow the direct use of structural drawings. If chemists were able to input a diagram in accordance with the way in which they think about problems, they (and others) would not need the elaborate training required at present to understand coding systems. The three schemes to be considered are (Barnard, 1987):

- a development of *DARC* to include generics called Markush *DARC*;
- the American Chemical Society's system for use with the *Chemical Abstracts* file;
- the work at Sheffield University.

MARKUSH DARC

This is being developed from the existing generic *DARC* system by its French suppliers Telesystemes in agreement with the French Patent office and Derwent Publications. The first version was launched commercially in January 1989 but only runs on post-1986 patents. Derwent will allow the new file, *WPID*, to run in parallel with *TOPFRAG*. A series of upgrades is planned and the older system will be relegated for use with the back-files only. There is no intention, at present, on the part of Telesystemes, to convert older files. However, the French Patent Office does intent to convert and plans to start with its oldest French patents online from the 1960s.

The Markush structures are represented in the computer by a single connection table concatenating all the partial structures for the constant part. Values are given for the variables and link descriptors to show the way they are connected. Generic groups are represented by a set of 21 'superatoms' e.g. 'halogens', 'carbocyclic aromatic' and 'polymer group' which can be qualified by various attributes; they seem to be as general as the claims in patent documents. The search involves a fragment screening stage followed by an exhaustive atom-by-atom match.

THE CHEMICAL ABSTRACTS SYSTEM

This system is similarly based on connection tables. Variables are defined by means of nomenclature as well as by reference to components of the structure diagram. The nomenclature employs terms such as 'alkyl'. With the diagrams, a hierarchy of generic group types may be defined and qualified by attributes such as element counts, ring sizes and number of rings. This system is protected by a software patent (American Chemical Society, 1987) and is planned to come into use in late 1989.

THE SHEFFIELD PROJECT – GENSAL

This work is still at the development stage at Sheffield University's Department of Library and Information Studies. *GENSAL* adopts a more radical approach and is based on a formal language similar to a programming language. It represents the structure by its topography. A test database has been created and commercial interest in this approach is high.

At present, these three systems represent the most advanced state of the art but this is a field where large sums of money are at stake and therefore research and innovation are bound to continue.

Product names

Both the active ingredient of a pharmaceutical and the specific products made from it are given names.

Generic names

This is a common, international non-proprietary name (INN) which is selected by national and international nomenclature committees. The INN is universally accepted and is important for communication in medical practice and for the labelling and advertising of medical products (Wehrli, 1986). Generic names look and sound much like trade marks but there is a major difference, the INNs use word stems to confer meaning. Thus, substances which are chemically and pharmacologically related share a common stem. Unfortunately INNs are not protected from imitation so that some very similar words are generated and registered as trade marks in order to capitalize on the high level of awareness of INNs. New trade marks should be checked against INNs and for this purpose the WHO makes available magnetic tapes listing its INNs. This possibility of conflict and confusion seems to be a recent problem. Generic names can be found in pharmacopoeias and tend to be used where non-brand drugs were manufactured and used in hospital pharmacies. A code of practice on trade marks has been drawn up by the British Pharmacopoeia Secretariat and released to the pharmaceutical industry for consultation. Trade marks would no longer be distinctive if names became too similar, so resolution of the problem would be of universal benefit. Ideally INNs should be included in trade mark registers.

Registered trade marks

These are the brand names given by companies to their products. If a name is associated with the product during the time that it is patented and the company is the sole supplier, then the reputation of that product will continue to live in that name even though the active agent is subsequently produced by others. Trade marks can stay in force for as long as they are used, thus affording a potentially indefinite form of protection.

The mark can consist of symbols and/or words although words are inevitably used because a product needs to be named when it is purchased. The words must be distinctive, i.e. they are attached to that product only, but do not describe the product. It follows that the best words are those which are totally made-up and have no other connotations. The terms attachment and use refer to the trade (in the case of pharmaceuticals this will be doctors, hospital staff, pharmacists and the like). The public are users for those substances which are sold directly to them e.g. aspirin. If a word enters common parlance and is applied to more than just the one brand, then it can be struck from the Register at the request of a competitor. For this reason, trade mark owners are very strict about their acknowledgements.

Products have been the main focus of this discourse and, as with all intellectual property, the laws are national and all countries have similar but differing laws covering the marking of products. Nearly all cover services as well as goods and would include, for instance, a pregnancy testing service. In the UK, service marks have been available for registration only since October 1986 whereas marks for goods began to be registered in 1876, over 100 years ago. This timing coincided with the tendency for trade to cover whole countries and regions and not be just local where suppliers would be known and unambiguous.

Marks can be applied for simultaneously in many different countries under the Paris Convention of 1883 and some later treaties. The pharmaceutical industry makes use of this opportunity because of its transnational trading patterns. Nevertheless, it is frequently advantageous to use words in the national language of the different national cultures.

Searching for trade marks

National trade marks registries have their official registers, search room and often a form of current awareness journal to notify the latest marks to be accepted. In Britain this is the *Official Journal (Trade Marks)*. Owners of marks monitor publications for marks which threaten their property. Anyone wishing to register a mark may search in the UK registry for all marks in force. There is a classification of goods into 34 divisions plus another eight for services. This entire process is currently manual but is in the process of being computerized. For commercial online searching, one can now use the very popular *UKTM* provided since 1986 by Pergamon Financial Data Services.

For international trade mark searching there is one major service, *Compumark*, based in Belgium; this provides the registered trade marks of 16 countries online and will also undertake manual searches (acting as an information broker) for other countries. *Trade Mark Scan* is a database from Derwent Publications and covers US trade marks only.

A particular advantage of online searching for trade marks is that it is easy to look for words deriving from one stem of phonetically similar words, e.g. 'skeeta' for mosquito. Whilst this requirement is recognized and suitable facilities are provided it does make the maintenance of online systems particularly difficult in this field. Searching the pictures of marks is also possible, but work still needs to be done to improve the software which is based on strings of shape classifications. Perhaps certain studies on chemical structure topology and on pattern recognition could be relevant here.

Design and get up

This topic concerns the packaging, presentation and appearance of a product; the main vehicle here is the informal trade mark (ITM). This is, perhaps, best defined in contrast to the registered trade mark. An ITM is a device linked to a product which can indicate its source of origin and help establish a 'reputation' for the manufacturer's product. Thus, the application of colour to half a capsule of a pharmaceutical product with the other half left transparent to display the contents was deemed registerable (Smith, Kline and French, 1976). When the presentation was green capsules irrespective of the contents then this could not be registered as a trade mark but the Court accepted the form of the capsules as a badge of origin rather than advertising. This allowed the company to claim a reputation for the capsules and prevent another from using the same capsules (Hoffman LaRoche, 1972). The mechanism is the same as for RTMs except that reputation has to be established at the time of action. No automatic right exists and each action has to be fought on its merits.

Text and drawings, all written or diagrammatic material about a product (whether issued as information or advertising) would normally have the protection of copyright for its exact format although not necessarily the contents. However, there has been an example of a patent being granted for a combination of product and information about it. This was for a pack providing daily dosages of two contraceptives together with instructions about the days on which each should be taken. It was held that the particular arrangement conveyed the information in a helpful way (Organon, 1970).

It can be seen from these examples that various forms of protection can be invoked for packaging and each case must be examined on its merits to find the most effective course of action. It follows that there are no routine sources of information on informal marks. One merely encounters them although they can be cited, for instance, in opposition and all countries allow different degres of rights to informal marks. In short, this is a minefield (Schricker and Stauder, 1986).

Acknowledgements

I thank Ms Kathleen Shenton of Derwent Publications for reading this chapter and making various suggestions for improvement. I also thank Dr. C. Oppenheim for help of the chemical structures section. I am grateful to the Comptroller of Her Majesty's Stationery Office for permission to reproduce the front page of British Application no. 2,182,658A.

References

American Chemical Society. (1987). *Storage and Retrieval of Generic Chemical Structure Representations* US Patent No. 4,642,762.

Barnard, J. M. (ed.). (1984). *Computer Handling of Generic Chemical Structures*, (University of Sheffield Conference, Gower).

Barnard, J. M. (1987). 'Online Graphical Searching of Markush Structures in Patents'. *Database*, **10**(3), 27–34.

Blick, A. R. (1979). 'Computer Assisted Chemical Synthesis Packages', *J. Information Science*, **1**, 227–119.

Battelle. (1987). *PATENTS-PC A Method and Personal Computer Based Software for Analyzing Published Patent Data* (Battelle Publications).

Dixon, M. D. and Oppenheim, C. (1982). 'Derwent Publications' Patent Information Services' *World Patent Information*, **4**(2), 60–65.

Eisenschitz, T. S. (1984). 'The Student Research Programme into Patent Information at the City University London' *World Patent Information*, **6**(3), 108–1140.

Eisenschitz, T. S., Lazard, A. M. and Willey, C. J. (1986). 'Patent Groups and their Relationships with Journal Literature' *J. Information Science*, **12**(1), 35–46.

Eisenschitz, T. S. (1987). *Patents, Trade Marks and Designs in Information Work* (Croom Helm).

EPO. (1986a). *How to Get a European Patent* (European Patent Office).

EPO. (1986b). *Annual Report 1986* (European Patent Office).

Goodier, J. (1988). *Informational Aspects of the Naming and Describing of Biological Organisms* (unpublished interim report at City University).

Hoffman LaRoche and Co versus DDSA Pharmaceuticals Ltd. (1972). *Reports of Patent Cases 1*.

Jackson, S. E. (1984). 'Experiences of a Patent Searcher'. In Barnard, J. M. *Computer Handling of Generic Chemical Structures*. (Gower), 30–37.

Jalloq, M. C. (1982). 'Use of Patent Literature by Academics' *British Library R&D Report 5770* (British Library).

Korah, V. (1986), *EEC Competition Law and Practice*. (ESC).

Mann, H. and Hellyer, A. (1980) 'Coverage of UK Patent Specifications by Abstracting Journals'. *World Patent Information*, **2**, 27–28.

Oppenheim, C. (1974). 'Patents Coverage by Chemical Abstracts' *Information Scientist*, **8**, 133–138.

Oppenheim, C. (1983). 'A MicroComputer Program for the Statistical Analysis of Patent Databases' *World Patent Information*, **5**(4), 209–212.

Organon's Application. (1970). *Reports of Patent Cases 574.*

OTAF. (1977). *Patents as a Technological Resource.* (8th OTAF Report, US Department of Commerce), 23–27.

Patent Office. (1986a). *Patents as a Source of Information* (Department of Trade and Industry).

Patent Office (1986b). *Introducing Patents* (Department of Trade and Industry).

Patent Office (1986c). *Annual Report of the Comptroller* (Department of Trade and Industry).

Trade Activities. *Patents Throughout the World* (Trade Activities Inc.)

Phillips, J. (ed.) (1985). *Patents in Perspective.* (ESC).

Piltch, W. and Wratschko, W. (1978). 'INPADOC – a Computerized Patent Documentation System', *J. Chem Information Comput. Sci.,* **18**(2), 69–75.

Schricker, G. and Stauder, D. (eds.). (1986). *Handbuch des Ausstattungsrechts.* (VCH).

Shenton, K. (1987). Personal communication.

Terragno, P. J. (1984). 'The GET Command' *World Patent Information* **6**(2), 69–73.

Smith Kline and French Laboratories' Application. (1976). *Reports of Patent Cases 511.*

Tewnion, L. (1982). *A Study of the Patent Citation Index.* (MSc. Thesis, unpublished, City University).

Wehrli, A. (1986). 'Pharmaceuticals: Trademarks versus Generic Names' *Trademark World,* **4**, 31–35.

WIPO. (1985a). *PCT Applicant's Guide,* (WIPO).

WIPO. (1985b). *International Patent Classification Schedules.* (4th Edition, WIPO).

CHAPTER SEVEN

Marketing and business information

S. GURNAH assisted by J. DE PASS

The role of marketing and/or commercial departments in the pharmaceutical industry is essentially to make recommendations and influence decisions affecting the profitability of their company. The specific activities of these departments may vary from company to company depending on the organizational structure and the various responsibilities of the departments. In the introduction the activities of these departments will be discussed with particular reference to the information requirements resulting from these activities. The remaining sections of the chapter will concentrate on describing the information categories which satisfy these requirements.

Introduction

The responsibilities of marketing/commercial departments in the industry can be summarized into the following categories:

- preparing forecasts for new and existing products to give an assessment of the potential income of the company;
- preparing and executing marketing plans/strategies for the launch and/ or promotion of the company's products;
- making recommendations to ensure that research and development and marketing resources are used in the most profitable directions;
- pursuing third party agreements and potential acquisitions to maintain as complete and profitable a product portfolio as possible.

These responsibilities require constant and comprehensive analysis of different aspects of the marketplace.

For an assessment of the sales potential of launched and pre-launch products in the portfolio, thorough analysis of the existing markets and the competitor activity within those markets is required. This analysis is also necessary for the preparation and successful execution of marketing strategies to achieve the forecasted sales potential.

Analysis of new and existing markets involves many different types of information. Details of sales and medical practices and their historical trends are the cornerstone of market analysis. These details involve information at the product/pack level of market activity and cover all aspects affecting pharmaceutical sales. Knowledge of competitor research and development activity and potential new therapies is also important as it provides insight into future trends of the market-place. The political and economic climates of the marketplace in the various countries also affect the pharmaceutical industry and therefore knowledge of these is required for thorough analysis. All this provides the information necessary for the portfolio analysis a company performs. It is also used to support any investment decisions, such as third party agreements and acquisitions.

Details of the company's products in relation to those of the competition, both launched and pre-launch, are also important indicators of sales potential and possible new developments. The details provide vital information concerning formulations, permitted labelling, potential indications, dosing, and other data concerning the product. All this information can be used for forecasting, market strategies, and improving the product portfolio of a company.

The information required to carry out the analyses described above must be extensive and detailed and is often useful in more than one aspect of these marketing/commercial responsibilities. However, it is possible to group the various information requirements into information categories or information types:

current awareness	– general information about all aspects of the industry and factors affecting it
sales statistics	– data on sales estimates of the industry; the companies, therapeutic areas and products
medical statistics	– data on diseases in terms of incidence, treatment and medical practice

product information – details of own and competitor product profiles; formulations, ingredients, dosing and clinical data

company information – information about companies; their marketing strategies, sales, potential plans, financial resources, and so on

country information – aspects of individual country markets which affect the industry, i.e. demographic, political and economic environments

research and development activity – information on research and development activities of competitors.

The remaining sections describe each of these information categories, the information they provide and the most important sources of that information.

Current awareness

Keeping up-to-date in all aspects of the pharmaceutical marketplace across the world is as important to marketing personnel as current awareness is in any profession. Journals and newspapers are common sources for current awareness in most areas and marketing is no exception.

Some periodicals adopt an editorial function in providing summaries of activities which are of interest to marketing personnel specifically in the pharmaceutical industry. PJB Publications Ltd. publishes one of the best known of this type of publication, *Scrip*. This biweekly journal covers all types of information mentioned above, as well as many others, and provides journalistic summaries of the most important events affecting the pharmaceutical industry. IMS International publish the other well known periodical; this is the weekly journal called *Marketletter* which performs a very similar function to *Scrip*.

Both report on meetings and symposia that have taken place around the world; they review articles of interest in the current medical journals; they report on companies' interim and end-of-year results, and company activities in general; they report on events in countries across the world which have an impact on the pharmaceutical industry; and they also have editorial articles on topical issues in pharmaceutical business. There is considerable overlap between the two journals, but as with all information sources, there are areas in both publications which are unique to one or other and therefore for completeness, it is necessary to use both. *Marketletter* has details of the share price and stock market performance of pharmaceutical companies across the world in every issue and it also has more sales-oriented information than *Scrip*, mainly because of the extensive international IMS network. *Scrip*, on the other hand, carries more advertisements for service agencies and also covers more product and research and development details than *Marketletter*.

ADIS Press Ltd. publishes a weekly publication which provides abstracts of pharmaceutically relevant articles in the medical and scientific journals. This publication, *Inpharma*, is very useful as a review of clinical trial results, both for new products in development and for existing products which are still being studied in clinical trials. ADIS also publish a series of monthly reports which review clinical trial data in specific therapy areas. These are the *Literature Monitoring Service (LMS)* reports and are intended primarily as a current awareness tool, summarizing articles on clinical developments in defined therapeutic areas. Another publication, *Bio-Inpharma*, performs the same function as *Inpharma*, but concentrates on biotechnology instead of chemotherapeutic agents.

F-D-C Reports Inc. in the United States publish several newsletters which report on the various activities of the Food and Drug Administration (FDA) and other industry matters. These are the *FDC Reports*. The most notable of these for marketing personnel are the *Pink Sheets* which report on the proceedings and outcome of the FDA sessions reviewing applications by the pharmaceutical industry for the various licences and rulings required for marketing of products in the US. They also provide details of all aspects of industry news on a weekly basis. Other *FDC Reports* include *Drug Research Reports (Blue Sheet)*, *Weekly Pharmacy Reports (Green Sheet)*, and *Medical Devices, Diagnostics and Instrumentation Reports (Gray Sheet)*. As the US is the largest and most important world market, this information provides a very useful understanding of legislative developments and their effect on the marketplace.

Other important publications which should be scanned for current awareness include the financial press, notably the *Financial Times* and the *Wall Street Journal*. A publication which concentrates on the pharmaceutical industry in Japan is the *Yano Report* which is published by International Pharma Consulting Ltd. It is edited by Yoshio Yano and has the subtitle, *Japanese Pharmaceutical Industry Quarterly*. Each edition features in-depth articles on events of significance from the Japanese viewpoint, as well as regular features on company activities, both domestic and overseas. Other general financial and economic publications will of course review the pharmaceutical industry and factors affecting them from time to time, and therefore these should be scanned regularly. These include publications such as the *Economist*.

Although many of the publications mentioned above perform a reviewing/abstracting function of the scientific and medical press for the industry, it is beneficial to scan the most prominent medical journals directly. The medical journals which could be considered as 'policy makers' because of their influence and prestige are the *British Medical Journal*, the *Lancet*, the *Journal of the American Medical Association* (*JAMA*) and the *New England Journal of Medicine*.

There are many other publications which are considered important enough to scan on a regular basis. That is, they may mention items of interest to the pharmaceutical industry on occasion and these should not be missed. However, they are too numerous to list here or indeed to scan manually on the off-chance of an interesting item. As with many other areas of information dissemination, the most convenient way of scanning numerous publications on a regular basis is to employ a current awareness profile or selective dissemination of information (SDI) on a computerized database.

Table 7.1 provides a summary of the textual databases which cover information relevant to marketing/commercial departments. The summary includes information on the company hosting and/or selling access to the database, the type of information provided and the use to which it can be put. Although most of these databases offer the ability to store individual search strategies which can be re-run on a regular basis, only some of them provide true SDI facilities, i.e. the ability to submit a search strategy and for it to be processed automatically after every update and the results to be automatically sent to the subscriber.

Two of the databases which provide this service are worth special mention. One is the Predicasts database, *Prompt*, available through Data-Star and Dialog. A weekly current awareness profile can be obtained from *Prompt* which is mailed to the client direct. A company that is part of the Predicasts group, Informat, also provides a tailored information service as frequently as daily. This is not included on *Table 7.1* as it is not an online system, and it is not possible for clients to interrogate the service; they receive a printed or electronic report tailored to their requirements. *IMSBASE* also provides an automatic SDI service which can be tailored to individual needs. This is called the 'Tracker' service and because of the publications covered by *IMSBASE*, this SDI service provides an excellent way of keeping up with developments

Table 7.1. Online Textual Database

Database	Producer	Host	Comments
IMSBASE	IMS International, UK	IMS	Contains the full text of many IMS hard copy publications with extensive indexing. It is therefore useful for *all* aspects of information mentioned in the chapter. Entire pharmaceutical coverage with good worldwide coverage
PROMPT	Predicasts Inc., US	Data-Star/Dialog	Provides informative summaries of key articles on market information, technology, company news and government news. Good worldwide and pharmaceutical coverage
Predicasts Forecasts	Predicast Inc., US	Data-Star/Dialog	Collects and displays numeric forecasts on products, industries, demographics and national incomes. Worldwide coverage
Predicasts Marketing and Advertising Services	Predicasts Inc., US	Data-Star/Dialog	Covers key publications on marketing and advertising strategies and provides informative abstracts. More consumer-goods oriented
Predicasts New Product Announcements	Predicasts Inc., US	Data-Star/Dialog	Provides full text of press releases before they appear in print. Less useful for ethical pharmaceutical products
Predicasts Annual Reports	Predicasts Inc., US	Data-Star/Dialog	Contains abstracts from annual reports of publicly held corporations. Provides financial, directory-type and strategic information on US and major international companies
PharmaProjects	PJB Publications, UK	Data-Star	Full text of the hard-copy publications. Useful for keeping up-to-date with research and development in pharmaceutical industry. Covers >3500 compounds and <500 companies worldwide

Name	Source	Host	Description
PHIND	PJB Publications, UK	Data-Star	Full text of PJB publications *SCRIP, CLINCA, ANIMAL PHARM* and *AGROW*. Split into three files for timeliness. *PHID* (daily file) is pre-publication material. Useful for recovering full text of important articles
Pharmaceutical News Index	Data Courier Inc., IS	Dialog	A comprehensive source for current news on companies, products and people worldwide in the pharmaceutical industry. Covers issues such as legislation, regulations, research progress, market analysis and future development
BIS Informat Newsfile	BIS Informat Ltd., UK	Data-Star	Provides concise summaries of articles from over 500 business newspapers and journals. Worldwide coverage. Health-care industry one of key sectors monitored
PAIS International	Public Affairs Information Service, US	Data-Star/ Dialog	Over 1200 journals and 8000 monographs indexed annually. No abstracts provided but indexing means accurate retrieval of references. Worldwide coverage. Topics covered include public policy, demography, economics and politics. Good for general country information
Kyodo Japanese News Service	Kyodo News Service, Japan	Data-Star	This is an English-language economic report service provided by a Japanese news wire agency. It concentrates on business and financial data. There are updates daily
Investext	Business Research Corporation, US	Data-Star/ Dialog	This contains full-text company and industry research reports in US, Europe and Japan, from the leading professional analysts. Useful for all company information, i.e. sales, earnings, research and development expenditure and business evaluation

Table 7.1. Continued

Database	Producer	Host	Comments
ABI/Inform	Data Courier, US	Data-Star/ Dialog	Contains long informative summaries of articles on management and administration. Covers many sources on Marketing with coverage of US and Europe
Dow Jones News Service	Dow Jones and Company Inc., US	Data-Star/ Dialog	This service provides abstracts and articles from the *Wall Street Journal* and *Barron's* as well as the *Dow Jones News Wire Services*. Covers all business areas with a US bias
Financial Times	Financial Times	Data-Star	Updated daily, it provides full text of all articles which refer to a company in the London and Frankfurt edition of the *Financial Times*. Like the *Dow Jones Service*, covers all businesses, but with a European bias
Harvard Business Review Online	John Wiley and Sons Inc., US	Data-Star/ Dialog	Full text (or summaries) of articles from the *Harvard Business Review*. The references cited by the articles are included in the record and so provide useful pointers to other key sources
East European Chemical Monitor	Business International, Austria	Data-Star	The articles provide data collected from East European popular, commercial and trade press on the chemical industry sector. One product group within the database is pharmaceuticals. Provides company information, market information and new product developments

Chemical Industry Notes	Chemical Abstracts Services, US	Data-Star/ Dialog	Articles are taken from journals, reviews, market reports, technical reports, etc. and cover production, pricing, products and corporate activities amongst others
Biobusiness	Bio Sciences Information Service, US	Data-Star/ Dialog	Monitors both business and life science literature for findings of laboratory and clinical research that have commercial applications. Over 1000 periodicals worldwide are covered
Chemical Business Newsbase	Royal Society of Chemistry, UK	Data-Star/ Dialog	Sources scanned include European, US and Japanese journals, company literature, annual reports, government literatures and market research/analyst reports. European bias but only deals with chemical industry, including pharmaceuticals
Abstracts in Biocommerce	Biocommerce Ltd., UK	Data-Star/ Dialog	Provides abstracts of commercial biotechnology news from trade journals, scientific periodicals and newspapers. Key facts are summarized with several sources cited for each abstract. Important for monitoring commercial significance of biotechnology progress
Frost and Sullivan Market Research	Frost and Sullivan Ltd., UK	Data-Star/ Dialog	Contains detailed summaries of over 200 market research reports. Worldwide coverage with emphasis on US and Europe. Useful for determining if a *Frost and Sullivan Report* covers areas of interest. Full report needs to be purchased
Foreign Trade and Economic Abstracts	Netherlands Foreign Trade Agency, Holland	Data-Star/ Dialog	Provides summaries of articles on international market trends, economic developments and management problems. The abstracts point to sources of data on trade, investment, import and export internationally

Table 7.1. Continued

Database	Producer	Host	Comments
Textline	Reuters, UK	Finsbury Data Services	Either full text or abstracts are provided taken from news publications all around the world. Single abstracts are provided for the same news item covered by several publications. Extensive indexing is undertaken. Many business sectors are covered as well as most business aspects. Newsline is the current awareness file for *Textline*
McCarthy Online	McCarthy Information Ltd., UK	Datasolve Ltd.	The full text of articles from leading business publications around the world are stored and indexed on the database. 1000 company annual reports are aslo summarized. Covers most business sectors

in certain therapeutic areas, both in terms of research and development and marketing developments.

Sales statistics

Information about sales revenue of competitors, market outlets, and products is of immense interest to all marketing departments in all industries. Although the pharmaceutical industry is not a traditional consumer industry (for ethical products, promotion can only be to the medical professional), it is one of the best supplied industries in terms of sales and market data.

Most important in the supply of market data to the pharmaceutical industry is Intercontinental Medical Statistics. Among the various data collected by IMS, are the sales audits of various pharmaceutical markets across the world. The word 'audit' is applied since IMS uses records of sales (invoices or computerized tapes) from either wholesalers or pharmacies to compile the statistics. For some of the more developed countries, the data collected are virtually census data, i.e. over 80% of the market is covered and consequently the data are extremely accurate.

The audits of pharmaceutical sales or pharmaceutical indices published and sold by IMS to the industry contain large amounts of detailed sales data. One pharmaceutical index is published for each country where IMS has a presence. This accounts for over 40 countries and this number continues to increase. The pharmaceutical index provides information about sales of manufacturers and their products within that country, of products grouped by therapy class and of recently launched products. The sales data include local currency sales of the latest month or quarter (depending on frequency of publication), for the year-to-date and also for the moving annual total (MAT – latest 12 months). It also includes the number of actual packs sold within those times, and the dollar sales for the MAT. Additional information contained in a pharmaceutical index includes pack details (i.e. form, strength and size of different product packs), details of manufacturers and their performance within therapy classes, launch dates and prices of products and individual packs, details of growth and market share down to pack level, and in some cases, details concerning prescription status of products.

IMS also publishes the same information for the hospital market of most of the major country markets. This market sector is dealt with separately from the pharmacy sector as it has many different aspects in terms of pharmaceutical sales; such as channels of distribution, therapy areas, products and price structures. The coverage of the hospital markets is not as comprehensive as the pharmacy markets, and because of direct tenders between manufacturer and hospitals, it is a difficult market to audit.

Another IMS publication, the *Index of Drug Chemicals (IDC)*, provides sales statistics of individual chemicals across the world. These statistics are collected in terms of quantities of active ingredients sold in each country. It is therefore possible to find out how many kilograms or tonnes of a particular chemical has sold in a particular country.

There are several annual publications which IMS publish which provide summaries of the sales data they collect. The *Pharmaceutical Market World Review* is a consolidated survey of the pharmaceutical markets across the world in terms of major companies, major therapy areas and major growth areas for the year. This is extremely useful in giving an overview of worldwide pharmaceutical sales. Another publication, *Leading International Products Review*, provides sales data of the leading products worldwide in several major currencies. *Leading International Generics Substances* provides sales data on the leading chemicals worldwide, regardless of brand name or marketing company. These publications are useful summaries of important market areas.

Another important source of sales statistics is from the manufacturers/companies themselves. Although the information that can be obtained direct from the companies about their sales is not as detailed as the information contained in the IMS audits because of company confidentiality, it does give a useful perspective on such collected data. Sales information from companies can be obtained either from annual reports or from stockbroker reports. Companies' annual reports will be discussed in more detail later. Some companies provide details of their performance to the various health-care financial analysts with their end-of-year or interim results, with a view to encouraging investment in their shares.

Health-care analysts publish these results in reports for their investors. Although these reports are only distributed to the list of investors who are customers of the stockbrokers, they do often find their way into marketing departments of the industry. Apart from publishing companies' performances, they are valuable because of the evaluation and opinion given in these reports. Financial houses which have a considerable interest in the pharmaceutical industry include Robert Fleming Securities Ltd., Barclays de Zoete Wedd, Wood Mackenzie and Co. and S G

Warburg in the UK; Merrill Lynch, Goldman Sachs, Morgan Stanley and Kidder Peabody in the US; and Nomura, Daiwa and Jardine Fleming in Japan.

Wood Mackenzie and Co. also provides an information service, *International Pharmaceutical Services.* An annual subscription to this service provides a 'database' of company profiles for Europe, Japan and the US. These company profiles are regularly updated and are automatically sent to all subscribers. Another part of the service are the regular reports on the industry as a whole. Each report also contains a specialized section which analyses a particular aspect such as a therapy area or industry area.

Kidder Peabody and Co. also provide a syndicated information service called *Pharmaceutical Perspectives.* As with Wood Mackenzie, there is an annual subscription to the service and this covers regular reports written by the Kidder Peabody analysts on various aspects of the industry. There are also company profiles which are constantly updated. There is a significant bias towards American companies and marketplace in the Kidder Peabody service.

Another source of sales statistics is market research reports. Some market research companies perform studies which are not commissioned by companies but are assessed to be of sufficient interest to justify publishing and selling to any interested party. Amongst other information, sales statistics are often reported. Wood Mackenzie publish such one-off studies. Frost and Sullivan is another market research company which is well known for their pharmaceutical reports. IMS also publish individual market surveys which utilize the IMS audit data as well as providing qualitative data and analyses. Because of the complexity and the original research required to produce these market research surveys, they can be very expensive. It is also possible for a company to commission an ad hoc market survey. As these are even more expensive, they are used more to provide medical data than sales statistics.

It is possible to glean some information about the sales of pharmaceutical companies and products from the financial press. These sales data tend to be scanty and so are most useful when specific information is sought rather than comprehensive market data. Consultation of *Table 7.1* will show which databases cover the financial press and it is possible to retrieve sales statistics from these databases.

Apart from these textual databases which provide access to reports, there are statistical databases which provide access to the actual data. A summary of the statistical databases is provided in *Table 7.2.* The most notable of these are the IMS databases which provide online access to the audit data. As can be seen from *Table 7.2*, there is an international database, *MIDAS*, (Multinational Integrated Data Analysis Service), which contains summary data for all IMS pharmaceutical indices. There are also several national databases, mainly for the major countries, which provide access to the detail audit data for one country only. These statistical databases have a very sophisticated enquiry language, allowing the data to be retrieved in many different and complex permutations. Purchase of the hard copy audits is a prerequisite to purchasing access to these databases.

Other services which provide access to sales statistics, as can be seen from *Table 7.2*, are the services provided by Standard and Poor's and *Datastream*, a Dun and Bradstreet Service. These services involve expensive subscriptions and also require specialized in-house software facilities. As they do not provide pharmaceutical data exclusively, they are an expensive option for marketing departments in the industry, and are more commonly used by financial analysts and stockbrokers who are concerned with a wider range of industries than just the pharmaceutical industry.

Medical statistics

As well as needing to know how much a product has sold, it is also important to know why a product has been successful, what it is used for and who is using it. In all industries, this is assessed by market research surveys, i.e. opinion surveys, either ad hoc or on a regular basis, of the buyers of the products.

In the pharmaceutical industry, this kind of opinion research is carried out with doctors to determine patterns of usage, potential markets, new therapy needs, and other relevant information. This kind of survey tends to be run in response to one or a few pharmaceutical companies who have an interest in a particular therapy or disease area. As surveys of this kind tend to be one-off studies and only commissioned for a few companies, they tend to be very expensive.

There are many market research agencies who specialize in this type of pharmaceutical market research, nationally and internationally. It is not within the scope of this chapter to discuss the relative merits of the various agencies and the type of market research that can be done. However, it is important to bear them in mind as a source of medical information.

As well as undertaking sales audits across the world, IMS also regularly produce medical surveys in several major countries. The

Table 7.2. On-line statistical databases

Database	Producer	Host	Comments
MIDAS	IMS International, UK	IMS	This database provides online access to 12 years of data from most of the IMS pharmaceutical sales audits produced internationally, as well as most of the IMS Medical Indices produced internationally. The enquiry language, while simple to use, is very flexible allowing complex relationships between the various data to be analysed. Worldwide coverage of exclusively pharmaceutical data
National Database	IMS International, UK	IMS	In-depth data for individual countries is available on a series of national databases. They employ the same enquiry language and access procedures as *MIDAS*, but provide monthly sales and promotional statistics, as well as in-depth medical data for one country. They are available for France, Germany, Italy, Spain, Netherlands, Austria, Portugal, South Africa, US, Japan, Finland, Switzerland, UK, Belgium, Australia, New Zealand and Canada
Dataline	Reuters, UK	Finsbury Data Services	Provides access to company accounts with the last five years of history of financial statements. The information comes in three levels of detail and there is a sophisticated modelling system which allows for forecasting. Has UK bias and covers most business sectors

Table 7.2. Continued

Database	Producer	Host	Comments
Datastream	Datastream International UK, US and Netherlands	Dun and Bradstreet	This service comprises many databases, each offering a specific type of service to the financial community. The system offers research of company financial statistics and data with programmes available for graphics, forecasting and time services analyses. Investment opportunities can be monitored and evaluated, with historical analyses plus up-to-date stock market information. It is designed essentially for the financial community
S and P Marketscope	Standard and Poor's Corporation, US	Standard and Poor's	This service provides information on over 4700 companies. While some of the information is textual, there is also extensive financial data and various manipulations are possible with the data. Although all business sectors are covered, there is a bias to the US and the services offered are of principal use to the financial community. Financial, rather than marketing, departments will make more use of this service

medical surveys consist of panels of doctors providing detailed information on a sample of patient visits. The information they provide includes the age and sex of the patient, the diagnoses made at that visit, the speciality of the doctor, the place of the visit, whether this is a first or subsequent visit for that diagnosis and details of the products prescribed during the visit. These data are collected by IMS and published in the *Medical Indices*, usually on a quarterly basis. A large amount of medical data can be obtained from these *Indices*. They provide information on trends in prescribing, treatments and diseases. From the marketing viewpoint, these are important factors to consider for market strategies, forecasts and new product development. In addition to the medical country indices, IMS also publish a summarized consolidation of this medical data in the *Medical Market World Review*. As with the *Pharmaceutical Market World Review*, this provides data on the major diseases and therapy areas worldwide.

These medical data are stored and are available on the IMS databases mentioned in *Table 7.2*. It can be seen that it is possible to correlate sales data with medical data on these databases, adding a further dimension to the analysis of these data, and providing a sophisticated and comprehensive service to users of the system.

The IMS *Medical Indices* provide a wealth of information on treatment of diseases, but not on the aetiology, epidemiology, morbidity and mortality of a disease. Another important aspect of medical information is the non-treatment or potential treatments of a disease (this applies equally to a new, hitherto untreated disease). Market analysis, particularly for new product development, involves comprehensive understanding of diseases, in terms of their treatment, the progression of the disease, the number of people suffering, the number of fatalities and the latest developments in medical research for that disease. This information provides a necessary background for all marketing staff in the pharmaceutical industry to develop appropriate market plans and strategies both for existing product marketing/positioning and also for new product development at the basic research and clinical stage of development.

Sources of morbidity/mortality data are official publications of governments or organizations. The World Health Organization (WHO) publishes various reports on morbidity and mortality statistics. There are the annual reports on infectious diseases which give statistics on the morbidity of all major communicable diseases for most countries of the world. The WHO *Quarterly Statistics Report* contains articles on various health issues and frequently these reviews contain morbidity or mortality data for several countries. The *Weekly Epidemiology Report* is another forum for publishing multinational disease statistics and the WHO *Technical Report* series also provide relevant data. Because of its unique position with regard to health issues across the world, most WHO publications contain useful medical information on a multinational level.

Other official sources of morbidity and mortality data are governmental bodies such as the European Economic Community (EEC) and the Centre for Disease Control (CDC). Other national official bodies (such as HMSO in the UK) publish morbidity and mortality statistics, often because it is required by law. The developing countries publish less of this kind of information, and it is for these countries that the WHO data are particularly useful.

For the aetiology and epidemiology of a disease, useful sources of information are national or international organizations specializing in a particular disease, such as the British Diabetic Association. These societies and organizations often fund epidemiological research and are closely involved with developments in treatment and research. They are usually very helpful and can be a surprisingly rich source of detailed medical data.

Other probably more obvious sources of medical statistics, such as epidemiology, morbidity and mortality, are the medical textbooks and journals. Standard medical texts, such as *Cyril Medical Textbook*, provide good background data about diseases and their treatment. The drawback of these as sources is that they become quickly out-dated. Medical journals, such as the *British Medical Journal* and the *Journal of the American Medical Association (JAMA)* publish articles on epidemiology etc., and so it is useful to search databases such as *MEDLINE* (this is the online source of *Index Medicus* available on Data-Star or Dialog), when looking for up-to-date studies in these areas.

Product information

Information concerning sales and use of pharmaceutical products has already been discussed, and an aspect of pharmaceutical products which has a significant impact on these parameters is the detailed profile of the individual product. The profile of a product includes information on the form and strength, the packaging, recommended doses, contraindications, interactions and the ingredients. Most of this information can be obtained from the manufacturer's data sheet which accompanies the product. However, it is not practical to obtain a sample of all products sold in all countries, and hence their data sheets.

One source of such information is a published collection of these data sheets. Most of the major countries publish these collections, often by the relevant Trade Association in that country. In the UK, there is the *ABPI Data Sheet Compendium*, published annually by the Association of British Pharmaceutical Industry. In Germany, the *Rote List* is published by the Bundesverband der Pharmazeutischen Industrie, also annually. In France, an alphabetical list of products is published in the *Dictionaire Vidal* by OVP. The Spanish equivalent is the *Vademecum International de Especialidades Farmaceuticas y Biologicas*, published annually by Daimon. In Italy, *L'informatore Farmaceutico* is published by the Organizzaziorie Editoriale Medico. In the US, there are two relevant publications. The *Physicians Desk Reference* (*PDR*) published by the Medical Economics Company Ltd. contains details on prescription (ethical) products, while another book, *PDR for Non-prescription Drugs* covers the non-prescription (OTC) products. *The Japanese Drug Directory* (also called the *JAPTA List*) is published in English by the Japanese Pharmaceutical Association. As with the other publications, these are published annually. There are other similar publications for other countries, but they are too numerous to mention.

A publication which contains information on therapeutic agents with references to the marketed products and which is relevant internationally is *Martindale – the Extra Pharmacopeia*, collated by the Pharmaceutical Society of Great Britain and published by the Pharmaceutical Press. This publication is updated regularly but not necessarily annually. As well as details mentioned above, it contains detailed information on adverse reactions and drug interactions together with therapeutic use of pharmaceutically-active molecules and references to original articles on clinical trials. All this information is also available on an online database hosted by Data-Star (see *Table 7.1*).

Another online source of product information is the IMS textual database, *IMSBASE* (see *Table 7.1*). This database contains information published in various IMS publications and includes details of ingredients, packaging, manufacturer, price, country availability and therapeutic use. The information available on *IMSBASE* has the advantage of the worldwide data collection by IMS and in addition to the coverage of the smaller pharmaceutical markets, there is an abundance of detail about individual products and packs.

The information discussed above deals with facts about the products and packaging, required by doctors and pharmacists, such as dosage, indications and warnings. Other product information that is useful to marketing staff for competitor analysis includes clinical trial results for the products. It is useful to know in what diseases a product has been studied and what the results were, which products were used in comparative clinical trials and what adverse effects were reported. Another increasingly important dimension of clinical trials is the economic analysis of treatment with various products, i.e. the cost-benefits or cost-effectiveness of a particular product against another product or therapy.

This clinical data can be gleaned from published trial results in the medical press. There are many journals publishing clinical data and as this is more relevant in non-marketing fields in the pharmaceutical industry, it is not proposed to cover them here. An online search on a database such as *MEDLINE* or *EXCERPTA MEDICA* will retrieve most of the required information.

As mentioned earlier, ADIS Press Ltd. publish a series called the *Literature Monitoring Service* (*LMS*). This series of reports is divided into separate therapy areas, such as antivirals or anti-cancers, and every month a report is published which abstracts all relevant published clinical data in that therapy area. All pertinent points of the trial data are summarized in an easily acessible form. It is possible to subscribe to the therapy area reports separately. This service is available online on *IMSBASE*, which adds to the value of *IMSBASE* in this information area.

PJB Publications Ltd. publish reports on specific products from time to time, usually just before or just after the first launch onto the market. They offer a summary of details on the current market and background information to allow assessment of the commercial potential of the product. These are called *New Product Reviews*. Another useful series of publications available from the PJB Group is the *Summary Basis of Approval* (*SBA*) reports. This document is based on the approval proceedings of the FDA in the US and it provides information on clinical studies, pharmacology, dosage, side effects, and other useful facts about an approved product. A drawback of these publications is that they apply only to the US.

Company information

Some aspects of company information have been considered earlier with a concentration on sales data, whether for companies, individual products or therapy areas. There is a lot of other information about companies which is relevant to marketing personnel, depending on the particular requirements of a situation. It is necessary to be able to predict what plans a competitor may have for future development of its business, in order to assess its potential for growth in various market areas. Issues which help in the assessment of these plans are research and development expenditure, promotional spending, investment and acquisition plans and potential licensing agreements.

It is important to know not just the current spending in these areas, but also the future intentions of a company regarding resource allocations. Other information about companies which will have an impact on marketing decisions includes company image, corporate ownership, internal organizational structure and share prices/profitability.

One important source for this information is the company annual report. Most of the major pharmaceutical companies belong to large, publicly quoted multinational corporations who, by virtue of their public status, must disclose a fair amount of financial information in their annual report and account and must publish them. There is often an exchange of annual reports between licensing personnel in the industry and most companies will send an annual report on request. Market analysts and stockbrokers have access, not only to such company information but also to the management and so reports they write provide an additional indication as to the likely strategic focus of these companies. The analysts most interested in the health-care industry were mentioned earlier and of particular note are the services provided by Wood Mackenzie and Kidder Peabody. They write detailed analyses of major pharmaceutical companies and keep them regularly updated.

PJB Publications Ltd. commission studies on individual companies from time to time. These *Company Profiles* provide an overview of the company and examines their strengths and weaknesses in terms of financial structure, product portfolio, market presence, research and development activity and company strategy. Other PJB publications on companies include *Scrip's Pharmaceutical Company League Tables* (an annual publication containing sales, growth, profitability, research and development expenditure, etc. for over 200 major pharmaceutical companies) and the *Directory of European Pharmaceutical Companies* (containing name, address, telephone numbers, etc. for manufacturers in East and West Europe).

IMS International also publishes several reports containing company information. The *Guide to Pharmaceutical Company Sales (GPCS)* contains information on companies, their subsidiaries and sales in the countries across the world. The *World Drug Market Manual (WDMM)* contains a company section which is divided by country with details of the major pharmaceutical manufacturers in that country. The details given include name, address, senior executives, sales turnover, top major products and corporate affiliations. These publications are available online on *IMSBASE* and in addition, *IMSBASE* contains reports not available in hard-copy, the Corporate Profiles. These are in-depth analyses of major companies covering most of the aspects of company activities mentioned earlier. Because of the structure of the database, it is also possible to get listings of all a company's products with all the relevant product details from *IMSBASE*. This makes *IMSBASE* an important source of company information.

A source which provides a very specific form of company information is Wilkerson's *Healthcare Strategist*. This book is updated annually and provides information on activities such as acquisitions, mergers, joint ventures and co-marketing agreements in the health-care market. It tends to have a bias towards the US, but is a useful tool for this kind of company activity, which has a significant impact on sales potential of a company.

There are sources of company information which are not specifically pharmaceutical. These include local directories, such as *Kelly's* in the UK, which list companies with details of products, etc. The *Extel* service provides a regular card index file of companies giving details of the company's operations, performance and investments. Professional agencies, such as Company House in the UK, which collect information on companies, will perform searches on request into a company.

Table 7.1 gives details of some of the databases where company information can be found. These databases are useful for searching publications that cover all industries, not specifically pharmaceuticals. Important publications covered include the financial press, and the *Predicasts* databases provide a good coverage of company information in the pharmaceutical industry.

Country information

Knowledge of the political and economic environment within a country is important in terms of business prospects in that country. The political environment can have an impact on pricing policies, returns on investments and marketing for multinational corporations, legal requirements for product licenses, patent protection, etc. The economic environment affects the profitability of marketing ventures and the business prospects within a country and therefore is an important consideration for marketing plans of a company. The health-care provision within a country clearly impacts considerably on pharmaceutical business, so the structure of health-care systems is another important consideration.

Information sources which provide this kind of information aimed specifically at the pharmaceutical industry include two IMS publications. The *World Drug Market Manual* (*WDMM*) provides a resumé of factors influencing the pharmaceutical market in a country. It discusses the political environment, pricing policies, health-care provision in terms of finance, quality and size, and economic factors. It covers countries in Europe, the Americas, Asia, Africa and Australasia and also provides some demographic data and summaries of the pharmaceutical markets in each country. The other IMS publication which provides information on countries is *Pharma Prognosis International* (PPI). This provides a more in-depth analysis of political and economic factors and concentrates on providing forecasts for the pharmaceutical environment within each country, both in political and economic terms.

Country yearbooks are useful reference books for reviewing the economic and political facts about a country with some historical perspective. They do not, however, provide information on how this affects the pharmaceutical market/industry. Similarly, government publications will provide some information and monitoring government publications for any proposed or implemented changes in legislation which may impact on the pharmaceutical market, (such as pricing, health-care system structures and reimbursements/insurance schemes) gives important insight to these areas.

The press provides relevant information in this area, and again the best way to do a one-off, retrospective search for these publications is on the databases listed in *Table 7.1*. *Table 7.1* also shows other journals which are covered by databases and provide political, economic or health-care system information.

Research and development

The final category of information identified is information concerning the details of the research and development activity of companies. This covers all stages of research and development, but the closer the product is to market, the more crucial the details are. Therefore, although it is relevant to know in what areas competitors are concentrating their resources in terms of chemical and biochemical research, preclinical (pharmacology and toxicology) and clinical research are of more interest to marketing departments. There are two major publications which provide this 'pipeline' information; one published by the PJB group called *PharmaProjects*, and one published by IMS called *Drug License Opportunities*.

PharmaProjects is an annual publication which is updated every month. It consists of resumés of compounds which are in the research and development pipeline of a company. The resumé contains brief details on the research carried out so far, the company conducting the research, the country where the research is taking place and any licensing prospects for the compound. Information on proposed names and stage of development is also included. The coverage is extensive in terms of country and company coverage. These resumés are organized both by company and by therapy area and there are several indexes, allowing access to the information from a variety of starting points. The information is collected from published papers, conference proceedings and also from individual company contact.

Drug License Opportunities is a weekly publication, which is collated into annual volumes. The coverage is similar to *Pharma-Projects* in terms of providing details of what is known about a compound's progress in the research and development pipeline and in terms of the countries and companies covered. The layout and content are somewhat different. There are product profiles for each compound covered and these profiles give details of research results available so far with information on molecular structure, research company, patent and licensing information and commercial forecasts for revenue potential. These profiles are updated regularly, but items of news such as licensing agreements, product licences or approvals, etc. are given in the section, *News Round-up*.

Each week's update also comes with an index to what is in the week's issue. Every month a list of forthcoming meetings to be held worldwide over the next few years is also included. As with *PharmaProjects*, the weekly updates are filed into large binders and there are several indexes which are also regularly updated.

Both *PharmaProjects* and *Drug License Opportunities* are available on online databases (*Table 7.1*). *PharmaProjects* is available through Data-Star and *DLO* is available through *IMSBASE*. Although the coverage of both publications is similar, because of the difficulty in obtaining accurate and up-to-date information on companies' portfolio, it is advisable to use both *PharmaProjects* and *DLO* as regular sources. It is then possible to build up a fairly complete picture, and it also helps to highlight areas which may be speculative, by their inclusion in one and not the other publication or by differing items of information in both publications. As is the case with *Scrip* and *IMS Marketletter*, they should be seen as complementary publications.

As mentioned above, conference proceedings are a very good source of research and development activity information. Details of research compounds are often reported at conferences, while not being published in articles, as companies see these conferences as a useful environment to start to stimulate medical interest in the product. Publishing an article in a scientific journal provides extensive details of research which may be of use to a competitor, but it may not be read by a doctor and therefore will not necessarily be of help in the marketing of the product. As many medics attend conferences, this is often a better forum for first disclosures of research compounds. Publications which report on conference proceedings are therefore good sources of research and development information. In a rapidly advancing area such as biotechnology, this is particularly important.

In addition to conference proceedings, another source of research and development activity are the articles in research journals which are published by researchers in the pharmaceutical market. Both clinical and preclinical research is reported in relevant journals and so regular reading/scanning of specialist journals such as the *Journal of Cardiology* are important ways of keeping up-to-date in a particular therapy area.

Gower Academic Press have recently started a monthly journal that surveys recently published patents in cardiovascular and antimicrobial therapeutic areas. They plan to expand the range into other therapeutic categories.

FDC Reports Inc. (publisher of the *Pink Sheet*, etc.) also publish an annual book, the *NDA Pipeline*. This gives details of all compounds submitted to the Food and Drug Administration (FDA) in the US for NDA approval. As the US market is very important to multinational pharmaceutical business, this publication provides important information on pre-launch compounds for that market.

Apart from details about the compound or agent, up-to-date information on the progress in the research and development pipeline is very important from the marketing point of view. This information can be considered as current awareness and many of the sources mentioned in that section are relevant here as well. *Table 7.1* gives details of quite a few useful databases in the area.

Summary

This chapter has discussed the various types of information useful to commercial/marketing personnel in the pharmaceutical industry, giving details of their uses and the most important sources of that information. Rather than provide an exhaustive list of information sources, the key publishers in this area of information have been described along with the key publications. As well as the publication mentioned, they publish a variety of ad-hoc books/reports and therefore it is important to regularly survey their catalogues in order to keep up-to-date.

Apart from the key publishers, notably PJB Group Publications, IMS International, Frost and Sullivan and FDC Reports Inc., there are two publishers who regularly publish ad-hoc reports in all the categories discussed. These are AD Little and Euromonitor. They provide extensive surveys on a variety of areas of the pharmaceutical marketplace and hence it is useful to constantly survey their publications lists also.

Another aspect of the information concerns of pharmaceutical marketing which was highlighted was the importance of online databases. This is the case in all areas of information, but as can be seen in *Tables 7.1* and *7.2*, there is a wealth of online sources for pharmaceutical business information. As well as the textual databases, statistical databanks are extensively used. As a large part of the marketing function is analysis of sales trends in the past and prediction of sales trends for the future, these databanks are invaluable and as a consequence have become very sophisticated. As is the case with most successful businesses, information technology is a vital part of pharmaceutical marketing.

CHAPTER EIGHT

Legal and regulatory constraints upon pharmaceutical business

I. HARRISON

Introduction

Pharmaceutical products have a capacity for harm as well as for good and great skill is required not only in their design and manufacture but also in their use if optimum benefit for the intended purpose is to be achieved. In an attempt to ensure the safety, quality and efficacy of medicines many countries have enacted legislation to control all activities involved in the manufacture and distribution of medicines. In the United Kingdom the framework of control is laid down in the Medicines Act 1968 and the detail may be found in numerous Orders and Regulations (collectively referred to as 'subordinate legislation') made under the Act. Directives of the Council of Ministers under the European Communities over-ride the laws of member states and most provisions of the Act have been modified to be compatible with the relevant directives. However, there are areas (such as the so-called 'high technology medicinal products') where EEC provisions exist in the absence of specific UK regulations. It is therefore necessary to keep up-to-date with EEC activity. All these legal constraints are supplemented with quasi-legal 'codes of practice' such as Good Laboratory Practice and Good Manufacturing Practice. The final outcome is the registration of products throughout the world.

In addition to protecting patients from damage caused by medicines, it is necessary to protect the workers who make them and the public living near factories in which they are made. This is achieved in the United Kingdom in several ways. The controls over the manufacture of medicines imposed by the Medicines Act require manufacturers to be licensed and to follow the procedures laid down in the Guide to Good Manufacturing Practice (GMP).

The Health and Safety at Work Act 1974 and the Control of Pollution Act 1974, also play a part in ensuring the safety of those who work in or live near such factories.

From the information viewpoint, awareness of the constraints and their interpretation is essential.

Safety, quality and efficacy

These three properties of pharmaceuticals are fundamental and deserve some explanation. They are closely interlinked but quality is probably the most important because when this is variable the safety and efficacy of a product become unpredictable. The quality of a medicinal product is dependent upon many factors including the purity of each of its ingredients and the nature of the impurities present. The choice of the excipients used and the care with which the ingredients have been mixed together are also important. The more closely these factors can be controlled the higher the quality of the product. Of equal importance is the uniformity of quality from batch to batch, and this can only be achieved by rigid adherence to the specifications of every ingredient and process used, coupled with constant quality control procedures at every stage of manufacture. In short, quality has to be built into the product. If a product of poor quality were to be permitted to be administered, it would probably be less efficacious than a product of superior quality, and the impurities present may themselves be dangerous.

Safety is defined in the Medicines Act [sect.132(2)] in a way which means more than simple freedom from hazard. The definition encompasses long-term effects upon the patient, the possibility that a medicine may disguise an underlying pathology and thus hinder diagnosis and, in relation to animal medicines, that they do not enter human food chains. However, no treatment is completely safe. All that can be done with respect to medicines is to test them as thoroughly as possible prior to marketing them.

Efficacy is related to a particular disease, condition or treatment. It is not solely a pharmacological property and can be affected by factors such as the particle size, the excipients and pressures used in tabletting and the purity of the ingredients. Also of importance here is the care with which the patient complies with the prescribed dosage regimen. Packaging, labelling and package inserts have an important role to play in improving compliance as does careful counselling by the doctor and pharmacist.

The three criteria differ insofar as they can be quantified. Quality is usually comparatively easy to measure especially when the ingredients of the product are chemical substances. Each of the important properties which together combine to constitute the quality of the product can usually be scientifically determined. It must be borne in mind however, that different assay procedures often give different results, hence the need for standard assay methods laid down in official compendia. Official standards are usually designed for testing products at the consumer level, i.e. some months after manufacture, and therefore have wider tolerances than would be permitted in the manufacturing specification.

Safety is extremely difficult to quantify and is, in fact, relative. Furthermore, most drugs have been found to cause damage to a few hypersensitive individuals even when used in the correct dosage. The majority of adverse reactions are either of minor significance in that they merely inconvenience the patient or disappear when treatment is stopped. Finally, safety has to be considered in relation to the intended use of the product so an element of subjectivity is involved. For example, a new analgesic should be at least as safe as existing analgesics but a lower level of safety might be acceptable in a new cytotoxic drug.

Efficacy is also difficult to quantify. Experience shows that even when a diagnosis is certain, a percentage of patients treated with a given drug will fail to respond to it while some patients recover spontaneously without any treatment. However, it could be argued that a product which is much less efficacious than existing products should not be marketed even though it might be 'safe', because it will be given to some patients who would benefit more from the existing products.

Legal and regulatory constraints

Relevant laws embrace a variety of measures to ensure the safety, quality, and efficacy of medicines. Typically, the legislation requires the manufacturer (or importer) of the product to obtain a marketing authorization for each product. This is granted after some expert group is satisfied with the specifications of the product and the biological and clinical tests that have been performed with it. When the product contains a new active substance (NAS), formerly called new chemical entity (NCE), details must be given of the pharmacological and toxicological tests which demonstrate its efficacy and safety. Normally the law leaves applicant free to use whatever tests he considers best for his purpose rather than specify the tests that must be carried out. Unfortunately, each country sets its own criteria for acceptance of a product, so applications (and the data required in them) vary from country to country.

The licensing (or marketing) authority will also need to be satisfied that the manufacturer has the facilities and expertise to produce batches of consistent quality. His premises are inspected and samples taken of products being manufactured to ensure that the processes are carried out properly. This system should result in the production of preparations of acceptable quality and safety. However, a product can be adversely affected by subsequent faulty storage so it is necessary to control wholesale dealers. As usual this is done by means of a licence to obtain which the dealer must show that he has the necessary facilities, both physical and in respect of record keeping, to permit drugs to be kept secure and in optimum conditions. A product's activity can be affected by its container so the law often controls containers. The container is also important in preventing children from gaining access to medicines and legislation provides for certain products to be packed in specified types of container.

The safety of a medicine depends upon its being used properly (e.g. in the correct dosage, and at appropriate times) so labelling provisions require that the necessary information and warnings appear on labels. Where a defective product is released for sale or a product deteriorates abnormally quickly, the affected batch can be traced rapidly and withdrawn from sale by means of the system of batch numbers which must appear on all labels including those on packing cases. The marketing authorization normally specifies whether the product may be sold freely, or from pharmacies only or on prescription only. It also specifies the clinical conditions for which it can be advertised and promoted. Deterioration can, of course, occur in the retail establishment and adulteration (either accidental or deliberate) can also occur in such places. To guard against this inspectors are employed to purchase samples of products from retail outlets and submit them for analysis.

Another aspect of proper use involves the promotion of sales of medicinal products. The Medicines Act controls these in several ways, it prohibits the advertisement to the public of specified products or those referring to specified diseases. It also controls promotional schemes aimed at practitioners by requiring that a data sheet containing a summary of all of the relevant points is issued to the practitioner before the product is promoted. Advertisements are therefore more truthful than they were and it is harder to gloss over the less desirable properties of the product. Advertisements directed at the public are also controlled.

Any such regulatory scheme requires the employment of large numbers of pharmacists, doctors and administrators and few 'third world' countries have these resources. Instead they may require proof that a product has been licensed for use in the country of origin. Difficulties can arise when a country wishes to import a product whose licence has been allowed to lapse because of poor sales in its country of origin.

This chapter will attempt to survey the legal and regulatory provisions applicable to pharmaceuticals and, where applicable, describe how these are modified by codes of practice. It will also indicate where the information can be located and give advice on its interpretation and implementation. The material will be considered in relation to UK law under the following headings:

(a) the definition of 'medicinal product'
(b) licensing of medicines
(c) manufacture, packaging and labelling of medicines
(d) controls over distribution of medicines
(e) sales promotion
(f) product liability
(g) health and safety of workers
(h) prevention of pollution.

The controls applicable in the US will then be briefly discussed emphasizing differences from British requirements. Finally, sources of information on laws and regulations of other countries will be mentioned.

United Kingdom law

The Medicines Act 1968 and its subordinate legislation controls most of the activities relating to medicines. In the account which follows, the section number of the Act will be quoted for reference while provisions contained in the subordinate legislation (Regulations and Orders) will be identified by the year and reference number of the appropriate statutory instrument.

(a) What is controlled under the Act?

(i) MEDICINAL PRODUCT

The Act chiefly controls products which fall within the definition of 'medicinal product' namely:

> 'any substance or article (not being an instrument, apparatus or appliance) which is manufactured, sold, supplied, im-

ported or exported for use wholly or mainly in either or both
of the following ways, that is to say –
(a) use by being administered to one or more human beings
or animals for a medicinal purpose;
(b) use, in circumstances to which this paragraph applies, as
an ingredient in the preparation of a substance or article
which is to be administered to one or more human beings or
animals for a medicinal purpose.' [sect. 130(1)].

The important point to note in the above definition is 'wholly or
mainly' but not necessarily exclusively. Thus a product which is
used mainly for a non-medicinal purpose but occasionally is used
medicinally (e.g. brandy) is not a medicinal product within the
meaning of the Act. It is also possible for a product to be both a
medicinal product and a food, or a medicinal product and a
cosmetic.

The definition includes products ready for administration to
patients and also ingredients that will be used in dispensing by
pharmacists or practitioners [sect.130(3)]. Ingredients used by
manufacturers of medicines are excluded from the definition
because their quality is controlled by the specifications laid down
in product licences.

"Medicinal purpose" includes treating or preventing disease,
diagnosing disease or ascertaining the existence, degree or extent
of a physiological condition, contraception, inducing anaesthesia,
and otherwise interfering with the normal operation of a physiological function [sect. 130(2)]. This broad definition covers a
multitude of effects including perspiration so that an anti-
perspirant intended to be applied to the body is a medicinal
product. For most purposes of the Act, such toilet or cosmetic
articles have been exempted from control.

Experimental substances which fulfil the criteria listed in section
130(4) of the Act are excluded from control. Consequently there is
no *statutory* control over tests on animals or on human volunteers
but companies and practitioners involved in such tests should
abide by the appropriate guidelines.

The EEC directives which relate to the free movement of
proprietary products throughout the Community have necessi-
tated the distinction between those which are proprietary products
(sold under a trade mark or patent name, or in a special pack) and
those which are not. The exclusion of vaccines, blood products and
homoeopathic products is a consequence of the fact that the EEC
definition of medicine does not include such products. [Sect.7(7),
1977/1052, 1977/1053, 1977/1055, 1982/1729, 1983/1724; 1983/1725;
1983/1726; 1983/1727; 1983/1728; 1983/1931].

Certain restrictions apply to imported proprietary products. In this context imported means that the product has been imported from a non-EEC country (1977/1052; 1977/1053; 1977/1056).

(ii) OTHER PRODUCTS SUBJECT TO CONTROL UNDER THE
MEDICINES ACT

The Ministers are empowered under the Act to control any class of articles or substances which fall outside the definition of 'medicinal product' but which are, or may be, (a) used for medicinal purposes (sect. 104), or (b) used in the manufacture of medicinal products or which are capable of damaging the health of humans or animals if used without proper safeguards (sect. 105). The controls are imposed by Orders which specify which parts of the Act affect the substance or article. The following relevant items have been brought within the scope of the Act by such orders.

Cyanogenetic substances have been advocated for use in the treatment of cancer under names such as amygdalin, laetrile or vitamin B17. While there is no evidence of efficacy, there is evidence of toxicity. Their sale, labelling and advertising is controlled (1984/187).

Various antibiotics and other biological products are substances whose purity and potency cannot be adequately tested by chemical means, and special precautions must be taken in their manufacture. Their labelling and distribution are also controlled. Among the substances controlled are certain antibiotics and a few hormones etc. intended for use as ingredients in the manufacture of parenteral products for human use. Also controlled are oxytetracycline and tetracycline for any product for human or animal use and dextrans for parenteral use in humans (1971/1200; 1982/425; 1985/1403).

Antimicrobial substances are products (not being a medicinal product) containing any antibiotic, sulphonamide, nitrofuran or any other substance having antimicrobial properties (1973/367).

The Ministers are not empowered by the Medicines Act to prohibit the administration of a product or class of products.

For the sake of brevity, the word 'medicine' will be used in this chapter to include all the above mentioned products.

(b) Licensing of medicines

(i) INTRODUCTION

The licensing system is central to the control of medicines. The system utilizes three types of licence; Product Licence (PL), Manufacturer's Licence (ML), Wholesale Dealer's Licence

(WDL), and two types of certificate, Clinical Trial Certificate (CTC) and Animal Test Certificate (ATC). Each is tailored to permit a person or company to engage in specified activities. The law relating to licences and certificates is contained in Part II of, and schedule 2 to, the Act and in several sets of regulations. The Medicines Control Agency have published a useful series of booklets (called *MALs*) which provide guidance on a variety of matters relating to the Act. It is an offence to engage in any of these activities unless one holds the appropriate licence(s) (sect. 45). Practitioners and pharmacists do not require licences for their ordinary professional activities (sects. 9,10,11).

The licences applicable to medicines for humans are issued on behalf of the government by the licensing authority [the 'Health Ministers', (sect. 6)] acting on the advice of professionals employed within the Medicines Control Agency at the Department of Health. Advice is also received from 'section 4' committees composed of independent experts in respect of applications for products possessing some novel feature such as a totally new active substance (NAS) or a new synthetic route or therapeutic indication for a known chemical.

The application for a product licence contains full details of the product's composition, specifications and procedures for manufacturing and quality assurance. These data are expensive to obtain and carefully guarded as trade secrets. To protect the applicant's interest in this information it is an offence (sect. 118) for a person having access to the information to disclose it to anyone not authorized to receive it. The confidentiality of information is so strictly respected that in 1976 the British Pharmacopoeia Commission complained that it was not allowed access to analytical data provided in applications for product licences.

The licences normally operate for five years but can be renewed (sect. 24). Product licences are examined by the Committee on the Review of Medicines before renewal. A product licence comprehensively describes the product's specifications, therapeutic indications, legal class and packaging and labelling requirements. It also contains standard provisions with which the holder must comply. A product licence can be 'varied" to take account of new information or standards. This may occur at the request of the licence holder (sect. 30) or by the licensing authority after consultation with an expert committee (sect. 28). Licences can be suspended or revoked for infringements of their provisions, if they were obtained by fraud or if new information indicates that the product is less safe or efficacious than was originally believed (sect. 28). A system of batch numbers marked on containers and packages and records kept by manufacturers and wholesalers enables defective batches of products to be traced and rapidly withdrawn from sale.

The licensing provisions of the Act are enforced by inspectors of the Medicines Control Agency who may enter the premises of licence holders, inspect processes, take samples of products or containers, inspect books and documents relating to the business and make copies of any entries. They may seize any products required as evidence and may require that containers, packages and vending machines be opened. They have similar powers of inspection over homecoming ships, and aircraft to prevent the importation of unlicensed medicines (sect. 112).

It is an offence to sell, supply, import or export an unlicensed medicine or to manufacture or assemble a product except in accordance with a product licence (sect. 7). Corresponding offences have been created in relation to products for use in clinical trials and medicinal tests on animals (sects. 31, 32). A number of other offences have been created by section 45, for example, it is an offence for a product licence holder to fail to inform the manufacturer of any variation to the licence, or to possess for sale etc. an unlicensed product (sect. 45).

(ii) ASSESSMENT OF APPLICATIONS

Within the Medicines Control Agency are teams each consisting of a physician, a pharmacist and administrators. The teams scrutinize written applications for certificates and licences. Applications relating to medicines which have no novel features and which are to be promoted for appropriate conditions are passed for licences to be drawn up. Where an application contains a novel feature the team send it for examination by a section 4 committee.

There are two expert committees which are relevant to the licensing of human medicines, namely, (a) the Committee on the Safety of Medicines (CSM) and (b) the Committee on the Review of Medicines (CRM).

These committees are composed of experts drawn mainly from academia and professional practice. The committees base their judgment on the evidence provided by the applicant. It is therefore essential that applicants provide full details of all experimental work carried out, both on the active constituents and on the finished product and present the data in the form required.

The duties of the CSM are to advise the authority with respect to the safety, quality and efficacy in relation to human use, of any substance or article to which the Act applies, and promote the collection and investigation of information relating to adverse reactions to medicines. It has sub-committees viz.: (1) Sub-committee on Safety, Efficacy and Adverse Reactions (SEAR); (2) Sub-committee on Biologicals; (3) Sub-committee on Chemistry, Pharmacy and Standards and (4) Joint Sub-committee on Adverse Reactions to Vaccines and Immunological Products. The sub-committees submit applications to detailed scrutiny and report to the main committee to enable it to advise the licensing authority on each application. Where the Committee feel that insufficient evidence has been provided by the applicant, it can request additional data or ask specific questions using a 'section 21(1) letter'.

The CSM also has the duty to collect, investigate and assess the importance of reports of adverse reactions to drugs. The Committee publicises its findings in *Current Problems* and in other ways and also gives advice to other expert committees.

When the CSM is satisfied that the product is safe, efficacious and of suitable quality, it will recommend the grant of a product licence. The licence will contain the appropriate Standard Provisions unless the applicant can show any of them to be unnecessary. If the CSM consider that the product should not be licensed or that the licence should contain certain provisions, they send a 'section 21(3)" letter to the applicant informing him of this.

The CRM gives advice as to the safety, quality and efficacy of a product where an application is made to renew its licence or to replace a licence of right with a full licence.

The Medicines Commission, advises the Ministers on matters relating to the safety, quality and efficacy of medicines and on the exercise of powers under the Act such as the appointment of committees. This advice may be given when requested or when the Commission considers it appropriate. It also has appellate functions. An applicant for a licence or certificate is entitled to make oral or written representations to the Commission where the Authority proposes either to refuse a licence or certificate on the grounds of safety, quality or efficacy, or to grant it subject to conditions which are unacceptable to the applicant. Appeal to the Commission also lies where a licence holder objects to a proposal by the Authority to suspend, revoke or vary a licence or certificate (sched. 2), or to make an order prohibiting the sale, supply or importation of a medicinal product (sect. 62).

When exercising its appellate function the Commission considers the application afresh and has access only to the original application and any relevant correspondence. The Commission reports its findings to the Authority, which sends the advice and its decision to the applicant.

(iii) THE LICENSING AUTHORITY

In dealing with an application for a product licence, the Authority must consider the safety of the product, its efficacy for the intended purpose and its quality as indicated by its specifications. A licence cannot be refused on the grounds that the product is less efficacious than other products for the same purpose except on safety grounds [sect. 19(1),(2)]. A licence may only be refused on grounds of efficacy, safety or quality after consultation with the appropriate committee [sect. 20(3)]. If the Authority after considering the views of the committee decides either to grant the licence with modifications or to refuse it, the applicant must be given the reasons for the decision [sect. 20(5)]. He may appeal against such a decision to the Commission [sect. 21(4)]. The Authority may NOT refuse a product licence on grounds relating to the price of the product nor may conditions as to price be imposed [sect. 20(2)]. The Authority has corresponding powers to satisfy itself as to the quality, safety and efficacy of imported products.

The Authority may suspend a licence or revoke it (sect. 28) on the grounds of safety, quality or efficacy. The appropriate committee or the Commission should first be consulted but where necessary on safety grounds the Authority may suspend (for 3 months) a licence or certificate immediately. If necessary, the suspension may be extended for another 3 months. Before giving any advice to the Authority, the committee or Commission must seek and take account of any oral or written representations made by the licensee (sched. 2).

Normally, the Authority accepts the advice of the Commission, but if it does not the applicant may apply for 'a hearing by a person appointed" [sect. 21(5),(7); sched. 2]. The Authority must appoint a person to hear the case. The applicant states his case giving any information he considers relevant and may with the consent of the person appointed, call witnesses. The person appointed reports to the Authority giving all the facts and points made by the applicant but does not make a recommendation. The Authority takes no part in the proceedings. The Authority considers the report and decides the issue. If the decision is adverse, the reason for it must be given. This procedure is seldom invoked.

On any matter relating to licences and certificates the decision of the licensing authority is final but an applicant may (within three months) appeal to the High Court on the grounds that the decision is not within the powers of the Act or that the requirements of the Act or regulations have not been complied with (sect. 107). The High Court may suspend the operation of the decision or quash it.

(iv) CLINICAL TRIALS

When an NAS (chemical or biological) has reached that stage of its development when it is necessary to determine its efficacy in a specified condition in patients the test is called a clinical trial and the NAS becomes a 'medicinal product' in law [sect. 130(1)]. Such a trial can only be carried out lawfully by obtaining a clinical trial certificate (CTC) and performing the trial in accordance with its provisions. The trial is designed to elicit the information required for the application for a product licence.

In his application for a CTC the person who wishes to arrange a trial submits details of the data obtained from laboratory studies on the NAS. This is examined by the CSM to ensure that the product is safe, of good quality and has real therapeutic potential.

Some years ago it took 4–8 months to process applications for CTCs in the United Kingdom. To expedite matters, the clinical trial exemption scheme (CTX) was established. This enables applicants to submit a certified summary (rather than a complete copy) of the laboratory data together with a copy of the protocols and the applicant's medical adviser must certify that the trial was reasonable. The Authority has 5 weeks in which to consider the application and notify the company of its decision though a 4 week extension period can be requested. The holder of an exemption must notify the authority of any change in the protocol, of any adverse reactions detected, of any information which casts doubt on the safety of the product or of objections raised by an Ethics Committee.

An applicant who is refused a CTX may apply for a CTC. This will involve him in the trouble and expense of providing multiple copies of all his data instead of merely summaries of it. It does however, entitle him to appeal against an adverse decision.

Clinical trial certificates contain standard provisions. The most important requires the applicant to keep samples of each batch of material used in a trial for at least 6 months after the batch expiry date and submit a sample and detailed protocols of the tests applied to that batch if requested by the authority. He must report any information which casts doubt on the validity of the data submitted with the application. The trial must be carried out in accordance with the protocol contained in the application, and the products may only be administered by a practitioner who has received a copy of the certificate and has been approved by the authority.

(v) PRODUCT LICENCES

Written application has to be made. In addition to such information as the company's name and address, the product's name, pharmaceutical form, composition and specifications the application must include all the scientific data (and the evidence on which this was based) required for the CTC, any additional laboratory work and the results of the clinical trials. This is bound in several volumes and includes Product Particulars, Pharmaceutical Data on the Dosage Form, Safety and Efficacy as detailed in *MAL 2*.

The application for a product licence (and for a clinical trial certificate) contains all available data which demonstrates the safety, efficacy and quality of the product. The manufacturer uses the tests and provides the evidence he considers best proves his case. Nevertheless, guidelines issued on behalf of the authority do indicate the types of data normally regarded as essential (e.g. *MAL 2*).

Abridged applications are permissible where the drug substance is well known, for example, it is the subject of a pharmacopoeial monograph or has been described in a previous application. However, if the substance has been obtained from a new source or by a new synthetic route, a full application is necessary.

A manufacturer can save considerable time and expense when drafting his applications for product licences if he uses the same excipients in several products. He can prepare and submit a manual describing the specifications for these materials and the analytical methods employed in their quality assurance. The analytical methods may also be applicable to other substances and future applications can refer to the manual when appropriate.

A product licence is required before an imported product can be marketed in the UK. The product licence is granted when the authority is satisfied that it is safe, efficacious and of suitable quality. The overseas manufacturer will be required to give an undertaking to permit inspection of his facilities overseas, to comply with any conditions imposed and to declare that the product has been manufactured in accordance with the law of the country of origin (1977/1038).

Special provisions apply to the parallel importation of proprietary products, that is those which are covered by a valid and current marketing authorization under Article 3 of EEC/65/65 by a member state of the EEC, and also by a valid and current marketing authorization issued in the UK.

(vi) HIGH-TECHNOLOGY MEDICINAL PRODUCTS

These are also referred to as 'biotechnology' products and are controlled under the Council Directive 87/22/EEC. This classifies them as List A products (which includes those developed by means of recombinant DNA technology or hybridomas and monoclonal antibody methods) or List B (which includes a variety of other types of product). Persons working in the field of genetic manipulation must obey the advice obtainable from the Health and Safety Executive (see later).

Medicinal products which fall within List A or B which it is intended to market in more than one member state need a marketing authorization issued by the Committee for Proprietary Medicinal Products (CPMP). The application is made to one state (which acts as Rapporteur) which assesses the product. This assessment is sent to the other states for comment and assessment. When all have reported, the applicant is notified of the finding and given the opportunity to provide additional information.

(c) Manufacture, packaging and labelling of medicines

The manufacture of a medicine can only be carried out by a person who holds a manufacturer's licence for the processes involved (sect. 8). In this context, the packaging and labelling of a product is called 'assembly" and constitutes 'manufacture'. The detailed specifications for all ingredients, processes and tests are contained in the product licence and it is a condition of the manufacturer's licence that each batch of every product is manufactured strictly in accordance with the product licence (sect. 23).

An additional licence, issued by the Home Office, is required if the medicine contains a 'controlled drug' i.e. one listed in schedule 2 (as amended) to the Misuse of Drugs Act 1971.

(i) MANUFACTURER'S LICENCES

There are five categories of manufacturing operation viz.:

(a) manufacture of 'biologicals', e.g. vaccines, blood products, hormones, enzymes
(b) manufacture of sterile products, e.g. powder or solution ampoules, eye preparations, some powders
(c) manufacture of unit dose forms, e.g. tablets, lozenges, capsules, pessaries, suppositories
(d) manufacture of multi-dose liquids, pastes or solids, e.g. mixtures, lotions, creams, gels, ointments
(e) manufacture of medicated dressings, e.g. lint, tulle, medicated plasters.

An application for a manufacturer's licence must be made in writing. It must specify each premises at which manufacturing processes occur and give details of the processes carried on and the facilities available at each premises. The names, addresses and qualifications of key personnel (e.g. production manager and quality controller at each premises) must be stated. The application must also contain information on a number of other matters, details of which are to be found in The Medicines (Applications for Manufacturer's and Wholesale Dealer's Licences) Regulations 1971 (as amended). Each premises is visited by inspectors from the Medicines Control Agency whose report is taken into account in deciding whether or not the licence should be granted.

The Medicines (Standard Provisions for Licences and Certificates) Regulations 1971 (as amended) require the licensee to provide the necessary staff and facilities for the operations covered by his licence and to notify the Agency of any proposed changes to them. Every batch of product must undergo the specified quality assurance procedures and durable records of all operations and tests must be available for inspection for at least five years from the date of manufacture. Premises may be inspected and documents or samples requested by the inspectors. Additional detailed and stringent standard provisions apply to manufacturer's licences relating to biological products.

(ii) PACKAGING OF MEDICINES

The container in which a medicinal product is packed protects the product and carries the label. It can also protect the user and other members of the public. Ribbed or fluted bottles have long been used to help the blind or visually handicapped distinguish poisons or liquid products for external use from those intended to be swallowed. More recently, containers have been designed to protect children from the dangers of ingesting brightly coloured and attractive tablets, capsules and similar solid dosage forms. The choice of container is clearly an important matter both in terms of effectiveness for the purpose and aesthetically. The material of construction of the container is important because it will be in intimate contact with the product with possible adverse effects.

The Act and regulations contain little detail relating to containers but the product licence normally describes the type and material of construction of the container to be used for the product. This permits a more appropriate form of control than would be possible by trying to draft regulations to cover all the eventualities. The *British Pharmacopoeia* contains standards for the glass and plastics to be used for packaging parenteral products and eyedrops (Appendix XIX).

It is an offence to sell or supply a product not in a container [sect. 85(4)] or enclose a product in container which does not comply with the requirements of the regulations [sect. 87(2)].

The Medicines (Child Safety) Regulations 1975 (as amended) require that tablets, capsules and cachets containing aspirin or paracetamol intended for retail sale for human use should be white and packed in child-resistant containers.

Liquid products for external use which contain substances specified in The Medicines (Fluted Bottles) Regulations 1978 must be packed in fluted bottles so as to be recognizable by touch. Exemption from this requirement applies to ear and eye preparations packed in plastic containers, products packed in bottles larger than 1.14l, and those for export or for scientific education or research.

(iii) LABELS AND LEAFLETS

The requirements contained in The Medicines (Labelling) Regulations 1976 (as amended) apply to all medicinal products. They are complemented by The Medicines (Leaflets) Regulations 1977 which control the information given in leaflets supplied with a proprietary medicinal product and which discuss the indications and dosage of the product. Leaflets which only contain instructions as to the method of use of the product are not controlled under these Regulations. Modified labelling requirements apply to products for clinical trial, import or export. It is an offence to apply a false or misleading label (or leaflet) with a medicine [sects. 85(1), 86(3)]. There are three main groups of labelling provisions namely, general, standard and warnings.

General provisions apply to all labels and are found in Regulation 17. Basically, these provisions require the label to be indelible, legible, fixed to the container, in English but other languages may be used in addition. Other general provisions relate to products containing aspirin or paracetamol and to the use of botanical names or references to the Act, its regulations, the Commission or to any of the Committees. Every container and every package immediately enclosing the container, of a medicinal product for retail sale must be labelled 'keep out of the reach of children" or similar phrase. Small containers and 'strip packs' are exempt from this requirement but the package must bear the warning (regs. 17, 18).

Standard provisions are necessary on containers and packages of products when sold, different provisions apply to dispensed products. The provisions are to be found in schedule 1 to the Regulations and include such details as the name of the product (i.e. either the 'non-proprietary name" or a proprietary name), its pharmaceutical form, the 'appropriate quantitative particulars" as defined in Regulation 3, the quantity of the product in the container, any contraindications, warnings or precautions to be taken in its use and the directions for use. In addition the label must state any special storage conditions, the expiry date, the batch number, product licence number and manufacturer's licence number.

Modified requirements apply to ampoules and other containers with a nominal volume not exceeding 10 ml, unit packs and 'bubble' packs (reg. 14E).

Warning labels must be used on certain medicines. Medicines which may only be sold from pharmacies must be labelled P, while those which may only be sold on prescription must be labelled POM (regs. 14B, 14C). Products containing aspirin, aloxiprin or paracetamol and for internal use must state the recommended dosage and 'If the symptoms persist, consult your doctor'. They must also be prominently labelled 'Contains aspirin' (or 'Contains an aspirin derivative' or 'Contains paracetamol') as appropriate. Paracetamol products must also be labelled 'Do not exceed the stated dose'. Warning phrases are also required for oral products containing antihistamines, for ephedrine products for use in asthma, and for products containing hexachlorophane. Furthermore, a product for internal use containing a POM substance which is exempt from POM control because of strength or dosage must be labelled 'Warning. Do not exceed the stated dose'. Further details on these requirements can be found in Regulations 14A, 14B, 14C and in schedules 5 and 6 to the Regulations.

Leaflets can be a valuable source of information for prescribers and patients, though it must be recognized that sometimes the information needs of the two classes are not identical. Where the leaflet is intended for practitioners it should be compatible with and supplementary to the data sheet. Leaflets intended for patients generally avoid mention of clinical indications for the product and concentrate upon helping the patient to use the product correctly. Such a leaflet is called a package insert or patient package insert (PPI) and it can be viewed as an ancillary label and is subject to analogous controls by The Medicines (Leaflets) Regulations 1977.

The leaflet must contain the same sort of information as appears on the label except expiry data and batch number. The particulars must be in English and consistent with the product licence. If the product is POM the leaflet need not contain details as to therapeutic indications or dosage. The required particulars shall be clearly separate from the remainder of the contents of the leaflet. Copies of the leaflet must be supplied to the licensing authority at least 6 weeks before it is first issued. If the authority considers that the leaflet should not be issued or requires modification it must inform the licensee accordingly within 21 days.

(d) Controls over the sale and distribution of medicines

The framework of these controls is provided in Part III of the Act, while the detail is contained in various sets of regulations and orders. The Ministers, after consultation with an expert committee or the Medicines Commission, have used powers specified in section 62 to prohibit the sale, supply and importation of products in the interests of safety.

(i) SALES OF MEDICINES

The Ministers have used powers conferred by section 61 to restrict the sale of medicines by manufacturers and wholesalers to specified classes of purchasers (1980/1923, reg. 5). This ensures that only persons lawfully entitled to sell or use the products can buy them in the trade.

For the purposes of sale and distribution, there are three classes of medicines namely, (a) general sale list medicines (GSL) (sect. 51), (b) pharmacy medicines [P] (sect. 52), and (c) prescription only medicines (POM) (sect. 58).

Medicines on the General Sale List (GSL). This is a list of ingredients which, when made into medicines, may be sold without the supervision of a pharmacist. The medicines must be sold from shops (or from automatic machines) in unopened containers which were packed and labelled elsewhere. There are two General Sale List Orders, one for human medicines (1984/769) and one for veterinary drugs (1984/768). For some products there are limitations as to dose, pharmaceutical form, route of administration or strength for certain substances, and some of these vary with the type and intended use of the product.

Pharmacy Medicines (P). These are medicines that contain one or more ingredients not listed in the GSL. They may only be sold by 'persons lawfully conducting a retail pharmacy business'. Such persons may only sell these medicines from registered pharmacies and each sale must be by or under the supervision of a pharmacist. There is no list of these products. A product which falls outside any of the GSL limits is a (P) medicine. Likewise some products are (P) because they are exempt from the POM provisions on the grounds of concentration, dose, use or route of administration. Other products are (P) because none of their ingredients are mentioned in either the GSL or the POM lists (e.g. promethazine).

Prescription Only Medicines (POMs). These are medicines which may only be sold or supplied to the public on the prescription of a practitioner. However, a pharmacist may supply such products for human use in an emergency at the request either of a doctor or the patient.

Medicines which contain an ingredient specified in The Medicines (Products Other Than Veterinary Drugs) (Prescription Only) Order, (1983/1212 as amended) are commonly referred to as POMs. The Order lists approximately 1200 substances but some of these are exempted from control when present in certain dosage forms, or for certain purposes or are labelled to show dosage regimens not exceeding specified levels. Included in the list are most of the therapeutically useful antibiotics and sulphonamides (except tyrothricin), all of the psychoactive drugs, most hormones, most alkaloids, and all the major drugs used in cardiovascular disease or in the promotion of diuresis. All parenteral products (except insulin injections) are POM. Local anaesthetics are POM when included in ophthalmic preparations but are exempt in other pharmaceutical forms. All products containing new chemical entities are made POM for a period of five years from first licensing unless there exists evidence of safety (sect. 59).

Controlled Drugs. All the drugs controlled under the Misuse of Drugs 1971 are POM. Under the Misuse of Drugs Regulations 1985 (as amended) some of these drugs are exempt from control when present in preparations containing less than specified concentrations. The same drugs are exempt from the POM restrictions when in preparations containing less than certain specified concentrations, but unfortunately different limits are set in the two documents.

The Misuse of Drugs Regulations 1985 specify the detailed control over 'controlled drugs'. The drugs are divided into five classes. Only persons holding a special licence from the Home Office may possess, prescribe, administer or supply the drugs specified in schedule 1 to the Regulations which include crude drugs such as coca leaf, and hallucinogens such as lysergide. Most of the medicinal drugs such as morphine, pethidine and cocaine are specified in schedule 2. Doctors, dentists and veterinarians registered in the United Kingdom are authorized to possess, prescribe, administer, sell or supply these drugs to their patients and pharmacists are authorized to possess, sell, supply and dispense them in the course of their business. Schedule 2 also contains all parenteral products containing any of the drugs in schedule 5, paragraph 1b (such as morphine, codeine, dihydrocodeine). Schedule 3 drugs include some barbiturates and again, practitioners and pharmacists are authorized to prescribe, sell, supply and dispense them under certain conditions. The drugs included in schedule 4 are the benzodiazepine tranquilizers while those in schedule 5 are preparations containing less than specified amounts of morphine, cocaine, codeine and certain formulations containing dextropropoxyphene, diphenoxylate, or propiram.

Registers must be kept by practitioners and pharmacists in which details of all incoming and outgoing supplies of the drugs in schedules 1 and 2 are recorded. Likewise manufacturers and wholesale dealers must keep records of all their dealings in the drugs. The drugs must also be kept in locked cupboards.

Prohibition Orders. Generally speaking, such orders are intended to prohibit the sale and supply of products containing certain hazardous substances. Sometimes, however, the restrictions fall short of a total prohibition. There are five such Orders. In the case of Bal Jivan Chamcho (a proprietary product of Indian origin intended for administration to children), the prohibition is complete due to the unacceptably high concentrations of lead. In the cases of chloroform, phenacetin and the various antimicrobial substances, sales to practitioners, pharmacists or hospitals are permitted.

(ii) CONTROLS OVER THE ADMINISTRATION OF MEDICINES

The administration of medicines to patients can be controlled by law (sect. 60) but there is no power under the Medicines Act 1968 to *prohibit* the administration of any product.

POMs may only be administered to human beings by doctors and dentists or persons acting in accordance with their directions (sect. 58). The restrictions over non-parenteral POMs were lifted (1983/1212 art. 7) so that only the parenteral administration of a POM and of any product controlled under the Misuse of Drugs Act is now restricted. The Ministers have also exempted the parenteral administration of adrenaline (and certain other specified products) in a life-threatening emergency (1983/1212, art. 5).

On the other hand, when necessary in the interests of safety, and where specialized knowledge or skills are required, additional restrictions may be imposed on the practitioners who may administer, prescribe or be supplied with certain medicines (sect. 60). Such restrictions have been imposed in respect of radioactive medicines (1978/1006).

(iii) QUALITY OF MEDICINES REACHING THE CONSUMER

The Medicines Act has applied pharmacopoeial (or other standards) to medicines including those supplied on prescription and it is an offence to supply (or possess for sale or supply) a product which does not comply with these standards (sect. 65). The *British Pharmacopoeia* (*BP*) contains 'relevant information' (i.e. descriptions of, standards for, or notes or other information) relating to such substances and articles or used in the manufacture of such products [sect. 99(7)]. The standards of the *European Pharmacopoeia* (*Ph.Eur.*) now take precedence over those of the *BP*.

A product is said to be adulterated when something has been added to it or an active constituent has been removed from it. Such practices were common with foods and drugs before modern analytical techniques facilitated their detection. It is an offence to adulterate medicinal products 'so as to affect injuriously the composition of the product with intent that the product be sold in that state' (sect. 63) or to sell, offer or expose for sale or to possess such a product for sale. Where the product is not 'injuriously affected' by an addition or removal (such as the addition of an antioxidant) no offence has been committed. A retailer may be convicted for selling or possessing for sale a product that was adulterated by someone else (e.g. a supplier).

A related but separate offence is to sell 'to the prejudice of the purchaser any medicinal product not of the nature or quality demanded' [sect. 64(1)]. Such an offence could arise by selling an adulterated product but might be easier to prove than adulteration. It is also an offence to supply such a product on a prescription.

(e) Promotion of sales of medicines

(i) INTRODUCTION

In section 92(1) of the Act 'advertisement' is defined in the broadest terms as including 'every form of advertising, whether in a publication, or by the display of any notice, or by means of any catalogue, price list, letter (whether circular or addressed to a particular person) or other document, or by words inscribed on any article, or by the exhibition of any photograph or a cinematograph film or by way of sound recording, sound broadcasting or television, or in any other way'.

Spoken words (except those forming part of a sound recording, broadcast or forming part of a sound track) are excluded from the definition of 'advertisement' [sect. 92(2)] but constitute a 'representation' and are controlled as such [sect. 92(5)].

Thus a document may constitute an advertisement even though it does not contain an inducement to buy or prescribe the product. Consequently, trade announcements about pack or price changes constitute advertisements within the meaning of the Act. The scope of the definition is shown by the fact that it has been necessary in one set of regulations to exempt from control company reports and circulars to shareholders giving details of the company's products [1978/41, reg. 8(2d)]. Because of the broad and all-embracing nature of the definition, it has been necessary to apply the restrictions over advertisements only to a 'commercially interested party' such as the holder of the product licence or anyone who manufactures or sells the product in the course of his business [sect. 92(4)].

Similarly the definition of 'representation' is broad enough to cover ordinary conversations between members of the public but again the legal restrictions apply only to those made in the course of a 'relevant business', i.e. one which consists of or includes the sale or supply of medicinal products [sect. 92(4).]

Many of the prohibitions relate to false or misleading statements, but some apply to the use of an 'unauthorized recommendation' which is 'a recommendation whereby a product of a description to which the licence is applicable is recommended to be used for purposes other than those specified in the licence' [sect. 93(10)].

The Standard Provisions attached to every product licence can be used by the Licensing Authority to control the advertising of a medicine in the most appropriate way. They operate only when the Authority invokes them by writing to the licence-holder explaining what is required. Three paragraphs of the Standard Provisions (numbers 9, 10 and 11) are involved. Paragraph 9 and 11 can only be used by the authority for the one or more of the following purposes: (a) to ensure that adequate information is given, (b) to prevent the giving of misleading information and (c) to promote safety.

Paragraph 9 of the Standard Provisions can be used to ensure that the advertisement contains the particulars specified in the licence. On the other hand, paragraph 10 enables the authority to request details as to contents and form, the means, media or medium by which it is to be issued and time of issue of planned advertisements prior to their issue. In effect, it allows the authority to 'pre-vet' advertisements for a product.

Paragraph 11 enables the authority to give a written direction that advertisements of a particular kind should not be issued, or reissued or that they should be modified or contain particular warnings or precautions. Such a direction can be used to prevent the issue of an advertisement which had been examined under paragraph 10.

(ii) CONTROLS OVER PROMOTION TO PRACTITIONERS

These aim to prevent practitioners being misled by promotional schemes, to prohibit the making of unwarranted claims for safety and efficacy and to ensure that practitioners obtain a balanced view of the product's usefulness and hazards.

Although the Act controls advertisements directed to all practitioners, to date the emphasis has been on controlling those directed to doctors and dentists.

At the centre of these controls is the data sheet, a document which is intended to provide an objective statement of the basic information about the product. This information must be consistent with that given in the product licence so the sheet may be viewed as that part of the product licence that is available to the health professions generally. Thus it is a yard-stick against which all subsequent claims can be assessed.

It is an offence to issue an advertisement which contains information inconsistent with that published in the data sheet [sect. 96(3b)]. All advertisements must state that a data sheet will be sent on request and it is an offence to fail to comply with such a request. Copies of data sheets must be sent at regular intervals to the licensing authority (1975/1326, reg. 2). The restrictions imposed upon advertisements in professional publications apply only to the advertiser, thus the editor and owners of the publication cannot be prosecuted for any infringement of the regulations (1978/1020, reg. 12).

Limitations on Advertisements. Advertisements may contain claims or suggestions that the product is safer, more efficacious, or has fewer side-effects than another product only when there is clear evidence for the claim. The word safe cannot be used without qualification. The product cannot be advertised as new more than a year after it became generally available in the United Kingdom. Similarly, claims for a new indication cannot be made after one year from the date of the first advertisement which mentioned that indication (1978/1020, reg. 8). In addition, graphs and tables included in advertisements must be relevant to claims made and give a fair and true view of the data (1978/1020, reg. 9).

None of the prohibitions and requirements apply to letters sent in reply to a specific enquiry, or to members of Parliament or to government departments. Nor do they apply to written notices of changes of packaging, warnings of adverse reactions or similar documents provided that no medical claims are made in the notice. Finally, they do not apply to an advertisement a copy of which was received by the authority but which the authority have not within 6 weeks, notified the holder should not be issued (1978/41, reg. 8; 1978/1020, reg. 5).

Data Sheets. A data sheet is a document relating to a particular product prepared by the holder of the product licence. It is only necessary when a licence holder intends to advertise or promote the product to practitioners, hence if a company does not wish to advertise or make representations about it, no data sheet is required for that product. The Agency takes the view that the issue of a price list or catalogue containing no recommendations as to the use of particular products or assertions as to their quality (e.g. a trade advertisement) does not in itself necessitate the issue of a data sheet. Most pharmaceutical companies ensure that practitioners receive data sheets annually by participating in the production of the relevant *ABPI Data Sheet Compendium.*

The Medicines (Data Sheet) Regulations 1972 (as amended) specify in detail the form of data sheets, their size, colour and typesetting, as well as the information to be included in each data sheet. The data sheet must contain all the specified headings and where there are no such particulars the heading shall be followed by the world Nil. There are restrictions upon the inclusion of other information.

The Regulations permit the publication of compendia of data sheets such as those published annually by the ABPI and lay down certain specifications for these.

Particulars Required in Advertisements Directed to Practitioners. Every advertisement sent to a doctor or dentist or contained in a professional publication (i.e. one sent only or mainly to doctors or dentists) must contain all the information specified in The Medicines (Advertising to Medical and Dental Practitioners) Regulations 1978. This includes the product licence number, the name and address of the licensee, a list of the active ingredients of the product (using official names where possible), one or more of the indications for the product consistent with the terms of product licence, a succinct statement of the data sheet entries relating to the side-effects, precautions and contraindications, dosage and method of use relevant to the indications shown, any warnings required to be included by the Authority and the cost (excluding VAT) of either a specified package or of a specified quantity or of the recommended daily dose. There are also provisions relating to the size of type and the location of certain of the pieces of information [1978/1020 reg. 1(2)].

Where an advertisement occupies more than one page (a multi-page advertisement), the above particulars must appear on at least one of the pages and ech of the other pages must identify on its outer edge the page on which they are to be found (reg. 10).

Modified requirements apply to 'abbreviated advertisements', 'promotional aids', 'audio-visual advertisements', 'reference advertisements' and 'trade advertisements'. Each of these terms is defined in the regulations as are the requirements applicable in each case.

(iii) ADVERTISEMENTS DIRECTED TO THE PUBLIC

The purpose of controlling these advertisements is two-fold, firstly to protect the public from the activities of unscrupulous advertisers of ineffective remedies, and secondly to protect the ignorant from attempting self-treatment for potentially serious conditions. Over the last 50 years the Proprietary Association of Great Britain (PAGB), has done much to improve the standards of advertising of medicines to the public. Legal force has recently been applied by The Medicines (Labelling and Advertising to the Public) Regulations 1978. This set of regulations is extremely detailed and, in effect, confines the advertising of medicines to the public to more or less trivial complaints. Medicines for human use which are POM or contain substances which are the subject of prohibition orders such as phenacetin and chloroform, cannot be advertised to the public (reg. 3). This, however, does not prevent the mention of such products in a company's Annual Report to shareholders (reg. 8(2)d).

The restrictions contained in the regulations apply to advertisements for medicines but those imposed on the diseases specified under 'prohibited advertisements" extend to 'any articles and substances" e.g. any non-medicinal substance or any device. Furthermore, some restrictions extend to labels and leaflets (regs. 2, 4). In addition, many of the restrictions apply to representations made by retailers. It should be noted that in many cases an offence is committed not only when a disease, disorder etc. is mentioned by name, but also when the advertisement (including labels and leaflets) 'is likely to lead to the use" of the product in treating or preventing the disease. It is necessary to frame such documents with care.

There is a total prohibition on the issue of advertisements to the public for any of the diseases or purposes specified in schedule 1 and Part V of schedule 2 to these regulations (reg. 2, 4).

The advertising of remedies for diseases, adverse conditions and body systems specified in schedule 2 is permitted to the extent specified in each case (reg. 4), and is often subject to conditions including the use of warning labels.

(iv) NON-STATUTORY CONTROLS – CODES OF ADVERTISING PRACTICE

There are several of these relating to the advertising of medicines. All are voluntary but persons or companies who fail to observe their provisions are subject to sanctions, including expulsion from the organization and possible boycott by the rest of the advertising industry. The codes can be divided into two classes, those which deal with advertisements to practitioners and those which deal with all other advertisements. All the Codes recognize the existence of the various statutory restrictions on the promotion of medicines, and are obviously compatible with them. It is usually the case that the law imposes the minimum acceptable standards, and that professional and similar bodies impose higher standards on their members. Copies of the Codes and additional information and advice regarding them can be obtained from the various bodies.

Association of the British Pharmaceutical Industry Codes (ABPI). The Association has two Codes, one applicable to human medicines and the other for veterinary products. The Codes do not merely control advertisements but extend to all methods of sales promotion such as frequency of calls by representatives, hospitality and gifts. The appropriate Code is now printed in the *ABPI Data Sheet Compendium* to enable the practitioner to read it and complain if he considers that the Code is not being obeyed. Each Code is enforced by a Committee having a legally qualified chairman.

The ABPI does not operate a pre-vetting scheme but its secretariat routinely scrutinizes advertisements for human medicines appearing in professional journals. Recently, a medical consultant has been employed to scrutinize the medical claims made in promotional material. When an advertisement is suspected of infringing the Code, the company is notified of the required changes and if these are made the matter ends. If the company disgrees with the ABPI's view, or refuses to comply with the request, the matter is referred to the Code of Practice Committee for investigation. The Committee gives its ruling, and if the company is found 'guilty" it is asked to sign a document agreeing to stop the offending advertisement (or other behaviour) at once and undertake to avoid repetition. Companies that renege on this agreement may be expelled from the Association, while if the company is not a member of the ABPI, the facts are reported to the Medicines Control Agency.

British Code of Advertizing Practice (BCAP). This is enforced by the Advertising Standards Authority (ASA), an independent body created by the advertising business. It has an independent chairman, who appoints 12 council members, 8 of whom have no links with the business. The members act as individuals, not as representatives of business or pressure groups. The ASA investigates complaints made in relation to advertisements and monitors advertisements to ensure compliance with the rules. The results are published as *Case Reports*. Where necessary, expert advice is obtained from scientific and technical consultants.

The Code has the support of about 20 organizations, including the Advertising Association, Newspaper Publishers' Association, Incorporated Society of British Advertisers, Institute of Practitioners in Advertising and the Proprietary Association of Great Britain. Representatives of the organizations constitute the Code of Advertising Practice (CAP) Committee. The Code applies to all advertisements except those broadcast on radio and television (which are subject to a similar code enforced by the Independent Broadcasting Authority), those published abroad, and those relating to medicines advertised to the medical and allied professions which are controlled by the ABPI. Sanctions applied to an advertiser or agency which refuses to amend or withdraw an advertisement include adverse publicity in ASA's *Case Reports*, the withholding of advertising, withdrawal of the agency's trading privileges and notification to other consumer protection agencies.

Proprietary Association of Great Britain (PAGB). This trade association represents the interests and views of manufacturers of non-prescription medicines, i.e. medicines which are sold to the public (sometimes referred to as OTC products). Strict compliance with the Code is a condition of membership of the Association. Companies must submit all advertising copy (except that issued in the trade or professional press) to the PAGB before the advertisement is issued. The copy is examined by a specialist staff of pharmacists and consultant physicians to ensure compliance with the Code. Approval lasts for two years so advertisements are kept in line with current developments in medical and public opinion. Companies must lodge a copy of the relevant parts of the product licence with the PAGB and all advertisements are compared with this to prevent the use of unauthorized claims.

The pre-vetting procedures used by the PAGB Code are obviously preferable to a procedure under which advertisements are only scrutinized after they have been issued, if only because they prevent the unscrupulous advertiser obtaining an advantage over his more conscientious rivals.

The Association regularly monitors publications in which the advertisements are published to ensure compliance with its requirements. The sanctions which could be used against those who fail to comply with the Code are those stated for the ASA.

The British Herbal Medicine Association and the Health Food Manufacturers Association also have a pre-publication vetting procedure.

Independent Broadcasting Authority Code (IBA). The IBA was required by the Broadcasting Act 1981 to formulate, maintain and enforce a code relating to advertising practice and methods. The Act also required the IBA to set up a Medical Advisory Panel consisting of representatives of general and specialist physicians, to advise on those advertisements which relate to medicines, medical and surgical treatments and applicances, and on any toilet articles for which therapeutic or prophylactic claims are made.

(f) Product liability

The Consumer Protection Act 1987 was intended to meet the requirements of the EEC Council Directive on Product Liability (85/374/EEC). Under this Act the producer of a product becomes liable for any damage caused by any defect (as defined in sect. 3) in it. The producer is defined (sect. 1) so as to include the manufacturer of a finished article, or of any raw material or of a component, or anyone who puts his name, trade mark or other distinguishing feature on the article, or who imports it into the Community. The injured person must prove the damage, the defect and the causal relationship between them but not that someone was negligent.

The Directive permitted each Member State to include in its legislation the 'development risk defence'. This enables a producer to avoid liability if the state of scientific and technical knowledge at the time the product was first marketed was such that the existence of the defect could not be discovered. Such a defence would be of great value to an innovative pharmaceutical company which developed a new class of drug or a new type of dosage form which subsequently proved to have a new type of hazard. The Act included such a defence [sect. 4(1)e] but the European Commission has complained that it is too lenient towards producers.

The Act has implications for practitioners who supply drugs to patients because in law they will be treated as producers unless the actual producer can be identified. Professional bodies have given advice to their members, for example, the General Medical Services Committee of the British Medical Association sent a letter to all members (24/3/88) containing measures to be taken to avoid liability under the Act.

Sources of information on medicines

(i) Statutes and statutory instruments

These legal documents may be purchased from HMSO. They state the law as originally enacted but many of them have been amended subsequently although the documents as purchased do not indicate this. Details of these alterations are published from time to time in Halsbury's *Statutes* and in Halsbury's *Statutory Instruments* (see below). The *Current Law Legislation Citator* together with the *Current Law Statutes Annotated Current Service* also gives full references to changes in statutes and statutory instruments but does not contain the text of the changes. Notices drawing attention to changes in medicines law are published periodically in *MAIL* (see below). The *Law on Medicines* (see below) contains the text of the Act and all Orders and Regulations as amended up to October 1985.

Medicines Act 1968

The Medicines

(Control of Substances for Manufacture) Order 1971	No.1200
(Extension to Antimicrobial Substances) Order 1973	No.367
(Control of Substances for Manufacture) Order 1982	No.425
(Control for Substances for Manufacture) Order 1985	No.1403
(Cyanogenetic Substances) Order 1984	No.187
(Manufacturer's Undertakings for Imported Products) Regulations 1977	No.1038
(Child Safety) Regulations 1975	No.2000
(Amendment) Regulations 1976	No.1643
(Fluted Bottles) Regulations 1978	No.40
(Labelling) Regulations 1976	No.1726
Amendment Regulations 1977	No.996
Amendment (No.2) Regulations 1977	No.2168
Amendment Regulations 1981	No.1791
Amendment Regulations 1983	No.1729
Amendment Regulations 1985	No.1558
(Leaflets) Regulations 1977	No.1055
(Data Sheet) Regulations 1972	No.2076
(Amendment) Regulations 1979	No.1760
(Amendment) Regulations 1981	No.1633
(Advertizing of Medicinal Products) Regulations 1975	No.298
(No.2) Regulations 1975	No.1326
(Labelling and Advertizing to the Public) Regulations 1978	No.41
(Advertizing to Medical and Dental Practitioners) Regulations 1978	No.1020
(Applications for Product Licences and Clinical Trial and Animal Test Certificates) Regulations 1971	No.973
(Amendment) Regulations 1972	No.1201
(Amendment) Regulations 1975	No.681
(Amendment) Regulations 1977	No.1051
(Amendment) Regulations 1983	No.1726
(Standard Provisions for Licences and Certificates) Regulations 1971	No.972
(Amendment) Regulations 1972	No.1226
(Amendment) Regulations 1974	No.1523

(Amendment) Regulations 1977	No.675
(Amendment) No.2 Regulations 1977	No.1039
(Amendment) No.3 Regulations 1977	No.1053
(Amendment) Regulations 1983	No.1730
(Exemption from Licences) (Clinical Trials) Order 1974	No.498
(Renewal Applications for Licences and Certificates) Regulations 1974	No.832
(Amendment) Regulations 1977	No.180
(Amendment) Regulations 1982	No.1789
(Fees Relating to Medicinal Products for Human Use) Regulations 1989	No.418
(Retail Sale and Supply of Herbal Remedies) Order 1977	No.2130
(Retail Sale and Supply of Herbal Remedies) Order 1977	No.2130
(Administration of Radioactive Substances) Regulations 1978	No.1006
(Chloroform) Prohibition Order 1979	No.382
(Phenacetin) Prohibition Order 1979	No.1181
(Sale or Supply) (Miscellaneous Provisions) Regulations 1980	No.1923
Amendment Regulations 1982	No.28
(Pharmacy and General Sale Exemption) Order 1980	No.1924
Amendment Order 1982	No.27
Products Other than Veterinary Drugs (Prescription Only) Order 1983	No.1212
Amendment Order 1984	No.756
Amendment Order 1986	No.586
Amendment Order 1987	No.674
Amendment Order 1987	No.1250
Products Other Than Veterinary Drugs-General Sale List Order 1984	No.769
Amendment Order 1985	No.1540
Amendment Order 1987	No.910

Misuse of Drugs Act 1971. Chapter 38. HMSO. The Act contains in schedule 2 a list of the drugs which are controlled under the Act. This list has been amended by a series of Orders as follows:

The Misuse of Drugs Act 1971 (Modification) Order 1973	No.771
1975	No.421
1977	No.1243
1979	No.229
1983	No.765
1984	No.859
1985	No.1995
1986	No.2230
The Misuse of Drugs Regulations 1985	No.2066
(Amendment) Regulations 1986	No.2330
(Amendment) Regulations 1988	No.916

(Notification and Supply to Addicts) Regulations 1973	No.799
(Amendment) Regulations 1983	No.1909
(Safe Custody) Regulations 1978	No.798
(Amendment) Regulations 1974	No.1449
(Amendment) Regulations 1975	No.294
(Amendment) Regulations 1984	No.1146
(Amendment) Regulations 1985	No.2067
The Radioactive Substances Act 1960 (Ionising Radiations Regulations) 1985	No.1333

(ii) Officially produced documents

The following documents have been published by the Medicines Control Agency of the Department of Health and, except where otherwise stated below, are supplied free on request. They are intended to explain in simple terms the legal provisions applicable to a given situation or type of product.

MAL 1 A Guide to the Licensing System (revised September 1984)

MAL 2 Guidance Notes on Applications for Product Licences. [HMSO ISBN 0 11 321024 8 £9.50 (revised February 1986) plus Supplement 1987 ISBN 0 11 321088 £7.50]

MAL 2(PI) Note on applications for Product Licences (Parallel Importing) (Medicines for Human Use) (revised 1987)

MAL 4 Guidance Notes on Application for Clinical Trials Certificates and Clinical Trial Exemptions (HMSO ISBN 0 11 321004 3 £8.75 and Supplement No.1 ISBN 0 11 321036 1 £2.25)

MAL 5 Notes on Applications for Manufacturer's Ordinary Licences (1971)

MAL 6 Notes on Applications for Wholesaler Dealer's Licences (revised 1979)

MAL 7 Licensing Fees (revised September 1987)

MAL 8 A guide to the Status of Borderline Preparations under the Act (revised July 1982)

MAL 18 Licensing Requirements involved in the Packaging and Labelling of Medicinal Products (revised 1980)

MAL 21 Notes on Licensing of Homoeopathic Products (revised 1982)

MAL 25 Notes on Data Sheets (revised May 1984) *MAL 30 A Guide to the Provisions affecting Doctors and Dentists* (revised June 1985)

MAL 32 Clinical Trials using Marketed Products (revised November 1981)

MAL 36 Notes for Guidance on Reproduction Studies (revised 1981)

MAL 39 Products containing Herbal Ingredients (revised 1981)

MAL 41 Additional Notes for Guidance – Biological Medicinal Products (revised May 1982)

MAL 42 Notes on the Medicines (Labelling) Regulations 1976 (revised 1983)

MAL 44 Implementation of the EEC Directives about the Marketing and Manufacture of Medicinal Products (1977)

MAL 45 Notes on the European Community Requirements about the 'Qualified Person' (revised 1982)

MAL 46 Notes on the European Community Requirements for the Importation of Proprietary Medicinal Products (for Human Use) (revised 1982)

MAL 55 Labelling and Advertising to the Public (1978)

MAL 57 Advertising to Medical and Dental Practitioners (revised Aug 1983)

MAL 58 Notes on the Preparation of Summaries of Information for Products subject to the Review Procedure

MAL 59 Hearings and Representations under Part II of the Medicines Act 1968 (revised September 1984)

MAL 60 Colouring Matters permitted in Medicinal Products (revised 1979)

Guide to Good Manufacturing Practice 1983 (HMSO £3.95). This booklet was written by members of the Medicines Inspectorate and contains practical advice on a variety of topics of importance to manufacturers.

Medicines Act Information Letters (MAILs). These are published quarterly by the Medicines Control Agency and sent to licence-holders and to other interested parties. They take the form of a newsletter in which impending changes in regulations or procedures are announced. They also contain a lot of interesting material relating to the work of the Agency.

Notice to Applicants for Marketing Authorizations for Proprietary Medicinal Products in the Member States of the EEC (Commission, Luxembourg 1989).

(iii) Codes of practice

Code of Practice for the Pharmaceutical Industry (ABPI, Revised 6th edition 1986).

Code of Pharmaceutical Marketing Practice 1985 (IFPMA).

British Code of Advertising Practice (BCAP Code) (1985)

Code of Standards of Advertising Practice (PAGB, 1986).

Code of Advertising Standards and Practice Appendix 3. The Advertising of Medicines and Treatments (IBA, 1987).

The Independent Television Companies Association (ITCA) booklet No.6, Medicines, Treatments and Health Claims (booklet, ITCA).

Report on Good Clinical Research Practice (ABPI, 1986).

(iv) Reference books

Harrison, I. H. (1985). *The Law on Medicines* (3 volumes, MTP, £200).

Volume 1. A Comprehensive Guide. This gives an overview of the legal and ethical controls over medicines in the UK.

Volume 2. Licensing and Manufacturing. This contains reprints of those sections of the Act and the Orders and Regulations made under them which relate to the administration of the Act, licensing of products and activities, the packaging and labelling of medicines and their sales promotion. Each reprint includes all the amendments made up to October 1st 1985.

Volume 3. Distribution and Selling. This contains reprints of the legal provisions relating to the sale, dispensing administration of medicines, the quality of medicines and the controls over pharmacies.

Dale, J. R. and Appelbe, G. E. (1989). *Pharmacy Law and Ethics* (4th edn, The Pharmaceutical Press, £14.95).

Halsbury's Statutes (4th edn, Butterworths, 1985, 49 volumes, especially volume 28).

Halsbury's Statutory Instruments (Grey Edition) (Butterworths, 23 volumes, especially volume 12).

International Federation of Pharmaceutical Manufacturers' Associations Summary of Registration Requirements 1988 (IFPMA, £200). This summarizes the legal requirements to be complied with by those who wish to market medicines in foreign countries.

International Drug GMPs (Interpharm Press, 1988, £150). This contains all the latest Good Manufacturing Practice Guidelines and Regulations published worldwide.

Steinborn, L. *Quality Assurance Manual for the Pharmaceutical and Medical Device Industries* (Interpharm Press, £115). This gives an FDA acceptable quality auditing programmeme, directions on maximizing the effectiveness of an audit, guidance on preparing for an audit and a training tool for new auditors.

Anderson, M. *GLP Quality Audit Manual* (Interpharm Press, £110). This provides both the theoretical considerations and practical applications in establishing a Good Laboratory Practice Quality Assurance Unit and in performing GLP Audits. It also contains six checklists which can be photocopied and used in different audit phases as well as specimen FDA acceptable master schedules, study inspection records and similar documents.

(g) Health and safety

The Health and Safety at Work Act 1974 supplemented but did not repeal two earlier statutes namely the Factories Act 1961, and the Offices, Shops and Railway Premises Act 1963. These Acts together with regulations made under one or other of them and various codes of practice are intended to protect workers and the general public from harm resulting from unsafe working methods and conditions. The provisions of these laws are part of the criminal law and both employers and employees may be prosecuted for failing to obey them, even if no one has been harmed. Usually, however, inspectors who detect hazardous conditions or practices use their powers to issue either an Improvement Notice (specifying the nature of the contravention and requiring the person to remedy it within a stated time) or, if the activity complained of involves the risk of serious injury a Prohibition Notice which directs the cessation of the activity.

For the purposes of enforcing this legislation The Health and Safety Executive (HSE) employ inspectors to visit various places of work. The local authorities also employ inspectors to visit catering establishments, offices, shops and certain other premises. Fire authorities also have powers to inspect various places of work.

The Act imposes upon every employer the duty to ensure, as far as is reasonably practicable, the health, safety and welfare at work of all his employees [sect. 2(1)]. This rather general duty is extended and amplified by particular requirements specified in section 2(2), such as the provision of plant and systems of work that are safe and without risk to health, proper instruction and training. He must also conduct his business in such a way that those NOT in his employment are not unduly exposed to risks to their health and safety (sect. 3). The emission of noxious or offensive substances into the atmosphere is controlled under section 5.

The Act imposes duties upon an employee, for example, section 7 requires him to take reasonable care of his own health and safety and that of others, and requires him to co-operate with the employer as regards any duty or requirement imposed upon the latter.

It is an offence for an employer or employee to fail to discharge any of the duties imposed upon him by the Act (sect. 33).

The potential for harm to the health of workers, visitors and consumers caused by unsafe working practices and conditions in pharmaceutical laboratories and factories is considerable and well recognized. The *Orange Guide* published by the Medicines Control Agency is an obvious starting point especially in respect to the safety of the consumer and of the worker. The Medicines Inspectorate will also point out deficiencies in processes and safety practices and suggest remedial measures.

RADIOACTIVE MEDICINES

Special additional provisions apply to these products. Every person who keeps or uses any radioactive material must register his premises with the Department of Environment under the Radioactive Substances Act 1960 (sect. 1). This Act requires that radioactive waste be disposed of only in accordance with an authorization issued by the Minister (sect. 6). Before a premises is registered under the Act, and while the registration is in force, the premises will be inspected by HM Inspectorate of Pollution.

The Ionizing Radiations Regulations 1985 require employers to protect employees and other persons against ionizing radiation arising from work with any source of radioactive substances or equipment. They specify the dose limits to which persons may be exposed, the designation of 'controlled" and 'supervised" areas, and of 'classified persons'. Employers are also required to appoint radiation protection advisers, and to make written rules to enable work to be performed in accordance with the Regulations. Employers must also ensure that all significant doses of radiation received by each employee are measured and recorded. The regulations also provide for the medical surveillance of workers, the safekeeping of radioactive substances, and the maintenance of records of quantities obtained and disposed of.

The transport of radioactive materials by road is governed by a Code of Practice published by the Department of Transport.

The prescription and administration of radioactive medicinal products is restricted to doctors and dentists holding a special certificate.

Sources of information

It would be impracticable in a book of this size to list all the sources of information which might be useful in every situation. Moreover, in addition to the Acts, the regulations made under them and the Codes of Practice mentioned above, there is a considerable body of case law which is very important in interpreting the legal provisions. The company's legal department can give valuable guidance as to the meaning of the various provisions and should always be consulted where there is doubt as to whether some proposed activity or procedure is safe. Specialist advice on health and safety can be obtained from the local inspectors of the HSE and from its Employment Medical Advisory Service (EMAS).

(i) STATUTES AND STATUTORY INSTRUMENTS

Factories Act 1961
Health and Safety at Work etc. Act 1974
Occupiers' Liability Acts 1957 and 1984
Offices, Shops and Railway Premises Act 1963
Radioactive Substances Act 1960

The following statutory instruments are likely to be of particular interest to readers:

Carcinogenic Substances Regulations 1967	No.879
Amendment 1973	No.36

Classification, Packaging and Labelling of Dangerous Substances	
Regulations 1984	No.1244
(Amendment) Regulations 1986	No.1922
(Amendment) Regulations 1988	No.766
Health and Safety (Dangerous Pathogens) Regulations 1981	No.1011
(Emissions into the Atmosphere) Regulations 1983	No.943
(Genetic Manipulation) Regulations 1978	No.752
Ionizing Radiations Regulations 1985	No.1333
(Amendment) Regulations 1986	No.392

(ii) OFFICIALLY PRODUCED DOCUMENTS

The Health and Safety Commission (HSC) have approved or authorized the publication of *Codes of Practice* which may be obtained from HMSO. In addition, the Executive publish *Guidance Notes* in five series namely, chemical series (CS), environmental hygiene series (EH), general series (GS), medical series (MS) and plant and machinery series (PM). These may be purchased either from the HSE or from HMSO. The Health and Safety (Guidance) Series and the Health and Safety (Regulations) Series may also be of interest in certain cases as may some of the *Toxicity Reviews* and *Occasional Papers*.

The HSC issue free of charge a series of leaflets (HSC series) on various aspects of the Act, and the HSE issue (also free of charge) two other series of leaflets [HSE series and IND(G) series].

Among documents of the types mentioned above, the following might be of particular interest to persons working in the pharmaceutical industry.

Approved Codes of Practice

Dangerous Substances – Classification and labelling of substances dangerous for supply
Ionizing radiation – The protection of persons against ionizing radiation arising from any work activity

Guidance Notes

Chemical series
CS/1 *Industrial use of flammable gas detectors*
CS/2 *The storage of highly flammable liquids*
Environmental Hygiene series
EH/18 *Toxic substances: a precautionary policy*
EH/42 *Monitoring strategies for toxic substances*

Medical series
MS/21 *Precautions for the safe handling of cytotoxic drugs*

Health and Safety (Regulations) Series
HS(R) 6 *A guide to the HSW Act*
HS(R) 7 *A guide to Safety Signs Regulations 1980*
HS(R) 11 *First Aid at Work*
HS(R) 12 *A guide to the Health and Safety (Dangerous Pathogens) Regulations 1981*
HS(R) 22 *A guide to the Classification, Packaging and Labelling of Dangerous Substances Regulations 1984*

HSC Leaflets (free)
HSC 2 *The Health and Safety at Work Act 1974: the Act outlined*
HSC 3 *The Health and Safety at Work Act 1974: advice to employers*
HSC 4 *The Health and Safety at Work Act 1974: advice to self-employed*
HSC 5 *The Health and Safety at Work Act 1974: advice to employees*
HSC 6 *Writing a safety policy*
HSC 7 *Regulations, codes of practice and guidance literature*
HSC 11 *Health and Safety at Work Act 1974 – your obligations to non-employees*

HSC Consultative document *Genetic Manipulation Regulations and Guidance Notes* ISBN 011–883–2026.

In addition, a series of *Guidance Notes* have been published by the Genetic Manipulation Advisory Group and later by the Advisory Committee on Genetic Manipulation.

(iii) REFERENCE BOOKS

Croner's Health and Safety at Work (Croner Publications Ltd.) This loose-leaf book is updated on alternate months. It is invaluable to those who have a special responsibility for health and safety within an organization. It contains an excellent summary of the Act, explains the functions of the HSC and HSE, the powers of the inspectors and the nature of the offences. It also contains lists of the statutory instruments in force, the leaflets, guidance notes and other documents published by the HSE. It also contains detailed entries on several hundred topics. Each entry identifies the relevant legal provisions (including any EEC requirements), describes their effects in simple terms and also refers the reader to other sources of information.

Dangerous Substances (Croner Publications Ltd.) This is a two volume loose leaf reference book which is updated bimonthly. It provides concise practical information and advice on both the UK and European requirements covering substance classification, packaging, labelling, storage, and transport. Volume 1 deals with explosives, gases, flammable liquids and solids, oxidizing agents and organic peroxide, and toxic substances. Volume 2 deals with radiations from all sources, corrosives and miscellaneous dangerous products which do not fall within the other classes.

(h) Pollution

This is closely related to the Health and Safety at Work Act when it results from the escape from a workplace of noise, noxious fumes or offensive smells, or toxic products which contaminate water supplies. However, pollution is also covered by other statutes such as the Control of Pollution Act 1974 which deals with the disposal of wastes [as defined in sect. 30(1)]. Special wastes are controlled under the Control of Pollution (Special Wastes) Regulations 1988. They are defined as being dangerous to health or having a flash point less than 21°C. Among these substances are mineral oils, halogenated hydrocarbon solvents, other solvents and wastes containing arsenic or mercury.

Substances used in the manufacture of pharmaceuticals need to be carefully disposed of when no longer needed; corrosives such as phenols are examples. These may cause damage to sewers, to persons working in sewers and may inhibit sewage treatment. They should not be poured down sinks and drains and spillages should be notified to the appropriate water authority.

The Health and Safety (Emissions into the Atmosphere) Regulations 1983 specifies the substances deemed to be 'noxious or offensive'.

Advice on matters relating to pollution can be obtained from the Environmental Health Departments of Local authorities, from the Department of Environment and from companies who specialize in the collection and disposal of waste products. In addition, written information is available in the forms mentioned below.

(i) STATUTES AND STATUTORY INSTRUMENTS

Clean Air Act 1956	
Clean Air Act 1968	
Control of Pollution Act 1974 (Special Waste)	
Regulations 1980	No.1709
(Amendment) Regulations 1985	No.1884
Collection and Disposal of Waste Regulations 1988	No.819

(ii) OFFICIALLY PRODUCED DOCUMENTS

The Department of Environment have published a series of *Waste Management Papers* (*WMP*). These may be purchased from HMSO.

WMP 19 *Wastes from the Manufacture of Pharmaceuticals, Toiletries and Cosmetics – A Technical Memorandum on Arisings, Treatment and Disposal including a Code of Practice*

WMP 23 *Special Wastes: a technical memorandum providing guidance on their definition*

WMP 25 *Clinical wastes*

Hazardous Waste Disposal Guide. (Croner Publications Ltd., £25). This contains information on the legal aspects of the disposal of hazardous wastes, the options available and on the contractors and consultants who may provide services and advice.

European Economic Community law

There are four sources of Community law, namely, the Treaties (especially the Treaty of Rome); the Acts of the various institutions, including regulations, directives and decisions; judgements of the Court of Justice of the Communities (these are not binding); legislation and decisions of municipal courts of member states on matters of Community law. The recommendations and opinions of the Council and the Commission are not legislative. Regulations are published in the *Official Journal* (*OJ*) and are directly applicable in all member states. Directives are binding upon the states to which they are directed but allow the state to choose its own method of implementation. Decisions are binding on those to whom they are directed, e.g. governments, companies. Directions and decisions are usually published in the *Official Journal*.

In addition, there are Council Recommendations but these are not binding upon governments, companies or individuals.

EEC law takes precedence over the laws of its member states who must ensure that their domestic requirements comply with those of the EEC. Nevertheless, there are slight differences in the requirements for marketing authorizations in the various member states.

It can be difficult to ascertain what EEC legislation has been enacted relating to a particular topic and to keep abreast of changes to it. Perhaps the best starting point is the *Directory* (see below) and the headings most likely to be of interest to readers would be *Proprietary Medicinal Products*, *Dangerous Substances* and *Ionising Radiations*. The *Directory* is revised annually and is printed in 2 volumes.

Directory of Community Legislation in Force and the Other Acts of the Community Institutions (11th edn, European Commissions, Brussels 1988. ISBN 92–77–36080–1 (the two volumes).

Documents of pharmaceutical interest

65/65/EEC Council Directive of 26 January 1965 on the approximation of provisions laid down by law, regulation or administrative action relating to proprietary medicinal products. The Directive was published in the *Official Journal* volume 22 of 9 February 1965, p. 369. It has been amended three times and the amendments may be found in *OJ* L 147, 9/6/75, p. 13 (the Second Directive, 75/319/EEC), *OJ* L 332, 28/11/83, p. 1 and *OJ* L 15, 17/1/87, p. 36.

75/318 EEC Council Directive of 20 May 1975 on the approximation of laws of Member States relating to analytical, pharmacotoxicological and clinical standards and protocols in respect of the testing of proprietary medicinal products. *Official Journal* L 147, **18**, 9 June 1975 9/6/75, pp. 1–22.

75/319/EEC Second Council Directive of 20 May 1975 on the approximation of provisions laid down by law, regulation or administrative action relating to proprietary medicinal products, *OJ* L 147, 9/6/75, p. 13, amended twice, 378 L 420 *OJ* L 123, 11/5/78 p. 26 and 383 L 570 *OJ* L 332, 28/11/83 p. 1.

75/320/EEC Council Decision of 20 May 1975 setting up a pharmaceutical committee, *OJ* L 147, 9/6/75, p. 23.

78/25/EEC Council Directive of 12 December 1977 on the approximation of the laws of Member States relating to the colouring matters which may be added to medicinal products, *OJ* L 011, 14/1/78, p. 18, amended, *OJ* L 183, 4/7/81, p. 53.

87/22/EEC Council Directive of 22 December 1986 on the approximation of national measures relating to the placing on the market of high-technology medicinal products, particularly those derived from bio-technology, *OJ* L, 17/1/87, p. 38.

67/548/EEC Council Directive of 27 June 1967 on the approximation of laws, regulations and administrative provisions relating to the classi-fication, packaging and labelling of dangerous substances, *OJ* L 196, 16/8/67, p. 1, amended many times.

76/759/Euratom Council Directive of 1 June 1967 laying down revised safety standards for health protection of the general public and workers against the dangers of ionizing radiations, *OJ* L 187, 12/7/76, p. 1, amended *OJ* L 83, 3/4/79, p. 18.

COUNCIL RECOMMENDATIONS

83/571/EEC Council recommendation of 26 October 1983 concerning tests relating to the placing on the market of proprietary medicinal products. This amplifies Directive 75/318.

87/176/EEC Council recommendations of 9 February 1987 concerning tests relating to the placing on the market of proprietary medicinal products.

OTHER PUBLICATIONS

Understanding Pharmaceuticals in the EEC (1987, Scrip, £80). Among numerous other matters of interest to manufacturers of pharma-ceuticals, this report describes the way in which Community institutions work and their influence upon the registration of drugs and high-technology products, GLP, GMP, product liability, information to doctors and consumers and OTC medicines.

O'Donnell P. (ed.) (1986). *Eurodrugs Step-by-step*. (Scrip, £60 PJP Publications). This manual explains the background and context of the EEC's new multistate drug regulation procedure. Designed to help companies prepare applications under this system, it contains all the official texts relating to the use of the system. It also contains a step-by-step guide to compiling and submitting applications, details of the new format and the role of experts and expert reports, a guide to the acceptable languages and other details of applications.
Product Liability in Europe (Scrip, £100).

European countries

The individual member states of the EEC have their own legal requirements and information on these can be obtained using the *IFPMA Summary*. More detailed information can sometimes be found, for example, Scrip produces English translations of various foreign requirements, such as: *French GMP Requirements* (£35, Scrip BS), *Austrian Medicines Act* (£50, Scrip BS) and *Austrian Enabling Decrees* (£55, Scrip BS).

United States

In the US the responsibility for approving the marketing of drugs (whether home-produced or imported) lies with the Food and Drug Administration (FDA) which also monitors the use of drugs after approval. The FDA acts in accordance with the Food, Drug, and Cosmetic Act 1938 and various regulations which are published in the Federal Register. The basic control is very similar to that described above for the United Kingdom but there are some quite important differences.

Whereas in the UK a potential product becomes subject to control only when it is to be tested on patients, in the US even trials on healthy volunteers are subject to control. In order to obtain permission for human volunteer tests the company must submit a form known as a Notice of Claimed Investigational Exemption for a New Drug and the drug is thereafter referred to as an Investigational New Drug (IND). The company must disclose on the form the composition of the product, its source, method of manufacture and results of all animal tests that have been performed to show that the product is potentially useful and not likely to expose the test subjects to unreasonable risk. The form must also include a protocol describing the plan for human testing. The company may begin these tests after the lapse of 30 days unless the FDA have requested a delay because of a potential safety problem.

The IND tests include those which in the UK would be covered by a clinical trial certificate. When the company is satisfied with the results of these tests, it applies for approval to market the drug by filing a New Drug Application (NDA). This contains all that the company knows about the drug and is accompanied by samples of the product and specimen labels. These are examined by the appropriate division in the Bureau of Drugs at the FDA. Each division includes pharmacists, chemists, doctors, and others skilled in assessing new drugs. There are six divisions each dealing with a class of drugs namely, cardio-renal, neuropharmacological, metabolism-endocrine, anti-infective, oncology and radiopharmaceutical, and surgical-dental. The divisions also use advisory committees of experts from outside the FDA.

The FDA approve the drug if it considers that when used properly, the benefits of the drug outweigh its risks.

Particular attention is paid to the labelling and the packaging insert. This latter describes in detail the nature, mode of action, therapeutic indications, adverse reactions, route and dose and other relevant information about the drug. It must be consistent with information given in the NDA and with the label. It is only removed from the pack when the drug is dispensed to the patient. Some products also contain a Patient Package Insert (PPI) which provides useful information for the patient, and some products such as the oral contraceptives are legally required to contain a PPI carrying certain warnings.

When the NDA has been approved, the product may be marketed. The manufacturer must keep batch production records and records of its safety and effectiveness. Any information which indicates that the product may cause unexpected problems must be reported to the FDA. The manufacturer may be required to perform post-marketing studies to determine the incidence of serious adverse reactions, or to ascertain the safety or efficacy of the product in a patient group not adequately studied.

The manufacturer must notify the FDA of any proposed change to the manufacturing process or to the labelling of the product. This is done by means of a supplemental NDA.

The FDA may withdraw approval from a product which is subsequently found to have serious unexpected side-effects or to be less safe or effective than was expected. The company is asked for its views before action is taken.

Antibiotics and also insulin are batch tested by the FDA before being released for sale.

The FDA controls the advertising of prescription drugs to doctors. All information supplied must be truthful and balanced. If promotional material is misleading, the product can be seized as 'misbranded' though this is seldom done. More frequent use is made of the remedial advertisement (of the same size, in the same journals for the same length of time as the original advertisement) or the remedial letter sent to all physicians.

The system described above applies to all medicines involved in interstate commerce. It is possible to manufacture and sell products not federally approved within a state if the Pharmacy Board of that state permit. Information as to what is permitted within a state can be obtained from the *Newsletter* of the relevant Pharmacy Board.

An 'orphan drug' is a drug or biological product intended for a disease or condition which occurs so infrequently in the United States that there is no reasonable expectation that the development costs could be recovered from sales in the United States. To encourage the exploitation of these drugs, the Food, Drug, and Cosmetic Act was amended by adding sections 525, 526 and 527 in 1986. Anyone interested in developing such a drug may ask the FDA to provide written recommendations for the animal and clinical studies necessary to permit marketing approval of the drug. Moreover, the manufacturer (or other sponsor) can request the FDA to designate the drug as an orphan and if this is granted, the applicant is entitled to tax credit for the clinical trials involved and a 7 year monopoly on its use for the purpose. Inquiries should be directed to the Director of Orphan Products Development, FDA, 5600 Fishers Lane, Rockville, MD 20857.

Sources of information

Code of Federal Regulations – Food and Drugs 1988, especially:
Parts 1–99 General matters
Parts 200–99 Drugs: general
Parts 300–499 Drugs for human use
Parts 600–680 Biologics
Parts 1300–16 Controlled substances
Neilsen, J. R. (1986). *Handbook of Federal Drug Laws* (Lea and Febiger). This is widely used as a text for pharmacy students and provides an excellent overview of the subject.
Allen, M. E. and Gerlis, L. (eds) (1987) *The Scrip Guide to US Good Clinical Practice* (£65). The FDA does not publish guidelines on good clinical practice (GCP) because many of the requirements are still developing. This report is intended to provide the information necessary to ensure that clinical trials meet the GCP requirements. It outlines the rules of GCP, instructs in the use of Standard Operating

Procedures (SOP) and provides a complete SOP for monitoring with specimen forms, gives advice on informed consent and ethical matters and on the reporting of ADRs.

Scrip Bookshop can also supply copies of the following:

US FDA Good Laboratory Practice Inspection Guidelines (£25).
Process Validation Guideline 1985 (£15).
Guidelines for the Clinical Evaluation of NSAI and Antirheumatic drugs (Adults and Children) 1986 (£20).
Pre-approval guidance for Drug Advertising and Labelling 1986 (£10).
Draft Guideline for Postmarketing Reporting of Adverse Drug Reactions (£20).
Draft Guideline for the Format and Content of the Human Pharmacokinetics Section of an Application (£15).
Draft Guideline for the Submission in Microfiche of the Archival Copy of an Application (£15).
Draft Guideline for Formatting, Assembling, and Submitting New Drug and Antibiotic Applications (£15).
Draft Guideline for the Format and Content of an Application Summary (£15).
Draft Guideline for the Format and Content of the Statistical Section of an Application (£15).
Draft Guideline for the Format and Content of the Microbiology Section of an Application (£10).
Draft Guideline for the Format and Content of the Nonclinical Pharmacology/Toxicology Section of an Application (£15).
Draft Guideline for the Format and the Chemistry, Manufacturing, and Controls Section of an Application (£10).
Draft Guideline for the Format and Content of the Clinical Data Section of an Application (£35).

Other countries

Information on the legal controls imposed upon medicines in other countries can be found in the IFPMA *Summary of Registration Requirements*. The Department of Trade and Industry can also provide information and help to those wishing to export medicines.

The Institute of Export, 64 Clifton St, London EC2A 4HB also supplies information to members and on a consultancy basis to non-members.

Other sources of advice

Advertising Association, Abford House, 15 Wilton Road, London SW1V 1NJ.
Association of the British Pharmaceutical Industry, 12 Whitehall, London SW1A 2DY.

Association of Information Officers in the Pharmaceutical Industry, Pharmaceuticals Ltd., Syntex House, St. Ives Road, Maidenhead, Berks. SL6 1RD.

British Herbal Medicine Association, PO BOX 304, Bournemouth, Dorset BH7 6JZ.

British Institute of Regulatory Affairs, Drayton House, 30 Gordon Street, London WC1H 0AX.

Department of Health, Medicines Division, Market Towers, 1 Nine Elms Lane, Vauxhall, London SW8 5NQ.

Department of Trade and Industry, Overseas Trade Division, 1 Victoria Street, London SW1H 0ET.

Home Office, Queen Anne's Gate, London SW1H 9AT.

The Royal Pharmaceutical Society of Great Britain, 1 Lambeth High Street, London.

Proprietary Articles Trade Association, 4 Margaret Street, London W1N 7LG.

Proprietary Association of Great Britain, Vernon House, Sicilian Avenue, London WC1A 2QH.

Overseas Pharmaceutical Organisations – see *Chemist and Druggist Directory 1989*, section D9 for addresses.

British Homoeopathic Manufacturers Association, 19A Cavendish Square, London W1M 9AD.

CHAPTER NINE

Post-marketing experience with drug products

H. TILSON, G. E. COLLINS and R. SIMPSON

Introduction

The completion of the pre-approval development programme
represents a milestone of great importance in the history of a new
pharmaceutical entity. However, by no means does it represent
the end of the scientific work on the material; indeed it marks the
launch of a whole new series of scientific challenges, not to
mention many of the other related tasks required for professional
support of a marketed product. There is a vast amount to be
learned about pharmaceuticals following their approval for
general clinical use: information which is gained through a wide
net of on-going formal and informal clinical experimental trials
(including trials not known to the manufacturer); through a
burgeoning literature of case histories, published studies, case-
series and other drug reports; through a continuing programme of
scientific endeavours from the manufacturers themselves; and
through the practical experience of prescribers and dispensers.
Capturing this information; analyzing it in a scientific and
responsible way; turning it around so that the people who need to
know can get the information they need in the best way for them;
and incorporating it into the on-going process of further drug
understanding . . . these are the tasks of the Drug Information
Manager in the pharmaceutical industry (*Figure 9.1*). This exciting
and rewarding challenge is portrayed in this chapter.

The 1980s have brought three trends which converge to make
information management a new and vitally important field: a
changing society which more and more values (perhaps even
demands) and which understands how to create accurate, timely
information; a changing climate which [in the face of a major
regulatory landmark, the 'NDA Rewrite' governing the require-

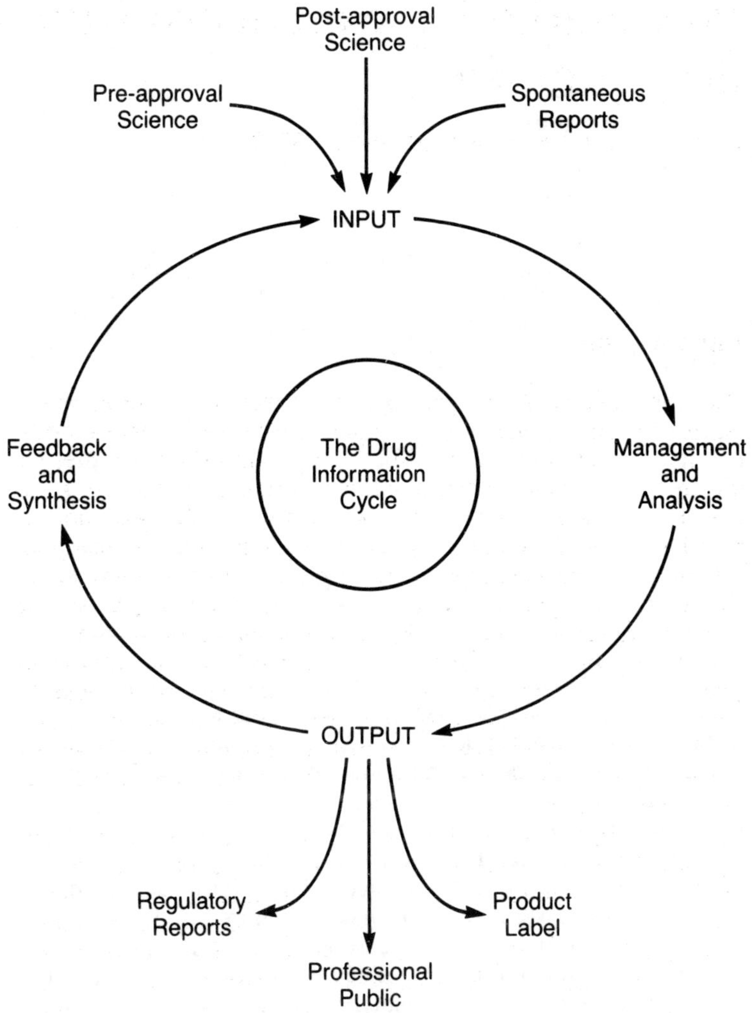

Post-approval
Science

Pre-approval
Science

Spontaneous
Reports

INPUT

Feedback
and
Synthesis

The Drug
Information
Cycle

Management
and
Analysis

OUTPUT

Regulatory
Reports

Product
Label

Professional
Public

Figure 9.1. The drug information cycle

Table 9.1. Translating the clinical experience

Inputs	Trials/studies
	  Internal
	  External
	Epidemiologic Studies/Post-marketing Surveillance
	Spontaneous reports
	Published literature
	Abstracting/Bibliographic Sources

Processes	Searching/Scanning/Case-finding Database creation Processing/Data management Analysis Synthesis Recommendations Communication

Outputs	Reports to
	  FDA
	  Investigators
	  Published Literature
	  File
	Letters
	Calls
	Label change
	Company research archive
	Lectures/Seminars

ments for professional support of drugs approved under the New Drug Approval System (NDA) in the United States and other parallel efforts abroad] demands and even provides guidelines for regular periodic reporting and updating of the World's insight into all marketed products from the manufacturers; and rapidly changing information management technology, rooted in computer storage and searching techniques and stimulated by the remarkable, fast and comprehensive in-house information analysis and reporting systems which make turnaround of enormous quantities of information both feasible and affordable.

The information in this chapter while drawing its examples from the experience of the American branch of one pharmaceutical manufacturer, the research-based Wellcome Foundation, should help drug information managers in industry, academia, government or practice in developing and operating the systems ('processes') needed in the 1990s to handle the various information 'inputs' and convert them into useful professional 'outputs' (*Table 9.1*).

The inputs

New challenges/new trials

Following the approval of a new chemical entity for marketing, there are still many more questions to be answered than the highly-focused milieu of the pre-approval clinical trial will permit. The process of drug development is a continuum in which approval to market is but a landmark, albeit a most important one.

Following approval, the nature, type and scope of clinical studies conducted by the manufacturer will include true experimental uses (e.g. new indications; new combinations, formulations and dosages; new populations, e.g. pediatric use). The programme of research will also address therapeutic utility, commercially useful direct experience surveys and comparisons with other therapies and other studies, most notably those which enhance understanding of the product for potential users and/or portray their cost effectiveness for competitive formulary decisions. Throughout this process, new information is learned which must be channelled into the drug information cycle properly and promptly. By the very nature of this information, however, it may be highly transitory, changing rapidly as new information unfolds. Often the primary 'input' into the cycle rests heavily (or even exclusively) upon the personal, professional colleague-to-colleague relationship between the Drug Information Manager and the research scientist.

Spontaneous voluntary adverse reaction reports

Perhaps the single most important area in which the incomplete information regarding a new drug at the time of approval may have major serious public policy ramifications is the rare but serious adverse event occurring in association with the clinical use of a drug. A typical clinical trial programme, comprising perhaps 3,000 patients, cannot be expected to detect rare, but possibly unacceptable drug toxicities. Following drug approval and for many (particularly much needed), highly effective novel compounds, the numbers of people receiving the drug will increase one, two, or even three orders of magnitude over a very few months. Further, persons with more complicated medical situations, (i.e. more concomitant medications, highly complicated, often severe concomitant illnesses) or even demographic populations who would not receive the drug in clinical trials (e.g. small children, pregnant women, the very old and infirm) will receive the new drug. The most powerful sentinel of a problem is the awareness of the health practitioner who, detecting something unusual in the therapeutic experience, has the wisdom and conscience to register the event through one of a series of drug information sensing systems. In the US, perhaps the most effective and extensively used of these is the spontaneous voluntary reporting to the pharmaceutical manufacturer. Over 80% of all new reports of adverse experiences following marketing in the US come first to the attention of society through a report to the manufacturer who, under FDA regulations and good public policy practices, will report these on to the regulatory authority and, as appropriate, include them in its external communications. Similar sensing systems which rely on spontaneous voluntary reporting of adverse drug experiences (ADEs) exist in many advanced nations. The best developed and most well regarded of these is the yellow card system in the UK. Under this system, physicians are exhorted to report all adverse experiences in association with new drugs and all potentially important adverse experiences with all drugs directly to the Committee on Safety of Medicines (CSM). Thus, in contrast to the US in which the industry is the central figure in this information gathering, in the UK it is the regulatory authority. Recently, through creative negotiation between industry and government, an active and timely method of notifying the manufacturer of the *ADE* database with the use of an 'anonymized single patient profile' (on microcomputer disks or hard copy) has been developed which can enhance the role of the industry-based Drug Information Manager with more current and useful information. Other nations have progressively followed suit in being more aggressive about soliciting reports directly from physicians though reporting rates from the field vary.

Albeit one of the most powerful signaling systems imaginable, sampling as it does from the entire 'universe' of treatments in the nation as the 'denominator', voluntary reporting has its weaknesses. For every health-care provider who does report, there are dozens who don't; for every instance which is reported, there are dozens that aren't; the pressures and stimuli (and therefore biases) which drive the reporting of one but not another reaction, under one but not another set of circumstances, by one but not another provider vary widely from drug to drug, patient population to patient population, time to time, nation to nation. Further, these biases are often subtle and difficult, if not impossible, to detect and/or quantify. This, in turn, precludes use of voluntary data to make quantitative comparisons across drugs, drug formulations or drugs and indications over time. This can only be done by proper epidemiologic studies. It is essential for the Drug Information Manager to understand these fundamentals to preclude transmission of this 'epidemiologic intelligence' as though it were an established set of morbidity and mortality rates, to know the differences between report ratio and true incidence!

Post-marketing surveillance epidemiologic studies

In addition to an ongoing programme of monitoring of spontaneous reports and clinical experimental studies, progressively an emerging source of important new drug information is the structured non-experimental observational epidemiologic study. The emergence of the era of pharmacoepidemiology in the 1980s has been made possible by the same expanding technology which has produced other drug information revolutions.[1] Proper epidemiologic studies of drug safety following approval require the establishment of a reliable drug history and an equally reliable set of data regarding health outcomes related to drug exposure in general clinical use. Such observational epidemiologic studies linking drug exposure to health outcomes are, of course, not new. However, in prior eras, these studies have been in relative disrepute because of high costs and difficulty in methodology. Following populations large enough over periods long enough to learn enough about drug/outcome associations to detect important but rare adverse events has been fraught with enormous costs and insurmountable biases. The emergence of the large automated database has changed all that. Much of this progress has occurred in the US because of its unique continuing emphasis upon non-governmental means of funding and organizing medical care services. During the 1970s and prominently in the 1980s, health plans, e.g. prepaid health plans and medical insurance schemes, have developed automated billing or accounting systems for diagnosis (e.g. all hospital discharge diagnoses), health plan membership (e.g. a single unduplicated plan number for each person), and pharmaceutical billing and management systems (i.e. a database with every prescription, dose, date and patient identifier). Through the use of a technique which allows the computer to link data from various sources, data regarding every member of large populations known to have received a prescription for a drug can be assembled, essentially without any but the most minor costs, from these automated pharmacy management records; and their subsequent medical experience likewise can be 'linked' (e.g. using a membership number) with hospital discharge diagnoses from their database. Through these linkage studies, drug firms, in collaboration with academic and regulatory researchers, can retrieve information about drug event associations in large populations and learn much earlier and much more accurately about unexpected excesses of important medical events, both adverse and beneficial.

The Drug Information Scientist will need to become progressively familiar with this emerging body of science, the growing body of scientists applying it, and the enlarging circle of resources for automated data upon which the science may be applied.

Perhaps the most noteworthy of the large automated linked datasets is that describing health-care transactions in a large, prepaid group practice (health maintenance organizations or HMO), the Group Health Co-operative of Puget Sound in Seattle, Washington. The database covers the medical experience of over 330,000 enrollees since the automation of the pharmacies in 1976, the hospital discharge diagnoses for the plan and the membership files having been automated some time prior. Perhaps the best known researcher in this field is the pioneer drug epidemiologist who developed the techniques and applied them at the Group Health Co-operative of Puget Sound, namely Professor Hershel Jick, the Director of the Boston Collaborative Drug Surveillance programme. This programme has issued hundreds of reports of associations between drug exposure and important medical outcomes.[2] Other important health-care organizations in which similar linkable data can be found in America include prominently the West Coast health maintenance organizations operated independently under the umbrella of Kaiser-Permanente (Portland, Oregon, Los Angeles and most recently, San Francisco, California) although similar data are available in other HMOs in America in Minneapolis, Boston, and elsewhere. Insurance payment schemes contain similar data regarding drug claims and medical experience claims paid. Notable among these is the province-wide data system for Saskatchewan in Canada, covering the experience of a million provincial residents again since the mid-1970s. Also of major interest are the databases from the medical claims payment systems of America's public-financed system for payment for medical care for the poor, the so-called Medicaid programme. Pioneered under leadership from (and financing by) the FDA, the *COMPASS* database is the largest and oldest of these systems, now describing medical experiences which can be linked with pharmaceutical exposure in a population of more than 6 million in more than 10 States.[3] More recently, several substantial investigators have determined the feasibility and desirability of working more closely with a single Medicaid programme, particularly because of the vital need for medical records to validate electronic signals of association and the requirement for a personal relationship between researcher and data resource in order to achieve this. Medicaid data from the states of Tennessee, California, Maine, New Jersey and Maryland are among those with individual researchers now collaborating. Researchers currently involved in the field using Medicaid data include Drs. Brian Strom, University of Pennsylvania; Alexander Walker; Jerry Avorn, Harvard University; Richard Moore, John Hopkins and Jeff Carson, New Jersey. Perhaps the best known and longest involved is Dr. Wayne Ray of Vanderbilt University with his colleagues.

More recently, active interest has developed in the UK and on the continent in harnessing similar computer payment and accounting systems to those users. Most notable is a promising MEMO system (Dundee Scotland), currently processing prescription data manually until suitable automation of prescription and patient identifiers can be arranged to permit linkage with existing automated morbidity information. Linkage of claim payment data is also under active discussion in Sweden, Germany and elsewhere.

Other epidemiologic resources are also important for the drug information specialist. For much as the large automated linked datasets contribute to the field, many questions can still only be answered using the 'traditonal' epidemiologic approach. Thus, the Slone Epidemiology Unit at Boston University under Dr. Syd Shapiro and many distinguished departments of epidemiology at universities particularly in the US, Canada, and the UK still conduct epidemiologic studies associating drug exposure with increased risk of adverse outcome. Several particularly creative approaches have emerged in the past decade. Notable among these are pharmacy-based patient registration programmes (e.g. Upjohn Pharmaceuticals, Purdue University); hospital-based drug information pharmacy networks (most notably co-ordinated through Buffalo University); automated medical record systems, in which prescription and diagnostic data relating to visits are all entered into a computerized database at the time of the physician's visit (Harvard Community Health Plan, Regenstreit System in Indiana, and the VAMP Network in the UK); and perhaps most notable, the unique prescription event monitoring system (PEM) of Dr. W. H. W. Inman at Southampton.[4] Professor Inman's system involves capturing all new prescriptions in the prescription pricing authority in the UK, notably for newly approved drugs. These data are entered along with the physician and patient identifier into a large ad hoc database. A note is sent to the participating physician requesting that he/she fill out a 'green card' describing significant events which may have occurred in that patient's life subsequent to receipt of a prescription, irrespective of the physician's index of suspicion regarding causality.

The published literature

An inventory of 'inputs' into the drug information process would, of course, scarcely be complete without a strong emphasis upon the published literature. From the very earliest phases of drug discovery (in which basic findings in pharmacology and toxicology relating to new products are extensively reported in the preclinical literature); through drug development (in which often dozens of papers will appear); through the period following approval (in which the numbers of papers increase by one or two orders of magnitude quickly after the initial dip following approval for 'successful' or 'interesting' products), the aggregate experience will unfold. Altogether the 'literature' in which an important article relating to a new pharmaceutical product might be found comprises thousands of journals with many thousands of published pages every year in dozens of languages from hundreds of countries. These are available to any pharmaceutical house directly on subscription and through multiple professional database abstracting services. These systems and services are extensively described elsewhere in this book.

The process

Building the database

Information vital to understanding and translating the experience with pharmaceuticals to the various audiences with a 'need to know' requires to be managed. The management of such information demands real organization, and has really only become possible through the advent of computers. Indeed in the 1970s, a spiralling of expectations occurred as computerization made available more information and more details faster and in a more manageable form; and this, in turn, created the expectation that more such information would be assembled in a manageable and managed form and would be reported to potential users in a more timely and comprehensive way. Under the provisions of the annual report requirements of the New Drug Application provisions of Federal Regulation (August 22, 1985) in the US, each year each company must summarize 'all important new information' regarding its products. This then requires a process throughout the year of accumulating in manageable form incremental progress reports on all on-going clinical trials and post-marketing surveillance studies. Under separate provisions of the same regulations, all adverse drug experiences reported spontaneously to companies must be analysed (quarterly for the first three years and annually thereafter) for possible increases in frequency in expected but serious complaints as well as important patterns in all complaints which might result in changes in company educational practices (including the official labelling). To support the last of these requirements (those regarding adverse experience reports), most of the major US companies have turned to automation, for example, Burroughs Wellcome Co. has assembled a powerful and extensive database using the database management system, *INQUIRE*. This system (*BW ADVERSE*) permits assembly of a dataset on each spontaneously reported adverse experience case comprehensive enough to complete officially mandated federal forms in the US (the FDA 1639) and, for official reports from elsewhere in the world, on the now more broadly accepted international adverse reaction report form (CIOMS Form). The systems have been described elsewhere.[5] An important emergent set of actors in the information management of pharmaceuticals is a growing body of epidemiologists in industry trained in managing large automated databases and recognizing, analysing, reporting and developing hypotheses from important epidemiologic signalling systems such as the spontaneous voluntary adverse reactions report.

Managing the enormity of the published literature, the PRODBIB system

During the 1970s and even more in the 1980s the modern pharmaceutical firm has created a large automated database to support its management of the burgeoning volume of published information. For example, in 1972 the Burroughs Wellcome Co.[6] used *INQUIRE*, a database management system from Infodata Systems Inc., to handle a bibliographic database into which bibliographic information, subject indexing, and text could be entered. Prior to its development the company had collected some 25,000 published product papers. These papers have been organized by drug name and then by author. There was no subject access. Since 1972 some 100,000 records (if two or more products are mentioned in a paper a record is created for each product) representing 75,000 papers have been added to the database. Only that part of the paper that is relevant to the product is indexed. A keyword-in-context approach is used, with indexing terms assigned to fields, (e.g. indication, concomitant therapy, therapeutic comparison, population, dosage, adverse experience, etc.). Numeric fields hold population information and occurrences of adverse clinical events. There are also free text fields for a therapeutic efficacy statement and an abstract.

Harnessing adverse reaction cases in the literature (looking for the bad to maximize the good)

The importance of the journal literature is without question. There is review by peers prior to publication and then there is opportunity for editorial comment and verification by others after publication.

Recent studies suggest that the leading form of continuing education is reading the journal literature.[7] Osiobe says that 'the use of medical journals take predominance over other sources as the first source of learning about advances in medicine and new drugs among health professionals'. Studies have also shown that most physicians become aware of new therapeutic advances via the journal literature (Osiobe, 1985). Kenny (1977) stated that 'collection from the world literature is an essential backup to the company's activities'.[8]

Identifying drug information in the literature via commercial bibliographic databases is dependent upon the indexing philosophy of the service and the significance of the drug in relation to the focus of the paper. Both the quality and quantity of references concerning pharmaceuticals is governed by these database producers. In order to assure proper coverage of the literature, pharmaceutical companies must scan the primary literature for references to their products and maintain in-house databases.

Commercial databases do not perform as well as an in-house database in handling a company's product literature and adverse clinical events. Of the papers published on a particular product around 33% are unique to an in-house database compared to up to 7% that would be unique in a commercial database. In addition to problems in identifying a particular drug there are also other problems that relate particularly to the reporting of adverse clinical events. Some of these are: the sheer volume of the literature; omission of basic details such as age, sex, dosage, symptom, etc.; multiple and pyramidal reports of the same patients; and no standard method for analyzing and reporting. In addition many literature reports are difficult to interpret. It is important to identify any event that coincides with the use of a particular drug and many times it is difficult to separate an adverse drug experience from the natural progression of a disease especially if the author(s) don't state their opinion regarding cause.[8] An in-house system must be finely tuned to record any event that may have only the slightest relationship to the drug therapy.

For example, the association of one Burroughs Wellcome drug in particular (we'll call it BMX) with the development of pancreatitis is an important and well known adverse clinical event from which much can be learned about literature signals. Of 137 papers currently reporting this association only 58 describe occurrences that had not been previously described earlier elsewhere. These are termed primary papers. The remaining 79 are previously reported occurrences or reports in reviews and surveys (i.e. secondary reports). Secondary reports in many instances may make a particular adverse clinical event appear more common than it actually is.

If one examines the reporting of the Drug X/pancreatitis association, it is found that the first association was published in 1966. This primary report was followed by three secondary reports which were published between 1967 and 1969. The second primary report was not published until 1970. From 1966 to 1988 there were 21 years in which a paper was published that reported the Drug X/ pancreatitis association (none in 1968), in 11 of which secondary reports outnumbered or equalled the primary reports. If world-wide spontaneous reports (not published but reported directly to a pharmaceutical company), are examined the first occurrence was noted in 1971 and 14 more in 1974, 1975, 1984, 1985, and 1986. Included in the 58 literature reports are 144 occurrences of pancreatitis compared to 15 individual spontaneously reported cases. Thus in looking at the two arms of voluntary reporting, literature and spontaneous, the literature arm produced almost 10 times more individual cases than the spontaneous arm.

The differences in scope and focus of an in-house database compared with commercial databases can be examined in the following way. Of the 58 primary papers reporting this association, only 35 could be retrieved from *MEDLINE, BIOSIS PRE-VIEWS*, or *EMBASE*. The remaining 23 papers lacked the drug name or pancreatitis or both in the databases' records. The remaining papers were identified by in-house scanning of the literature and by the detailed examination of papers that mentioned Drug X.

The FDA has obliged pharmaceutical companies to examine the literature for serious adverse experiences that are not included in the product's labelling. In a one-month period in 1988 59 papers were identified in the literature that meet this requirement. Of the 59 papers, 53 were identified by commercial sources. The remaining 6 were identified solely by in-house scanning of the literature. Of the 53 that were identified by commercial sources, the drug appeared in the title in 73% of the cases. 68% of the papers were single patient case reports. Venning[9] examined the types of studies in which 18 important adverse drug experiences were initially identified. Of the 18 only three were from cohort studies; the remainder were case studies or anecdotal reports. Venning goes on to criticize the UK yellow card system and hospital schemes such as 'Pre-Group Health' the Boston Collaborative Drug Surveillance program, for failure to identify first alerts. Eight of the 18 adverse reactions were identified after the UK yellow card system and the Boston programme were in effect (1964 and 1966 respectively) without the benefit of these schemes. This is in agreement with Lortie who stated that 'anecdotal reports are the backbone of an adverse surveillance system'.

Being able to identify only 61% of serious labelled clinical events (Drug X/pancreatitis) and failing to identify 10% of the serious unlabelled clinical events using commercial databases demonstrates the differences in indexing philosophy and scope that the pharmaceutical industry requires. A company's product literature system cannot ethically rely only on the commercial databases for product information.

Epidemiologic techniques for analyzing the published literature

'SIGNALLING'

A fundamental obligation of the reader of published literature, an obligation which is generalized to the pharmaceutical industry, is to recognize the important problems with medications described in an article; to identify, analyze, and obtain more data if necessary; and to report important new problems just as though the information had occurred in a spontaneous voluntary adverse reaction report telephoned into the pharmaceutical company. Indeed, the FDA rules and, progressively, rules from around the world, require such vigilance. In the US, the 15-day alert rule applies equally to published literature as it would to any spontaneous voluntary report. However, under this rule, the company is required to read all published literature worldwide within a reasonable period and to identify instances in which an adverse experience has occurred in association with use of the product which has resulted in an outcome which meets the criteria for 'serious' and which is outside the approved labelling ('unexpected'). Additionally, the reader (company) is expected to identify whether in the author's mind there is a reasonable possibility that the drug might have played a causal part in the event. Stated in the negative, articles are not required to be reported, no matter how dire the outcomes, if the drug in question is mentioned as an incidental fact in comprehensive case description. These requirements, elaborated in the NDA rewrite of August 22, 1985, may represent the single most sweeping and complex new requirement of the FDA in monitoring the published literature. They may also represent simultaneously one of the most potentially useful, if confusing, requirements; useful because the published literature (like other spontaneous, voluntary adverse reaction reports), describes the comprehensive world experience; confusing because it is often difficult to divine what was, in fact, in the author's mind and time-consuming and frustrating to attempt to find out through direct communication.

MONITORING THE LITERATURE FOR INCREASED FREQUENCY

When such serious reports (even if they describe 'labelled/ expected' events) apparently occur at an unexpected frequency in the literature, they too must be reported immediately using a narrative report to the FDA. This increased frequency reporting requirement generally involves the same criteria as that for spontaneous voluntary reports, that is to say, a 'doubling' over the expected rate in a published case series. Yet to be developed, and perhaps this is an impossible, if not undefinable task, are the ground rules for how to monitor the aggregate published literature over time to ascertain evolving rates of serious adverse reaction reporting and to 'flag' increases in frequency. Many have tried but, to this point, none has succeeded in developing a set of principles which can control the vagaries of what gets reported, and guide how and when to do pattern analysis over time, no matter how tempting. Conceptually, the task is easy: monitor all patients mentioned in a published article with a specific event, or all events, or all deaths as the 'numerator' and monitor all 'some-things' to calculate a denominator or surrogate denominator. Among the 'somethings' nominated have been all persons des-cribed in similar articles; or a system which only looks at articles which describe case series or populations; or the use of sales figures worldwide. The problems with this approach are myriad. That which gets into the numerator is influenced by the biases of that which gets into the published literature in general. Specific-ally, in the literature, the 'first report' of almost anything is likely to be published (even if spurious) while recurring reports of known events, however important, tend to be regarded as 'no longer newsworthy'. Likewise, articles which report unusual frequencies of occurrences of important adverse events will be published, whereas those which observe no increases or potentially decreases from the 'expected rate' will not be. This is not to detract from the usefulness of rigorous and extensive (if not comprehensive) review of patterns from the published literature to understand as much as can be known as quickly and as accurately as it can be known. It is rather to warn the information manager and transmitter against over-interpreting the data or over-applying the method.

Professional analysis

Of course, all the computers, databases, automated keyword searches, and technical information managers are no substitute for individual pharmaceutically-trained professionals within pharmaceutical houses doing their own professional reading and self-development, and, in the process, selecting important articles and information to add to this bibliography. The bibliography having been assembled, there is further, no substitute for the professional in the pharmaceutical house choosing, from among articles nominated, those needing extensive review and/or interpretation because of clinical merit (including potential reportability in official reports to regulatory authorities). The 'system' works because it includes in it the concerned and dedicated body of scientists who can think while they process and create the focus needed to make manageable order from information chaos.

The outputs

The pharmaceutical industry is recognized primarily for the discovery and development of innovative drugs, but it has a major responsibility for the rapid, complete and accurate dissemination of information to health professionals using its products to ensure the safe and effective use of these drugs. Traditionally, the pharmaceutical sales staff has been the primary provider of information from the company. However, the industry must also be prepared to accommodate the specialized needs of health professionals who must assimilate the increasingly complex database on drug therapy. To accomplish this, pharmaceutical companies must develop systems to capture, analyse and disseminate necessary information. These new systems should complement and expand the traditional methods of information dissemination employed by the industry.

Drug information services

The pharmaceutical industry has many ways of disseminating the information to health professionals in the US and many companies have established organized drug information services. Personnel in these groups have specific training in processing and communicating scientific information and they usually have a background in clinical pharmacy; they thus possess the skills necessary to relate the available drug data to specific patient care problems in an objective, unbiased manner.

While the pharmaceutical industry is recognized as an important source of drug information,[10,11] there have been few studies conducted to evaluate how well this service is provided. Knight and Baumgartener wrote to pharmaceutical manufacturers requesting quality control information about seven drugs and the responses of brand and generic companies were compared.[12] Brand companies had an acceptable responses rate of 30%, generic companies 25%, while the total acceptable responses rate for the study was only 27%. Tse and Mansur evaluated response time and completeness of replies received from 59 pharmaceutical companies by a drug information centre during an eight-month period.[13] Response time to telephone and written requests varied. However, some written requests required up to 2 months for a reply and only 57% and 65% of telephone and written responses, respectively, were considered complete.

One group of investigators evaluated the pharmaceutical industry's response to a question about a clinical problem.[14] 40 pharmaceutical companies marketing tetracycline were queried concerning the concurrent administration of tetracycline and cimetidine. Brand manufacturers performed better than generic companies. However, results indicated that few manufacturers are organized to answer drug information requests involving clinical problems in response to a telephone enquiry.

Two studies examined the opinions of users of information from pharmaceutical manufacturers concerning the information they received. In the first study, health professionals contacting the medical department of a single pharmaceutical company for information were surveyed.[15] They found that 83% of the respondents felt their question was answered completely or more than adequately, and 88% thought the response time was fast or about right.

Drug information and clinical pharmacists were surveyed to determine the problems encountered when attempting to obtain information from the pharmaceutical industry.[7,16] 120 respondents indicated that the most difficult information to obtain from the pharmaceutical industry pertained to stability/compatibility and the effects of drugs in pregnancy/nursing infants. Telephone transferring within the company was reported frequently and often the individual with the needed information was not available. However, once the appropriate individual was reached, the necessary information was forthcoming. Written replies, although useful, were frequently (44%) not provided in a timely manner. 66% preferred written follow-up to all verbal responses, while

Table 9.2. Automated product information resources

Product bibliography (*PRODBIB*)
Published adverse reactions (*BW ADVERSE*)
Spontaneous adverse events
Product master formula database
Internal scientific document database
Package insert labelling database
Drug information request database (*DISBASE*)

26% felt that copies of cited references should be automatically included. Many respondents suggested that several services would encourage more pharmacists to request information from pharmaceutical companies. These included the availability of a 24-hour-a-day, toll-free telephone service, the establishment of a centralized information centre, and access to computerized, on-line information services.

To function appropriately, such a group would have to harness the available information databases within the company as well as establish its own database to capture the information the group receives daily from health professionals. In many companies drug information personnel have access to all company databases. These include the special databases for the published literature described previously and other automated files. As an example, the automated databases available at Burroughs Wellcome Co. are listed in *Table 9.2.*

Several companies have developed unique databases to capture information they receive from health professionals contacting the company for medical information. These databases contain information concerning all enquiries and can be searched using many data fields which include: (1) product name; (2) adverse reaction case number (if product complaint); (3) requestor's name; (4) state where requester is located; (5) type of contact; (6) category of contact; (7) nature of request; (8) type of request; (9) index term (key word).

These systems allow rapid retrieval of past responses to similar enquiries and assist in developing timely and accurate replies to similar requests or in providing follow-up to a previous requester. Periodically, reports can be generated which allow the staff to summarize the number and types of requests received. This information can guide the development of educational material for the sales staff and provide feedback to the clinical research staff so that research programmes can be fine-tuned to answer important clinical questions.

The data gathered and analysed by the pharmaceutical company is useful only if it is communicated rapidly and accurately to health professionals and regulatory bodies. The ability of a company to communicate information about its products contributes greatly to the success of its research and development programme. Because the company which develops a product has the most comprehensive and in-depth knowledge about that product, it has a considerable social obligation to provide complete, objective information.

Medical letters from the manufacturer

The pharmaceutical company frequently provides unsolicited information directly to health professionals in the form of promotional material and letters. Although promotional material is often prepared by the Advertising and Marketing Divisions of the company, the medical content must be accurate and balanced, presenting both the merits of the product as well as fairly portraying the drug's safety profile. As the company learns more about the drug, it may be necessary to communicate directly to the medical community rather than relying on the infrequent visits of the sales staff. Whether through promotional material or letters directly to the health professional, all drug information must be reviewed and approved by the medical professional in the pharmaceutical company.

Annual and periodic reports

In the US, each pharmaceutical company must provide to the FDA an annual summary of all new scientific information received during the past year for each marketed product. In addition to summarizing studies completed by the company, the report must also review findings published in the world's medical literature and adverse events reported spontaneously to the company. While this is a formidable task for many products, it is an excellent opportunity for the company to analyse how the new information might impact on the content of its educational campaigns, the product's labelling and the direction of its current research programme.

Understanding uncertainty

Ultimately, the responsibility of the information manager is not simply to capture information but to translate it into useful, clinical frames of reference. Intelligent informed decision-making is the goal. At the patient-treatment level, this requires informed choice, application and monitoring of treatments, monitoring of risk factors, and vigilance for early signs of problems which might indicate need to discontinue the drug to minimize the risk of impact of undesired effects. To achieve that objective, the prescriber, dispenser and/or monitor need to know not simply that which is known, but also that which is not(!). Therefore, the role of the Drug Information Manager must include a keen sense of the epidemiology of the situation, always recognizing the possibility that rare things do happen rarely and given the weakness of current monitoring techniques and the likelihood of missing such rare things, there will never be a substitute for clinical judgement in an individual case. One of the greatest limits of current drug information science is our ability to develop a synthetic estimate which summarizes the extent of our knowledge of benefits into some form of 'aggregated social usefulness number', and of risks into some form of 'aggregated problem index.' Regulators, educators, prescribers and marketeers alike speak glibly of the risk-to-benefit ratio of one chemical entity or another as though we knew what we were talking about. But, the reality is that we cannot. Extending beyond this quandary, often the regulator, the creator of a formulary, and the individual therapist wish to develop a hierarchy of interventions ('first-line' drugs, 'back-up' therapies and 'medications of last resort'). Our tools for creating such classifications are primitive at best and, among groups of drugs all of which 'work' and all of which have relatively 'clean' side-effect profiles, e.g. potent antibiotics or non-steroidal anti-inflammatory drugs, still far from scientifically satisfactory. The responsibility of the Drug Information Manager is to understand the state of knowledge and attempt to translate this to the decision maker who, in the last analysis, must make a professional judgement in the face of that uncertainty. And the job of the contributor to drug information, scholarly researcher and observant clinician alike, is to be specific enough to allow the world to understand the place which the additional information they are providing might have in reducing the level of social uncertainty. Thus, it is far more useful to say not simply, 'this is the first reported case of X with this drug' but to add the statement 'with over Y thousands/millions of users'. Often, individual authors won't know the 'denominator' but a responsible editor can solicit such data (e.g. from the manufacturer) for an editor's note in association with such a published case.

All actors in The Drug Information Cycle have the recurring responsibility to ensure that what is known (or believed to be known) is factored in to the evolving information database. As new information develops, it is often the trigger for further information needs, priorities for further research, and imperatives for new communication.

Summary

Following approval of a new drug, the information available increases by several orders of magnitude and does so very quickly. The task of the information manager is to detect, compile, sift, sort, analyse, format, translate and communicate. The job is to know as much as we can, as fast and as accurately as we can about new drugs so that lessons learned can be applied to achieving the greatest possible benefit while reducing risks to their irreducible minimum. The costs of managing information are not inconsiderable; however, they are minor indeed compared with the costs of our failure to do so!

References

1. Hartgema et al. *On Drug Information Pharmagia.*
2. *Pharmacotherapy* (BCDSP Bibliography).
3. *ETIP Report.*
4. Inman, W. H. W. (1986). *Monitoring for Drug Safety* (2nd edn, MTP Press).
5. Doi, P., et al. (1988). *Drug Information Association Workshop.* (Drug Information Journal in process).
6. Freedman, B. C. (1978). 'The *PRODBIB* database: Retrieval of product information from the published literature.' *American Chemical Society Symposium Series No.84. Retrieval of Medicinal Chemical Information* (eds. W. J. Howe, M. M. Milne, and A. F. Pennell).
7. Osiobe, S. A. (1985). 'Use of information resources by health professionals: a review of the literature.' *Soc.Sci.Med.*, 21(9) 965–973.
8. Kenny, E. C. (1977). 'Information centres in the pharmaceutical industry – the prime source of drug information.' *Postgrad.Med.J* 53 562–565.
9. Venning, G. R. (1983). 'Identification of adverse reactions to new drugs. III: alerting processes and early warning systems.' *Br. Med. J. (Clin. Res.).* 286 289–292.
10. Stinson, E. R. and Mueller, M. A. (1980). 'Survey of health professionals' information habits and needs.' *J. Am. Med. Assoc.*, 243, 140–143.
11. Manning, P. R. and Denson, T. A. (1980). 'How internists learned about cimetidine.' *Ann. Int. Med.* 92, 690–692.
12. Knight, J. L. and Baumgartner, R. P. (1974). 'Quality control information – its availability from pharmaceutical manufacturers.' *Hosp. Formul. Manage.*, 9, 35–52.
13. Tse, C. S. T. and Mansur, J. M. (1984). 'Drug industry response time.' *Drug Intell. Clin. Pharm.*, 18, 325.
14. Generali, J. A. and Hogan, L. (1983). 'A comparison of pharmaceutical manufacturers as a source of drug information to a telephone inquiry: generic vs brand.' *Drug Inform J.* 17, 195–204.
15. Bainbridge, C. V. et al. (1982). 'Pharmaceutical manufacturer's response to drug information inquiries.' *Am. J. Hosp. Pharm.*, 39, 1532–1534.
16. Collins, G. E. et al. (1986). *Audit of hospital pharmacists' utilization of the*

pharmaceutical industry as a source of drug information. (ASHP Mid-Year Clinical Meeting, Las Vegas, 1986).
17. Lortie, F. M. (1986). 'Postmarketing surveillance of adverse drug reactions: problems and solutions.' *Can. Med. Assoc. J.*, **135**(1), 27–32.

PART II

CHAPTER TEN

Publicly-available information on marketed drugs

C. BARRY

The 1980s gave rise to a remarkable increase in the amount of drug information being made available to the public and health professionals in developed countries such as the US and the UK. Not only has the number of marketed drugs proliferated markedly since World War 2, but also the complexity of government regulations and the pressures of health-care costs have heightened the awareness of patients, physicians, pharmacists and other health-care providers about the need for facts and figures in making effective decisions. This chapter will address, among other things, the desire of an increasingly sophisticated public to know about commercial drugs. Busy physicians need authoritative sources to consult for information they can use to select appropriate prescriptions and that they can pass on to their patients. While studies have failed to show a clearcut association between patient education and improved compliance (Masur, F. T., III (1981) 'Adherence to health care regimens,' in *Medical Psychology; Contributions to Behavioral Medicine*. eds. C. K. Prokop and L. A. Bradley, 441–470), health-care professionals intuitively believe in its value and, inadvertently or deliberately, patients know more about drugs than they did even 10 years ago. In an increasingly expensive and ethically ambiguous health-care environment, both care recipients and providers need information upon which to base their choices. Pharmaceutical industry representatives and marketers want to know what the competition is selling and how their own products compare with other drugs available. Reporters, industry analysts, government officials, lobbyists, and economists need basic information to assess industry claims and the potential success of new products and therapies in the market place. Hardly a day passes that newspapers or television do not carry a story related to prescription drugs and the pharmaceutical industry. Drug tampering incidents, new regulations, personal success in the stock market and product liability cases have made the public aware of the drug business in all its multifaceted complexity.

Part I covered the wide variety of drug information sources compiled by many different organizations for different consumer groups. These formally organized efforts target their products in a variety of formats to physicians, research scientists, industrial institutions, pharmacists and educators. The US is as well served as any society in this respect and consequently is a good model. This chapter selects and pulls together aspects of many drug information channels that find their way into electronic and printed media available to the public.

Drug handbooks and compendia

There is no intention here of covering to any great extent the guides and compilations available internationally but the better known ones are very important. *Martindale: The Extra Pharmacopoeia* (Pharmaceutical Press, 1982) provides the most comprehensive, detailed information in a single volume, including different trade names of products sold in a variety of countries. *The Merck Index* (Merck and Co., 1983), also available as a semiannually updated database on Dialog, BRS and Questel, provides encyclopedic coverage with brief entries on chemical names, structures, patents, manufacturers, and therapeutic uses on drugs, chemicals and biological substances.

Most developed countries publish a compendium of their marketed products. *Rote Liste*, *Vidal Dictionnaire* from France and *ABPI Data Sheet Compendium* from the UK are representative of these reference tools. In the US, besides manufacturer catalogues like the *Physicians' Desk Reference*, many publications provide the lay public and the specialist with comprehensive lists. Usually products are cross-referenced by generic and brand name, and vary in depth and completeness. The following are typical of this class of information guides: *American Drug Index* (J. B. Lippincott, annual), *Consumer Drug Digest* (American Society of Hospital Pharmacists), *Facts on File, 1982, Top 200* (R. J. Tallarida), a compendium of pharmacologic and therapeutic information of the most widely prescribed drugs in the US, *Drugs of Choice* (ed. W. Modell, C. V. Mosby, revised every two years), and *The Essential Guide to Prescription Drugs*, (J. W. Long, Harper and Row, 1982).

American Hospital Formulary Service Drug Information (American Society of Hospital Pharmacists, annual) and *USP DI, Vol. I, Drug Information for the Health Care Professional* and *USP DI, Vol. II, Advice for the Patient: Drug Information in Lay Language*, (United States Pharmacopeial Convention, annual) are relatively new publications in their current form and deserve special mention. The former, previously published as a looseleaf service, and the latter, a new approach, consist of in-depth drug monographs written and continually reviewed by a large number of medical specialists. The *Advice to the Patient* volume has been especially well received and should improve the quality of patient care and intelligent use of medications.

Periodicals

Physicians, pharmacists and the lay public who are concerned about current therapeutic approaches, new drugs, and the status of marketed drugs may turn to a variety of non-technical, non-research journals, newsletters and periodicals. The publications mentioned here are typical of those in the US that enjoy wide circulation and a reputation for unbiased, accurate reporting.

The *New England Journal of Medicine*, the *Journal of the American Medical Association* (*JAMA*), and the *British Medical Journal*, all weekly publications, include a significant amount of up-to-date drug information in the form of letters, short reports and significant reviews in addition to scientific reports of clinical studies. Papers submitted to these publications are subjected to stringent refereeing which maximizes the validity and accuracy of the data presented. A regular reader of any of these publications could hardly fail to be cognizant of new and recommended therapies or adverse experiences. Consumer groups, legislative offices, newspaper reporters, and many sophisticated citizens and patients (especially AIDS sufferers, for instance) scan these publications as soon as they appear.

A few newsletters typically abstract articles appearing in specialized journals. These summaries are intended for readers who may not have access to many clinical periodicals but nevertheless need to keep up with potential benefits and hazards in drug prescribing. *Clin-Alert* deals with adverse reactions and interactions, and the *Medical Letter on Drugs and Therapeutics*, also biweekly, reports on drug experiences and, in addition, draws conclusions and makes evaluations on its findings. Other alerting bulletins in wide use are the *International Drug Therapy Newsletter*, monthly, and the *Adverse Drug Reaction Bulletin*, bimonthly, published in several languages besides English.

The public reads about new drugs in many general magazines, but a few publications are of special interest to patients and health-care personnel. Among notable ones in the US is the *FDA Consumer*, an official, inexpensive magazine put out by the US Food and Drug Administration (FDA) as part of its mission to inform the public about marketed pharmaceuticals: product recalls, court actions, tampering and contamination, new approvals and substantive articles on drug abuse, health, vitamins and quack remedies are part of its coverage. Another valuable source of information is a bimonthly from the US Pharmaceutical Convention entitled *About Your Medicines*.

News about drugs

A revolutionary aspect of most major newspapers in recent years is the emergence of coverage devoted to pharmaceutical discovery, development and marketing. *The Wall Street Journal* (Dow Jones), the *New York Times*, the *Financial Times*, and important papers in all major cities now have special science reporters and pharmaceutical news analysts who cover the 'drug beat'. Undoubtedly the public's increasing awareness of issues related to the drug industry has sparked the insatiable demand in the developed countries for technologically sophisticated analysis and immediate access to research breakthroughs. In recent years, the AIDS epidemic has been on the front page almost every day, making celebrities of researchers and, in turn, raising and dashing the hopes of sufferers with items about potential therapies from huge research organizations and tiny laboratories alike. The financial success of many pharmaceutical companies has attracted investors in great numbers and progress now tends to be publicized in press releases as a commercial strategy long before definitive therapeutic advantage is proven in clinical trials and reported in scholarly journals.

Still a vastly under-utilized resource, press releases and local stories, most of which are never published, make their way into newspapers' offices over international news wires. *AP News, UPI* and *Reuters News Service* are also available to the public via computerized database vendors subject to a day or two delay. Full-text databases of the *Washington Post*, the *Financial Times*, *The Wall Street Journal*, the *New York Times*, *The Guardian* and the

Japan Economic Journal are just a few of those available on NEXIS (Mead Data Central). Still too expensive for most private individuals, NEXIS, introduced in 1980, is the most widely used online vendor in the world, but other online vendors such as Dialog are now offering many of the same publications for less money as well as the full text of many regional US newspapers that contain news stories about small, often privately owned firms and new ventures that specialize in biotechnology.

Besides daily newspapers, and other purveyors of international news about pharmaceutical development such as TV and radio, several general scientific periodicals cover research and regulatory issues on a global basis. *Science* (American Association for the Advancement of Science) and *Nature*, the prestigious UK weekly magazine, report important scientific discoveries, especially in molecular biology, biochemistry and biotechnology, and analyze important therapeutic advances wherever they occur. Major clinical successes reported at scientific conferences and meetings are covered by these publications and often special in-depth reviews are devoted to new therapies. *New Scientist* and *Chemical Engineering News* are also excellent sources of news and analysis of scientific progress in the pharmaceutical field.

Similarly, newsletters satisfy the needs of industry and public alike for prompt information about product launches, financial movements, research breakthroughs, and technological advances. Much information is too non-technical to satisfy the scientist or specialist, but newsletters provide a useful adjunct to patents and scholarly journals for keeping up with discoveries and break-throughs in fields related to pharmaceutical development such as genetic engineering, computerized drug design, and regulatory issues. Much material is based on press releases and pronounce-ments by public relations officers in the industry or at research institutes, but journalists and market analysts seem to consume them *ad libitum* despite errors, exaggerated claims and premature announcements.

Scrip (PJB Publications) and *FDC Reports* ('*The Pink Sheet*'), specializing in news on legislation and regulations, are both available in full-text for computerized retrieval. *Genetic Engineering News* and *Generic Line* (Scitec Services, Inc.), provide timely coverage of pharmaceutical news that varies in timeliness and global coverage, but it is difficult to keep up with relevant events day-to-day without them.

Public access to many newspapers and news periodicals is facilitated by printed indexes such as the *Reader's Guide to Periodical Literature* and the *New York Times Index*, as well as online database indexes, such as the *National Newspaper Index*,

Magazine Index, and *Pharmaceutical News Index* (PNI) available on BRS and Dialog.

Popular guide books

Consumer-oriented books and popular home health guides fill whole ranges of shelves in general bookstores. The public has become an active participant in its own health care, seeking reliable information with which to make decisions about medications, often confronting their physicians with choices and decisions based on material they have read. The American Association of Retired Persons is a typical influential consumer organization that provides a wealth of information for its members, emphasizing especially the needs of the elderly. *AARP Pharmacy Service Prescription Drug Handbook* (AARP, 1988) lists over 1000 drugs with descriptions of proper administration, safe use and special precautions and restrictions for persons over 65. The AARP Pharmacy service also provides the largest mail order source of low-cost prescription drugs in the US.

Other popular guides found in public libraries include the *Good Housekeeping Family Guide to Medications and Dictionary of Prescription Drugs* (J. K. Jones, Hearst Books, 1980), and *Drug Information for the Consumer* (Consumer Reports Editors, Consumer Reports Books, 1987).

Book departments in major department stores, discount stores and even airports make available many self-help guides mostly in paperback form. Some examples include *How to Save Dollars with Generic Drugs: A Consumer's Guide to High-quality, Low-priced Medicines* (M. A. Ferm and B. Ferm, Morrow, 1984) and *Patient's Guide to Medicine: From the Drugstore through the Hospital* (ed. W. J. Brown, Aero-Medical, 1987).

A phenomenon in the field of consumer drug information is pharmacologist Joe Graedon, whose best-selling books, *The People's Pharmacy* (Avon Books, 1976), *The People's Pharmacy-2* (Avon Books, 1980) and *Joe Graedon's The New People's Pharmacy* (Bantam Books, 1985), in conjunction with his syndicated column appearing weekly in newspapers across the US, has probably contributed more to the public's knowledge of marketed drugs and the drug industry than any other single source. Despite his slangy, sensationalist style, the information he disseminates is authoritative and respected even by drug companies subject to his occasionally unfavourable commentary.

Drug companies as drug information producers

Patients, health professionals, journalists and financial analysts increasingly address their questions about marketed products to the manufacturer. To respond to this demand for information about adverse effects, indications and available formulations of their drugs, many companies have set up Consumer Information Units or Drug Information Centres charged with the responsibility of providing accurate, authoritative facts about their own products. Companies have found it more efficient, cost-effective and ultimately profitable in terms of public relations as well as monetary return to centralize this function in a single administrative group that has been trained in both the scientific aspects of drug therapy and appropriate communications skills. To keep the information up-to-date and readily retrievable, many of these units call upon search specialists to retrieve literature and internal reports about their company's and competitive products on a regular, usually monthly, basis. Such Centres may compile their own in-house files and databases of published literature references, and data about dosage forms, prices, contra-indications, and precautions. Authoritatively selected and systematically maintained, this well organized function works to ensure that the company's products are used correctly and to make a favourable impact on the public's perception of the company and the industry as a whole.

Besides in-house product information, US companies are required by the FDA to furnish officially approved information along with the product, usually as a 'package insert'. The most widely distributed compendium of package inserts on marketed products in the United States is the *Physicians' Desk Reference (PDR)* (Medical Economics Company, annual). Pharmaceutical manufacturers provide the content of these package inserts; many marketed products, over 2000, are included, indexed by manufacturer, brand name, generic name, chemical name and product class. Product descriptions include the name, address and phone number of each company, pictures of capsules and tablets, data on its composition, actions, warnings about side-effects and forms supplied. Although originally intended exclusively for physicians, it is now available in public bookstores. It is well documented that more patients consult the *PDR* than doctors. Unfortunately public reliance on this reference is probably misguided since it contains only material approved by the FDA and, while it gives extensive coverage of adverse effects, no evaluation is made on their severity or frequency.

Drug Topics Red Book (Medical Economics Company, annual), and *American Druggist Blue Book* (Hearst Corp., annual) are other compendia of manufacturer-based information, especially useful for determining prices and forms in which products are sold.

Professional and trade associations as information providers

Many professional organizations provide information for their members and publish handbooks, directories and drug evaluations for health practitioners. Examples of such authoritative sources are the *American Hospital Formulary Service Drug Information* (American Society of Hospital Pharmacists, annual) and the *Handbook of Nonprescription Drugs* (American Pharmaceutical Association and National Professional Society of Pharmacists). *AMA Drug Evaluations* (American Medical Association Department of Drugs and the American Society for Clinical Pharmacology and Therapeutics, 1986) brings in current scientific findings pertinent to marketed drugs that go beyond the official indications and dosages sanctioned by the FDA on the package insert. Side-effects and caveats about drug interactions with other medications and diet are also comprehensively covered in this volume.

The Pharmaceutical Manufacturers Association, a trade organization of ethical and biological products manufacturers, disseminates information about government regulations and compiles a *Statistical Fact Book* on the industry which is of interest to the media and financial analyst. The National Association of Retail Druggists and the Generic Pharmaceutical Industry Association publish periodicals and promote their particular concerns to the public often by responding to queries, putting out press releases and lobbying at federal and state legislative offices.

Sources on poisoning and toxicity

One reads in news or business magazines that approximately 65–75% of the US population is taking one or more prescription drugs at one time or another. Most of these prescriptions are being taken by non-hospitalized patients, meaning that capsules, tablets and syrups are sitting on bedside tables and in medicine chests. In addition a great number of the over-the-counter medications are purchased and self-administered. Besides commercial drugs, 'street drugs" have become a pervasive fact. Drug hot-lines, detoxification units, emergency rooms and police headquarters need ready access to information sources about drugs and chemicals. Adverse drug reaction emergencies and overdosage occur so commonly that a programme of poison control centres became established in the US in the 1950s. Originally the growth of these centres occurred haphazardly and resulted in a lack of standards and depth of expertise. Currently over 500 hospitals list a poison control telephone number for emergencies in the *Health and Medical Directory* (Yellow Pages of America, 1988). Today, the function of providing emergency information to the public is more reliably carried out by approximately 36 regional centres which pool resources, often including a library and access to computerized databases, as well as funding. An American Association of Poison Control Centres (see the current *Encyclopedia of Associations* for Headquarters) maintains protocols and standards for the organized network of centres, conducts educational programmes, prepares guides, maintains a database of statistics, etc. Pharmacists usually staff these centres on a 24-hour per day basis to respond to emergency calls with immediate information on identifying potential poisons and recommendations on antidotes and treatments. In the US the National Institute on Drug Abuse, part of the US Department of Health and Human Services, makes available a National Clearinghouse on Drug Abuse aimed at health professionals and organizations dealing with these problems.

Drug information centres (DICs), which began to spring up in the US in the 1960s, acquire and disseminate information on drugs but with a broader scope and different mission than poison control centres. Like them, however, DICs tend to be affiliated with medical centres, hospitals or schools of pharmacy, but because responses do not need to be made in such a narrow time frame, DICs offer more elaborate services such as computer services and specialized literature analysis for health professionals who call in. Typical requests addressed to a DICs include therapeutic use, dosage, identification, side-effects, availability and compatibility with other medications. With improvements in organization and information management, the two types of information centres often merge to combine resources (*Am. J. Hosp. Pharm.*, **40**, 1219–1222).

Typically, the most widely used information source at these centres is the National Library of Medicine's MEDLINE and TOXLINE databases. Other sources include *DRUGDEX*, a computer-generated microfiche system put out by Micromedix, Inc., which also publishes the *De Haen Drug Information Systems* on microfiche and online with Dialog. *POISINDEX*, a service listing the ingredients of many thousands of commercial products, is available by subscription on microfiche as well in computerized form, from the Rocky Mountain Poison Center. Besides standard reference compendia like the *PDR*, particularly valuable for drug identification with its coloured illustrations of marketed capsules and tablets, popular sources are usually geared to ease of use and timeliness. *Clinical Toxicology of Commercial Products* (ed. R. E. Gosselin et al., Williams and Wilkins, 1984), *Meyler's Side Effects of Drugs: An Encyclopedia of Adverse Reactions and Interactions* (ed. M. N. G. Dukes, Elsevier, 1984) and *Handbook of Poisoning: Prevention, Diagnosis and Treatment* (R. H. Dreisbach, Lange Medical Publications, 1983) are among the printed sources such centres have on hand for ready reference.

Secondary sources: abstracts, indexes and computerized databases

Until the 1970s access to specialized scientific and technical information about drugs was limited to information specialists, usually in university health sciences libraries or schools of pharmacy, with comprehensive collections of printed indexes and abstracts such as *Index Medicus* (US National Library of Medicine), *Excerpta Medica* (Elsevier Science Publishers) or *International Pharmaceutical Abstracts* (American Society of Hospital Pharmacists). Users now connect by national packet switching networks to databases on hosts such as Dialog, SDC, NLM or BRS in the US and Data-Star or DIMDI in Europe and key in appropriate search terms in a series of logical combinations to retrieve references that satisfy their information needs. To search most effectively requires extensive training and experience as well as financial resources usually beyond the reach of the average consumer or individual health practitioner.

However in the 1980s a new phenomenon broadened the access of non-information professionals to drug information. Database producers have designed and marketed computerized search systems targeted to the occasional user for whom extensive training and continuing education is undesirable and unrealistic. With the spread of microcomputers, 'do-it-yourself' searching has become available to physicians, pharmacists, and information-minded consumers in their offices, homes and public libraries. These systems often have menus, natural language, and online guidance to help the user find the information he or she seeks. For example, *Drug Information Fulltext*, available on Dialog in the US, corresponds to two printed publications, *The American*

Hospital Formulary Service Drug Information and *The Handbook on Injectable Drugs*, and is representative for the comprehensive, authoritative sources widely used in advanced countries for immediate, timely information. The public information consumer and health practitioner usually do not have many current periodicals at their fingertips. Full-text databases have filled this information gap and their availability and comprehensive coverage continue to expand.

In the US the American Medical Association has developed a Medical Information Network in conjunction with the General Telephone and Electronics Corporation that distributes the products and services. MINET is aimed at the private physician who does not have ready access to the resources of health sciences libraries and includes news, a bulletin board, electronic mail, a continuing education calendar, a link to patient data management, document ordering services, direct access to government Advisories such as ones put out by the Surgeon General's Office and simple retrieval systems for clinical literature databases including the full text of the *AMA's Drug Evaluations*.

Traditional indexing and abstracting services are marketing medical and pharmaceutical databases in a wide variety of products. One of the most widely touted forms is CD-ROM; an extremely helpful publication for locating and evaluating these products is *CD-ROM Librarian* (Meckler). MEDLINE is available from six CD-ROM vendors including Silver Platter, Cambridge, and Dialog. Each producer packages the database in different formats and subsets. The search software rarely matches the online version in quality and sophistication, and the documentation is often clumsy and confusing, although these drawbacks are being reduced as experience and feedback from customers take effect in updated versions. Other CD-ROM products of interest that have recently arrived on the market are *Martindale, DRUGDEX* and *POISINDEX* from Micromedix and the *PDR* from Medical Economics. The advantages of using CD-ROM discs over online searching include being able to avoid unreliable telecommunications, ease of distribution, and a fixed price. The disadvantages include less timely updates, often unwieldy search software in the name of simplicity, and the fact that unless CD-ROMs are available via microcomputer network, only a single user can access them at one time.

Online systems aimed at the occasional user or health professional have been appearing over the past few years in bewildering multiplicity. One of the least expensive and most satisfactory products is *GRATEFUL MED*, produced by the US National Library of Medicine (NLM), which provides access to *MEDLINE* and the other NLM databases. *GRATEFUL MED* is a front-end which guides the user in formulating a search, calls the NLM computer, executes the search online, downloads the retrieval and logs off. The system allows the user to examine the results of his search, reformulate the terms, and log back on for further searching if desired. Also widely used is BRS/Saunders' *Colleague*, an integrated information resource produced by BRS and W. B. Saunders, which includes the full text of over 50 medical journals, over 30 up-to-date medical textbooks and access to most of the major biomedical bibliographic databases online. Another front-end system that is widely used in the US is *EASY-NET*, the software developed by Telebase Systems, Inc., available in several versions including Infomaster from Western Union, I-Quest from Compuserve, and, soon, as Direct Connect which will feature a dedicated connection. The system provides, theoretically, a seamless, transparent, uniform gateway to over 900 databases available from more than a dozen major database vendors. Users pay a flat fee per simple search and first ten citations retrieved. A series of menus and choice points leads the user to an appropriate database and helps him select relevant terms. Extra charges are levied for printing abstracts or fulltext articles. An information specialist may be consulted online for help. Students and casual users of this gateway report favourably on their experience with it. Information specialists find it cumbersome and costly although the number of databases and ease of use make it attractive for searching rarely used sources without having to establish an account or learn the search software. Keeping up with systems targeted to the end-user is difficult. The publication *Database Searcher* is a valuable source of product news and evaluations and attending one of the annual online conferences is recommended for sharing experiences and previewing new products. When full-text pharmaceutical and medical periodicals published on optical disks become economically and technically within the grasp of the public, we can anticipate a massive redistribution of drug information in the form of locally published and customized products.

This chapter has described sources of information on marketed drug products. Much of this information is specific to the US and is well known to pharmaceutical industry personnel, especially librarians and information specialists. However, consumers, health professionals and many interested groups who report on the drug industry are seeking information on their own and need guidance to find accurate, timely sources written in language they can understand and interpret. In keeping with this trend, information vendors have directed their efforts to analysing new markets and designing products to fit an expanding niche. Regrettably, many new formats have been extravagantly oversold, resulting in disappointment and wasted investment in systems that fail to fulfill the expectations of the uninitiated information consumer. However, as users explore the world of self-service information retrieval, a new awareness of the vast possibilities is taking place.

At the same time, rapidly developing information technology and expertise is improving access and global reach. The most important feature of technological improvement is the flexibility it affords. With multiple formats, subsets, distribution channels, and retrieval software, users have a great variety of ways to gather and use relevant data and published material. As will be obvious to readers of this book, groups who need drug information represent a broad range of sophistication and motivation. The individual physician's requirements are extremely different from those of the regulatory agency and those of the newspaper reporter. And, without doubt, economics will always be a powerful force affecting how important drug information finds its way to populations who need it.

Evaluating and selecting information products and educating users turns out to be an even more demanding responsibility of the information professional than the 'simple" computer searching function of 10 years ago. Besides the overwhelming growth in the amount of electronically stored material, its diversity and quality is even more bewildering. The next generation of information products will depend on 'value-added' features based on past experience and will be tailored for very specific consumer groups. It is hoped that readers of this chapter have gained an overview of how publicly available information on marketed drugs is obtained in the United States. The situation changes daily. Finding ways to select, package, and deliver this important commodity promises information professionals a complex and challenging future and consumers hitherto unimaginable opportunities.

CHAPTER ELEVEN

The performance of the pharmaceutical industry in providing scientific and technical information

S. E. WARD

Introduction

Previous chapters have clearly demonstrated the vast range and quantities of information required and produced throughout the drug research, development and marketing life cycle. It is evident that the pharmaceutical industry is an information intensive industry. Information is its lifeblood and information is its principal product. For research, development, production and marketing, effective gathering of external information and the streamlined organization of proprietary, i.e. company-generated, information is a critical activity, essential if the company is to maintain its rate of innovation and its sales performance (*Figure 11.1*).

The industry utilizes the patent to protect its intellectual property in new inventions – the patent will normally be the first publication on a novel compound of potential therapeutic interest. Permission to take a drug into clinical trials and the granting of a product licence require the systematic and regulated creation of an organized dossier of experimental data covering all aspects of the drug's clinical performance, safety and manufacturing and its submission to regulatory authorities worldwide. The responsibility of the industry to advise regulatory authorities of information on the drug continues into marketing with post-marketing surveillance activity involving monitoring literature, contact with clinicians prescribing the drug, and further clinical trials. In

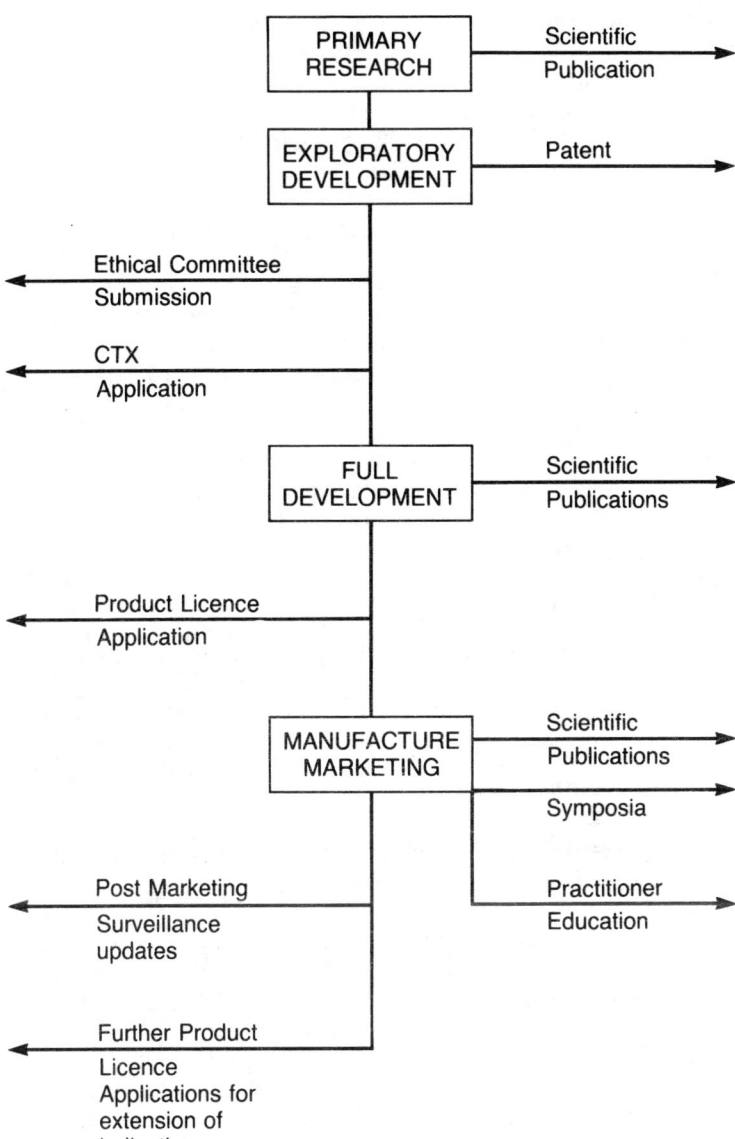

Figure 11.1 Information landmarks in the drug lifecycle

addition, the industry relies heavily on the publication of papers in the primary scientific literature, journals, monographs and so on, both to disseminate its achievements in research and development, and for the promulgation of information on the clinical performance of its products, to encourage their prescription.

It is therefore not surprising that the industry has made a considerable investment in information management and that it employs large numbers of information professionals. It must also be obvious that the industry is particularly well placed to be a source of information on its products. Where these are the result of its own research and development in fact, a company's 'dossier' of information will be both comprehensive and unique. As a product moves onto the market this information can and should be exploited by those involved in prescription and administration of drugs within the health-care field. 'The effective use of drugs depends on the accurate and comprehensive communication and understanding of information about them' (Hoff, 1983).

This chapter will concentrate on the role of the industry as a provider of information on medicines through its product information services, representative activity, and publication and education programmes. It will also consider, albeit briefly, the industry's contribution to information service developments. The industry's research and development information activity has been considered elsewhere.

Why is industry potentially such an important source of information?

The volume of biomedical information currently being published is substantial and it continues to grow. At least 20,000 biomedical journals are currently published; articles on medicines will be scattered across many of these, in many different languages. Even for just one medicine, publication volumes can be substantial (*Table 11.1*). Sources such as online databases will facilitate the identification of useful references. However, for comprehensive coverage, more than one database will need to be consulted, as key references will be distributed across more than one database.

The pharmaceutical industry will, for its own and for regulatory purposes, evaluate all publications on its medicines to create its internal knowledge base. The industry can therefore be regarded as a centralized source of evaluated published information on its own products, complemented by its own proprietary information. Much of this information is available to health-care professionals.

Table 11.1. Ranitidine publication volumes

Language	1989	1988	1987	1986	1985	1984	1983	1982	1981	1980	1979	Total	%
Chinese	1	0	1	0	0	0	0	0	0	0	0	2	1
Czech	0	0	0	0	1	0	0	0	0	0	0	1	1
Danish	0	0	2	2	1	0	2	0	0	0	0	7	1
Dutch	3	0	1	1	4	1	2	1	0	0	0	13	1
English	237	594	509	509	408	454	287	355	163	76	15	3607	81
Finnish	0	0	0	0	0	1	0	0	0	0	0	1	1
French	2	11	11	13	25	17	17	9	10	2	2	119	3
German	10	28	35	43	45	46	37	32	20	7	3	306	7
Greek	0	0	0	0	0	3	0	0	0	0	0	3	1
Hebrew	0	0	0	0	1	0	0	0	0	0	0	2	1
Hungarian	0	0	0	0	3	8	0	0	0	0	0	11	1
Italian	0	12	6	20	30	42	54	9	4	1	0	178	4
Japanese	2	15	12	26	18	26	17	2	3	0	0	121	3
Norwegian	0	1	0	0	0	0	1	0	0	0	0	2	1
Polish	0	0	0	1	1	1	0	1	0	0	0	4	1
Rumanian	0	0	0	0	1	0	0	0	0	0	0	1	1
Russian	0	0	1	0	0	1	0	0	0	0	0	2	1
Serbo Croat	0	0	0	1	0	0	0	0	0	0	0	1	1
Spanish	1	13	11	4	8	8	9	2	0	0	0	56	1
Swedish	0	0	0	0	0	1	0	0	0	0	0	1	1
Total	256	674	589	621	546	609	426	411	200	86	20	4438	

The 'regulation' of industry information services

The marketing of a medicine inevitably involves the provision, by a manufacturer, of information on the product in the context of its promotion. Within the UK, promotional activites of most pharmaceutical companies are voluntarily regulated by the *'Code of Practice for the Pharmaceutical Industry'* produced by the Association of the British Pharmaceutical Industry, (ABPI). In its introduction (ABPI 1988), the code 'emphasises the importance in the public interest of providing the medical and allied professions with accurate, fair, and objective information on medical products so that rational prescribing decisions can be made. Moreover, the Code establishes the principle that such information should be presented in a form and by ways and means which conform not only to legal requirements but also to ethical standards and canons of good taste.' The code also reflects the obligations of the industry to provide information to the pharmaceutical profession.

Many of the specific provisions of the code are concerned with the quality and context of promotional material (advertisements, mailings, and presentational information) and with the background and activities of the medical representative. In addition, specific responsibilities are laid on member companies to provide high quality and objective information on their medical products. Extracts from clause 4 of the code which defines these responsibilities are given below:

> '4.1 Upon reasonable request, the company concerned shall promptly provide members of the medical profession with accurate and relevant information about the medical products which the company markets.
> 4.2 Information about medical products should accurately reflect current knowledge or responsible opinion.
> 4.3 Information about medical products must be accurate, balanced and must not mislead either directly or by implication.
> 4.4 Information must be capable of substantiation, such substantiation being provided without delay at the request of the medical profession.'

The *Code of Pharmaceutical Marketing Practices* of the International Federation of Pharmaceutical Manufacturers Association (IFPMA) places similar responsibilities on the various National Pharmaceutical Manufacturers Associations who have endorsed it (currently 46 associations are signatories). The *Code* outlines the industry's obligation to base the claims for substances and formulations on valid and current scientific evidence; to provide scientific information with objectivity, scrupulous regard for truth, and clear statements with respect to indications, contraindications, tolerance, and toxicity; and to use complete candour in dealing with public health officials, health care professionals, and the public. It also stresses that companies must provide accurate, fair, objective and current information on their products (IFPMA).

The company Medical or Product Information Service is the vehicle through which most companies discharge these responsibilities.

Information services provided by the industry

Traditionally the pharmaceutical industry has featured two types of information function, the research and development based information service providing support to company employees and the medical or drug information service, providing information not only internally but also outside the company. These established units are now increasingly complemented by the business information specialist or analyst supporting market research, strategic planning, public relations and also the procurement function. Information units supporting the manufacturing process will also exist in many companies.

Quantitative information on the level of investment in information services staffing is patchy. Certainly in the UK research and development sector, approximately 3% of the research and development workforce is a common level of investment, leading to departments of over 100 staff in some companies. In the USA, a recent survey (Nelson, 1989) shows an investment of roughly 2.8 information staff per 100 researchers served; of these an equivalent of 1.8 are directed to published information provision, the remainder dealing with proprietary information, that is internal reports, compound information and sample management, biological records, microfilming and computer modelling.

Within medical information units in UK companies a survey in 1984 (Williams 1984) showed staffing levels of up to the mid-twenties. Certainly units of ten staff are not uncommon. An American survey (Salter 1986) conducted at a similar date showed the overall average of information professionals per drug company as 6.2, with an average of four non-professional support staff members.

For both research and development and medical information services, staff investment includes a considerable proportion of professionally qualified scientists with first degrees or doctorates (Nelson, 1989; Williams, 1984). This is essential: staff in a scientific information service function must have both good scientific qualifications and a thorough grounding in information science if they are to fulfil their role effectively.

The medical information service

Roles and responsibilities

The medical information service exists both to serve company staff involved in product support, marketing, and product development and to provide information on marketed products outside the company. Internally, its customers will include company physicians and other staff involved in the conduct of volunteer studies and clinical trials, product managers, regulatory affairs staff, marketing personnel and sales representatives. Externally, services will be provided to doctors, pharmacists, and other health care professionals. Within the UK, a particularly active customer group are the drug information pharmacists, responsible for drug information provision within NHS regions, districts or working at the local hospital level. For companies with over-the-counter (OTC) medicines (drugs available through retail pharmacies, without prescriptions) an information service may need to be provided to the general public. For prescription-only-medicines (POMs), companies will generally not answer queries from patients directly except where the enquiry is a simple one of fact, for example, the appropriate timing of a medication. In fact, in the UK the ABPI *Code of Practice* (ABPI, 1988) requires the industry to refuse requests from individual members of the public for information or advice on personal medical matters and to recommend that the enquirer consults his or her own doctor. The doctor-patient relationship is thereby protected. Similar practices are not necessarily followed in other countries.

The medical information service will normally but not invariably report to a medical director; if it does, its organizational context lends authority to its role in information provision. Its staff will include qualified pharmacists and biomedical information scientists with support personnel; the professional staff will be experts in sources of published information relevant to their company's drugs and the related therapeutic areas. They will have access to considerable in-house information on the drug and will have developed substantial expertise on the product, often being regarded as an in-house expert.

Most companies have a well developed medical information service. A UK survey in 1984 showed at least 50% of companies having a dedicated unit, others handling the service via medical advisors or marketing staff (Williams, 1984).

Given the responsibilities laid on companies by the ABPI *Code of Practice,* it is hardly surprising that the answering of enquiries is central to the role of the medical information service. Nevertheless, the complete service does include other functions, which can be classified into the following groups: enabling (i.e. activities which assist the service to function effectively), services to internal groups, and services outside the company.

Table 11.2 classifies services most likely to be found in medical information groups into these categories. Depending on the organizational framework, medical information services personnel

Table 11.2. Functions performed by medical information services (Williams 1984; Bassin 1986; Bauman 1989)

Enabling services	Monitoring published and internal information on company drugs. Maintenance of collections of product information. Development of approved responses to enquiries. Monitoring competitor information
Internal services Sales and marketing	Training of sales representatives, Preparation of training booklets. Approval of product advertisements, labelling and press releases. Preparation of technical and product monographs. Commentary on competitor advertisements. Preparation of commentary on new key papers
Regulatory affairs	Support for preparation of prescribing information documents, patient guidance leaflets.Compilation of bibliographies for product registration submissions and updates
Clinical research	Editing company papers for publication.Preparation of slides for external meetings. Literature reviews to provide background for proposed studies. Review of study protocols
Drug surveillance	Published literature monitoring for adverse events. Ensuring that external enquiries relating to adverse events are formally reported
External services	Handling enquiries on all aspects of company products and related areas. Product complaints. Contributing to symposia and conferences via lectures and also via search services. Developing good relationships with external information personnel. Providing submissions to Hospital and Regional Formulary Committees.

may also be directly involved in product registration activities and in the conduct of clinical trials.

Enquiry answering

Of the range of activities which make up the work of a medical information service *in toto,* enquiry answering is the most significant (Williams, 1984). Numbers of enquiries from outside the company are substantial. In some companies and countries toll free telephone numbers are used. Volumes of work of the order of 20,000–100,000 enquiries per year are quoted by large companies such as Lederle (Bassin, 1986), Glaxo Inc, (Bauman, 1989), Roche (Bayland, 1989) and others (Salter, 1981). Enquiries will be received directly or via the sales force, by telephone or by letter. The enquiries will arise principally from physicians, both general practitioners and those working in hospitals, pharmacists in hospital and retail environments, the general public, the representatives themselves, research workers and nurses. Within the UK hospital pharmacists form the highest proportion of company information enquirers – although often they will be acting as intermediaries (Williams, 1984).

Enquiries received fall into several categories (Williams, 1984; Bauman, 1989; Rumore, 1989):

- basic information related to the proper use of the company's products, often required speedily and sometimes in an emergency e.g. dosage and administration in particular circumstances;
- information for clinical investigators or for academic research projects utilizing the company's products, often concerning non-approved indications;
- adverse drug reactions;
- toxicology;
- biopharmaceutics and pharmacokinetics;
- drug identification;
- drug interactions from a patient or laboratory perspective;
- drug stability;
- product ingredient information, often related to the avoidance of allergy;
- requests for drug samples.

Much of the information to be provided is factual. However, medical opinion is also often required and the medical information service will have access to a medically qualified advisor for this type of enquiry. Replies may be verbal or written and can require considerable effort in their creation. Enquiries on products continue throughout their lifetime on the market. However, it is not surprising that enquiry loads are particularly high in the period surrounding the launch of a new product or following significant publications on the company's drug or that of a major competitor (Collins, 1987).

In most cases, the company's information service will focus by and large on its own products. However, some companies do establish information services covering a general area of therapy, for instance, Beecham's infectious diseases information service providing a 'round the clock' service on infection and anti-infective drugs in general (Scrip, 1989).

Many of the enquiries can be answered using published literature and reprints of papers will often be supplied with the response. Sometimes enquiries will require the release of information which is proprietary. Most reputable companies will provide proprietary data if a serious treatment problem exists with a patient. However, the enquirer will be asked not to publish such information without written permission and it will be emphasized to the enquirer that the information is being given for humanitarian reasons, for that particular patient. Often the company will be asked to comment on the use of the drug outside the approved indications, e.g. the use of the drug beyond its expiry date. The company's response here will always be conservative and the enquiring professional referred to the datasheet or package insert (Boyland, 1983). However, the company will provide details of relevant literature reports which refer to non-approved uses.

The drug information service can provide the stimulus for further company research – the receipt of a group of enquiries may suggest the conduct of further studies to provide data to meet the information need (Bauman, 1989) or may provide new ideas for extension of the use of existing products (Boyland, 1983, 1989; Hoff, 1983). Frequently, questions asked may highlight the need to include more information in promotional materials and this will be followed through by the medical information service with company colleagues.

Inactive ingredients

Non-therapeutic ingredients can in some patients cause a reaction to a medicine. Where a patient has a known history of sensitivity to certain substances, knowledge of the ingredients of a medicine will enable the doctor to avoid likely sensitizers when prescribing. It is commonly accepted practice within many developed countries that inactive ingredients are disclosed; regulations vary from country to country on factors such as which substances should be disclosed (all drugs and preservatives only, specific bioactives); whether information should be qualitative or quantitative; where the information should be displayed (on the label, the pack, or on the doctor or patient information leaflet). Within the UK, however, no statutory regulations have been developed to cover information provision on ingredient data. As a result, previous UK practice has been for pharmaceutical companies to give the requisite information to *bona fide* professional enquiries in response to specific requests. The public do not have direct access to this information.

The UK pharmaceutical industry via the ABPI has introduced voluntary guidelines to increase information provision on inactive ingredients both to the patient and to the doctor and pharmacist in response to growing public demand for the release of such information evidenced by both enquiries received by companies and consumer pressure groups (ABPI, 1988). The ABPI's proposals recommend qualitative information on specific ingredients for most medicines with full qualitative disclosure for injectables, inhalants, topicals and ophthalmic preparations. Declaration would ideally coincide with the introduction of patient information leaflets and original pack dispensing in the UK.

The ABPI's recommendations were published at the same time as the UK's Department of Health produced a consultation document on information on medicines which includes proposals for full qualitative disclosure for all medicinal products. The debate continues but the initiative of the industry in responding to consumer demand reflects its personal commitment to improving the provision of information on its products. Current proposals from the European Commission are in line with the selective ABPI policy of disclosing those inactive ingredients which are of significant interest.

Emergency and out-of-hours service provision

Most pharmaceutical companies recognize the need for procedures to handle information requests received after normal working hours. Such emergency services recognize the need for enquirers faced with overdose problems or the need for critical dosing information to be able to contact company experts around the clock. These services often function 24 hours per day and 7 days per week and throughout holiday periods (Boyland, 1983). Methods for handling such services will vary between companies and can include use of company security personnel to arrange for the enquirer to be contacted by company staff from their home, automatic switching of telephones to home numbers after hours, or referral of calls to local information centres such as poisons information centres. The ABPI encourages member companies to publish their emergency number in the Data Sheet Compendium (See later).

The tools of the medical information service

The most important facility required by the medical information service is current information on company and competitor drugs. Product information specialists will have access to unpublished product information in the form of internal research reports, product licence and ingredient data, correspondence, and drug surveillance information. In addition, all product information specialists will be expected to be completely *au fait* with the published literature on products and will normally contribute to its collection through scanning of primary journals and monitoring of online databases. The drug information specialist will also be responsible for the evaluation, indexing and storing of such information.

It is common to find internal databases which index all papers on company products for later retrieval. In the large international companies such databases are often developed by the parent company and are available online to subsidiaries to ensure that all companies in the group have access to the core database, irrespective of location and local resources. The effort involved in sustaining such systems is substantial. However, the inadequacies of commercial databases from the perspective of comprehensiveness and drug indexing and timeliness compared with the benefits of a comprehensive and current online collection of company specific information provide the basis for cost-justification of internal systems. If a paper deals with a company's product only marginally then the commercial databases may well ignore this in indexing. However, to the company this information is important and the paper will be fully indexed in the in-house collection. Indexing in these internal collections is detailed: in addition to the expected information on authors, their affiliation and location, language and type of paper (editorial, letter, review, journal article and so on) systems such as the Glaxo database, *Glaxoline,* will index all aspects of information recorded on a Glaxo drug as shown in *Table 11.3.*

Such databases provide for precise and rapid searching when urgent enquiries are received and will be supplemented by the commercial online databases when required.

As well as computerized collections of source data, companies will also maintain collections of responses to enquiries, often in electronic form. Responses take time to prepare since every effort

Table 11.3. Typical indexing features in company product information systems

Drug action	Number of patients
Disease or organism	Biochemical/physiological parameters affected
Route of administration	Biopharmaceutical parameters measured
Formulation	Type of study
Dosage schedule	Adverse events
Drug interactions	Drugs use in combination or concurrent therapy

is made to ensure that the information provided is accurate, fair, objective and based on current scientific data, both external and internal. Master enquiry files maintaining current and approved responses to regularly received enquiries or prepared in advance to anticipated questions prior to a product launch can facilitate the turnround of enquiries considerably. Again, in international companies, approved corporate responses can be shared across a number of subsidiaries via a central and accessible database.

The characteristics of the enquiries received by a pharmaceutical company provide useful management information. Computerized monitoring systems which record product, enquiry type, details of enquirer, method of enquiry – phone, letter, etc. – and response time can yield useful information not only on the performance of the information service but on trends in drug-related issues. The effective information service can then exploit this information to brief sales representatives fully on matters of current medical concern.

The role of the industry in information provision at external meetings – or the use of information for company promotion

The development of good quality medical information services is in itself a promotional activity and companies recognize the benefits that this can bring. Extensions to this service can be noted as companies take their enquiry service to the customer through provision of information-dominated exhibits at major and often international medical meetings. The service provided normally centres around computerized literature searching utilising either online or local services such as CD-ROM. Companies do not limit searches to their products but will willingly retrieve information on competitor products, the work of particular researchers, in short on any scientific topic, related or not to the topic of the conference. Many of the requestors will have little or no access to computerized literature retrieval facilities in their own country or will be unaware of the benefits of modern techniques for information retrieval. Services of this type therefore not only promote the image of the company as one committed to the provision of information services of the highest quality but meet a real information need from the attending physicians and researchers (Anderson, 1988). They are on the increase.

Prescribing information documents

The provision of a full medical information service on their products is a facility that pharmaceutical companies voluntarily determine to provide. The publication of prescribing information documents is generally mandatory in those countries with regulatory authorities. Prescribing information documents have already been referred to in Chapter 8. The objective of such documents is to provide accurate and objective information on medicinal products so that rational prescribing decisions can be made by the clinician or general practitioner. The information will be a distillation of the proprietary data accumulated by the manufacturer in preparing for the product licence submission. Prescribing information documents can take a number of forms and titles, depending on country – the product data sheet, the product information leaflet, the package insert, etc. The document, before distribution, must be approved by the local regulatory authority.

In the UK, the manufacturer must produce a data sheet for every product which it is intending to promote – via advertisement, or representative activity for instance. Data sheets must be provided to all doctors and the vehicle normally chosen to do this is the ABPI *Data Sheet Compendium*. Changes to the data sheet which may be made between editions will be notified to the prescriber directly by companies.

In the US, the Food and Drug Act requires that all drugs must have adequate directions for use. For OTC drugs, these instructions for the patient are incorporated into the label. For POMs, where the instructions are for the practitioner, the information can accompany the drug either on or within its package (Fisher,1987). The *Physicians Desk Reference* is an independently produced compilation of technical information on US drugs, which contains manufacturers' information on their products and which is produced with their co-operation. Similar national arrangements and compendia exist in many other countries.

The categories of information contained in the prescribing information document will vary from country to country. Drug name, presentation(s), active ingredients, drug uses, dosage and routes, method, and frequency of administration, contraindications, warnings regarding the drug's use in pregnancy, lactation, while driving or using machinery, overdosage, drug interactions, pharmaceutical properties and precautions, e.g. storage conditions, shelf-life, package quantities, and the product licence number are featured on the UK data sheet. The information in the prescribing information document must, in most countries, be consistent with that in the approved product licence submission. The depth of presentation required does, however, vary from brief information to a very detailed technical statement. The prescribing information document provides the route for promotional control; advertisements and other literature must be consistent with the content of the prescribing information document.

The prescribing information document is dynamic and will be edited and updated as new information on the product becomes available to the manufacturer. For international companies, and most pharmaceutical companies are now international, considerable effort may be made to ensure that the information provided in prescribing information documents is consistent from country to country. Companies will choose a variety of mechanisms for ensuring such consistency – central vetting of all prescribing information documents, or a core prescribing information document which will be provided to all territories for the minimum of local adaptation to national regulatory needs. Some pharmaceutical companies will require all their subsidiaries to produce prescribing information documents, whether or not there is a local legal requirement so to do. This is likely to be an increasing trend; the embarrassment to a large company of being accused of inconsistent claims in different countries is considerable.

The prescribing information document, where legally required, is intended to ensure that the prescriber has a core unbiased summary of information on the drug to balance against other more biased information they will receive. In practice considerable effort is expended by the pharmaceutical industry in their production but it is an effort that a responsible industry regards as being very worthwhile. In the past, the industry has been criticized for inconsistent claims for drugs worldwide; consistent prescribing information is one mechanism for avoiding this.

Electronic provision of information

The vast majority of information which the industry provides is provided on paper, through the telephone, and, to a far lesser extent, face to face contact with the prescriber. There have been few attempts to capitalise on other methods of information provision but noteworthy here is the *Philex* database and various Viewdata services.

Philex (Scrip, 1989) is a database, containing a condensed version of the data sheet for each drug, prepared by the manufacturer. All core technical data is included in a form suitable for economic screen display. The database is marketed with sotfware to assist prescribing, which covers the whole of the *British National Formulary,* and with prescription and practice management software. The industry interest in the project was fostered by the ABPI and currently approximately 70 companies subscribe. The database is produced by Exeter Data Base Systems.

Within the UK, pharmaceutical companies have used Viewdata services as a method of transmitting product and educational material to doctors via closed user groups. Educational material has included details of forthcoming symposia, case studies, and journal contents pages. Little information is available on the success of such services in providing information; several companies do, however, continue to sponsor this type of information provision. Further systems are under development. *Medicinet,* a system developed in Holland, combines information provision through electronic bulletin boards (meetings and electronic journals) with software for capturing and transmitting clinical trials, post marketing surveillance facilities and access to online biomedical databases.

Other industry information activities

Industry as a publisher

The pharmaceutical industry is a significant contributor to the primary literature and, to a lesser extent, the tertiary literature. Scientists and clinicians, and other professionals, such as information scientists, statisticians, bio-engineers are, despite their industrial base, part of the scientific communication chain. Frequently, these workers will be members of relevant scientific societies, will participate in scientific meetings and will publish in learned scientific journals. As for professionals with their base in the academic sector, this publishing activity will stem from the desire to gain peer-group recognition, coupled with the responsibility felt by many to make new information widely available through the vehicle of publication. Publishing is encouraged by most companies who recognize the value of peer-group contact and it is only restricted by the obvious need to secure patent protection for new ideas prior to publication.

A summary of the volumes of primary literature published by the top ten pharmaceutical companies in 1988 is given in *Table 11.4*. The articles cover a wide range of topics – chemical synthesis, chemical and biological analytical techniques, biological mechanisms, *in vitro* performance of new compound classes, etc. Only a small proportion of this literature, less than 20%, discusses the clinical performance of medicines. Much of the clinical research is done by external clinicians who will publish their results in their own right. This publishing activity is encouraged by companies and many companies will support this activity through the provision of scientific editing services or medical writing services. Internal writers, skilled in the creation of scientific papers, will provide assistance to the clinician in the production of the paper. This service, while creating the larger benefit of facilitating the publication of new research work, undoubtedly

Table 11.4. Journal articles published by some pharmaceutical companies in 1986–1988[a]

Company	1986	1987	1988
Merck & Co.	647	707	721[b] (729)
Glaxo Holdings plc	142	174	190
Ciba Geigy AG	516	640	527[b] (604)
Hoechst AG	254	399	278[b] (368)
American Home Products Corp.	125	123	92
Smith Kline Beckmann	548	608	587
Pfizer Corp.	173	192	169
Sandoz AG	381	430	402
Bayer AG	322	280	312
Eli Lilly & Co.	314	367	427

a Figures taken from a corporate author search on *Science Citation Index*. b Figures corrected by elimination of papers on non-pharmaceutical business of the company.

benefits the company by ensuring that results on trials on its drugs are published as rapidly as possible. It should be noted, however, that in a UK report on relationships between physicians and the pharmaceutical industry some research workers expressed concern on their right to publish data *sometimes* being restricted by the companies with whom they have collaborated (Royal College of Physicians, 1986). This was addressed by the ABPI in its own subsequent report on relationships (ABPI 1988).

TERTIARY LITERATURE – BOOKS, MONOGRAPHS

The pharmaceutical industry will also encourage its workers to contribute to the tertiary literature in their field of expertise. In addition, companies will commission monographs on relevant topics, often drug-related, and will sponsor the publication of proceedings of conferences with which they have been associated or which deal with therapeutic areas with which the company is concerned. In the main these are often of high quality. However, where symposia proceedings are published as supplements to a journal, concern has been expressed that these papers are not always subject to the peer-review process (Royal College of Physicians, 1986). The ABPI has itself strongly recommended that reports should only be published in journals with an independent refereeing system and has identified those journals which have such a policy.

Industry's contribution to medical education

The pharmaceutical industry continues to sponsor scientific meetings and conferences and postgraduate and graduate educational events. Product launches will often be preceded by a programme of educational events whose speakers will include clinicians of international repute, who may have been involved in the clinical development of the product, and industry speakers. Conferences and meetings will often be of a high scientific value. Doctors claim to favour *inter alia* company-sponsored meetings and symposia, discussions with other doctors and postgraduate educational meetings as the most useful information sources when prescribing a medicine for the first time (PJB Publications Ltd 1984). Companies are investing significant resources in education programmes utilizing, in the developed world, techniques such as video conferencing to bring together large numbers of doctors. Video cassettes are also being used for dissemination of educational material and some companies are beginning to make a moderate charge for educational material and training opportunities to reinforce their educational rather than promotional content. However, it must be said that marketing via medical education is one area of activity which earns the industry criticisms of attempting to influence prescribing opinion through meetings in attractive venues with over-elaborate hospitality.

The role of the medical representative

The medical representative continues to provide a focus for transmission of information on products to health-care personnel, particularly on new products. Medical representatives still account for the largest proportion of promotional expenditure by the industry (PJB Publications Ltd., 1984). The industry spends significant time and effort in training representatives in accordance with codes of practice for their conduct and representatives continue to feature as a significant source of information in many surveys, though less important than professional medical journals, contact with other doctors, postgraduate education programmes and conferences. (Dewayne Caldwell, 1986; Abate, 1987; Peay, 1984). The representative is often denigrated (Vere, 1983; Osiobe, 1988) yet s/he does provide a direct communication link between industry and the practitioner; in addition to the representative's role in disseminating information, many of the enquiries received by the medical information service are transmitted by the medical representative and the representative plays his or her part in gathering information on prescribing attitudes and possible adverse events.

Information for patients

It seems self-evident that patients will benefit from knowing more about their drugs and certainly there is nowadays considerable interest in drug information on the part of the consumer. The industry with its comprehensive product information files should have a logical role in such provision. Despite this, the attitudes and regulations relating to patient information vary from country to country (see *Table 11.5* for examples). Industry's role in information provision must of course reflect the need to avoid interference in the relationship between patients and medical practitioners and industry activity has therefore tended to focus on the supply of printed information to be handed on to the patient by the pharmacist or the doctor.

Within Europe, improvement and harmonization of information to patients is a current priority and the European Commission hopes to have harmonized patient information provision by 1993. Groups such as EFPIA (European Federation of Pharmaceutical

Table 11.5. Patient information provision, ABPI (1987)

Country	Current practice
US	Patient package inserts (PPI) required for certain products only. In 1980, US Pharmacopoeia – Dispensing Information System (USP-DI) introduced and contains a compendium for consumers. FDA Committee on Patient Education established
France	PPI or information on package is compulsory. PPIs established for 19 drug categories by the SNIP, the French Consumer's Union, GPs, and pharmacists
Sweden	A patient information compendium (Patient – FASS) was first produced in 1983
Denmark	PPIs not encouraged except for oral contraceptives. A patient compendium is available
Finland	A patient compendium is available
Greece/Portugal	PPIs are produced for the combined use of patients and physicians
Belgium	Since 1986, all new product applications must include a package insert
W. Germany	From 1976, package inserts are mandatory, but need not be patient oriented
Italy	PPIs are produced separately for patients. The Italian Health Industry Ministry is experimenting with simplifying patient information leaflets

Industries Association) and CRIOC (Centre of Research and Information for Consumer Organisation) are now supplying detailed evidence to the Commission in response to a comprehensive proposal to amend much of the existing directive which applied to patient information leaflets.

The ABPI in the UK has also invested considerable effort in considering industry's role in information provision to patients. Its recent guidelines on patient information (ABPI, 1987, 1988) have been developed following consultation with representatives of consumer organizations and interested professions, and have utilized patient information surveys to obtain a view on desirability, content, design, and comprehensibility. The ABPI initiative precedes the introduction of original pack dispensing in the UK and encourages not only the introduction of patient information leaflets at this stage but also the general introduction of such information provision for all medicines.

The role of the industry in providing patient information is therefore becoming more firmly established. In undertaking this activity the industry sees its role as reinforcing and amplifying information which may already have been provided by the doctor or the pharmacist. Its prime objective is to improve patient understanding with respect to the use of their medicines. In preparing leaflets, the industry is at pains to avoid accusations of product promotion, inaccuracy, and inappropriateness of information. The ABPI guidelines on the preparation of patient information leaflets reinforce this stance: 'leaflets must be consistent with the product licence; the leaflet should not be promotional in content or intent; companies should be encouraged to liaise with others producing leaflets for similar or identical products in order to achieve compatibility of layout and text; the leaflet must be reviewed by a doctor or pharmacist as part of the company approval mechanism.' Considerable thought has been given to leaflet design: a front sheet reinforcing key information e.g. the product name, the type of medicine, and up to five key points on the product; a reverse sheet providing more detailed information e.g. the general purpose of the medicine, precautions, how and when to take the medicine, duration of treatment, possible side effects, and how to store and dispose of the medicine (ABPI, 1988).

The industry's progress in providing patients with simple factual information on products is to be welcomed as an opportunity for patients to receive good quality information of a pre-defined and internally regulated standard. It does not represent, however, the sum total of industry activity with respect to patients. The provision of educational booklets on particular diseases treated with pharmaceutical products and on how to use drug administration devices such as inhalers effectively via patient organizations, in this instance the Asthma Society, is relatively common.

Industry sponsorship is also extending to more active patient information programmes; for example the SHARING programme for education in ulcer disease (Anon, 1988) and AIRLINK, a pilot information service for patients with asthma (Chaplin, 1988). In the US where direct consumer advertising by the industry is allowed, companies are beginning to advertise in support of their prescription products (Scrip 1988). Here, however, the use of industry-sponsored advertising to encourage patients to seek information packs on a particular disease which will provide product samples and refund coupons is hardly likely to enhance the industry's reputation as an objective information provider.

Information for developing countries

As in the developed world, the industry acts as a source of information on its products via the medical representative and other activities to the developing countries. The IFPMA *Code of Practice* imposes the same standards of information provision on member companies *wherever* they market their products and stresses that information provided in developing countries must be consistent with that provided in the developed world. However, this commitment has developed following considerable criticism of the industry's promotional activities in developing countries.

IFPMA, mindful that many developing countries are not aware of the indications, contraindications, and side effects of drugs that have been accepted on the market in developed countries, has offered since 1982 to supply copies of current standard compendia, free of charge to government health departments of these countries e.g. the ABPI's *Data Sheet Compendium (UK)*, the *Physicians Desk Reference (US)*, the *Rote Liste* (Federal Republic of Germany) and the *Dictionnaire Vidal* (France) (IFPMA).

The industry's initiatives to improve health-care in developing countries focus on many activities other than product promotions. They include projects to improve procurement, storage, and distribution of medicines, projects to improve quality control testing of the generic products used in many countries, and initiatives to train and educate health-care workers. The pharmaceutical industry provides funding and practical experience for many of these projects. For instance, in East Africa the ABPI is working with the African Medical and Research Foundation (AMREF) to encourage rational use of medicines through a diagnostic/information service (Scrip 1986). A central and portable diagnostic laboratory will provide the focus for field work and training. Technology transfer is the aim of another initiative, established by WHO and IFPMA to offer training in industry quality control laboratories to technical workers involved in government control laboratories and inspection services (Scrip, 1987). This type of activity is far more relevant for the countries in question than the medical information activities common in the developed world.

From the viewpoint of the western observer, the impact of these activities locally is unclear. However, a recent survey of physicians' and pharmacists' drug information services in Nigeria suggests that the industry is certainly not regarded with a high degree of credibility (Osiobe 1988). Journals, monographs and textbooks, personal contacts, private files and pharmacy reference works were cited as principal sources of drug information. In fact, the researcher noted the low use of pharmaceutical industry-based sources of information as one gratifying aspect of the study, noting previously published criticisms of the industry's lack of objectivity in view of its commercial motives. Undoubtedly, the industry's promotional approaches in developing countries in the past have not always been of a particularly high standard. The industry now must realize its part in encouraging developing countries to use their limited resources for health care as wisely as possible.

Information on generic medicines

This chapter is not the place to comment on the pros and cons of generic prescribing; this debate will undoubtedly occur elsewhere. It is, however, appropriate to note an information side-effect of increased prescribing of generics, stimulated by economically driven measures such as hospital or regional formularies and, in the UK, by the introduction of the *'Limited List'* in 1984.

The ability and commitment of a company to provide a professional information service to back up its product marketing will obviously be related to its size and willingness to invest in information provision, and also to the company's involvement in research and development, which will largely determine the information in its possession. Companies which exist to market generic products which are manufactured by other suppliers are unlikely to have either the resources or expertise to provide post-marketing information. Where the original manufacturer continues to market the generic equivalent of the original branded product it is likely that medical information support will continue, albeit on a reduced scale (Griffin, 1986). However, the original manufacturer is likely to become increasingly wary of information requests on the product which may derive from users of generic equivalents. The success of the *Philex* project has been hindered by controversy on whether information on generics should be included alongside information on branded drugs.

For long established products, information will be available from standard texts and monographs although this is unlikely to satisfy all possible enquirers. It is therefore important that generic manufacturers develop medical information services of an adequate standard to provide the required product support. The industry in general via its trade associations must monitor and encourage professional standards in generic product support – generic houses so far show few signs of providing the type of medical information service described above.

Perspective on the pharmaceutical industry's information provision: the question of bias

Drug information is so fundamental to health care systems that in the advanced nations of the world, drug information networks have been developed as an integral part of the health service system. Within the US, drug information centres have existed for more than 20 years (Amerson, 1986). Within the UK their advent is much more recent (George, 1983). Their roles, however, are similar – services provided by drug information centres will include (Rumore, 1989):

- query answering from pharmacists, physicians, and nursing staff – aimed at providing unbiased and good quality information on drugs and other aspects of therapy;
- participation in therapeutics committees concerned with the development of local formularies and the preparation of drug evaluations;
- active dissemination of information on drugs through newsletters, bulletins, and journals;
- new drug reviews;
- reviews of drug use;
- involvement in regulatory reporting programmes, e.g. for adverse events;
- provision of poison information;
- education and training;
- and, in some countries the provision of information to patients.

Drug information centres in the UK make heavy use of the industry's information services (Williams, 1984). Yet the industry itself is still just barely recognized as a significant contributor to the health information chain –although the pharmaceutical representative features as a recognized though low-level information source in surveys on sources of information on drugs (Osiobe, 1988; Abate, 1989). And in one US study representatives were found to be the principal source of information on new products for community pharmacists (Dewayne Caldwell, 1986). Information from the industry is, of course, inevitably tarnished with the term 'bias'. The charge of bias is difficult to refute since it is rooted in the firm view that if a company wishes to sell products, any information it provides on them must be subjective and extravagant in its claims. Survey information which provides quantitative information on company performance in the provision of medical information services is limited and ambivalent (Hibberd, 1981; Bainbridge, 1982; Tse 1982). From the published evidence, therefore, the industry's performance in terms of bias is unproven; its performance in terms of efficiency of turnround of requests and comprehensiveness of replies is adequate but no more than adequate. Yet its services are used and used again by the very professionals who levy the charge of subjectivity.

Professional information organizations in the pharmaceutical industry

Information professionals working within the drug industry recognize the importance of setting and maintaining the highest possible standards of professionalism. They also recognise the value of contact between colleagues throughout the industry to the development of professional standards and information practice.

In many countries therefore pharmaceutical information professionals have formed organizations to facilitate communication between professionals on current activities and new developments and to provide pressure where necessary for the improvement of the services and tools which are essential to their work.

Some of these groups are composed of individual information professionals seeking a focus, e.g. The Association of Information Officers (AIOPI) in the UK, and the Library and Documentation Group of the Drug Industry (LIDOK) in Sweden. The Gruppo Italiano Documentalisti Industria Farmaceutica e Istituti di Ricerca Biomedica (GIDIF-RBM) in Italy is a group which includes both drug industry staff and workers in research institutes and health units. The Drug Information Association is an international association focusing on information systems for handling drug information throughout the drug development process and it also includes industry, academic, and government personnel among its membership.

Other organizations have been established under the auspices of the pharmaceutical industry trade association for that particular country, e.g. the Science Information Sub-section of the Pharmaceutical Manufacturers'Association in the US. The ABPI in the UK has established a Medical and Scientific Information Committee to advise it on information matters. This group maintains a watching brief on industry information activities and investment in information services.

Company-based associations include IMPI (Information Managers in the Pharmaceutical Industry), an informal network of UK information managers in large research and development based pharmaceutical companies, and the Pharma Documentation Ring (PDR). The latter consortium was founded in 1959, has an extensive European membership, and was formed to provide a framework for specific collaborative projects, discussed elsewhere in this Chapter.

Despite their different constitutional arrangement, all these organizations provide a focus for professional development through seminars, conferences, newsletters and other publications. Often working parties will be set up to study new information developments in depth in order to create a knowledge base for the membership, e.g. AIOPI's working parties on records management, information technology, and adverse event monitoring. Training is of particular interest. In the UK, a key training scheme for many new entrants to the pharmaceutical industry information departments is provided by AIOPI. Run as a co-operative self-help scheme, the AIOPI training course comprises a series of 10 or so visits to different UK pharmaceutical companies during a 12 month period. The scheme has been very successful, has run for more than 10 years and continues to attract large numbers of applicants. One session is run jointly for industry staff and drug information pharmacists in the health service and is a particularly important contribution to fostering good relationships between the two groups.

All these various information organizations are dynamic and developing, reflecting the interest of the industry as a whole and the commitment of its information workers to improving the industry's status and their own standards of operation.

The contribution of the industry to the general development of information services

As an extensive user of external information sources, the pharmaceutical industry has undoubtedly provided a stimulus for improvements in the available services. Likewise, its search for effective methods of handling its proprietary information has stimulated developments in information handling methods whose benefits extend well beyond the industry.

Developments in sources

The Pharma Documentation Ring (PDR) was established in 1958 as a vehicle for co-operation between European pharmaceutical companies (Ash and Hyde, 1975). One of their purposes was the development of a chemical indexing system, *Ringcode,* to encode structural information and biological activity data from the chemical and drug literature. The resultant database, *Ringdoc,* was subsequently transferred in 1964 to Derwent Publications who continue to offer it both as a commercial search service and as an abstracting and indexing service with the regular publication of *Ringdoc* profile booklets. The *Chemical Reaction Documentation Service* produced by Derwent Publications is also derived from the PDR's initiative.

The Royal Society of Chemistry's *Hazards Information Service* was developed as a result of the UK pharmaceutical industry's interest and commitment to the improved availability of health and safety information on both laboratory based and industrial hazards.

Developments in information handling methods

The importance of chemical structure handling to the industry has resulted in significant contributions to information management techniques. The ICI *CROSSBOW* software developed in the 1970s to provide sub-structure searching facilities for large chemical files based on Wiswesser Line Notation (WLN) was a particularly important development. ICI's decision to make the *CROSSBOW* software commercially available provided many UK companies with the opportunity to take a quantum leap in improving access to their large in-house chemical information collections. ICI's release of the text handling package, *Assassin,* provided a large number of companies with the opportunity to purchase software for indexing and retrieval of textual information developed with the practical problems of the industrial information unit in mind. The software has undergone major development from its original batch searching version to a version which competes effectively with other more recent text handling packages.

The *CROSSBOW* software provided the mechanism by which a number of UK pharmaceutical companies in the 1970s collaborated to create a database of structural information on commercially available organic chemicals. *CAOCI (Commercially Available Organic Chemicals Index)* took approximately three years to develop, was subsequently handed over to Fraser Williams Scientific Systems Ltd, to manage under the name *ChemQuest* and is now owned by Maxwell Communications Corporation (North, 1989).

Currently, the most important source of published chemical information is *CAS ONLINE.* The structure-searching techniques on which this service is based were introduced by the Basel Information Centre, a consortium of Swiss pharmaceutical companies. Their experience and commitment to improving access to *Chemical Abstracts* files undoubtedly stimulated CAS to launch a similar development themselves.

The impetus for this type of activity in the 1960s and 1970s appears to have been replaced in the 1980s by the collective lobbying of information providers and software houses to develop or improve services and systems for use by the pharmaceutical industry; a possible reflection on the scale of the information industry sector now available to be influenced and the high costs of information systems development.

Conclusions and forward directions

The pharmaceutical industry recognizes several key responsibilities in the provision of information on drugs (Boyland, 1983):

- to identify all published and unpublished information on its products;
- to evaluate, index and store such information;
- to ensure that this information is accurate;
- to communicate this information both pro-actively and on request;
- to provide a fair balance in responding to requests for information.

The industry has made and continues to extend its considerable investment in information services. It compiles the most comprehensive information source on its products and devotes a considerable resource to the communication of this information. Its initiative in areas such as patient information and medical education deserve credit as a contribution to the safe and effective use of drugs.

The industry's information services are used by health-care professionals and by researchers and, through educational initiatives and patient information publications, are increasingly being used indirectly by the general public. Surprisingly, since resources for information provision outside the industry are often severely limited by lack of funding, the industry is still not regarded credibly as an information source and industry's aim must surely be to secure a recognized and respected role in this respect. However, realization of this aim is not something which industry can accomplish alone – there needs to be a corresponding appreciation and acceptance of what industry has to offer by the drug prescribers and consumers.

What should industry do to improve its services? As a first step, it must review the effectiveness of its information provision to drug information users on a regular basis. At the moment, companies know that their information services are used but they do not know what value those services are accorded. Industry should continue to promote its information facilities at conferences, exhibitions, and in publications focused at potential users and use these contacts for evaluative feedback.

Industry must also continue to improve standards of information provision, including the accuracy and objectivity of information provided, and must reinforce the need for international consistency of product information throughout companies and their overseas subsidiaries. Recipients of information must neither be overloaded nor deceived: both accusations are frequently levelled at the industry. More positive approaches to collaboration with its information consumers should be considered, including actively contributing to independent sources of product information and comparative assessments of product performance. As electronic access to information by doctors becomes more common, the industry will have to consider whether it will allow limited access to its own information databases.

The emerging themes are, therefore, quality control and co-operation. It is vital to the well being of the pharmaceutical industry that its services continue to develop and gain respect since it is partially through the use of such services that the industry receives feed-back on the use of its medicines. The proposed partnership is equally important to the consumer since making effective use of the primary source of information on medicines is logically the most appropriate and least costly method of obtaining information relevant to improved health-care.

References

Abate, M. A., Jacknowitz, A. I., Shumway, J. M. (1989). 'Information sources used by private practice and university physicians,' *Drug Information Journal*, **23**, 309-19.

ABPI Code of Practice for the Pharmaceutical Industry, (December 1988). *7th ed* (Data Pharm Publications Ltd., 7th edn).

ABPI (1987). *Information to patients on medicines* (ABPI).

ABPI (1988). *Guidelines on the disclosure of inactive ingredients and the evolution of these guidelines* (ABPI).

ABPI (1988). *Relationships between the Medical Profession and the Pharmaceutical Industry* (ABPI).

ABPI (1988). *Patient information: advice on the drafting of leaflets* (ABPI).

Amerson, A. B. (1986). 'Drug information centres: An overview,' *Drug Information Journal*, **20**, 173-8.

Anderson, T. (1988). 'Database marketing: Squibb's educational promotion,' *Pharmaceutical Executive*, 8/10, 46.

Anon (1989), 'Beecham's infection information service in France,' *Scrip*, **1420**, 12.

Anon (1988), 'Glaxo's SHARING (Self-help advice Resource in Gastroenterology) initiative,' *Scrip*, **1353**, 11.

Ash, J. E. and Hyde, E. (1975). *Chemical information systems* (Ellis Horwood Ltd., 247).

Bainbridge, C. V., D'Ambrosio, G. G., Petrick, R. J., and Weiss, E.M. (1982). 'Pharmaceutical manufacturer's responses to drug information inquiries,' *American Journal of Hospital Pharmacy*, **39**, 1532-4.

Bassin, L. G. and Bilhuber, P. A. (1986). 'Functioning of a medical information unit within a large pharmaceutical organisation; a methodological approach and review,' *Drug Information Journal,* **20,** 165-172.

Bauman, J. H. and Fuentes, R. J. (1989). 'Drug Information at Glaxo Inc.,' *Drug Information Journal,* **23,** 95-103.

Boyland, J. I. and Brdlik, M. F. (1989). 'One critical hour: handling drug information calls at Roche,' *Pharmaceutical Executive,* April, 46-9.

Boyland, J. I. (1983). 'Providing drug information to health professionals and consumers: role and responsibility of the pharmaceutical manufacturer,' *Hospital Pharmacy,* **18,** 211-6.

Chaplin, S. (1988). 'Patient information, from leaflets to TV?,' *MIMS Magazine,* 15th June, 23-4.

Collins, G. E., Frieden, C. S., Cloutier, G., and Cato, A. (1987). 'An integrated approach for processing inquiries at the time of a new product launch,' *Drug Information Journal,* **21,** 165-71.

Dewayne Caldwell, R., Generali, J. A., Scott, B. E., White, S. J., and Gerald, K. (1986). Comparison of Drug Information Resources observed in community pharmacies, *Drug Information Journal,* **20,** 77-82.

Fisher, M. P. (1987), 'Preparation and maintenance of prescription drug labelling,' *Drug Information Journal,* **21,** 153-7.

George, C. F. and Hands, D. E. (1983). 'Drug and therapeutics committees and information pharmacy services: the United Kingdom,' *World Development,* **1113,** 229-36.

Griffin, J. (1986). *Generic pharmaceuticals and the provision of information* (AIOPI, Proceedings of the 13th Annual Conference, Harrogate, 1986, 64-77).

Hibberd, P. L. and Meadows, A. J. (1981). 'The pharmaceutical industry as a provider of drug information,' *Journal of Research Communication Studies,* **3,** 289-97.

Hoff, L. C. (1983). 'The role of the innovation-based pharmaceutical industry in international drug information communication,' *Drug Information Journal,* **17,** 271-6.

Hyams, P. (1989). 'Deeply Coding with CAIRS,' *Information World Review,* **34,** 16.

IFPMA Code of Pharmaceutical Marketing Practices (FIIM/IFPMA, Geneva, Switzerland).

Medicinet (De Medicus, P.O. Box 182, 2350 A.D. Leiderdorp, The Netherlands).

Nelson R. P. (1989). 'Organization of information in the U.S. pharmaceutical industry,' *Drug Information Journal,* **23,** 151-64.

North, S. A., Skidmore, J. F., Tarr, I. J., and Ward, S. E. in 'Sources of information, chapter 4-14' in 'Volume 1, Comprehensive Medicinal Chemistry' (Pergamon Press, 1989, ed. C. Hansch).

Osiobe, S. A. (1988). 'Physicians' and pharmacists' drug information sources in Nigeria,' *Drug Information Journal,* **22,** 553-63.

Peay, M. Y. and Peay, E.R. (1984), 'Differences among practitioners in patterns of preference for information sources in the adoption of new drugs,' *Social Science in Medicine,* **18,** 12, 1019-25.

PJB Publications Ltd. (1984). *The pharmaceutical market in the United Kingdom.*

Royal College of Physicians (1986), 'The relationship between the physician and the pharmaceutical industry, *Journal of the Royal College of Physicians of London,* **20,** 4, 3-10.

Rumore, M. M. and Rosenberg, J. M. (1989). 'Comparison of drug information practice in hospitals and industry,' *Drug Information Journal,* **23,** 273-83.

Salter, F. J., Kawano, J. C., and Pherson, D. A. (1986). 'Management of drug information inquiries from health care professionals: An industry wide survey,' *Drug Information Journal,* **20,** 187-94.

Scrip (1988). 'Smith Kline and French launches TV ulcer campaign,' *Scrip*, **1297**, 8.
Scrip (1988). 'Glaxo standardises product information,' *Scrip*, **1303**, 12.
Scrip (1986). 'ABPI's Third World projects,' *Scrip*, **1101**, 17.
Scrip (1987). 'Association of the British Pharmaceutical Industry's Third World projects continue,' *Scrip*, **1265**, 3.
Tse, C. S. T. (1982). 'Disclosure of product information by manufacturers,' *American Journal of Hospital Pharmacy*, **39**, 645-6.
Vere, D. (1983). 'Drug information,' *Journal of Royal College of Physicians of London*, **17**, 3, 159-160.
Weiss, P. (1986). 'Health and biomedical information in Europe,' (WHO Regional Office for Europe, Public Health in Europe 27).
Williams, G. A., Beaumont, C., Roach, M, Ward, S. E., and Wallis, D. A. (1984). 'Information services in the development and use of medicines. Part 1,' *Pharmaceutical Journal*, **233**, 381-3.
Williams, G. A., Beaumont, C., Roach, M., Ward, S. E., and Wallis, D. A., (1984). Information services in the development and use of medicines. Part 2,' *Pharmaceutical Journal*, **233**, 500-2.

CHAPTER TWELVE

The role of the World Health Organization

J. F. DUNNE

This chapter, which is published with the permission of the United Nations agency, the World Health Organization (WHO), is based upon material prepared for the Conference of Experts on the Rational Use of Drugs in Nairobi, Kenya, 25–29 November 1985, and included in *The Rational Use of Drugs* (World Health Organization, 1987).

Introduction

Within the terms of its constitution WHO acts as the directing and co-ordinating authority on international health work. Among its functions it is required to assist governments, upon request, in strengthening health services, and to provide information, counsel and assistance in the field of health. Virtually from its inception in 1948, the governing bodies of the Organization identified drug regulation and control as a field in which these responsibilities could be applied to useful effect. Over the years, many formal resolutions have been adopted by the World Health Assembly which call for the establishment of international norms, exchange of information and multilateral collaboration to support the drug regulatory apparatus of national governments.

Whereas the cumulative rate of expansion of the scientific medical literature is prodigious and much of this is related to drug therapy, little of this output directly influences the prescribing practices of doctors. The original literature is largely inaccessible to the busy generalist and over the past two decades an appreciation has developed that greater efforts are needed to provide prescribers with readily assimilated, independent and objective information that will keep them adequately informed of changes in therapeutic practice throughout their professional careers. The problem is evident in both developed and developing countries. It is a product of the current and unparalleled rate of therapeutic innovation, and it is exacerbated in market economy countries by the promotional activities of competing pharmaceutical manufacturers. This, in turn, has resulted in varying measures of governmental and self-imposed control over the content and presentation of advertising material by pharmaceutical manufacturers and has also stimulated governments and the medical profession to take a variety of initiatives in the supply of independent prescribing information.

To some extent the prescriber's need for information on drug products has been alleviated by the institution of national drug regulatory authorities, particularly in those industrialized countries where drug innovation is largely concentrated, since their influence has resulted in the elaboration of independent and authoritative standards of quality, safety and efficacy in marketed products. It is beyond the capacity and the competence of prescribers to assess at first hand the potential risks and benefits of the drugs that they use. Thus, the necessity of creating independent multidisciplinary bodies at national level to adjudge the acceptability of new products for general marketing and to subject existing products to systematic review, would ultimately have become apparent without the emotive stimulus of the thalidomide tragedy. Regulatory authorities in market economy countries are not, however, constituted to develop as primary sources of drug information. Although several authorities are becoming more active in this regard their terms of reference typically invest the licence holder — usually the drug manufacturer — with the prerogative and responsibility of informing and advising prescribers on the use of the relevant product. The informational role of the regulatory authority is limited, in these circumstances, to ensuring that the product is advertised in a manner that is consonant with the terms of the product licence.

Countries that have yet to introduce comprehensive provisions for drug regulation can draw from a diversity of national systems in determining their own requirements. Nonetheless, problems in establishing drug control in developing countries have, too often, resulted from the adoption of legislative provisions successful elsewhere, but of a complexity that precludes their effective implementation with the resources available. As an alternative to adopting regulatory systems devised for countries with different economic, commercial and social circumstances, scope exists for developing countries to consider whether statutory recognition might be accorded to existing international systems for exchange of information such as, for instance, an essential drugs policy as embodied in the various WHO reports on the *Model List of Essential Drugs*[1] and the *WHO Certification Scheme on the Quality of Pharmaceutical Products Moving in International Commerce*[2].

However, in order to place the task of the Organization in perspective, it is necessary to review briefly how current trends in drug control and drug development determine the structure of national markets.

The structure and control of national drug markets

The administrative control of pharmaceutical products

Contemporary approaches to the control of biological and pharmaceutical products have evolved in highly developed countries over more than 50 years. The course of the evolutionary path has been determined by the fact that the beneficial, therapeutic effects of marketed drugs are, in general, more readily demonstrable than their attendant hazards. Whereas it is incontestable that innovative development of new products has transformed the practice of medicine, drugs are biologically active substances with an innate potential to exert adverse as well as beneficial effects at pharmacological dosage, and that therapeutic progress is inherently associated with risk. Evaluation of the costs, benefits, and hazards of drug treatment has consequently assumed a complexity that was formerly unappreciated. Already the spectrum of controls applied to the research-based pharmaceutical industry is singularly comprehensive.

The greatest emphasis has been accorded to the need for exhaustive technical assessment of each new product prior to marketing. However, adequate assurance of quality, efficacy and safety is contingent on many other safeguards including the implementation of good manufacturing practices, efficient distribution and storage, informed use of products, and systematic collation and analysis of experience with marketed drugs. In highly developed countries controls are consequently exercised over manufacture and packaging, labelling and promotion, distribution, sale and use, and the reporting of post-marketing experience.

In turn, this requires not only a central agency to determine the conditions under which each product is accepted for registration, to oversee advertising and promotional practices, and to ensure adequate post-marketing surveillance, but also a highly educated cadre of doctors, nurses, and pharmacists qualified to prescribe and dispense the products with understanding and discretion, and a body of inspectors and chemical analysts to assure the quality of products throughout the distribution chain. The cost of the required administrative apparatus is burdensome even to the most affluent of countries and it is reflected in the prices of pharmaceutical products whether they are destined for domestic use or for export. Thus developing countries support an infrastructure of control, as well as a research base, in exporting countries that they are unable, for lack of resources, to institute themselves.

However, the need for drugs is no more forcefully apparent than in the developing world. To deny populations access to the benefits of medical technology for want of administrative capacity is inadmissible in concept. The problem has to be relieved by ensuring that national approaches to drug control and the provision of associated information to the profession and the public are optimally adapted to local circumstances. The extent to which success can be achieved is contingent upon the support afforded by governments of exporting countries and by the manufacturers.

Governments in exporting countries can ensure that their statutory provisions and regulations provide for adequate control of exported products without impeding the delivery or development of drugs legitimately needed elsewhere for which there is no domestic market. They can also assure the effective operation and further evolution of the *WHO Certification Scheme on the Quality of Pharmaceutical Products Moving in International Commerce*[2] with a view to providing importing countries not only with an attestation of the quality of the product but also with authenticated information on its safe and effective use.

Finally, a complementary role exists for WHO in promoting contact and understanding between national drug regulatory authorities; in effecting the exchange of objective, unbiased technical information selected and presented in a manner that sustains countries in their quest for self-sufficiency; and in fostering and co-ordinating therapeutic research on an international basis, particularly within the developing world.

The technical assessment of pharmaceutical products

Whereas the *WHO Model List of Essential Drugs* contains less than 250 pharmaceutical substances, the number included in registered pharmaceutical products in some countries exceeds 3000. The total number of products available on these national markets is many times greater since most of these substances are offered in a variety of formulations – either singly or in combination – both as branded and as generic items. Traditional and unorthodox systems of medicine retain recognition and considerable patronage in many highly developed countries, but in the rural areas of some developing countries it is the traditional healers that provide the most readily available resource for the development of primary health-care services. Also at issue, however, is an extensive and highly remunerative illicit trade in fraudulent or spurious medicines which is not confined solely to countries with weak regulatory systems.

Comprehensive national licensing systems that require marketed drug products to meet independently determined standards of quality, efficacy, and safety have been instituted only within the past two decades and priority has consistently been accorded to the assessment of newly developed products. Few national regulatory authorities have, as yet, reviewed every product currently available on their domestic market, but many are committed to do so within the next few years. The recent demonstration that a substance as widely contained in herbal remedies as aristolochic acid possesses a potent carcinogenic potential has disposed of the complacent assumption that prolonged use is, of itself, sufficient to establish the safety of a medicinal product.

The rational use of drugs evolves from a profound and secure understanding of their clinical performance. At present, much of this knowledge is superficial and incomplete. However, the emergence of epidemiologically based approaches to the investigation of drug performance seems destined to create a shift of emphasis in drug assessment to a more balanced combination of pre-marketing evaluation and post-marketing surveillance. The resources and organizational effort required to undertake such studies are daunting. Nonetheless, the development of computerized databases of prescribing information on the one hand, and of patients' hospital records on the other, renders feasible the collection and analysis of more information on drug performance than has previously been possible. Governments, because of their general responsibility for their health services and pharmaceutical companies, because of their responsibility and liability for the safety of their products, have a direct interest in fostering such studies and in exploring their logistic, financial and ethical implications. Regardless of the approaches ultimately adopted, the efficiency with which the work is executed and the extent to which it influences the use of drugs globally will depend greatly on international co-ordination of efforts and timely exchange of information.

Economic factors as determinants of drug development and use

The concept of the rational use of drugs implies a need for socially oriented approaches to new drug development. The commercially determined and competitive structure of the research-based pharmaceutical industry require research expenditure to be determined by the prospective return on investment as well as by medical need. The problem is compounded by other perceived disincentives to innovative research: the lack of secure scientific leads for new forms of therapy; rising costs of drug development at a time of general economic recession: stringent price controls imposed within the public sector: and short effective periods of patent protection coupled with selective encouragement of generic manufacture and prescribing.

The consequential pattern of drug development, with its emphasis on treatment and prevention of the common diseases of affluent communities, draws criticism as being ill adapted to global therapeutic needs. Thus the *Sixth Cumulative List of International Nonproprietary Names for Pharmaceutical Substances* published by WHO in 1982,[3] shows that no fewer than 78 benzodiazepines and closely related compounds had been taken to an advanced stage of development. The inclusion in the list of 67 beta-adrenergic blocking agents, 53 non-steroidal anti-inflammatory agents of the ibufenac and ibuprofen groups, 57 penicillins and 42 cephalosporins offers revealing insight into the prevalence of repetitive research patterns. Nevertheless, the new perspectives that have been opened by recombinant DNA technology already exemplified by the commercial availability of new hepatitis B vaccines, thrombolytic agents and erythropoietin belie much pessimistic prognostication about the pace of therapeutic progress. Moreover, the ongoing clinical development of drugs such as ivermectin, mefloquine, halofantrine, praziquantel and polyamide inhibitors provides evidence of the commitment of some research-based companies to the advancement of tropical medicine. One disturbing trend is, nonetheless, already apparent: in recent years several leading manufacturers of vaccines have opted to withdraw from the field pleading reduced profitability and sharply increased liability for product-induced injury. At a time when the emergence of AIDS and legionnaires' disease provides a salutary reminder that infectious disease will never remain a static target; when vaccination holds encouraging prospects in the management of parasitic, as well as bacterial and viral infections; and when bioengineering techniques offer important new approaches to vaccine manufacture, this is an ominous portent.

Whether action is now required to redress these trends is an issue that is critical to the future of drug development. Some governments are either contemplating, or have already introduced, measures that offer a degree of protection or inducement to companies prepared to address societal responsibilities. Several governments have also provided direct financial incentives and rights to prolonged periods of commercial exclusivity to stimulate the development and production of 'orphan' drugs for rare diseases that would otherwise offer no commercial attraction.

At the global level the call to increase the responsiveness of pharmaceutical research to contemporary needs brings the problems of the developing world into sharp relief. The prevalence of malaria has probably doubled over the past decade as both the parasites and the vectors have developed resistance against the available drugs and insecticides. Other conditions such as measles and diarrhoeal diseases, which rarely result in death in affluent countries, remain major causes of infant mortality in developing countries. Even so, national populations are increasing in these countries, as elsewhere, at a rate that is imposing an intolerable strain on economic resources already extended to breaking point.

It is against this background that WHO has initiated major research programmes in the fields of tropical and diarrhoeal diseases and in human reproduction,[4-6] in the expectation that much may be done on budgets that are modest by commercial standards to rationalize existing therapy and even to develop new therapeutic tools, both by co-ordinating and building upon resources that already exist within the academic world and industry and by judicious financial support of promising research.

Coincidentally with the early development of these research programmes, the World Health Assembly endorsed, in 1975, a report of the Director-General[7] pointing to the need for comprehensive, centrally directed drug policies within developing countries as a prerequisite to satisfying the basic health needs of under-served populations. Reference was made to experience gained in countries where schemes of basic or essential drugs[8] had been implemented and WHO was called upon to advise Member States on the selection and procurement, at reasonable cost, of essential drugs of established quality meeting their national health needs. It remains the responsibility of governments to determine the extent to which selective procurement policies are implemented and to adapt the *WHO Model List of Essential Drugs* to specific local needs and policies. Wide contrasts in national circumstances render it impossible to draw up a drug list of general applicability and acceptability.

Since the essential drugs concept was first elaborated it has gained wide recognition. It has provided a rational yet flexible basis for systematizing drug procurement and use at national level and for establishing drug needs at specific points, or for specific purposes, within the health-care system.

Whether it is primarily a function of government or of professional self-discipline to constrain prescribing costs is a political issue. Many national drug regulatory authorities are required, in their licensing function, to confine their attention to matters of quality, safety and efficacy. Others are empowered also to consider, as a condition of registration, whether a product meets a perceived medical need. Nonetheless, public expenditure on drugs is everywhere identified as an important and potentially negotiable element in the overall cost of public health services. Increasingly, governmemts reveal a determination to reduce drug costs, not only through direct price controls and selective registration but also through selective reimbursement of prescription costs, compulsory generic licensing, or promotion of generic prescribing and dispensing. The dilemma that emerges for all governments is to reduce public expenditure on drugs as far as is practicable without eroding the standards of the health services they provide and yet assuring socially productive investment in new drug development.

The exportation of pharmaceutical products

A majority of Member States import large quantities of both finished pharmaceutical products and bulk substances. Many do not, however, possess the technical and financial resources to undertake a comprehensive and independent assessment of the drugs on which they depend. The major drug-exporting countries have adopted divergent positions in their legislative responses to this situation. One approach, which derives from the doctrine of state responsibility and the concept of international minimum standards, is to disallow the export of pharmaceutical products that have not been approved for domestic sale. The other, based upon the principle of comity of nations, is to accept the right of each sovereign government to decide what medicines it will import having regard to its own assessment of its particular health needs, the diseases and health-related characteristics of its population, the nature of its health-care delivery systems, the availability of treatment and its own evaluation of benefits and risks. Patterns of disease and the structure of medical care vary within wide limits from country to country. The balance of risk and benefit in using medicines varies correspondingly. The unacceptable burden of disease in many developing countries results from infections that either do not occur in developed countries or that are effectively contained where highly evolved medical services are available. In consequence, the administrations of many exporting countries consider that they are inadequately informed to make value judgements on the safety and efficacy of medicinal products that are to be used under circumstances, and for conditions, that are alien to their experience. Developing countries, nonetheless, rely upon drug-exporting countries to develop legislative and administrative mechanisms that will effectively block the shipment of unacceptable or substandard products. Such provisions should neither impede the movement of valuable medicines where they are most urgently needed, nor provide a disincentive to the development of new drugs for diseases endemic exclusively in developing countries.

Both these principles are embodied in a resolution of the UN General Assembly (CA37/137) adopted in 1982 on problems of international trade in hazardous products, with particular reference to pharmaceuticals. It records that products either banned or withdrawn on grounds of human health and safety within the country of origin have subsequently remained available in export markets, and that newly developed products intended – but not yet approved – for domestic use have been released prematurely for export. The resolution also acknowledges that a product unauthorized in one country may subserve a legitimate need elsewhere. It therefore calls upon governments to prevent the exportation of any pharmaceutical product not authorized for domestic use, save at the specific request of the competent authority within the importing country or when evidence is available that its use is officially sanctioned within the importing country. In either case governments are requested to ensure that the supporting information and labelling are adequate to provide for the safe and effective use of the product. To assist importing countries to identify imports that present undue or exceptional hazards the General Assembly called for the publication and regular updating of a consolidated list of products that have been banned, severely restricted or not approved in the country of origin.[9] This listing complements the *Certification Scheme on the Quality of Pharmaceutical Products Moving in International Commerce*, which was established by WHO in 1975. The scheme provides an administrative mechanism whereby regulatory authorities in developing countries may obtain, on request, details of the regulatory status of any imported product in the country of origin. It requests the competent authority of the country of manufacture to certify pharmaceutical products intended for export by supplying, at the request of the importing country, assurances, first, that the exporting country has approved the product for domestic sale (or if not, to state why not) and, secondly, that the manufacturing premises are subject to regular inspection and conform with standards set by WHO in its principles of good manufacturing practice.

International exchange of information

Mechanisms of exchange of information

The complexity of the international drug market and the urgency with which messages sometimes need to be conveyed leave no doubt about the need to develop efficient international channels of communication between national drug regulatory authorities. The development of such channels also needs to take account of the existence of profound differences both in the structure of national drug markets, and in the circumstances in which drugs are used. A decision taken in one country may not be immediately applicable elsewhere. For the most part, however, developing countries are constrained to use products that have been designed for use within totally different systems of medical care. In such circumstances the process of drug control should embrace not only consideration of evidence generated elsewhere – and most frequently in highly developed countries – on the performance of candidate drugs, but also evidence of their relevance to health-care within the national context.

International systems of exchange of information relieve regulatory authorities in developing countries of the need to undertake fully indpendent assessments of the drugs registered under their aegis. They are thus enabled to focus their attention on the acceptability of products within the prevailing health-care infrastructure. No mechanisms for international exchange of regulatory information can operate effectively, however, where there is no indigenous system of drug registration. Many developing countries have yet to create such a system. This may have been impracticable in the past, but advances in informational science in recent years have yielded data storage and retrieval techniques that bring effective data management within the reach of virtually every country. International dissemination of technical information needs to be complemented by the development of information systems that can be readily adapted to the registration requirements of every national authority.

The purpose of central administration is to serve an effective peripheral infrastructure. This demands not only a flow of appropriate information to drug prescribers and the public but also the carefully planned, assured and controlled delivery of appropriate drugs at every level of the health-care system. In any situation, rational use of drugs is contingent on effective administration of health services and effective education of health-care providers.

A number of regulatory authorities reserve the right to inspect the premises of manufacturers of imported products. Some countries have entered into bilateral or multilateral agreements to recognize and accept each others' inspection provisions. WHO, in establishing global standards for *Good Practices in the Manufacture and Quality Control of Drugs*, which are now recognized by 127 Member States, has created a basis for extending mutual recognition of inspection procedures to all countries. This is the essence of the *WHO Certification Scheme on the Quality of Pharmaceutical Products Moving in International Commerce.*

Harmonization of regulatory requirements

In the late 1960s, when many countries were first instituting statutory systems of drug registration, WHO frequently provided a forum for the discussion and elaboration of norms employed in the technical assessment of drugs. Many basic recommendations on the pharmaceutical, toxicological and clinical aspects of drug evaluation were issued under the auspices of the Organization at that time. This tradition is now continued from within the WHO Regional Office for Europe which has issued an extensive series of guidelines for the clinical evaluation of specific classes of drugs. Overall, however, WHO is now less involved than formerly in the provision of didactic technical guidance for drug regulators. Until the principles of drug assessment are further advanced, the divergences now apparent in national policies and practices are unlikely to yield to attempts to forge an international consensus. Within a purely scientific context WHO now promotes collaborative approaches to the validation of the methodological basis of drug control. The multicentre validation of analytical techniques described in the *International Pharmacopoeia* has recently been complemented in the toxicological field by a collaborative study on mutagenicity testing performed under the auspices of the International Programme on Chemical Safety.[10]

There is a case for comparative analyses of toxicological data generated by pharmaceutical companies. Indeed, a databank containing comprehensive animal toxicological data from pharmaceutical companies operating in the United Kingdom has been set up under the auspices of the industry to provide for this need and to enable a more rational approach to animal studies to be developed.[11] Similar initiatives have been taken in other countries. These will operate most effectively to the advantage of public health – and, possibly, to the commercial interests involved – if both the data and the conclusions are made generally available. The case for increased accessibility to pre-marketing safety and efficacy data applies, with the same validity, to the results of clinical studies. New drugs are typically registered for marketing on the basis of their performance in a small number of time-limited, controlled, comparative clinical trials. Even in the case of indisputably innovative products intended for prolonged use in man, therapeutic potential will rarely be assessed in more than 1000 patients, few of whom are likely to have received the product for more than a year. Prescribers disposed to use a new drug would have an opportunity to gain a deeper insight into its properties if, in addition to the information carried on the labelling concerning indications, contra-indications, precautions and warnings, they were informed of the clinical data on which the marketing approval is based and of any hypothetical risks identified by the results of toxicological testing.

In this specific area, WHO has retained the initiative to develop and update guidelines relevant to drug assessement that are of unquestioned global relevance. They concern the need to safeguard the basic human rights, the safety and the welfare of human subjects involved in biomedical research. Clinical research is undertaken in many countries. If, in particular, the conquest of tropical disease is to be advanced, drug performance must be assessed in endemic areas, and even in countries that have not previously had cause to develop relevant regulations. WHO is, itself, a sponsor of such research. It has, therefore, collaborated with the Council for International Organizations of Medical Sciences in broadly based consultations that have resulted in the publication of two sets of guidelines: *Proposed International Guidelines for Biomedical Research Involving Human Subjects*, and *Safety Requirements for the First Use of New Drugs and Diagnostic Agents to Man*.

Promotion of international collaboration and harmonization also underlies the concept of the biennial International Conferences of Drug Regulatory Authorities. The conferences were originally devised to offer drug regulators from all WHO Member States opportunity to exchange views and experience on the administrative and technical aspects of their responsibilities and to advance interagency communication. The first, which was jointly sponsored by WHO and the United States Food and Drug Administration, was held in Annapolis, Maryland, US, in 1980 [12] and subsequent meetings have been convened, in Rome (1982), [13] in Stockholm (1984), [14] and most recently in Tokyo (1986). A prime concern at all of these meetings has been to improve the flow of information between regulatory agencies and to promote the effective utilization of the *WHO Certification Scheme on the Quality of Pharmaceutical Products Moving in International Commerce*. However, regulatory policy and administration also raise issues for discussion. Without mutual understanding of administrative systems, and without a collective will to approximate methods of work, international exchange of technical information between regulatory authorities will remain   inadequate   and   unacceptably   vulnerable   to misinterpretation.

The advisory functions of WHO

Technical reports

Many of the WHO's technical reports bear upon the management and treatment of conditions of prime importance to community health standards. They reflect an internationally based consensus of expert opinion and, as such, they influence the formulation of relevant national health policy in many countries. De facto, and as a direct consequence of its constitutional mandate, the Organization is cast in an advisory role on issues that impinge directly on drug marketing and the regulatory process. Typically, these issues relate to the major health problems of the developing world. They are exemplified by the promotion of oral rehydration therapy in the treatment of infantile diarrhoea; the advocacy of combination chemotherapy in leprosy in the face of increasing dapsone resistance; and proposed constraints on the use of newly introduced antimalarial, mefloquine, to impede the emergence of resistant *faliciparum* malaria.

The concept of essential drugs

In 1975 the Organization's mandate to provide advice on the socioeconomic aspects of drug use was considerably strengthened. In the light of a report submitted by the Director-General calling for rationalization and extension of primary health-care services, particularly in rural communities, the World Health Assembly adopted a series of resolutions requesting WHO, inter alia, to co-operate with Member States in formulating drug policies and management programmes relevant to the health needs of populations, and to identify drugs and vaccines which, in the light of scientific knowledge, are indispensable for primary health care and control of diseases prevalent in the population at large. Following wide consultation, an initial model list of essential drugs was issued in 1977 within the first report of the WHO Expert Committee on the Selection of Essential Drugs. This has subsequently been revised and updated in four further reports. The objective has been to retain under review a comprehensive yet limited array of drugs of proven value in the prophylaxis and treatment of commonly occurring conditions and to demonstrate that rationalization of procurement can hold advantage in terms of both economy and efficiency in any health-care setting. A list that is consonant with local needs and policies must be compiled, as appropriate, at national, or even institutional level. Nonetheless, the WHO list has exerted considerable influence in several respects: it has served as a stimulus to all countries to consider the available options for establishing cost-effective drug policies in the public sector; it has provided a systematized approach to drug selection that is applicable in a wide variety of situations, and it has highlighted the outstanding yet discriminatory nature of therapeutic innovation. The emergence of the essential drugs philosophy and of effective primary health care programmes within developing countries offers a renewed challenge and an incentive to pharmaceutical companies disposed to remain vigilant to global health issues.

The *WHO Model List* is comprehensive in its scope. Some of the listed drugs are intended exclusively for use in specialized hospital departments. Some can be prescribed safely only within a relatively sophisticated system of medical care, and many others can be used effectively only when continuity of treatment and availability of experienced supervision can be assured. Special attention is accorded, nonetheless, to drugs required in a primary health-care setting that can be used safely and effectively by individuals with little formal medical knowledge. The Committee, however, emphasizes that the range of drugs supplied to community health workers must be determined at national level having regard, not least, to force of circumstance. In an ideal situation antibiotics, for instance, should be used only by individuals with advanced diagnostic skills with access to appropriate microbiological facilities. In practice, the lives of many children now dying from pneumonia in the developing world could be saved if injectable procaine penicillin were available at the time of their initial presentation. Health care would also be ameliorated by the availability of accurate and objective information understandable to each category of prescriber.

The generation of information that is attuned to local needs and circumstances is, inherently, a national responsibility. WHO, however, has given priority in several of its programmes to the development of resource material included for local adaptation. Training material on case management and on control strategies that has been generated within several of the specialized technical units of the Organization is now being collated and extended to provide the basis for promulgating Model Prescribing Information. The more flexible format of a formulary has been preferred to drug information sheets for two reasons: it averts the difficulty of ensuring that the WHO material is consonant with officially approved product information already settled between manufacturers and national drug regulatory authorities, and it focuses attention on practical aspects of case management in different clinical settings rather than on the innate properties of individual drugs. It is vital that such information be available to complement governmental efforts to develop and improve primary health-care services.

Inter-governmental exchange of information

Mechanisms for collaboration

In giving expression to the need for rational use of drugs, the World Health Assembly has assigned a key role to the Organization in developing activities at national, regional and global levels that will improve the provision of unbiased and complete information about drugs to the health profession and the public. At the same time, the United Nations General Assembly, in a series of resolutions relating to the export of hazardous products and to consumer protection, has recently emphasized the continued importance of the inter-governmental systems of information long established by WHO within the pharmaceutical field. The broader charge now placed upon the Organization to provide technical information on drug use directly to the end-users must devolve from its existing inter-governmental remit if its message is to remain aligned with and responsive to nationally determined policies and decisions. In general terms, the various resolutions call upon the Director-General to support national drug regulatory authorities by arranging for interchange of information both on the registration of new products, and on the withdrawal or restriction of existing products on grounds of safety, and to collaborate with these authorities in the collation, analysis and interpretation of reports of suspected adverse drug reactions.

National drug regulatory authorities operate within a rigid statutory framework constructed to control the domestic drug market. Much of their work is undertaken in camera since it demands access to privileged, commercially valuable information submitted by pharmaceutical manufacturers specifically and exclusively for regulatory purposes. The need to maintain confidentiality, however, is counter-balanced by an underlying commitment to safeguard health and, in particular, to ensure that sufficient information enters the public domain to permit the safe and effective use of registered drugs. In response to these factors and also, in some cases, to a general relaxation of governmental policy on classification of confidential data, several more highly evolved authorities have consciously expanded their informational role. This extends not only to the provision of officially approved prescribing information but, in some cases, to summaries of the evidence on which licences are granted. Many authorities now regularly publish news sheets and bulletins on matters of current therapeutic concern. These changes have not only promoted a flow of information into the public domain, they have fostered discussion and collaboration between regulatory authorities which has operated to the benefit of all countries.

Exchange of information on regulatory decisions

In 1963, the World Health Assembly requested Member States to communicate immediately to WHO any decision to prohibit or limit the availability of a drug already in use: any decision to refuse the approval of a new drug; and any approval for general use of a new drug accompanied by important restrictions, whenever these decisions are taken as a result of serious reactions. The Assembly has reaffirmed the importance it places on the efficient exchange of this information in several subsequent resolutions which additionally call for the transmission of decisions to withdraw or restrict the availability of a drug already in use on grounds of lack of effectiveness, and data on the scientific basis and the conditions of the registration of individual drugs. Effective implementation of the open-ended collaboration effort that these resolutions require has proved to be dependent upon the creation of an organizational infrastructure. This is now provided through a chain of designated network information officers responsible for assuring efficient lines of communication with the Organization [13] and the biennial International Conference of Drug Regulatory Authorities (ICDRA). Nonetheless, logistic difficulties and considerations of confidentiality have frustrated the full implementation of the Assembly's aspirations.

Initially, governments were simply invited by the Organization to notify restrictive regulatory decisions of international relevance. These were then transmitted to all Member States in *WHO Drug Circulars*. For many years this scheme fell short of its potential, partly through default and partly because the removal of drugs from national markets by voluntary agreement with manufacturers rather than by enforcement of statutory controls were rarely reported, even when safety was at issue. Over a period of some 16 years extending up to 1979 a total of only 199 notifications was received by WHO, of which about one half was provided by the US Food and Drug Administration. The situation has since been greatly improved, both because national authorities have become more perceptive of the need to exchange information and because of intensified promotion of this need by WHO. In 1984 alone, WHO transmitted information on more than 800 regulatory decisions received from 37 countries. These are conveyed in monthly *Pharmaceutical Newsletters* addressed to information officers in every Member State. The flexibility and informality of this channel of communication has, on several occasions, also proven its value as a means of canvassing opinion on a broad international basis on regulatory issues of immediate concern. However, formal understandings on confidentiality between manufacturers and regulatory authorities still impose significant constraint on exchange of information. Very few regulatory authorities are prepared to disclose information on rejected marketing applications, even when approval is withheld on grounds of safety.

Similarly, with the notable exception of the material published by the US Food and Drug Administration, little information has become available internationally on the systematic reviews of efficacy that many authorities are committed to undertake on all currently marketed products. This situation persists not withstanding a recommendation adopted by consensus at the Third ICDRA, 1984, to the effect that:

'Effective international exchange of this information will directly assist regulatory authorities in their responsibility of removing products from national markets that do not conform with prevailing standards of efficacy and safety. It will also substantially reduce the technical and administrative burden on national administrations that derives from this responsibility, since it will reduce the need for independent and repetitive national assessments.'

Although information on newly registered drugs is more readily available, its presentation in an international communications system raises both logistic and policy issues. In particular, there is a strong case for selectivity. The same product is frequently licensed in many countries under virtually identical conditions; many newly registered products are not innovative, many being new brands or new dosage forms of existing drugs; and relevant information is already generally available to importing countries as an adjunct to the *WHO Certification Scheme*. Nonetheless, the assessment of new products accounts for a high proportion of the total work load within many regulatory authorities, and many countries appreciate sight of a regular updating of new products registered elsewhere. Some 20 highly developed countries regularly offer this information to WHO and a selective annotated listing is included in the quarterly publication *WHO Drug Information*. The process of selection is directed exclusively to eliminating undue repetition and non-innovative and combination products that do not accord with criteria of rationality proposed by the WHO Expert Committee on the Selection of Essential Drugs. However annotation does not in any way indicate approval of the product by WHO, and each entry is limited to a brief description of the pharmacological class of the active component together with the licenced indications, contra-indications, warnings, precautions and serious known adverse effects.

Arrangements have also been made within the past few years to provide every competent national authority with comprehensive independent and authenticated information in the English language on marketed drugs. Through the good offices and generosity of several governmental authorities, of the International Federation of Pharmaceutical Manufacturers Associations, and of various national organizations within these countries, the following material is distributed on a complementary basis to all Member States:

Belgium: *Répertoire commenté des médicaments* (Centre Belge d'Information Pharmacothérapeutique); *Folia Pharmacotherapeutica* (Centre Belge d'Information Pharmacothérapeutique).
France: *Dictionnaire Vidal* (OVP Paris).
UK: *British National Formulary* (British Medical Association/ Pharmaceutical Society of Great Britain); *Data sheet compendium* [Association of the British Pharmaceutical Industry (information in compliance with regulations of the UK Department of Health)]; *Martindale's Extra Pharmacopoeia* (Pharmaceutical Society of Great Britain, 1982).
US: *Drug Evaluations* (American Medical Association); *Drug Information* (American Hospital Formulary Service); *Physicians Desk Reference* [Medical Economics Co. (information in compliance with regulations of the US Food and Drug Administration)]; *Summary Bases for Approval* (US Food and Drug Administration).

Negotiations are now in hand to explore whether this service can be extended to material published in other widely used languages. Meanwhile, a compendium of publications and documents prepared by national drug regulatory authorities, or by professional and consumer organizations is provided at six-monthly intervals to information officers, while relevant statutory instruments and regulations are either translated into the working languages of WHO in the quarterly *Digest of Health Legislation* or indexed by the WHO Collaborating Centre on Drug Information, Hungary.

Monitoring of adverse drug reactions

The governing bodies of WHO have identified a need not only for international exchange of information on regulatory decisions but also for central collation of the reports of suspected adverse reactions from which many restrictive decisions devolve. In 1967 the Director-General was requested to establish an international system of monitoring reports of adverse reactions to drugs using information derived from national centres. This was inspired by an expectation that infrequent and unanticipated drug-related hazards could be identified with greater efficiency by pooling case-reports submitted to individual national centres. Over 20 countries have contributed consistently to the scheme virtually from its inception in 1968, and the databank, which is now housed in a WHO Collaborating Centre located within the Swedish Department of Drugs, currently contains over 500,000 case reports of suspected adverse reactions to drugs notified spontaneously by clinicians. Nonetheless, the data need to be interpreted with caution: probably only a small proportion of drug-induced adverse reactions are reported to national centres and there is no assurance that these are reliably representative of the true hazards of treatment. In most instances, spontaneous reports merely provide an alert to the possible existence of a hazard that requires independent investigation for its confirmation. It is for this reason that information in the database is, as yet, confidential to the national centres participating in the scheme.

The emergence of epidemiologically based approaches to drug monitoring has created possibilities for new approaches to post-marketing surveillance and the feasibility of developing a comprehensive array of surveillance mechanisms adequate for drug regulatory purposes is an issue that has preoccupied drug regulatory authorities and manufacturers for some years. An ideal solution is possibly unattainable, but without a stimulus for international consultation and collaboration involving research-based pharmaceutical companies, university departments of epidemiology, toxicological centres and responsible government agencies, practicable possibilities of utilizing the diverse facilities that already exist to best advantage will remain unrealized. WHO, working together with the Council for International Organizations of Medical Sciences, is already engaged in promoting this dialogue and in exploring the extent to which various forms of surveillance might be feasible and cost-effective in generating information on drug-related hazards in developing as well as developed countries.

Evaluated information on regulatory decisions

In 1975 the World Health Assembly requested the Director-General to disseminate evaluated information on drugs to Member States. This brief has now been broadened and given a different focus with the adoption, in 1984, of a further resolution which requests the Director-General 'to continue to develop activities at national, regional and global levels aiming at the improvement of use of drugs and of prescription practices and the provision of unbiased and complete information to the health profession and the public'. At present, the norm in many countries is to place greater emphasis on the pre-marketing assessment of new products than on the information that will subsequently assure their responsible use. Given that inter-agency communications are improving, that more effective use might be made of the WHO *Certification Scheme*, and that adoption of the essential drugs concept offers a basis for rationalizing registration policies, many regulatory authorities, particularly in developing countries, might now review the balance of their activities. Whereas they might reasonably, for the most part, rely on pharmaceutical or toxicological assessments undertaken by other national authorities or, in some instances, by WHO, they are uniquely placed to determine the form and content of the information required both by health professionals and by patients to assure the effective use of the drug under prevailing national conditions.

WHO with a view to supporting national authorities in their informational responsibilities, has already embarked upon a three-fold strategy:

- production of a WHO Model Prescribing Information to complement the model list of essential drugs;
- preparation of discursive commentaries on regulatory policies and decisions within the subscription journal *WHO Drug Information*;
- renewed promotion of the WHO *Certification Scheme*, and consideration of its possible extension to provide for exchange of all nationally authenticated product information.

Work in developing the Prescribing Information is already in hand. It is envisaged as a handbook of treatment relevant to the first referral level of medical care. It will also embody teaching material for primary health-care workers and the public that can readily be adapted to local educational standards and cultural precepts. Although it will place greatest emphasis on drug treatment, the information will be organized having regard to practical case management. To the greatest possible extent the information will be drawn from advice already issued within the technical reports of the Organization. Completed drafts are being submitted in consultative form, section by section, to designated national information officers, members of relevant WHO expert advisory panels, interested pharmaceutical manufacturers and non-governmental organizations.

Having regard to the wide circulation of the WHO Drug Information, the many requests that have been received from Member States for permission to translate its contents, and the supportive references made within the governing bodies of WHO, its preparation retains high priority. Its presentation will be changed to lend it greater appeal and it will be forcefully promoted. Emphasis will be accorded, as in the past, to discursive comparative commentaries on the regulatory status of essential drugs and other important products in various countries, and the present consultative procedures will be maintained to ensure that interested regulatory authorities and manufacturers receive sight of material in draft.

In order that these documents remain responsive to the needs of Member States the designated information officers will be asked to arrange that they be kept under review and to indicate how they are used and with what effect. This information will be submitted to the ICDRA which regularly receives a status report from WHO detailing progress in all aspects of the Organization's collaboration with national drug regulatory authorities. In turn, a short report of the proceedings of each Conference highlighting the adopted recommendations will subsequently be made available to the governing bodies of WHO.

Presentation of information to prescribers and the public

There is no stereotyped national pattern by which information is presented to the end user of drugs. In many market economy countries the manufacturers, the government, representative professional bodies, technical journals, the media and, to a growing extent, consumerist organizations, have each assumed a role. In centrally planned economies, in contrast, where industry is an arm of government, domestic drug advertising is designed to provide factual information rather than to promote sales. In these circumstances information flows primarily from governmental and professional sources. A composite list of the mechanisms by which information is regulated and conveyed around the world would include the following.

The product licence

This is the instrument issued by the national licencing authority to the licence holder authorizing the distribution and sale of a specific pharmaceutical product within the jurisdiction of the authority. With the exception of any approved promotional (data sheet) and labelling material, the product licence is a confidential document containing commercially sensitive information that is necessary for the precise registration and effective control of the product. In a few countries it is considered mandatory that all doctors should receive objective, officially approved information about the properties and use of each new prescription medicine before it is launched on the market. The data sheet, which also provides the basis of the package insert, serves this purpose and it must be posted individually to all registered medical practitioners within a specified period before marketing. It sets out the approved indications for the product, recommended dosage regimens, contra-indications, precautions, warnings, details of packaging, pack size and optimal storage conditions. Although a data sheet defines the proposed usage of the product in terms acceptable to the licencing authority, it provides no description of the preclinical or clinical data on which these proposals were founded. To provide prescribers with an account of the nature and quality of these data the US Food and Drug Administration now publishes summaries of the pharmacological, toxicological and clinical studies on which the marketing authorization was based. References to preclinical data are similarly included in product monographs that are updated from time to time by the Canadian Health Protection Branch, and in drug profiles produced by the Australian National Drug Information Service.

Pharmacopoeial monographs

Having regard to the potent biological activity of drugs and to their crucial importance to the welfare of patients, clinical users need every reasonable assurance that their quality conforms to specification not only at the time of manufacture but throughout their shelf-life. Publicly accessible monographs defining these specifications are published both by national pharmacopoeial commissions and by WHO. The *International Pharmacopoeia* produced by WHO is intended particularly to subserve the needs of developing countries.[15] It is now complemented by a series of basic tests of identity of drugs included in the *WHO Model List of Essential Drugs*, simple tests of gross degradation and a compendium of stability data on these drugs under tropical climatic conditions. The objective is to provide peripheral health workers with some ability to test the quality of their stocks without recourse to laboratory facilities, and to be able to refer doubtful material for further analysis.

Manufacturers' promotional activities

The basic right to promote the products of private enterprise is conferred by statutory or even constitutional provisions in some countries. The advertising of pharmaceutical products is, nonetheless, extensively controlled in developed market economy countries both by regulation and by voluntary codes of practice, and some regulatory authorities routinely monitor advertising material. These restrictions place constraints on the types of products that can be advertised direct to the public, and on the format and presentation of printed advertising copy and promotional material used in film and the electronic media. Particularly in countries where access to journals is limited, company representatives are widely deployed to promote products. When these representatives are highly trained, they are in a position to offer constructive and practical advice. Their mission, however, is to promote the products of the company for which they work and it is not feasible to subject their activities to effective monitoring. At international level, the Twenty-first World Health Assembly, in 1968, adopted a series of ethical and scientific criteria for pharmaceutical advertising[16] and, in 1982, the International Federation of Pharmaceutical Manufacturers Associations produced its own *Code of Marketing Practices*.[17]

Notwithstanding the existence of these constraints, reservations are still expressed about the basic premise of treating drugs as normal commercial entities and soliciting their use through competitive advertising rather than objective evidence. New products, in particular, are intensively promoted when they are introduced to the market. At this time they have been subjected to limited clinical use. Their advantages and disadvantages with respect to alternative drugs are, at best, incompletely defined, and no secure judgement can be offered as to whether they will ultimately become established in routine practice. Moreover, with the exception of those authorities that are required to assess whether a proposed new product satisfies a medical need, national drug regulatory authorities are not empowered to undertake comparative assessments of products. In the latter case, no test is applied at the time of registration as to whether a new product is more or less efficacious than others already available.

Lists of reimbursable products

Because of large price differentials that commonly exist between competing products and particularly the high cost of sophisticated dosage forms, such as slow-release preparations, several national administrations notify doctors and pharmacists that the charges for specific products are excluded from permitted reimbursements to patients participating in national social security systems.

Officially-sponsored prescribing advice

To advise doctors further on efficient and cost-effective drug use some national health authorities also provide tables or bar charts indicating the comparative costs of interchangeable products as well as authoritative prescribing advice issued on a regular or intermittent basis. The content and format of officially sponsored national formularies is also being changed in some countries to accord with the same general objective: entries are selective rather than comprehensive and they are prefaced and accompanied by didactic prescribing advice.

Drug compendia

Collated sets of data sheets are issued in many countries, as regularly updated drug compendia, and, in some, they are accessible online through desk-top computers. Some are published privately, while others are compiled by national associations of pharmaceutical manufacturers. In some cases, however, entries are included that have not been approved by national authorities. Whereas these are readily distinguishable in some publications, this is not invariably the case. Some of the most widely available compendia are sustained on advertising revenues – and issued gratis to doctors. In this case there may be no assurance that any of the entries confirm to authorized texts, for the controls applied to information issued by manufacturers may not apply to information issued by independent publishers. Other compendia, such as those produced by the American Hospital Formulary Service, the American Medical Association, the British Medical Association and the Pharmaceutical Society of Great Britain, are clearly independent, authoritative and encyclopaedic in their coverage.

Medical journals

Refereed technical journals provide the ultimate, original source of much information on the properties and uses of drugs. Articles relevant to any one topic are so widely dispersed, however, that an extensive library with specialized search facilities is essential for adequate access to the literature. Several professional abstracting services now provide weekly or monthly updates on articles relating to a given specialized field that are published in major technical journals. The preparation of this information, however, is a labour-intensive task: the output is directed to a relatively small, specialized readership and, in consequence, it is often prohibitively expensive for individual subscribers, even in highly affluent countries. Faced with this situation, editors of many national and international medical journals now assume more responsibility for educating – as opposed to informing – their readers. Authoritative basic reviews of clinical and scientific topics, question and answer features, reports of the activities of professional governmental committees, and expanding correspondence columns challenge the traditional dominance of the original research paper, and leading articles have become concerned more with practical medicine and less with erudition.

Review journals and drug bulletins

A particularly encouraging feature of the past five years is the number of bulletins that have appeared and sometimes become established in developing countries. The information, however, is often drawn largely from international sources and, as such, this is sometimes of questionable priority – or even relevance – to local needs.

Textbooks

Textbooks of repute, particularly those that are the product of a single author or editor, hold attraction for the undergraduate student because they are cohesive, comprehensive, yet concise. Yet what is striking is the lack of textbooks written from within the perspective of the developing world. Even standard texts of tropical medicine remain dominantly a product of specialized institutions in developing countries. In large measure, this situation reflects economic constraints. Textbooks are not commercially viable unless high volume sales can be guaranteed. This is not the case in smaller developing countries which are left no option but to resort to importation of books at prices that are now virtually prohibitive. The domain of therapeutics presents an exceptional challenge to authors. It is becoming an overwhelming task, however, for a small editorial group, no matter how highly motivated, to keep abreast of current therapeutic information in an authoritative and comprehensive manner.

The media

In some countries public television time has been accorded in off-peak hours to post-graduate education for doctors and, in recent years, the media have also become more attentive to the presentation of health issues to the general public. Under the influence of consumerist groups with the active involvement of doctors, the mass media have created a greater awareness of the importance of lifestyle to health, of the values and limitations of self-medication, and of the strengths and shortcomings of the public health services. Through these activities lobbying pressures have also been exerted on drug manufacturers and governments to bring them to re-appraise their performance and attitudes in the promotion and control of drugs. Undoubted successes have been achieved in obtaining or accelerating the withdrawal of several marketed products and in stemming unacceptable promotion of several drugs in countries lacking well established regulatory systems.

However, bias in media reporting, even when it is unintended, can readily undermine public confidence in established practice. In the early 1970s, for example, reservations were expressed about the safety of whooping-cough vaccine and particularly about its propensity occasionally to induce encephalopathy and permanent brain damage. Subsequent controversy caused pertussis immunization to be discontinued in some countries and discouraged elsewhere; major epidemics of the disease followed resulting in many deaths. The implementation of community health programmes commonly presents governments, public health officials and pharmaceutical manufacturers with onerous ethical responsibilities including the need to make equitable provision for the rare and unfortunate victims of injuries induced by drugs or vaccines. To dramatize remote yet tragic risk when this is the inevitable price of community protection against the unacceptable burden of serious communicable disease could mean that society, particularly in developing countries, will forego potential benefits of new drugs and vaccines and it may jeopardize established programmes of health-care.

Evaluation of the influence of drug information

The significance of given factors on prescribing practices of doctors is not readily assessed when a variety of messages competes for their attention. However, the considerable sums invested by pharmaceutical companies in advertising their products are presumably known to be cost effective. Much of this expenditure is directed to establishing and maintaining products in key markets in which companies are interlocked in commercial competition. This expenditure adds to the cost of health care without offering tangible benefit and, unconstrained, it threatens to damage the performance and even the stability of the industry itself. Even in the light of authoritative objective evidence doctors are not always readily induced to alter their established prescribing habits. For example, short-term chemotherapy for pulmonary tuberculosis, which has been shown to offer important economic as well as therapeutic and social advantage in developing countries, has so far not gained wide acceptance in many developed countries despite the persuasions of committed and authoritative exponents.[18, 19]

Busy doctors, with onerous clinical responsbilities, particularly when they are working outside large institutions, tend to become isolated in their work. Without the opportunity of formal in-post training or even of informal discussion with colleagues, they need to remain highly self-critical if they are to adjust effectively to the steady evolution of therapeutic practice and the pressures of competitive advertising. The failure of doctors to seek out objective information on newly introduced drugs[20] contrasts strikingly with the acknowledged success of local committees, particularly in hospitals, to institute and sustain efficient and cost-effective prescribing practices in both developed and developing countries. Collective discussion and decision-making among professional peers is evidently a potent stimulus to rational prescribing practices.

Rational prescribing devolves from sound education and not simply from access to objective information. In developing countries, the use of a limited number of essential drugs and their appropriate indications is a part of the basic information every young medical graduate and every health worker should possess. To achieve this, pervasive problems of communication have to be overcome in those countries with a multiplicity of cultures, local languages and dialects. Community health workers, no less than other prescribers within the health infrastructure, need the support of instruction and discussion to remain up to date in their responsibilities. The staff of first-referral hospitals are well placed to assume the role of educator, and arrangements could be made to outpost staff from time to time.

National drug regulatory authorities or associated national committees are also appropriately placed to provide independent and objective information on products registered under their aegis. They are competent to determine labelling requirements for medicines not only for doctors but for other cadres of personnel including community health workers and traditional birth attendants. They can also ensure that information on pharmaceuticals is placed in reasonable perspective with regard to the overall social, economic and educational development of the community. National authorities commonly have a large measure of assurance from regulatory decisions taken in other countries about the quality, safety and efficacy of products available in international commerce. More demanding than the decision to license such products is the challenge of assuring, through appropriate control of labelling and distribution, that each product is used to best advantage within the national context. In the absence of a basic system of drug registration that addresses this need, a call for the rational use of drugs can never be securely founded.

References

(1) WHO Technical Report Series, No. 772 (1985). *The use of essential drugs: report of a WHO Expert Committee* (WHO).
(2) *WHO Official Records, No. 226 (1975). Good practices in the manufacture and quality control of drugs* and *Certification scheme on the quality of pharmaceutical products moving in international commerce.*
(3) *International nonproprietary names (INN) for pharmaceutical substances. Cumulative list no. 6,* (WHO, 1982).
(4) *Special programme for research and training in tropical diseases. Sixth programme report, 1 January 1983 – 31 December 1984.* (UNDP/World Bank/WHO).

(5)   *Programme for control of diarrhoeal diseases. Fourth programme report 1983–1984,* (WHO).
(6)   *Special programme of research development and research training in human reproduction. 13th Annual Report, December 1984* (WHO).
(7)   WHO Official Records, No. 226 (1975). Annex 13, pp. 96–110.
(8)   WHO Technical Report Series, No. 685 (1983). *The use of essential drugs* (report of a WHO Expert Committee).
(9)   *Consolidated list of products whose consumption and/or sale have been banned, withdrawn, severely restricted or not approved by governments* (UN, 1983).
(10) 'Evaluation of short-term tests for carcinogesis'. *Prog. Mutation Res.,* **5** (published on behalf of the International Programme on Chemical Safety (WHO, ILO, UNEP), Elsevier, 1985).
(11) 'A toxicology data bank based on animal safety evaluation studies on pharmaceutical compounds'. *Human Toxicology:* 1985, **4**, 447–460.
(12) *Proceedings of the International Conference of Drug Regulatory Authorities. 28–31 October 1980, Annapolis, Maryland, US* (FDA/WHO, 1981).
(13) E. D. Poggiolini (1983). *Second International Conference of Drug Regulatory Authorities.* (Raven Press).
(14) *Proceedings of the Third International Conference of Drug Regulatory Authorities, 11–15 July 1984, Stockholm, Sweden.* (Swedish National Board of Health and Welfare/WHO, in press).
(15) *The International Pharmacopoeia,* 3rd edn, vol. 1, 1979: vol. 2, 1981 (WHO).
(16) Ethical and scientific criteria for pharmaceutical advertising. World Health Assembly Resolution WHA21.41, May 1968.
(17) *IFPMA Code of Pharmaceutical Marketing Practices.* (International Federation of Pharmaceutical Manufacturers Associations, 1982).
(18) Fox, W. (1983). 'Compliance of patients and physicians: Experience and lessons from tuberculosis II.' *British Med. J.,* **287**, 101–105.
(19) Cooke, N. J. (1985). 'Treatment of tuberculosis'. *British Med. J.,* **291** 497–498.
(20) *Who receives NDIS profiles?* (Australian Drug Information Service. Department of Health Newsletter, September 1985).

PART III

CHAPTER THIRTEEN

Information needs and availability in the United States

T. ANDREWS, N. S. HEWISON AND S. A. WILDER

Background

For many years, health professionals, researchers, and information specialists have been coping with an information explosion. The phenomenal growth of the US National Library of Medicine's *MEDLINE* database illustrates this. When *MEDLINE* first went online in 1971, it indexed 236 biomedical journals and included references to 147,000 articles. By 1986, it contained over 5,000,000 citations, with current citations drawn from 3400 journals. In 1988, *MEDLINE* was reported to be growing at the rate of 23,000 citations a month. Initially, *MEDLINE* was available to 25 users, who had access to the database 4 hours a day. In 1986, domestic users alone numbered 6746 (not including clients of 15 non-US *MEDLARS* centres and those searching *MEDLINE* via commercial online services), and the database was available virtually 24 hours a day (*NLM Technical Bulletin*, Sept. 1986, supp. 1).

Among the general public, a growing consumer health movement has been seen. There has been increased interest in nutrition, physical fitness, and the participation of the patient in making medical care decisions. Patient education programmes have been established in hospitals. Informed consent is now required for many surgical procedures. Many patients have adopted the role of consumer and given up their belief in the physician as God.

In 1986, national health expenditure in the US (including medical research and the construction of medical facilities) totalled $458.2 billion. This figure was an 8.4% increase over 1985 and represented 10.9% of the gross national product. Approximately 87% of US citizens have some form of health insurance (private, government sponsored programmes for the aged and/or indigent, or military), and patients themselves plus their insurance plans spent $30.6 billion on drugs and drug sundries in 1986. Profits of the pharmaceutical firms among the country's 500 largest industrial corporations increased 20.3% in that year (*World Almanac*, 1988).

The major causes of death in the US in 1986 were heart diseases, cancer, cerebrovascular disorders, and accidents. The average lifespan increased by 4 years from 1970 to 1986, so that 12.1% of the population is over the age of 65. Thus, geriatric health-care has become a significant specialization and there has been increased attention to the development and prescribing of drugs for the elderly.

The cost of developing new drugs in the US is very high, mainly because of the high standards of efficacy and safety required by federal regulations administered by the US Food and Drug Administration. Labour costs also increase the cost of drug development and production. This burden of cost has created orphan drugs, those for which research and development costs are out of proportion to the number of people whose rare diseases require them. The government is seeking ways in which to fund or supplement such research and production.

Competition in the marketplace affects the pharmaceutical industry in the US. One of the largest developments has been the increase in availability of generic drugs. In many instances, these may be substituted for the brand name item, thus decreasing sales for the companies who have invested much time and money in researching and developing the original drug. Such intense competition increases the necessity to keep and protect patents, research ideas, and manufacturing processes.

Extent and nature of needs

Information centres, including libraries, play a most important role in satisfying the ever-increasing needs for information. Drug information sources are needed for use by health sciences practitioners as well as manufacturers and researchers, and there has also been an increasing demand for information which can be used by the patient.

Information activities of industrial information centres and academic pharmaceutical libraries are well developed in the US, but needs are still expanding. This chapter deals primarily with these needs, how they are being satisfied, and how information centres are providing pharmaceutical information to their primary users.

Rapid access to information is of critical importance to research scientists, especially those in the drug industry, and industrial information departments have been known for their progressive, innovative, and co-operative activities. Academic libraries, on the other hand, where development of new products and return on investment is not of such primary importance, do not find rapid access quite so critical, but they are more likely to hold long runs of back volumes of periodicals of particular interest to all researchers. They make these available to industrial libraries through co-operative arrangements such as inter-library loan and document delivery by telefacsimile reproduction. There is, thus, a developing library network throughout the US.

Academic institutions that offer advanced degrees in the pharmaceutical sciences, and some institutes, conduct research which resembles that done in industry. For many years large research grants have been awarded to professors in universities from both industry and governmental agencies. Recently there has been even greater emphasis on research in certain diseases, for example, cancer and AIDS.

A special need has developed for information dealing with adverse drug reactions and drug interactions because the public has become more aware of possible risks as new, more potent drugs are introduced.

Lest the human element be forgotten, one of the most important sources of information in the pharmaceutical field is the information practitioner who is available to assist in the use of materials as well as their collection and organization. In addition, the practising pharmacist is an important source of information for the drug consumer, the physician and other health professionals.

These sophisticated needs have directed the development of the information sources and systems that are available in the US and which are increasingly being made available worldwide.

Constraints

There are many constraints that prevent the capture of relevant, needed information. First, there is the quantity and complexity of information. There has been a tendency toward scientific specialization, but at the same time, the pharmaceutical information that is sought is interdisciplinary in nature. Further, information science itself is interdisciplinary, involving business, computer science, law, communications, liberal arts, library science, and engineering, making it a difficult, complex field in which to operate. The development of information technology has probably been too rapid, so that even in a wealthy country such as the US, information centres and libraries have been unable to keep pace with the 'opportunities' that are offered. And, of course, there are limits to what technology can do.

The greatest constraint, however, is the cost factor. Studies are desperately needed to show whether the benefits of certain of the new technologies really outweigh the costs. Hasty action may turn out to be expensive if systems which rapidly become obsolete may be replaced (Doughtery, R. M. (1987). 'The Pace of Change,' *J. Academic Librarianship*, **12**(6), 339). The most obvious costs are the purchasing of computers and other hardware, software, and systems. But, in addition, there are ongoing costs such as fees for database searching, funds to buy periodicals and monographs, and the cost of reference services (staff as well as database access), and the training of information specialists.

A proliferation of machine-readable information sources is available through well developed communication networks, and many of them are important to pharmaceutical scientists even though the services overlap considerably in coverage. Cuadra/Elsevier, publishers of the *Directory of Online Databases*, reports that 468 new databases went online during 1987, bringing the total number available to 3700 (*Information Today*, **5**(6) 1988, 1). It is possibly encouraging to note, however, that there has been a slowdown in the number of new online databases becoming available. This may indicate there is a trend toward consolidation and standardization.

Bibliographic databases, in providing references, have stimulated the need for access to the publications they index, particularly periodical articles. This has resulted in increased use of inter-library loan services, the development of formal networks to facilitate inter-institutional borrowing, and various means of rapid document delivery.

A disturbing result of extensive inter-library borrowing is that costs of materials have risen more rapidly than they might have otherwise. Librarians, particularly those in academic institutions, feel that prices of materials, especially periodicals, have reached critical levels. Many university professors and administrators feel that recent price increases are outrageous. *Science* (**239**, 1988, 81) reports that the average technical book in the natural sciences cost $32.70 in 1977, and and risen to $59.06 in 1988. Journal prices show even greater increases; in the last two years they increased at four times the general rate of inflation (Houbeck, R. L., Jr. (1987) 'If Present Trends Continue: Responding to Journal Price Increases.' *J. Academic Librarianship*, **13**(4), 214-220). The relatively low exchange rate of the American dollar compared with currencies in other countries is responsible for some of this price increase.

Although journal prices have been making up a large percentage of library budgets, librarians, faculty members, and researchers feel that universities must provide literature resources just as they provide other services. This is difficult, because at the present time there is a good deal of pressure in the US to keep the costs of higher education from escalating further.

Another aspect of the funding problem is that the federal government no longer furnishes as many subsidies as it once did for publication grants, page charges, and the like. Commercial publishers have taken over most of the role of disseminating information. Incidentally, many professional and scientific organizations have begun to behave like commercial publishers with correspondingly high prices. One hopes that new, less costly systems of scholarly communication will evolve, such as electronic publishing, but as yet this has not developed to a widely accepted degree.

Those working in the information field are sometimes bothered by the secrecy that surrounds some research; they feel that information should be available to all who seek it. Pharmaceutical companies, however, do not want competitors to know what they are developing and, thus, do not publish their research results until their intellectual property is adequately protected. Industry contributes substantially, of course, to the general pool of scientific knowledge. The federal government, also, has been criticized of late because it does not make the information generated through its subsidies and grants readily available at low cost to the scientific community. The charge is that private enterprise is taking over this role and is making unreasonable profits from it.

The use of any complex indexing or abstracting service can be a complicated task, whether it be online or on paper. Many information workers feel that users do not always make the best use of these tools, and that they need a good deal of assistance for best results. Until relatively recently, librarians or information specialists did almost all the computerized database searching for their clients. Whether or not they achieved the best results is open to question, too, although the professionals providing these services have been instructed frequently by vendors and others about the fine points of a search and how the databases are constructed. In any case, there is now a trend toward the users themselves performing their own searches. Suppliers claim that their computerized services are more 'user-friendly' now than formerly. Many of them present the user with menus of search and display options rather than relying upon memorized commands, and some provide assistance in the formulation of search strategies. Again, it is open to question whether the 'end-user' will really find this to be an improved way of obtaining information. The cost of instruction in the best use of services is yet another factor to be dealt with, another barrier that comes between the user and the information.

What is being done

Printed sources currently used in the US

This section discusses some of the printed sources of information that have proved valuable to the pharmaceutical scientist of the US. A list of indexing and abstracting publications can be found in the General Appendix. Most of them are also available for searching via the computer as well as in hard copy.

Publications that show how to use some of these indexes are listed in the General Appendix. Representative titles are *How to Use Index Medicus and Excerpta Medica* by B. Strickland-Hodge (Gower, 1986) and *How to Use Chemical Abstracts, Current Abstracts of Chemistry, and Index Chemicus* by B. Livesey (Gower, 1987). In addition, numerous guides to the literature of the fields of chemistry, medicine, and life sciences, etc., provide information of this kind.

PERIODICALS

The number of journals of interest to the pharmaceutical scientist grows very rapidly. The individual who wants to keep abreast of the literature is therefore obliged to read more and more journals. In addition to pharmaceutical journals, journals of clinical medicine are important, partly because they report adverse drug reactions. Many newsletters of interest are published, and virtually every state has a journal published by the state pharmaceutical association. Schools of pharmacy also publish periodicals, but these ordinarily are news publications only, with little or no scientific information included.

T. Andrews (1986) in *Guide to the Literature of Pharmacy and the Pharmaceutical Sciences* (see General Appendix) lists about 250 periodicals that are exclusively pharmaceutical in nature; publications in ancillary fields are not included, nor are state pharmaceutical society publications or those of schools of pharmacy. The second edition of *A Basic Booklist and Core Journals for Pharmaceutical Education*, by P. S. Piermatti and B. M. Hill (see General Appendix) lists about 75 core journals. An early compilation by T. Andrews and J. Oslet ('World List of Pharmacy Periodicals, Revised and Enlarged Edition, 1975.' *Amer. J. Hospital Pharmacy*, **32**, 85–124) lists approximately 2000 periodicals. However, this included titles that have ceased publication and some that are quite obscure. The compilers made an attempt to get as complete a list as possible. A large research-based pharmaceutical company's library fairly typically subscribes to about 1000 periodicals.

DIRECTORIES

There are directories available that list people or pharmaceutical manufacturers or provide other information of interest in this field. A recent one is *Pharmaceutical Manufacturers of the United States* (4th edn ed. D. J. De Renzo, Noyes Data Corporation, 1987), which provides addresses and a description of product line, annual sales, number of employees (where available), and a directory of key executives. One with international scope is *International Directory of Pharmaceutical Companies (Scrip)* (PJB Publications, 1984). Another very useful publication, though not limited to pharmaceutical firms, is *Directory of American Research and Technology*, (20th edn R. R. Bowker Co., 1986). A brief publication, *Drugs and Pharmaceuticals Worldwide: Trade List* (US Department of Commerce, International Trade Administration, US and Foreign Commercial Services, 1982) provides export market information for drugs and health-care services. *World Pharmaceutical Directory* (Unlisted Drugs, 1988) provides worldwide information on compounds under investigation and marketed pharmaceuticals, and also contains a good list of corporate names and addresses. The *Exectuive Directory of the U. S. Pharmaceutical Industry* (Chemical Economics Services, 1966 to date) annually provides a list of companies (with the ownership indicated if a subsidiary or division of another company), corporate officers and other personnel, annual sales, and product information.

A list of retail drug stores in the United States, *Hayes Druggist Directory* (Edward N. Hayes, 1912 to date), is published yearly.

A biographical directory of interest, *Pharmacology and Pharmacologists: An International Directory* (Oxford University Press, 1981) lists about 3200 pharmacologists and also includes international societies.

REVIEWS AND SURVEYS

Serial publications such as scientific reviews or surveys have in common the fact that they are a means of bringing the reader up to date on topics of current interest. They point out significant developments in a special area of the pharmaceutical sciences and summarize research. The titles of the publications usually include such words and phrases as 'advances', 'review', 'report', 'progress in', 'methods', 'annual review', 'recent advances', and 'yearbook', indicating their content. Such publications are frequently published annually, but some are irregular. Specific titles are listed in the General Appendix.

PHARMACOPEIAS AND STANDARDS

Pharmacopeias are publications that provide assurance of the quality of medicines that reach the consumer. They are collections of standards designed to protect the patient. Pharmacopeias are 'official' publications, and nearly every country has one, has adopted that of another, or has collaborated to produce one. Of recent years the practising pharmacist in the US has made little use of the official pharmacopeia, although it is of value to the pharmaceutical manufacturer and those interested in analytical procedures. The current *United States Pharmacopeia* (21st revision) and *The National Formulary* (16th edn) (U. S. Pharmacopeial Convention, Inc., 1984) are now produced together in one volume. These two works are the legally recognized compendia of standards for drug strength, quality, purity, packaging, labelling, and storage for most US drugs. The publication is kept up to date between revisions by cumulative supplements.

Another important work that provides standards is *Official Methods of Analysis of the Association of Analytical Chemists* (14th edn, Association of Official Analytical Chemists, 1984). It has never been designated as an official compendium by law or by regulation, but it has earned the legal status of an authoritative text and has been accepted as such in court actions.

Official publications of other countries of the world are discussed elsewhere in the book.

The role of organizations

Organizations have an important role in the dissemination of information. They put individuals of like interests in touch with each other, and they are important generators of information.

There are two organizations for librarians and information scientists that have divisions or sections for those interested in the pharmaceutical sciences: the Special Libraries Association and the Medical Library Association. The American Society of Information Science is of interest also, particularly to those who are involved with automated systems. In addition, there is a Drug Information Association whose membership is made up of those who are concerned with drug information in any way.

There are many organizations for pharmaceutical scientists. Among those are the following: the American Association of Colleges of Pharmacy (which has a section for librarians), the American Pharmaceutical Association, the American Association of Pharmaceutical Scientists, American Society of Hospital Pharmacists, American Society for Pharmacology and Experimental Therapeutics, and the American Society of Pharmacognosy. In addition there are organizations for retail pharmacists such as the National Association of Retail Druggists and the National Association of Chain Drug Stores. There are organizations primarily for manufacturers of pharmaceuticals such as the Pharmaceutical Manufacturers Association, the Proprietary Association, and the Parenteral Drug Association.

Virtually all the organizations mentioned above publish material, notably a journal.

There is one federal agency of much importance to those involved with drugs, the US Food and Drug Administration. It is the principal consumer health protection agency of the federal government; it is responsible for ensuring that foods, drugs, and cosmetics are safe and, in the case of drugs, efficacious.

Computerized information and new technologies

Computers and telecommunications play significant roles in information flow in the US. Computerized indexing and abstracting sources, listed in the General Appendix, are important tools for searching the literature, and for retrieving data and factual information. Equally critical are the document delivery mechanisms designed to secure the actual document once its existence has been discovered.

The concept of the electronic desktop workstation has great appeal for information professionals as well as for their clients. A microcomputer is a multifunctional tool which allows activities such as wordprocessing, database management, the construction of personal information files, and the production of spreadsheets, plus (if a modem is added to permit telecommunications with remote computers) access to computerized databases, automated library catalogues, and electronic messaging systems. In short, a microcomputer workstation allows not only access to a wide range of information services, but also enables retrieved material to be further refined, packaged, and transferred. Concern has arisen that intellectual property rights may be violated by some of the capabilities introduced by advances in communications technologies, and the US Library of Congress Network Advisory Committee has held several meetings to examine whether the copyright law needs revision in this new technological environment.

COMPUTERIZED DATABASES

Many computerized databases are bibliographic; that is, they contain references to the literature where information can be found; they may also include abstracts. Other types of databases, informational rather than bibliographic, include dictionaries of chemical substances, and files which contain the complete text of journal articles or books. Examples of the latter are the *Comprehensive Core Medical Library* (BRS Information Technologies), a database containing the complete text of a group of medical books, textbooks, and journals, and the *CJACS File* (American Chemical Society), which contains the full text of articles, notes, communications, etc., from 19 printed journals, including *J. Medicinal Chemistry*. Many computerized databases have printed counterparts, e.g., *BIOSIS* (*Biological Abstracts* and *Biological Abstracts/RRM*), *The Merck Index Online*, and *International Pharmaceutical Abstracts* (*IPA*). The General Appendix lists some of the databases important for the pharmaceutical sciences.

Computerized databases got their start in the US in the 1960s when the National Library of Medicine(NLM) and other organizations began to use computers to typeset their printed sources. NLM's *MEDLINE* database (which is the computerized equivalent to *Index Medicus* and related indexes) was initially searchable by computer only in batch mode at NLM. Search requests were submitted by mail to NLM and the resulting bibliographies were mailed back. Beginning in 1971, librarians across the United States were able to search the most current years of *MEDLINE* interactively from their own libraries by using, first, teletype machines and, then, computer terminals and modems which utilized telephone lines to establish a link with the host computer at NLM. Eventually all the *MEDLINE* backfiles were also made available online, and 15 non-US *MEDLINE* centres began to provide access in various parts of the world.

The number of commercially available databases grew from 300 in the mid-1970s to approximately 600 in 1980 and more than 3000 in 1985, according to M. E. Williams' paper in the *Proceedings of the 1987 National Online Meeting* (Learned Information, Inc., 1987). Bibliographic, full-text, directory, and other non-numeric databases were searched 18,000,000 times in 1985, up from 6,000,000 in 1980 and 1,000,000 in 1975. These databases had revenues in excess of $260 million in 1985.

During the 1980s, librarians began to use microcomputers to search online databases and to formulate search strategies in advance (before making the connection with an online service which would charge fees based partly on time connected to the host computer), save the strategies, and then 'upload' them. Similarly, it became common practice to 'download' the records retrieved in the search by copying them onto the floppy diskettes or the hard disk used by the microcomputer for data storage. This step allows for post-search processing of the results before giving them to the requestor as a printout or disk file.

Information staff have acted as intermediary searchers for most of the history of computerized databases. Changes in this trend have occurred in the 1980s with the development of a number of means of access designed for the 'end-user'. These have included search software with such 'user-friendly' features as menus of search choices (rather than complicated command languages), context-sensitive help screens, and programmes that choose the database for the requestor. Some of this software is mounted on the host computer; other software packages may be purchased for the user's microcomputer.

CD-ROM AND OTHER LOCAL-USE PRODUCTS

Other developments include the use of CD-ROM technology to store databases, or subsets of databases, on compact discs which are accessed via special players connected to the user's microcomputer. Coupled with user-friendly search software, the immense storage capacity of CD-ROM discs is thus utilized to make computerized databases searchable locally, without the use of telecommunications. Disadvantages of CD-ROM products include the present costs which represent an extreme constraint for academic and many other libraries. For example, the annual *Physician's Desk Reference* (*PDR*) (Medical Economics Company Inc.) costs $400 on CD-ROM as compared to $40 for the printed version. A further constraint is that the number of simultaneous users is severely restricted, which is not true of online searching. Currency of information on the disc is another issue, as CD-ROM producers typically issue updated discs only quarterly. It is anticipated that many searches of CD-ROM databases may be followed up by a search of the corresponding online database to retrieve the most recent information.

One advantage of CD-ROM is that the annual lease fee covers unlimited access, thus eliminating the per-minute and per-record charges typical of online services. Others are the ability to manipulate the data in ways not possible in the printed version (e.g., to search for a particular side effect across all the drug entries in *PDR*), the user-friendly software mentioned above, plus special features, such as the inclusion on the *PDR* CD-ROM of records from related sources. Information professionals working with CD-ROM products need to evaluate a number of factors, including frequency of use, cost, and comparison with printed and online counterparts.

Many producers and vendors also make magnetic tapes of their databases available for mounting on the customer's computer under a lease arrangement; a company or university library may thus make particular databases available for unlimited local searching.

CURRENT AWARENESS SERVICES

Current awareness services, designed to provide regular updates on current literature, are available in several forms. An information centre may photocopy in-house the tables of contents of selected journals and send them to interested persons who scan them to identify articles of interest. (However, the copyright issue surrounding this is still in question.) The dissemination of tables of contents may also take place via electronic mail.

A subscription to a printed *Current Contents* edition from the Institute for Scientific Information (ISI), provides a weekly compilation of contents pages from journals in a given area, for example, *Current Contents: Clinical Practice*. The ISI current awareness journals are also available online and on CD-ROM, leased tapes and diskettes.

Many information centres offer SDI (selective dissemination of information) services in which search profiles are stored in computerized form and executed regularly against the latest updates of specified databases.

DOCUMENT DELIVERY

Once desired documents have been identified, libraries can assist in obtaining even those that are not included in their own collections. The National Library of Medicine has implemented DOCLINE, a national automated inter-library loan system, and many libraries also participate in local consortia which facilitate inter-library borrowing and lending. The *OCLC* database (Online Computer Library Center, Inc.), a computerized record of the holdings of 6000 libraries in the US and 16 other countries, makes it possible to locate books and journals and to request that they be loaned following established inter-library loan practices.

The delivery of the actual document is most often accomplished through the US Postal Service or private carriers such as United

Parcel Service. For faster delivery, the use of telefacsimile machines allows the rapid transmittal of documents over the telephone lines.

In addition to information centres, there are private firms which specialize in the rapid and reliable delivery of journal articles, government documents, and other material; some of these can respond to requests sent electronically as well as by mail and telephone calls.

INDIVIDUAL DATABASES

Information on pharmaceutical topics may be found in databases from a wide variety of subject areas including pharmaceutical sciences, biomedicine, chemistry, the pharmacy industry, regulations, and patents. The General Appendix lists important databases in each area, their producers, and printed counterparts if available.

What the future holds

It is difficult to examine and predict the future, although it is perhaps reasonable to assume that since revolutionary changes have been taking place in health-care delivery and information technology in the US, certain trends will continue to be evident. Changes are affecting physicians, hospitals, and all those in the business of providing health care, including the pharmaceutical industry. The changes have been brought about to a great extent by pressures from the federal government to keep health-care costs under control. This comes at a time when biomedical research and drug development continue to be more complex, time-consuming, and costly.

No definite answers are available to show in what direction health-care planning will go. A universal health-care system has not evolved, and no easy solutions are in sight. However, the health-care marketplace today is driven by cost containment, and likely will remain so for some time to come. It appears that there will be no return to the days before government regulation, even if those days did bring us a vast array of 'wonder drugs' from the drug industry.

Information centres will continue to play a most important role in providing pharmaceutical information to those who need it. Many feel, however, that the information revolution is near the beginning, rather than the end, at the present time. It is reasonable to assume that there will be incentives to develop newer information technologies that are more cost-effective and that make it easier to get the information to the user. If costs make it possible, information specialists will soon be using more powerful computers and optical disc-based devices. A trend of the future may be an increase in the levying of fees for service, especially in public and academic libraries. In addition, there will likely be more online whole-text searching done to speed the process of obtaining information from documents, and information may be disseminated electronically making use of the 'electronic journal'. Document delivery services of varying types are being initiated, and will continue to develop.

Another problem is information security, the need to protect sensitive but not necessarily government-classified material. It is possible for computer 'viruses', or unauthorized programmes, to migrate unchecked through an entire computer system from one database to another. Frost and Sullivan predict that the $498 million spent on overall computer security in 1987 will increase to $588 million in 1988, and will go to over $1 billion by 1993 (*Frost and Sullivan Forecast*, Report #A1743).

In some sectors CD-ROM is seen as a threat to online revenues. However, information specialists have overwhelmingly called for a reduction in prices to make CD-ROM systems more commonly available in information centres, especially libraries. They have ranked this issue as more critical than the need for multi-user systems, for networking capabilities, and for standard search protocols. It appears at present that prices for CD-ROM versions of print or online systems are prohibitive for all but very heavy users of the other media. However, almost all information specialists view the CD-ROM technology as highly acceptable for their workplace, so it seems reasonable to predict that a way to make it cost-effective will be found in the future (N. M. Nelson. (1988). 'CD-ROM: Librarians and Vendors,' *Information Today*, **5**, 17–18).

Another problem area now emerging is the protection of intellectual property rights as they relate to advances in communication technologies. There is evidence that this will continue to be a problem in the future.

The newsletter *Information Hotline* frequently publishes projections of costs that will be incurred in the information industry in the near future. It is estimated that electronic services, including interactive communications, information processing, and entertainment services may become a $100 billion market by 1995. At the present time the information technology industry performs over one-fifth of all industrial research and development. Database services reached a level of $1.9 billion in 1986 (*Information Hotline*, **19**, 1, 11, 1988). It is projected that this will double in the next five years, although as yet there has been no slowing of paper-based publishing.

There has been speculation that by the year 2000 the role of libraries may change. Academic and public libraries will likely still be a combination of print and paper and electronic materials. They will develop as storehouses of materials, but not necessarily of materials used day-to-day. The actual role of industrial information centres may not change so much, but it is likely that they will rely more heavily on electronic materials.

CHAPTER FOURTEEN

Information needs and availability in Western Europe with special reference to the EEC

N. VAN PUTTE

Introduction

This chapter describes the information scenario for pharmaceuticals in Western Europe. For several reasons this is not an easy task. In the first place such a scenario has so many aspects that these cannot all be discussed in sufficient detail; a selection of items has to be made. Secondly, the scenario is changing so rapidly, that only an impression of the situation can be given at the time of writing. Thirdly, one should realize that Western Europe is large and heterogeneous so that it is not possible to describe the scenario for the whole region. Consequently, this chapter will concentrate on the situation in the major West European countries.

Western Europe

Western Europe consists of more than 16 countries of different size, which usually have high density of population. Most countries are strongly industrialized, but each has its own distinct history and its own political, economic and cultural level. Most countries also have their own, unique language, so that Western Europe is a multilingual community.

After the region had been stricken by two world wars this century, the years following World War 2 brought an unparalleled expansion of wealth and productivity. In 1957 the European Economic Community (EEC or EC) was founded by the Treaty of Rome and now comprises the majority of West European countries. In spite of the depression, which followed the period of economic growth, there is a high level of welfare in almost the whole region, although there are still considerable differences in gross domestic product between certain countries. In 1987 the Single European Act came into force. This act forms part of the Treaty of Rome and defines the single market which should be achieved by the end of 1992. That market will consist of 320 million people in 12 countries.

Health care

After World War 2 there was enormous progress in health care in Western Europe. The present and future patterns are influenced by a number of factors (R. B. Tabor, 1986). There is, for example, a changing pattern of disease towards chronic illness and handicap associated with ageing and often aggravated by lifestyle factors. Circulatory disorders and cancer have become the major causes of death, followed by accidents and violence on the one hand and respiratory diseases on the other hand. Infectious diseases have been controlled to a large extent. The average life expectancy is over 75 years for women and over 70 years for men in most countries.

Advances in medical terminology introducing far more sophisticated patterns of diagnosis and care and extending the scope of medical care have also changed the state of health care. There are rising public and professional expectations connected with technological advance and increased reliance on formal health services for alleviation and comfort, rather than on informal coping mechanisms in the family and the community.

Some factors are regarded as counterbalancing these demands, such as (1) increasing awareness of, and freedom to choose, alternative therapies outside the boundaries of orthodox medicine (2) rising demand by the consumer to understand and to influence medical decisions, accompanied by initiatives towards self-help, and (3) shifting in the balance of health care from the hospital towards the community (R. B. Tabor, 1986).

The health services have been organized in different ways in the different countries, but the state has a strong influence on all systems, especially in countries where national health services exist such as in the UK and Italy. There are many physicians and the number of hospitals and pharmacies is usually high.

The WHO Regional Office for Europe has been responsible for pulling together the many strands of government, medical and expert opinion leading to the realization that a whole new health strategy is needed for the region. It will, in the same way, help in the formulation of the plans and programmes needed to give effect to that strategy.

More information about the role of the WHO in Europe can be

found in the book *Health Crisis 2000* (P. O'Neill, 1983) and in the WHO publication *Targets for Health for All* (1985).

The pharmaceutical industry

The pharmaceutical industry plays an essential role in the provision and use of pharmaceutical information in Western Europe. Many of the world's leading pharmaceutical companies can be found in Western Europe. Some of the largest are located in Switzerland, West Germany and the UK. Drug companies are often involved in activities outside the field of pharmaceuticals. They may form part of larger organizations and are e.g. the pharmaceutical arms of large chemical conglomerates. The European Federation of Pharmaceutical Industries' Associations (EFPIA) is the grouping of national associations representing the pharmaceutical industry in the region.

Many pharmaceutical companies in Western Europe are innovative and have high expenditure on research and development. This expenditure has increased under the influence of increasing demands by the health authorities in the different countries during the last 20–30 years.

The Single European Act of 1987 will have important consequences for pharmaceutical companies in the member states of the EC. The Act will particularly affect registration or licensing, pricing and marketing, but it will also have an impact on regulatory procedures, product liability, patents, developments in biotechnology, over the counter medicines and information products/services. In marketing, the EC approach has been to harmonize existing member states' rules for quality, safety and efficacy. All medicinal products should be the subject of specific marketing rules by 1992 (Edit, 1988). There are plans for a Community post-marketing surveillance system. As far as registration is concerned, discussion centres round the question of mutual recognition of national regulatory decisions versus a central 'pan-European' regulatory agency, or a system embracing both approaches.

At the moment the number of pharmaceutical products on the market of the different countries in Western Europe still varies considerably. However, there is a target date of 1990 for the national authorities of the EC member states to review all existing products according to new criteria, which are the result of harmonization of national standards into binding legislation at EC level (D. Macarthur, 1987). This harmonization will also include a 10-year period of protection for an innovator against a second applicant.

Patent protection is important for the innovative pharmaceutical industry because of the high costs involved in the development of a new drug. Since the European Patent Convention or Munich Convention came into force in 1978, it is possible to file a single application covering a number of European countries. It is hoped that by 1992 all the provisions of the Convention will have been agreed (Edit, 1988). The pharmaceutical industry in Western Europe is faced with the fact that during the last few years the use of generics has increased. This has been encouraged by European governments in order to reduce total health expenditure, but the pharmaceutical industry fears that this may have a negative effect on the output of research.

Users of pharmaceutical information

The different categories of users of pharmaceutical information in the region are briefly reviewed. In the first place, of course, physicians should be mentioned. They prescribe drugs to their patients and therefore need to be well informed. In general, however, medical practitioners use relatively little scientific information. They have little time or motivation (P. Weiss, 1986). They are dependent for their information on the pharmaceutical industry, pharmacists, health authorities, contact with colleagues, training courses and the scientific literature.

Pharmacists form another group of users of pharmaceutical information. Their role as specialists in the clinical use of pharmaceutical products has increased, but also patients (consumers) in modern West European society want to know more about the drugs prescribed to them and to be better informed than in the past.

Health authorities have special responsibility with regard to the registration of new drugs, surveillance of existing drugs and information provided to physicians.

In general, however, research workers use information most and seek it most actively (P. Weiss, 1986). These researchers can be found in research institutes, at universities and in pharmaceutical companies.

All the above mentioned persons have their own information needs. This makes it interesting to review, as far as possible in the context of this chapter, the sources and services available to them and to indicate, where possible, to what extent these fulfil their information needs.

Information sources and services

Without doubt it can be said that Western Europe is rich in sources and services for pharmaceutical information such as books, journals, retrieval systems, publishing houses, libraries and the like. For a long time the sources containing the information have been available almost exclusively in printed form, but during the last two decades there has been a considerable growth in the amount of information available in electronic form. The use of online databases has shown rapid development and in several other ways the introduction of the computer has changed the process of information transfer.

One can, however, observe that in spite of these developments the popularity of paper for the transfer of information in Western Europe has not declined. In many cases electronic information is in addition to printed information and has not replaced it. The printed journal has remained the main source of primary pharmaceutical information and the number of these journals is still growing, so that some people (e.g. P. Weiss, 1986) are of the opinion that there are now too many journals.

Journals

In many cases journals are the official publication of a professional society. The abundance of learned societies is therefore one of the main reasons for the large number of journals in Western Europe. The first scientific journal was *Philosophical Transactions of the Royal Society* in the UK, which appeared in 1665. Nowadays there are thousands of different scientific journals of West European origin. Some of them have become very famous and are used worldwide such as the *British Medical Journal*, others are mainly of local value. In this connection the language plays an important role. Only when the journal is written in English, it has a real chance of success in the international scientific community, both inside and outside Europe. A summary in English of each article in it is usually not sufficient.

During the last twenty years a number of scientific journals of a more European nature has appeared such as the *European Journal of Pharmacology*, which was first published in 1967. These are of course all written in English. They may belong to a European learned society or to a European federation of learned societies. Their aim is to report scientific developments in the whole area, but contributions from non-European countries may be included as well.

All these primary journals are published by a number of different publishing houses in Europe. Well known publishers of pharmaceutical information are e.g. Elsevier in Holland, Blackwell in the UK, and Springer and Thieme in Germany. On the other hand, American journals are equally important for many users of pharmaceutical information in Europe, especially for research workers.

Other primary sources

Apart from journals there is a great variety of other printed sources of primary pharmaceutical information available in Western Europe such as dissertations, congress proceedings, reports, patents and the like. Unfortunately part of this information belongs to 'grey literature' or even to 'dark literature', which means that access is poor or non-existent (P. Weiss, 1986). Books or monographs are another important source of information. The significance of books has not declined, but seems to be overshadowed somewhat by the increased role of journals as a source of information. In general, books play a minor role as a source of information in natural sciences (P. Weiss, 1986).

New developments are often reported at frequent congresses and symposia. Therefore it must be regretted that the results of these meetings are not always published fully and quickly. Equally, congress proceedings are covered to a less extent than primary journals by the retrieval systems.

Secondary information sources

Beside the primary information sources there is also a great number of secondary sources of pharmaceutical information available on the European market, both in printed form and online. In this case the growth of the online services has reduced the significance and use of the printed products considerably. Literature retrieval is now carried out in Western Europe to a great extent by means of computer systems. Only smaller libraries do not have the necessary equipment and have to rely on printed products only.

The different forms of pharmaceutical information used in Western Europe are briefly discussed with the emphasis on the secondary information sources. These may be used for current awareness, retrospective searching or both purposes.

BIOMEDICAL INFORMATION

Pharmaceutical information is in the first place often biomedical information: the effect of drugs in animals and man. A great variety of services in this field can be found in Western Europe. The popularity of a service differs more or less from one user group to the other (researchers, physicians, hospital pharmacists, etc.). One of the most popular services is *Index Medicus* of the National Library of Medicine (NLM) in the US, which is used in many libraries with biomedical literature. Its online version, *MEDLINE*, has also become very popular; it is available now through several European host organizations.

In general, *MEDLINE* is more frequently used than its European counterpart, *EMBASE* (Excerpta Medica), which is available through two major European hosts (DIMDI and Data-Star) only. However, European biomedical literature is better covered by *EMBASE* than by *MEDLINE*.

Excerpta Medica pays special attention to the coverage of drug literature, but for the majority of European pharmaceutical companies with research capacity the *Ringdoc* system of Derwent Publications in London is still the main source of information for drug literature. This is due to the fact that *Ringdoc* was originally developed by a group of European companies (the Pharma Documentation Ring or PDR) and that it is maintained by the publisher in close co-operation with its users. New developments always receive much attention at the annual European subscriber meeting.

Not only pharmaceutical companies have their own system. Hospital pharmacists in the UK have also developed a system, called *Pharmline*. This database is now available online through Data-Star. It is less sophisticated than *Ringdoc*, which is only available through two American hosts (Orbit and Dialog).

Several other biomedical information services are used in Europe, the most important one undoubtedly being *Biological Abstracts* or *Biosis Previews* (Biosciences Information Services). Online access is possible through several European hosts. New developments are discussed annually at the European Field Representatives Meeting and the organization opened its European help desk in 1987.

Biomedical databases are enhanced if new publications are included as soon as possible. The short delay in *Ringdoc* for instance is important for European pharmaceutical companies because of the intensive use of this system as a current awareness service. The faster turnaround introduced recently in *Embase Plus* is welcomed by European users. Such developments will further reduce the need for (manual) journal screening to prepare in-house alerting bulletins.

One of the most rapid and in Europe also one of the most popular current awareness services in the biomedical field is *Current Contents Life Sciences* (ISI); this in spite of the fact that only bibliographic data and no summaries of articles are supplied. There are libraries which maintain a similar service by distributing sets of contents of their own journals.

CHEMICAL INFORMATION

While there is a variety of biomedical databases, the whole chemical world depends almost entirely on one non-European system for chemical literature: *Chemical Abstracts* (American Chemical Society). Chemical Abstracts Services (CAS) still publish collective indexes, but these have become very expensive because of their size. This is one of the reasons why European and other users have become more and more dependent on the online system. However, topological systems have also brought a considerable improvement in online chemical literature searching. Beside *CAS Online*, developed by CAS, in France another system has become available, called *DARC*. Both systems are used in Western Europe. *DARC* is available through the French host Télésystèmes Questel, *CAS Online* through the European STN node in Karlsruhe (Germany). STN is also the only European host that offers the abstracts online. Other hosts for *Chemical Abstracts* in Western Europe are ESA-IRS and Data-Star. Some big European pharmaceutical companies use *Chemical Abstracts* tapes in house.

New developments in the system are discussed at the annual European User Council meetings. Political items are of particular interest because of the rather monopolistic position of *Chemical Abstracts*. The recent decisions of CAS about their price policy and the new licensing plans have received strong opposition in Europe.

Another important source of chemical information for European pharmaceutical companies and others is Beilstein's *Handbook of Organic Chemistry*. A project to create an online database is being supported by the West German government. Agreements have been signed with some host organizations and the first part of the online database became available through STN at the end of 1988. Another agreement has been announced between the German publishing house Springer and Télésystèmes for the marketing of Beilstein structural files with *DARC* software for in-house systems.

Other notable chemical information services in Western Europe are those of the Royal Society of Chemistry (RSC) in the UK. *Analytical Abstracts* is a well known service of RSC and the Mass Spectrometry Data Centre, a centre d'excellence that collects and disseminates mass spectra, is also an RSC product. The Society is the representative for CAS products in some European countries.

During the last few years graphic systems have become available for the storage and retrieval of structural chemical information. *MACCS* and *DARC* are the most frequently used graphic systems used by West European pharmaceutical companies for their own compounds. These companies would also like to have the chemical structures in *Ringdoc* searchable with their own system. Therefore the first step recently made by Derwent in the direction of such a topological system for *Ringdoc* was welcomed by the European subscribers. European subscribers are also waiting for a graphic system for another Derwent service, the *Chemical Reactions Documentation Service*. Hyams (1987) discusses the commercially available software packages designed specifically for in-house storage and retrieval of chemical reactions based on graphics input and output in his paper *Information for Chemists – a European Viewpoint*. He reviews the available software and databases, and indicates that *ORAC* of Leeds University in the UK may become the most popular system in Europe.

In the field of chemical patent information the co-operation between Derwent, Télésystèmes and the French patent office INPI is seen as an interesting development. This has resulted in a generic structure retrieval system based on topological coding, the *Markush-DARC* file. IDC in Frankfurt (the International Documentation Centre for Chemistry, a consortium of mainly German chemical companies) is the only other organization coding *Markush* structures. They have decided to continue with their *Gremas* code, but to use graphics input to generate these complex codes more easily (Hyams 1987).

BIOTECHNOLOGY INFORMATION

The increased interest in biotechnology of the last 10 years has resulted in a number of special information sources in this field. The most popular bibliographic system among European users is at this moment probably *Biotechnology Abstracts* (Derwent), which covers both the patent and journal literature. It was launched in 1982 and is used in printed and online form. Online access is possible through the same two American hosts used for *Ringdoc*.

Another service of British origin is *Current Biotechnology Abstracts*. It was started in 1983 by the Royal Society of Chemistry and is used both in printed and online form.

Beside these bibliographic databases some databanks have been developed in Western Europe which are of special interest in relation to pharmaceutical information, viz. DNA and protein sequence databanks and online microbiological culture collections. In 1981 a DNA sequence databank was developed at the European Molecular Biology Laboratory (EMBL) in Heidelberg (W. Germany) and in 1985 the European node of the international *Hybridoma* databank was established in Nice in France (CERDIC, the European Centre for Documentary Research on Immunoclones), both supported by the EC.

As far as the culture collections are concerned, there are several national databases in Europe. MINE, the Microbiological Information Network Europe, is an EC-funded operation to link these collections.

The significance of biotechnological information for Europe is illustrated by the fact that the British Library has a special Biotechnology Information Service. This has resulted from the European Biotechnology Information Project, which was supported by the EC.

DRUG INFORMATION

In the field of drug information an enormous variety of sources can be found. A special survey of these sources has been compiled by J. P. Revill (1984) in his book *Drug Information Sources*. A good overview is also given by D. H. Calam in chapter 9 (*Pharmacology and Therapeutics*) of the Butterworths guide *Information Sources in the Medical Sciences* (Morton and Godbolt, 1984). In both books the majority of drug information sources available in Western Europe can be found. Only the most recent developments are missing.

Revill has subdivided the sources into international, regional and country sources, while online databases are mentioned in an appendix. His book shows the wealth of sources available in different European countries as well as some special European sources. Calam makes a distinction between different types of sources: indexing and abstracting services, primary sources, reviews, monographs, pharmacopoeias and drug indexes and compendia.

There are various drug indexes and compendia which contain factual rather than bibliographic information. The factual information may consist of a great variety of data such as chemical structures, trade names, manufacturers, clinical indications, side-effects and many other drug-related data. Certain sources are of special interest for the innovative pharmaceutical industry, other sources mainly for physicians and pharmacists.

One of the most popular sources of drug information among European pharmaceutical companies is now *Pharmaprojects* (V and O Publications, UK). It contains detailed information on potential new drugs and is appreciated because of its speed in providing information that is often unpublished and obtained directly from the industry. It has been available online on Data-Star for a few years. There are companies which would like to load the database on their own computer, but this is not yet possible.

A popular journal in the same field is *Drugs of the Future* (ed. J. R. Prous, Barcelona, Spain). For a more comprehensive coverage of drug information *Drug Data Report* by the same publisher and American systems such as *Unlisted Drugs* and the systems of Paul de Haen are used.

In spite of the availability of these and other sources, pharmaceutical companies in Europe usually also have their own internal file with drug information. One of the reasons for this is that none of the existing systems fully covers the field of interest of a company. Moreover, the majority of systems are not yet available online. Further, the available systems, with the exception of *The Merck Index*, do not allow structure-activity searches. Many companies have a special interest in this type of search, especially those with their own software for substructure retrieval. For this reason Derwent is expanding the *Standard Drug File*, now a part of the *Ringdoc* system, into a more comprehensive drug information system allowing substructure searches with the different software packages used by European pharmaceutical companies such as *DARC* and *MACCS*.

Information on available drugs is primarily of value to physicians and pharmacists. It is supplied by different organizations in different ways in Europe. In this field the pharmaceutical industry, pharmaceutical societies, hospital pharmacists, health authorities, publishing houses and others are active.

The pharmaceutical industry has its mailings and medical representatives. Meetings are organized, e.g. for the launch of a new drug. Medical information departments answer questions of practitioners about the products of their organizations, for which purpose internal databases are maintained, e.g. containing the literature about their products. On behalf of the industry, compendia are published with information on products available in a country such as the *Rote Liste* in Germany and *Dictionnaire Vidal* in France.

Some pharmaceutical societies have their own databases and publications with pharmacotherapeutic information. The famous *Martindale Pharmacopoeia* for instance is published by the Pharmaceutical Society of Great Britain. This clinical pharmacopoeia is now also available online at Data-Star. The society of German pharmacists (ABDA) maintains another extensive online file, which is available via DIMDI. It is an example of a sophisticated non-bibliographic databank.

The health authorities in several European countries contribute to correct drug information by distributing drug bulletins. They have a special interest in the safety of drugs and collect reports on side-effects of drugs. In this field there is also co-operation with the pharmaceutical industry, where similar files are maintained.

Many other sources of drug information could be mentioned. Once again the reader is referred to the books of Revill and of Morton and Godbolt.

Finally reference should be made to the development of viewdata systems with pharmaceutical information, which are now available in some European countries. An example of a national viewdata service is *Prestel* (British Telecom), which already has many receivers and information providers. Medical information for doctors is supplied by different providers including a number of pharmaceutical companies. Besides drug information, mailbox services, training and the like are given.

HOST ORGANIZATIONS

In connection with the increased use of information in electronic format a large number of computer hosts has become available in Europe, which offer numerous online databases. *Dianeguide*, the directory of ECHO (the European Community Host Organization) with publicly accessible databases, shows that there are now more than 80 hosts in both Community and non-Community European countries. At present, the most important ones in Western Europe for pharmaceutical information are Data-Star, DIMDI, ESA-IRS, STN and Télésystèmes-Questel. Each host has its own mixture of bibliographic databases, full-text databases and databanks with factual information. These are mainly of American and European origin.

Data-Star is one of the most popular hosts in Europe for biomedical information. It originates from the American host BRS (both have the same software) and has its head office and computer facilities in Switzerland, but its marketing organization in London. On 1 January 1988 Radio Suisse, the owners of the service, became a subsidiary of a private company. The host is also strong in business information. It is developing its own network.

DIMDI, the Institute for Medical Documentation and Information in Cologne, is the German centre for biomedical information. It is not just a host, but also the information resource and consultancy for a large German government department. It started its operations in 1974 and has mainly German customers, but strives for a position as a European biomedical information centre.

ESA-IRS is the Information Retrieval Service of the European Space Agency in Frascati (Italy), an intergovernmental organization. It was the first to introduce online searching in Europe and has databases in different fields with the exception of biomedical sciences and patents.

STN is the fastest growing host in Europe. The Scientific and Technical Information Network is a worldwide co-operation between three organizations in the field of online information: CAS in the US, the FIZ Centre in Karlsruhe (Germany) and the Japan Information Center for Science and Technology in Tokyo. All three organizations supply databases for this network. One of the results is that STN has become the most important host for *Chemical Abstracts* in Europe.

The French host Télésystèmes-Questel was primarily oriented towards chemistry and patents, but for the last few years it has also concentrated on biomedical information. It has a special relation with Derwent Publications and is trying to expand its position outside France. Recently, it has upgraded its search software capabilities both for chemical structures (*DARC*) and bibliographic data (*Questel plus*).

Infoline in the UK was another European host for scientific databases, but since the takeover of the American host SDC with the Orbit service by Pergamon, there has been a reorganization, which has resulted in two new host organizations: Pergamon Orbit Infoline with technical and scientific databases in the US and Pergamon Financial Data Services for financial and commercial information in the UK. Recently, Orbitnet, Orbit's new telecommunications link between Europe and the US has become operational.

A recent trend in the world of the host organizations is the development of gateways from one host to the other. This enables the online searcher to access the databases of a second host through his connection with the first one. Existing gateways are e.g. ESA-Infoline, Infoline-BRS and DIMDI-ECHO. In fact, STN has created its own gateways.

Of course, there is much overlap in databases between the different hosts, but each host has also its unique files. Preference for a certain host is also determined by its price structure, database mixture and command language. DIANE (Direct Information Access Network for Europe), the information service of the EC, once developed the Common Command Language (CCL), but this never became a success. European online users no longer need American host organizations for most searches as in the beginning of the online era. However, the American service Dialog is still very popular in Europe, mainly because of its great number of databases and databanks.

EHOG is the European Host Operators Group. Recently this group has joined forces with another group, EURIPA, the European association of information producers, to form a new association: the European Information Industry Association (EIIA). The new group hopes to create a stronger voice for 1992.

LIBRARIES AND DOCUMENT DELIVERY

Libraries are one of the main sources of pharmaceutical information in Western Europe. They are widely available in most countries. In the Community the total number of libraries is estimated at about 90,000. It is clear that not all these libraries play a role in the process of pharmaceutical information. The most important ones are national and university libraries, as well as those of research institutes and pharmaceutical companies.

Collaboration between different libraries can be very useful in view of inter-library loan and document delivery. However, collaboration between libraries in European countries is often limited and at a European level it is only in its infancy. For good co-operation computer systems are needed, but only about 10% of the libraries in the EC have computer systems and these systems are often not compatible. In general, there is a lack of standardization and there is a language barrier, because existing systems are only available in the language of the country. This means that the number of library networks in Europe is limited and that such networks are often only at a local level. There is e.g. a local network in the area of Rome. On the other hand Italy is one of the countries building a national network.

According to a preliminary report of the EC in 1987 (*Plan of Action for Libraries in the EC*) the above-mentioned situation is probably due to many factors such as the progressive imbalance between increasing running costs and decreasing real resources, a certain traditional and institutional rigidity in the library sector and the fact that automation in libraries has taken place on the whole in an unco-ordinated and fragmentary way. In order to improve the situation the Council of Ministers adopted a resolution in 1985 on collaboration between libraries in the EC. The aim is not only to achieve the interconnection across Europe of computerized catalogues but also to promote libraries as an active contributor to the innovation process. An action plan in this direction is under discussion. In the meantime one can observe that there is indeed a tendency in Europe to transform libraries in more active information centres with different kinds of information services. The libraries of many pharmaceutical companies have already become a part of an integrated information unit in their company.

The current situation means that inter-library lending and document delivery are patchy in Europe. This is the conclusion of D. Russon (1987), director of the British Library Document Supply Centre (BLDSC), the largest document supply centre in Western Europe. He remarks that some countries have well used systems which perform reasonably well and that others have relatively small demand and/or poor performance. The demand varies between over 3,000,000 requests per annum in the UK to 80,000 per annum in Spain. The majority of titles is in the subject area of life sciences. The satisfaction rate seems to vary from 60% in Italy to 90% in the UK.

Russon also describes how technology and other developments can assist in providing better document delivery services in Europe. One of the developments is the ADONIS project, the experimental document delivery service of some European publishers which supplies copies of articles from 219 biomedical journals held on CD-ROM. The aim of the project is quicker, better and cheaper document delivery. It receives support from the EC Commission. Test sites in Europe are BLDSC, its French equivalent CDST (Centre de Documentation Scientifique et Technique), the library of the KNAW in Amsterdam (the central Dutch library for biomedical literature) and the Medical Library in Cologne (Germany).

Another method mentioned by Russon for improvement of document delivery is electronic delivery. He refers to a project under the name APOLLO (Article Procurement with online Local Ordering), an experiment of the European Commission and ESA for electronic document delivery between West European countries via satellites.

More experiments could be mentioned such as TRANSDOC in France, one of the EC's DOCDEL (document delivery) projects for the electronic storage and delivery of documents. These developments show the significance of technology for the process of pharmaceutical information transfer. Therefore attention is now turned to the technical environment in Western Europe.

Technical environment

Technical developments have had an enormous impact on the process of information transfer in Western Europe during the last 20–30 years. This is evident when one looks for instance at the techniques of literature searching. In 1964, when *Ringdoc* was taken over by Derwent from the Pharma Documentation Ring, it had a chemical and biological code based on the punched card. Literature searches were time consuming because many punched cards had to be put through a sorting machine. Nowadays there are computers with efficient programmes for in-house and external retrieval of information. Large databases can be scanned quickly, even if they are on the other side of the world. Initially, terminals were used for information retrieval, but gradually they are or have been replaced by personal computers (PCs). The PC has become the workstation for the information specialist and many others. It is used for purposes like database management, information retrieval, uploading, downloading, file conversion, statistics, terminal emulation and word processing. For all these applications many programmes are available. Graphics have become a normal feature.

The Information Resources Centre at Aslib, the Association for Information Management, has carried out a survey of the use of information technology within the library and information community, as represented by its membership (G. Sippings, 1987). The survey covered microcomputers, VDU, print and videotex terminals, printers, modems, CD-ROM drives and microform equipment. Computers, PCs, printers and especially microform equipment appeared to be widely used. The number of respondents using some form of software for information retrieval purposes was lower than expected (about 50%). Electronic mail and videotex were frequently mentioned and a large number of organizations was found to be using fax, although this is a rather new facility.

TELECOMMUNICATION

Good data communication networks became necessary in Europe for the international transfer of information. Such networks are essential for the access of online commercial databases in the different countries, for parent companies to provide product information to overseas operating companies, electronic mail, electronic document transmission and the like. Therefore a network to collect and disseminate scientific and technical information was set up by the PTTs of the EC member countries. It had to be a network with location independence and uniform tariffs. This Euronet has been in operation for some years. In 1983 the Euronet DIANE service offered more than 500 databases and databanks on some 50 hosts. However, the plan was not entirely successful. According to M. Hyams (1987) this was due to the following reasons. (1) Hosts could not agree on who should have the most profitable databases. (2) No suitable software was at hand, and insufficient funds were available for development. (3) The language barrier made training and promotional effort difficult between countries. (4) There was not enough critical mass of databases available for cross-file searching, or to justify learning a new command language. (5) Euronet communications lacked sufficient power and reliability, and the various national PTTs were most unco-operative in making concessions. Hyams concludes that the various factors contributing to the downfall of Euronet are still present, so that the European online market is fragmented and in his opinion still dominated by the Americans.

Online users in Europe are now dependent on the national telecommunication networks, the Packet Switching Services (PSS) which are interconnected in the IPSS. The development of these networks has been stimulated by Euronet. In the new situation most European users are only a local call away from a central node that connects them to all the other major networks in the world. However, the online users see that their telecommunication costs may differ considerably from one country to another. Moreover, surveys of EUSIDIC, the European Association of Information Services, have shown during the last few years that 25–30% of all attempted online connections failed because of failures by the telecommunication system. The monitoring was carried out in 12 European countries and showed only a slight improvement in 1988 compared with 1986 and 1987. Competition on the market is considered by EUSIDIC as the only viable, permanent solution of telecommunication problems. Most European countries now also have an academic network allowing communication between universities and like institutes. A good example is JANET (the Joint Academic Network), a wide-area network linking together computers and users in British universities, polytechnics, establishments of the research councils, and the British Library. RARE (Reseaux Associés pour la Recherche Européene; associated networks for European research) has been established to co-ordinate developmental activities amoung research networks and hopes to harmonize electronic communications to ensure that scientists all over Europe can communicate.

USE OF ONLINE INFORMATION

The use of online information differs from one country to the other. A user panel has been created for the European Commission to help monitor trends in the demand for electronic information services. According to a report prepared by BIS Mackintosh (1988) on behalf of the EC about this panel (*Electronic Information Users in Europe*), the extent of use is approximately in line with economic development and is particularly aligned with the status of the telecommunications infrastructure. The UK dominates the professional electronic information services market due to the high concentration of financial information services and the accessibility of English-language US online databases. The French market is particularly important in view of the overwhelming success of the Minitel videotex service, a popular service for the end-user and also used to access Télésystèmes Questel.

According to the report the more conservative approach to the provision of electronic information services has hindered their development in West Germany, where information gathering is concentrated on central information departments rather than on end-users. In Italy the telecommunications infrastructure is of a relatively underdeveloped nature. This means that the new technology of CD-ROM is now more extensively used in Italy than in any other European country.

CD-ROM

Much attention has been paid in Western Europe to the CD-ROM during the last few years. Online databases and reference works became available on optical disc, often as experimental systems, but in general publishers in Europe have been more cautious so far than their colleagues in the US. Reasons for this caution can be found in a report, entitled *Market Opportunities for CD-ROM 1987–1992*, from Knowledge Research, a London-based market research company. There is a fear that CD-ROM will erode revenues from existing alternative publications and there is an unwillingness to commit to the large up-front investments needed. Hardware manufacturers in particular have shown little interest in entering the market. According to the report the strongest commitment to CD-ROM has come from relatively small software publishers, but in Europe only 20% of online databases are being converted to CD-ROM compared with 40% in the US, where the penetration of PCs is about three times as high as in Europe. Some projects in Europe are supported by the Commission of the EC. One of them is *BIO-ROM*, i.e. *Biotechnology Abstracts* (Derwent) on CD-ROM. Another is *Medata-ROM*. This product contains two-monthly updates of *MEDLINE*, *Embase* and *Pascal*, the French database for scientific literature. It is a joint project among a few French organizations. The ADONIS project of some European publishers for document delivery from biomedical journals is also supported by the European Commission.

Not only publishers but also users in Europe are cautious as far as the CD-ROM is concerned. The number of CD-ROM drives in Western Europe is still rather low. The new medium is tested at many places. *MEDLINE* is one of the most frequently used files for these tests, but *Excerpta Medica* has a test file for *Embase* as well.

In the meantime CD-ROM has come into use in several libraries, e.g. in universities and pharmaceutical companies. In these cases not so much the information specialists but the end-users, the scientific workers, have become the users of this new facility, although they are often trained and supported by information specialists. In this way CD-ROM is a replacement of hardcopy indexing and abstracting services rather than a replacement of online searching.

ACTIVITIES OF THE EC

Within the European Commission, Directorate General XIII (DG XIII) deals with telecommunications, information industries and innovation. Its responsibility is to put forward and carry out activities that will give the Community as a whole the ability to control its own future in all aspects of information. Two important current technological activities of DG XIII are ESPRIT and RACE.

ESPRIT, the European Programme of Research and Development in Information Technology, began its pilot phase in 1983. The main programme, 1984–1989, had a budget of 750 million ecus on a 50% cost sharing basis with companies, universities and research centres. The major aims were to improve Europe's information technology capability, to reinforce technological co-operation and pave the way to internationally accepted standards. Action areas were patent information, biotechnology information, electronic publishing, library co-operation and telecommunication. Projects such as ADONIS, APOLLO and *BIO-ROM* have already been mentioned. The second phase of ESPRIT has been given the go-ahead by the Council of Ministers. It will have a budget of 1,600 million ecus.

RACE, Research and Development in Advanced Communications in Europe, is designed to help create the telecommunications channels and infrastructure for the mid-1990s. Such a broadband network will allow the digital transmission of both images and data.

Discussion and conclusions; the future

As stated in the introduction it is impossible to discuss all aspects of pharmaceutical information in Western Europe. Special attention has been paid to the needs of the pharmaceutical industry, which plays an essential role in the use and transfer of pharmaceutical information in Western Europe. Special attention has also been paid to developments in the EC because these are important for the future of the information provision in general in Western Europe.

Within the EC there has been a growing awareness of the value of information. Recently the EC Council of Ministers has approved an action plan for a European services market. The aim of the action plan is to help set up an internal information services market by the end of 1992 and to stimulate and reinforce the competitive capacity of the suppliers of information services in Europe. ECHO is expected to play an important role in the plan. Professional electronic information services will be a decisive factor in achieving the Community's 1992 objective of a unified internal market.

Western Europe has for a long time been relying on traditional paper-based methods for the transfer of pharmaceutical and other forms of information, but is now using information in electronic format more often. However, while the number of databases of European and American origin is comparable, the number of European databanks is still lower than the number of American databanks.

The EC attempts to solve the problem of grey literature by means of SIGLE, the online system for information on grey literature in Europe.

In 1992 there should be one common market in the EC for telecommunication products and services. Rules for the supply of these products and services have been laid down by the European Commission in the report *Towards a Community-wide Telecom-*

munications Market in 1992. The role of the national PTTs is also described in this report; they should be commercial organizations in a free market.

One may hope that the libraries in Europe will also benefit from improved telecommunications, so that better collaboration between libraries in the different European countries may lead to a better library service.

Technological developments play an essential role in the future of the information scenario for pharmaceuticals in Western Europe. However, technological development cannot be isolated from economic, political, social and other factors (P. Weiss, 1986). The development of information and communication in the future therefore depends on their adequate organization.

In general, pharmaceutical information is of course to a large extent an international affair and therefore the free flow of information, not only in Western Europe, is essential. Prof. Luckenbach (1988), director of the Beilstein Institute in Germany, mentions a number of political, technical, economic and legal factors which may prevent this free flow of information, such as the non-availability and nonaccessibility of essential services, lack of education, language problems and monopolism. One may hope that these barriers, as far as they still exist in Western Europe, may disappear so that in the future pharmaceutical information will always be available where it is needed.

Acknowledgements

I thank Marlou Prevoo and Paul Landman for their assistance in preparing the manuscript and some European colleagues for their contributions.

References

Calam, D. H. (1984). 'Pharmacology and therapeutics'. In *Information Sources in the Medical Sciences*, eds. L. T. Morton and S. Godbolt, 3rd edn, pp. 188–207 (Butterworths).

Edit (1988). '1992 The single European market'. *The Pharmaceutical Journal*, **241**, 76.

Hyams, H. J. (1987). 'Information for chemists – a European viewpoint'. *Aslib Proceedings*, **39**, 169–181.

Luckenbach, R. (1988). 'The free flow of information: a Utopia? Ways to improve scientific and technological information and its international exchange'. *J. Chem. Inf. Comp. Sci.*, **28**, 94–99.

Macarthur, D. (1987). *The EEC environment for medicines – a progress towards harmonisation*. (PJB Publications).

O'Neill, P. (1983). *Health crisis 2000*. (Heinemann).

Revill, J. P. (1984). *Drug Information Sources* 2nd edn (Gothard House Publications Ltd.).

Russon, D. (1987). 'Document delivery to medical libraries in Europe: technology, policies and practices'. In *Medical libraries: cooperation and new technologies*, eds. C. Deschamps and M. Walckiers, pp. 33–38 (North-Holland).

Sippings, G., Ramsolen, H., and Turpie, G. (1987). The use of information technology by information services: the Aslib information technology survey 1987. In *Online Information 87. Proceedings of the 11th International Online Information meeting London 8–10 December 1987*, pp. 247–252 (Learned Information).

Tabor, R. B. (1987). 'Integrated information for health care'. In *Medical libraries: cooperation and new technologies*, eds. C. Deschamps and M. Walckiers, pp. 121–126 (North-Holland).

Weiss, P. (1986). *Health and biomedical information in Europe* (WHO).

World Health Organization. (1985). *Targets for health for all. Targets in support of the European regional strategy for health for all.* (WHO).

CHAPTER FIFTEEN

Information needs and availability in Japan

T. SASAKAWA

Background to the Japanese pharmaceutical industry

As an East Asian country, Japan is a neighbour of China, Korea and the USSR. Surrounded by the Pacific Ocean, the Sea of Okhotsk, the Sea of Japan and the East China Sea, Japan consists of some 3,000 islands, the four major ones being Honshu, Hokkaido, Shikoku and Kyushu. The total area is 377,835 square kilometres (around 1/25th of the land mass of the US); the population, 121,672,000 (1986 census), is about half that of the US. Much of the land area varies from hilly to very mountainous. The plains regions of Japan are relatively small but of vital importance, because the population is so heavily concentrated there.

Disease profile

According to the 1986 census, the mortality level in Japan that year was 751,000, a decrease of 2,000 over the previous year. The major causes of death in both sexes were malignant neoplasms together with cardiovascular and cerebrovascular diseases. About 61% of deaths are caused by diseases in adults.

According to the national health survey performed by the Ministry of Health and Welfare (MHW), the main cause of morbidity is cardiovascular disease followed by gastrointestinal and respiratory diseases. Japan has become one of the leading countries in terms of longevity and the proportion of aged people in the population is still on the increase.

Mortality and morbidity data are much in demand by the pharmaceutical industry for strategic planning purposes. Typical mortality and morbidity information sources are:

Kokumin Kenko Chosa (National Health Survey), compiled by MHW, published by Kosei Tokei Kyokai, annual
Jinko Dotai Tokei (Vital Statistics), compiled by MHW, published by Kosei Tokei Kyokai, annual
Kosei Tokei Yoran (Health Statistics Handbook), compiled by MHW, published by Kosei Tokei Kyokai, annual.

Status of the industry

Approximately 360 Japanese pharmaceutical companies are listed in *Pharmaceutical Manufacturers of Japan 1988–89* (Yakugyo Jiho Co., Ltd.). According to *Yakuji Kogyo Seisan Dotai Tokei Nenpo (Pharmaceutical Industry Production Trend Statistics)*, Yakugyo Keizai Kenkyujo, the production value of pharmaceuticals manufactured in Japan during fiscal 1986 was 4.8 trillion yen, an increase of 12.7% over the previous year. Another source of Japanese pharmaceutical market information is *Japan Pharmaceutical Market* (IMS Japan Ltd.). *Table 15.1* shows values of ethical drug production by therapeutic use. Antibiotics are still ranked first in

Table 15.1. Ethical drug sales by category

Rank	Therapeutic category	Production value	Percentage of total
1	Antibiotics	683,361	16.0
2	Cardiovascular agents	563,301	13.2
3	Central nervous system agents	430,516	10.1
4	Gastrointestinal agents	372,044	8.7
5	Other metabolic agents	328,629	7.7
6	Dermatologicals	261,030	6.1
7	Vitamins	243,712	5.7
8	Respiratory agents	166,904	3.9
9	Antitumor agents	160,822	3.8
10	Biologicals	157,887	3.7
11	Others	912,526	21.1
	Total	4,280,732	100.0

Source: *Pharmaceutical Industry Production Trend Statistics* (Yakugyo Keizai Kenkyujo).

the ethical drug market but their share by sales value has declined due to recent drug price cuts. Other top selling drugs in Japan are cardiovascular drugs, central nervous system agents, and gastrointestinal agents (which is to be expected in view of the disease profile of Japan). The Japanese pharmaceutical market is the second largest in the world after the US and, in fact, only about 3% of the pharmaceuticals manufactured in Japan are exported. This is a very low level compared with other developed countries. In direct contrast to licensing trends, approximately 2.5 times as many pharmaceuticals are imported as exported.

Health-care delivery

In 1961 the Japanese government established a Health Care Insurance programme with the aim of providing insurance coverage for the entire population. Consequently, according to statistics in the 1985 Report *Developments in National Health Care* (MHW) 99.5% of the Japanese population was covered by Health Care Insurance.

Table 15.2 shows expenditure on health care and, in particular, elderly health care alongside average national income. Note that the 1986 data are estimates and 1987 data forecasts. The increase in health care expenditure is a major political issue and has been brought about by the ageing of the population and the use of highly sophisticated medical equipment and drugs.

In 1981, the Japanese government began a policy of systematically reducing official drug reimbursement prices. In 1987, the government introduced a new pricing formula and on 1 April, 1988, the first price cuts under the new pharmaceutical pricing system occurred. In Japan, reimbursement prices are set by the government for all ethical drugs but because pharmaceutical companies compete with one another for sales, the actual price paid for a drug is generally lower than the official listed price. The objective

Table 15.2. Health care expenditure

	1983	1984	1985	1986	1987
Health care expenditure	14.5	15.1	16.0	17.1	18.0
Elderly health care expenses	3.3	3.6	4.0	4.4	4.8
National income	227.9	239.7	254.5	264.5	275.6

Source: **Yakuji Handbook '88** (Yakugyo Jiho Co., Ltd.).

of the government's price revision system has been to lessen this gap. A useful compendium of drug prices is *pdp Japanese Ethical Pharmaceutical Tariff* which covers the official prices for over 6,000 ethical drugs.

Research and development in pharmaceuticals

Research and development costs in the pharmaceutical industry are higher than in other industries and these costs have been increasing steadily in proportion to sales. Foreign pharmaceutical companies and domestic non-pharmaceutical companies have been entering the Japanese market in greater numbers recently. This has increased competition thus making the development of new drugs even more important to the survival of the established pharmaceutical companies. Sources for research and development in the Japanese pharmaceutical industry are: *Annual Survey of Investigational Drugs in Japan 1987* (16th edn Drug Business Research Co., Ltd.) and *Pharmacast* (Technomic).

In Japan the new drug approval process takes 10–16 years from the time a new compound is discovered until the drug is finally marketed. A period of 2–3 years is spent in preliminary research and screening. This is followed by 3–5 years of preclinical animal tests. After submitting notification of the completion of preclinical tests and receiving governmental permission to continue, another 3–5 years of clinical trials are conducted. Then a New Drug Application (NDA) is filed: 2–3 years is required for evaluation by the MHW before the NDA may be approved. Once approval has been received the drug must be included in the drug reimbursement price listing before it can be marketed. The entire procedure requires the expenditure of approximately 8–10 billion yen by the pharmaceutical company. The Japanese government formally decided recently on a revision of the patent law, and extended the patent period for newly developed drugs. The revised law came into effect on 1 January, 1988, and should improve the prospects of the patented drug market and encourage innovative research.

Japanese pharmaceutical regulations are influenced by those of the US. The Good Manufacturing Practice (GMP) and the Good Laboratory Practice (GLP) regulations will be followed in the near future by the Good Clinical Practice (GCP) and the Good Post Marketing Surveillance Practice (GPSP) regulations.

Outline of library and information scene

Japan has been steadily developing as a technological nation since World War 2. This policy was stimulated by the lack of natural resources and the large, highly concentrated population. According to the *Handbook of Science and Technology* (JICST, 1986), the development of policy on science and technology information in Japan has proceeded in four steps.

The first step was the period of economic rehabilitation after the War. Major organizations were established in the field of scientific information. The National Diet Library was founded in 1948. In 1950, the Library Law came into effect, establishing a basis for the introduction of modern library organization and management practices in Japan. In 1956, the Science and Technology Agency was established, followed by The Japan Information Center of Science and Technology (JICST) in 1957.

The second step was a high-growth period in the Japanese economy. Research laboratories were organized in many private companies, the need for scientific information increased, and JICST expanded. But these organizations had developed separately, so the need for total regulation and the establishment of an organizational policy was recognized, particularly in the light of the US President's Science Advisory Committee Report *Science, Government, Information* (the Weinberg Report). Therefore in 1969 the Government created the Science and Technology Council which submitted an interim report on the National Information System for Science and Technology (NIST) Plan. The objective of the NIST Plan was to improve substantially and consistently the co-ordination of scientific information activities in the future.

The third step was the realization of the NIST Plan. In the late 1960s computer technology developed rapidly, and many information systems and special information centres were established in universities, governmental offices, national research laboratories and private companies.

In 1974, the Science and Technology Agency offered a final report detailing the measures required for the ultimate realization of the NIST Plan. Since then a number of special information centres (e.g. Japan Association for International Chemical Information (JAICI), International Medical Information Center (IMIC), Japan Pharmaceutical Information Center (JAPIC), etc.) have been established.

The fourth step, which is in progress now, is a period of internationalization and widespread information flow in science and technology using optical fibre, optical discs, local networks and artificial intelligence. An example of internationalization is the establishment of the Scientific and Technical Information Network (STN) international network in co-operation with the US and West Germany.

In 1986, the National Center for Science Information Systems (NACSIS) was established. This centre produces databases of books and periodicals. It has also provided an online service, NACSIS-IR since 1986. As of March 1988, 62 university libraries were connected with this network. However, these bibliographic utilities are not available for use by private companies.

Information sources

Secondary sources

In the 1970s only the largest chemical companies were able to use the major magnetic tape secondary sources (e.g. *CA Search, Excerpta Medica*) because at that time commercial online services were not available and a company needed a large computer and specialized software in order to mount the large expensive databases. But during the 1980s, online systems have developed rapidly and small companies are able to access many commercial services via relatively inexpensive online vendors. The rapid development of computer technology and storage materials (e.g. magnetic discs) has led to changes in the production methods and processing of secondary sources. In Japan it led to the ability to handle 'Kanji' (ideographs) and consequently Japanese databases using Kanji were produced and marketed. In recent years, most printed secondary sources have become available as databases through commercial online systems. The foreign medical and pharmaceutical databases most frequently used in Japan are *MEDLINE* and *Chemical Abstracts*.

These and other foreign secondary sources commonly used in Japan are shown in the General Appendix. Japanese secondary sources are listed in the General Appendix and, since they are unique and likely to be of general interest, they are briefly described below.

JAPANESE SECONDARY SOURCES MOST FREQUENTLY USED BY THE PHARMACEUTICAL INDUSTRY

Igaku Chuo Zasshi (Japana Centra Revuo Medicina) was first published in 1903 and is intended primarily for individual physicians. There are 36 issues a year produced by Igaku Chuo Zasshi Kankokai (The Japan Medical Abstracts Society) in Tokyo. This abstracts journal covers the whole field of medical subjects and consists of more than 200,000 citations from 1,917 journals (1,770 Japanese journals and 147 English language journals published in Japan). Each volume has a cumulated subject index and name index. In 1981, input commenced of part of *Igaku Chuo Zasshi* into one of the Japanese medical databases produced by the Japan

Information Center of Science and Technology (JICST), the *JICST File on Medical Science in Japan (JMEDICINE)*.Some 40,000 articles per year from about 700 journals were input into this database over the next two years. From 1983, the society initiated computerised publication and henceforward all data in *Igaku Chuo Zasshi* have been input into *JMEDICINE*. Approximately 140,000 articles per year are now input into *JMEDICINE* and the database incorporates about 220,000 citations annually. An English version of *JMEDICINE* is available outside Japan through STN international. In 1986 *Igaku Chuo Zasshi* initiated its own online service entitled *Ichushi Title Guide* in order to supply clinical information more rapidly.

Nihon Iyaku Bunken Shorokushu (Abstracts Journal for Japanese Pharmaceutical Literature) published by the Japan Pharmaceutical Information Center (JAPIC) indexes and abstracts drug information extracted from more than 321 journals. Approximately 13,000 abstracts are published each year. In 1987 there were 8,933 original articles and 1,897 reviews. It is available as an online database (*JAPICDOC*) through Tokyo Information System (TIS) and Japan Information Processing Service (JIP) online systems.

Kagaku Gijutsu Bunken Sokuho (Current Bibliography on Science and Technology) is published by JICST. It covers about 5,600 Japanese journals, 7,500 foreign journals published in 50 countries, technical reports, proceedings and so on. The abstracts are arranged in 12 sections. The abstracts are available online in the *JICST File*, through JOIS (JICST Online Information System), and are updated at the rate of 521,000 abstracts per year. An English version, *JICST File on Science and Technology*, is searchable through STN International in European countries and the US.

Asu no Shinyaku (Pharmacast) is published by Technomic Information Service, Inc. Results of trials with novel experimental drugs from pre-clinical studies through to phase III clinical trials are sorted according to manufacturer. The publication supplies current research information abstracted from about 350 published journals, conferences, and pharmaceutical newspapers. The editing of this information is enhanced by direct contact with the pharmaceutical industry. It is published annually with a quarterly update. Data provided includes drug names, use, manufacturer, clinical stage, CAS registry number, structural formula, molecular formula, synonyms and chemical names. There is also provided detailed background information on each drug. An English version of *Pharmacast* is published. The online version is updated monthly and is available through JIP.

The publication *CONTENTS* provides copies of the table of contents pages of approximately 400 current Japanese journals in pharmaceutical, medical and library sciences. Each issue also previews forthcoming meetings in the medical and pharmaceutical sciences, as well as preprints and abstracts of such meetings.

Primary sources

Journals and reference books are listed in each section. They are divided into two parts: those published outside Japan and those published in Japan.

JOURNALS

Journals published outside Japan which, according to the Union Periodical Catalogue of the Japan Pharmaceutical Libraries Association, are most often used by the Japanese pharmaceutical industry are listed in the General Appendix.

Journals published in Japan which according to the Union Periodical Catalogue of the Japan Pharmaceutical Library Association and the journal *CONTENTS* are most often used by the Japanese pharmaceutical industry are described below.

JAPANESE JOURNALS IN ENGLISH

Chemical and Pharmaceutical Bulletin (Tokyo, The Pharmaceutical Society of Japan, 1953-, monthly, ISSN 0009–2363), an official journal of the Pharmaceutical Society of Japan, is published monthly. It covers all aspects of drug research, namely studies on physical chemistry, organic chemistry, medicinal chemistry, pharmacognosy, analytical chemistry, biological science, pharmaceutical science and pharmacological science.

The Japanese Journal of Pharmacology (Kyoto, The Japanese Pharmacological Society, 1951-, monthly, ISSN 0221–5198) is an official English language journal of the Japanese Pharmacological Society. Articles are included on all aspects of pharmacology. Proceedings of meetings of the Japanese Pharmacological Society are published in supplementary issues.

The Journal of Pharmacobio-Dynamics (Tokyo, The Pharmaceutical Society of Japan, 1978-, monthly, ISSN 0386–846X) is an English language journal which publishes research articles on dynamic interaction between pharmaceuticals and biological systems. It also covers abstracts of the symposia of the Pharmaceutical Society of Japan.

The Journal of Toxicological Sciences (Tokyo, The Japanese Society of Toxicological Sciences, 1976-, quarterly, ISSN 0388–1350), the official English language journal of the Japanese Society of Toxicological Sciences, provides research papers on toxicology and papers relating to toxic substances and their metabolites.

RESEARCH JOURNALS

Byoin Yakugaku (The Japanese Journal of Hospital Pharmacy) (Tokyo, Japanese Society of Hospital Pharmacists, 1975-, bimonthly, ISSN 0389–9098), the official journal of the Japanese Society of Hospital Pharmacists, publishes reports on original research bimonthly.

Drug Delivery Systems (Osaka, Iyaku Journal Co., Ltd., 1986-, semi-annually, ISSN 0913–5006) is intended for physicians, researchers in the pharmaceutical industry and laboratories. This journal covers papers on drug delivery systems.

Gan to Kagaku Ryoho (The Japanese Journal of Cancer and Chemotherapy) (Tokyo, Gan to Kagaku Ryohosha, 1974-, monthly, ISSN 0385–0684), is intended for physicians and medical doctors, and covers articles on cancer treatments and experimental research. Each issue has a special feature.

Gendai Iryo (Modern Medical Treatment) (Tokyo, Gendai Iryosha, 1969-, monthly, ISSN 0533–7259) focuses mainly on clinical and therapeutic research. In addition there are sometimes commentaries and review articles.

Hinyokika Kiyo (Acta Urologica Japonica) (Kyoto, The editorial board of Hinyokika Kiyo, 1955-, monthly, ISSN 0018-1994) is intended primrily for medical doctors of urology. It provides articles on orginal research and clinical research.

Iyaku Journal (Medicine and Drug Journal) (Osaka, Iyaku Journal Co., Ltd., 1965-, monthly, ISSN 0287–4741) is a semi-technical journal for hospital pharmacists.

Nippon Kagaku Ryoho Gakkai Zasshi (Chemotherapy, Tokyo) (Tokyo, Nippon Kagaku Ryoho Gakkai, 1952-, monthly, ISSN 0009-3165) is the official journal of the Nippon Kagaku Ryoho Gakkai and publishes articles for medical doctors on chemotherapy. It covers many papers on pharmacology, pharmacokinetics and clinical trials of new antibiotics.

Nippon Yakurigaku Zasshi (Folia Pharmacologica Japonica) (Kyoto, The Japanese Pharmacological Society, 1925-, monthly, ISSN 0015–5691) is the official journal of the Japanese Pharmacological Society and publishes articles on original methods in pharmacology and other aspects of pharmacology.

Nippon Yakuzaishikai Zasshi (Journal of the Japan Pharmaceutical Association) (Tokyo, Japan Pharmaceutical Association, 1949-, monthly, ISSN 0369–674X) is published for the Japan Pharmaceutical Association. This journal provides research papers and reports on issues and events that influence the pharmacist. There is also news as well as comments and papers on the safety and efficacy of new drugs, legislation, pharmaceutical education and public health.

Oyo Yakuri (Pharmacometrics) (Sendai, Oyo Yakuri Ken-kyukai, 1967-, monthly, ISSN 0300–8533) publishes original papers dealing with the clinical and pharmacologic effects of new drugs. Most papers relate to the evaluation of new drugs in patients or healthy volunteers.

Pharma Tech Japan (Tokyo, Yakugyo Jiho Co., Ltd., 1985-, monthly, ISSN 0910–4739) is a semi-technical journal for pharmacists and quality controllers. Most papers are derived from *Pharmaceutical Technology* (Aster Publishing), and translated into Japanese.

Rinsho Yakuri (Japanese Journal of Clinical Pharmacology and Therapeutics) (Tokyo, Nippon Rinsho Yakuri Gakkai, 1970-, quarterly, ISSN 0388–1601) publishes original papers dealing with the clinical and pharmacologic effects of drugs. There are a great many papers related to the evaluation of drugs in patients or healthy volunteers.

Toxicology Forum (Tokyo, Science Forum, 1977-, bimonthly, ISSN 0287–8712) publishes original papers covering all aspects of toxicology. It contains information and comments on the development of new methods of investigation used in toxicology, and short articles on the activities of the pharmaceutical industry and related societies.

Yakubutsu Dotai (Xenobiotic Metabolism and Disposition) [Tokyo, Nippon Yakubutsu Dotai Gakkai (Japanese Society for the Study of Xenobiotics), 1986-, monthly] is an official journal which covers information on the biotransformation and the mechanisms of toxic action of several drugs. Most papers concentrate on the actions of exogenous chemicals in biological systems.

Yakugaku Zasshi (Journal of the Pharmaceutical Society of Japan) (Tokyo, The Pharmaceutical Society of Japan, 1881-, monthly, ISSN 0031–6903) is an official journal which publishes papers and reviews on all aspects of pharmaceutical science (e.g. organic chemistry, biological chemistry, medicinal chemistry and pharmacology). It is primarily intended for pharmaceutical researchers.

Yakuri to Chiryo (Japanese Pharmacology and Therapeutics) (Tokyo, Life Science Publications, 1973-, monthly, ISSN 0386–3603) a technical journal covering papers on pharmacological tests and clinical trials.

Yakuzaigaku (Journal of Pharmaceutical Science and Technology) [Tokyo, Nippon Yakuzai Gakkai (The Academy of Pharmaceutical Science and Technology), 1940-, quarterly, ISSN 0372–7629) publishes original and review articles in the broad field of pharmaceutical science and technology in Japanese or in English. There is also a short letters column. This journal had been published for 45 years by the Pharmaceutical Society of Japan before the Academy of Pharmaceutical Science and Technology was founded in 1985.

CLINICAL JOURNALS

Chiryo (*The Journal of Therapy*) (Tokyo, Nanzando, 1920-, monthly, ISSN 0022–5207) publishes articles on trends in the medical field as well as advances and developments in diagnosis and therapy.

Kiso to Rinsho (*The Clinical Report*) (Tokyo, Yubunsha, 1967-, monthly, ISSN 0385–2806) is a journal for physicians. It gives priority to publishing articles on drug effects in human trials and clinical uses of new drugs.

Naika (*Internal Medicine*) (Tokyo, Nankodo, 1958-, monthly, ISSN 0022–1961) is a medical journal, but it contains a lot of articles on drug therapy. Intended for the use of physicians, each issue has a special clinical topic on treatment, diagnosis and diseases.

Rinsho Hyoka (*Clinical Evaluation*) (Tokyo, Rinsho Hyoka Kankokai, 1972-, quarterly, ISSN 0300–3051) covers papers on clinical uses, especially double blind tests, of new drugs.

Rinsho Iyaku (*Journal of Clinical Therapeutics and Medicine*) (Tokyo, Rinsho Iyaku Kenkyu Kyokai, 1985-, monthly, ISSN 0910–8211) publishes critical reviews and papers on clinical trials and clinical pharmacokinetics of new drugs.

Rinsho Kagaku (*The Journal of Clinical Science*) (Osaka, Sekai Hoken Tsushin Co., Ltd., 1965-, monthly, ISSN 0385-0323) provides comments on clinical trials and pharmacology of drugs. Each issue has a special theme on clinical science.

Rinsho to Kenkyu (*The Japanese Journal of Clinical and Experimental Medicine*) (Fukuoka, Daido Gakkan, 1924-, monthly, ISSN 0021–4965) covers all aspects of clinical medicine for general practitioners.

Saishin Igaku (*Medicine Today*) (Osaka, Saishin Igakusha, 1946-, monthly, ISSN 0370–8241) provides comments on recent advances in diagnosis, treatment and clinical cases.

Shindan to Chiryo (*Diagnosis and Treatment*) (Tokyo, Shindan to Chiryosha, 1914-, monthly, ISSN 0370–999X) is published for physicians to provide information and comments on clinical cases, diagnosis and treatment.

Shinryo to Shinyaku (*Medical Consultation and New Remedies*) (Tokyo, Iji Shuppansha, 1963-, monthly, ISSN 0037-380X) publishes information on new drugs of all therapeutic categories and papers on clinical therapeutics and trials.

Shinyaku to Rinsho (*Journal of New Remedies and Clinics*)

(Tokyo, Iyaku Joho Kenkyujo, 1952-, monthly, ISSN 0559–8672) publishes papers and comments on pharmacology and clinical therapeutics of new drugs.

Sogo Rinsho (*Clinic All-Round*) (Osaka, Nagai Shoten, 1952-, monthly, ISSN 0371–1900) is a clinical journal for physicians. It covers papers and comments on all aspects of recent clinical problems.

Tadashii Chiryo to Kusuri (*The Informed Prescriber: Critical Choices in Drugs and Therapeutic Alternatives*) (Tokyo, Iyakuhin Chiryo Kenkyukai, 1986-, monthly, ISSN 0914-434X) provides unbiased information on adverse drug reactions and case reports.

Other clinical journals are *Gan to Kagaku Ryoho* (*Japanese Journal of Cancer and Chemotherapy*), *Gendai Iryo* (*Modern Medical Treatment*), (*Hinyokika Kiyo* (*Acta Urologica Japonica*), *Nippon Kagaku Ryoho Gakkai Zasshi* (*Chemotherapy, Tokyo*), *Oyo Yakuri* (*Pharmacometrics*), *Rinsho Yakuri* (*Japanese Journal of Clinical Pharmacology and Therapeutics*) and *Yakuri to Chiryo* (*Japanese Pharmacology and Therapeutics*).

MARKETING JOURNALS

Diteruman (*Detailman Information*) (Tokyo, Mikusu, 1973-, monthly, ISSN 0288–559X) publishes papers dealing with the marketing of drugs and the development of new drugs.

Kokusai Iyakuhin Joho (*International Pharmaceutical Intelligence*) (Tokyo, Kokusai Shogyo Shuppan, 1972-, semi-monthly, ISSN 0388–211X) contains news and information on marketing, sales earnings, stock reports, prescription prices, legislation and regulations and so on.

Iyaku Journal (*Medicine and Drug Journal*) also contains information on marketing.

PRODUCTS AND FORMULATIONS

Information on products and formulations can be found in *Diteruman*, *Kokusai Iyakuhin Joho* (*International Pharmaceutical Intelligence*), and *Gekkan Yakuji* (*Pharmaceuticals Monthly*) (Tokyo, Yakugyo Jiho Co., Ltd., 1959-, monthly, ISSN 0016–5980), which covers papers on new drugs and their side-effects, and comments on government regulations.

REGULATIONS

Articles on regulations can be found in *Gekkan Yakuji, Kokusai Iyakuhin Joho, Nippon Yakuzaishikai Zasshi* and *Pharma Tech Japan*.

GENERAL

Information on general aspects of medical and pharmaceutical sciences can be found in *Farumashia* (*Pharmacy*) (Tokyo, The Pharmaceutical Society of Japan, 1965-, monthly, ISSN 0014–8601) a monthly newsletter of the Pharmaceutical Society of Japan covering all aspects of pharmacy. It is not really a newsletter but a substantial news magazine containing topical items plus feature articles on drugs.

Igaku no Ayumi (*Journal of Clinical and Experimental Medicine*) (Tokyo, Ishiyaku Shuppan, 1946-, weekly, ISSN 0039-2359) is an outstanding general scientific journal that covers all aspects of medicine. It often provides good critical reviews on medical science.

Yakkyoku (*The Journal of Practical Pharmacy*) (Tokyo, Nanzando, 1950-, monthly, ISSN 0044–0035) covers reviews and comments on all aspects of the pharmaceutical industry. There is also coverage of major issues and trends in the pharmaceutical field, and developments in this area.

Other journals containing information on general aspects of medical and pharmaceutical sciences are *Chiryo* and *Gekkan Yakuji*.

INFORMATION SCIENCE

Joho Kanri (*Journal of Information Processing and Management*) (Tokyo, The Japan Information Center of Science and Technology, 1958-, monthly, ISSN 0021-7298) publishes papers on the theory and practice of documentation, and news and comments on information sciences. It is of interest to information specialists, academics and librarians.

Joho no Kagaku to Gijutsu (*Journal of Information Science and Technology Association*) (Tokyo, Joho Kagaku Gijutsu Kyokai, 1950-, monthly, ISSN 0913–3801) was until 1986, called *Documentation Kenkyu* (*The Journal of the Japan Documentation Society*). Its purpose is to disseminate information on the theory and practice of documentation and library science.

Yakugaku Toshokan (*Pharmaceutical Library Bulletin*) (Tokyo, Nippon Yakugaku Toshokan Kyogikai, 1956-, quarterly, ISSN 0386–2062) covers articles on library science and information science in the pharmaceutical field. There are also literature guides to drug reference books, pharmaceutical secondary sources and pharmaceutical journals.

Igaku Toshokan (*Journal of the Japan Medical Library Association*) (Tokyo, The Japan Medical Library Association, 1954-, quarterly, ISSN 0445-2429) provides information on all aspects of medical library practices. Most contributors are medical librarians.

Kagaku Gijutsu Bunken Service (*Science and Technology Information Service*) (Tokyo, National Diet Library, 1962-, quarterly, ISSN 0022–7633) introduces pharmaceutical and medical reference books, and provides information on all aspects of information sciences in science and technology.

OTHER PERIODICALS

Newsletters provide coverage of regulations, scientific findings, activities of the pharmaceutical industry, and recent developments and events. They also present viewpoints, research and development and new product announcements.

Typical Japanese newsletters in the pharmaceutical sciences are *Yakuji Nippo*, *Yakugyo Jiho* and *Nikkan Yakugyo*.

The following foreign magazines and newsletters for current awareness are most often used in the Japanese pharmaceutical industry. They cover world-wide or American pharmaceutical news of significant developments, company performance, new products, side-effects and regulations: (1) *SCRIP* (PJB Publications Ltd.), (2) *In Pharma* (ADIS Press), (3) *Reactions* (ADIS Press), (4) *FDC Reports; Prescription and OTC Pharmaceuticals (Pink Sheet)* (FDC Reports Inc.), (5) *FDC Reports; Health Policy and Biomedical Research* (*Blue Sheets*) (FDC Reports Inc.), (6) *Chemical Regulation Reporter* (Bureau of National Affairs Inc.), (7) *Food Chemical News* (Food Chemical News Inc.), (8) *Market Letter* (IMS Publications).

Other Japanese periodicals are *Joho* (Tokyo, Japan Pharmaceutical Information Centre, 1973-, weekly) and *Farumashia Review* (Tokyo, Nippon Yakugakkai, irregular). *Joho* is only distributed to members of the Japan Pharmaceutical Information Centre. It provides current abstracts on clinical pharmacology, side-effects, interactions and toxicity. It also covers regulations of Japan and the US, new product introductions inside and outside Japan, and listings of government reports. The purpose of *Farumashia Review* is to present review articles on all aspects of pharmaceutical sciences. The targeted readership includes all pharmacologists, pharmacists and scientists in related fields working in academia and industry. As of 1987, 24 issues have been published.

REFERENCE BOOKS

In Japan, there are more guides to the literature in pharmaceutical sciences than in other subjects with the exception of medical science. These guides provide information on journals, abstracts, indexes, databases and reference books. Unfortunately they are not often revised and therefore the currency of information is uncertain. Typical guides to the literature in the pharmaceutical science are:

Yakugaku Joho Kagaku Gairon (Yosoji Ito, Chijin Shokan, 1974)
Iyakuhin Joho (Mitsuo Sasamoto, Hirokawa Shoten, 1981)
Drug Information, Rinsho Yakugaku Koza 12 (Taro Saito, Chijin Shokan, 1980)
Iyakuhin Kaihatsu Gairon, Iyakuhin Kaihatsu Kiso Koza 1 (Kyosuke Tsuda et al., Chijin Shokan, 1970)
Kusuri to Joho no Dynamics, Farumashia Review 11 [Farmuashia Review Henshu Iinkai, Nippon Yakugakkai (The Pharmaceutical Society of Japan), 1983].

This section provides the formularies, drug compendia, regulations, directories and dictionaries which are most often used in the Japanese pharmaceutical industry.

The reference books published outside Japan which are most often used by the Japanese pharmaceutical industry are listed in the General Appendix.

Reference books published in Japan which are most often used by the Japanese pharmaceutical industry are described below according to subject matter.

JAPANESE REFERENCE BOOKS IN ENGLISH

Hoken Yaku Jiten (April 1988 edition) (Yakugyo Kenkyukai, Tokyo, Yakugyo Jiho Co., Ltd., 1988) provides the recognized prices for over 6,000 major prescription drugs (excluding Chinese medicines and dental drugs) covered by the Japanese health insurance programmes. The pharmaceutical specialities are classified according to 125 pharmacological/therapeutic groups with manufacturer, composition, presentations, dosage, retail price, therapeutic use and indications. Between editions it is kept up to date with supplements. With every official price change this book is revised. In 1988 an English version was also published under the title: *pdp Japanese Ethical Pharmaceutical Tariff 1988.4.*

Iyakuhin Seizo Shishin 1987 (Nippon Koteisho Kyokai, Tokyo, Yakugyo Jiho Co., Ltd., 1987) is intended for manufacturers to facilitate the procedures required for filling applications for approval and licensing of drugs in Japan. It also addresses topics such as GLP, efficacy re-evaluations and drug safety measures. Additional information on this subject is provided in *Seiyaku Kankei Tsuchisyu 1987*. An English version is also published entitled *Drug Approval and Licensing Procedures in Japan 1987*, but this book is a translation of the 1986 Japanese version.

Nippon Kosei Busshitsu Iyakuhin Kijun Kaisetsu 1986 (Koseisho

Yakumukyoku, Tokyo, Yakugyo Jiho Co., Ltd., 1986) provides requirements for licensing of antibiotic products in Japan. It is authorized by the Ministry of Health and Welfare in accordance with the specifications in the Pharmaceutical Affairs Law. It consists of 38 articles on General Notices, 19 articles on General Rules for Preparations, 29 articles on General Tests plus tables and covers 528 products. An English version was also published in 1986, entitled *Minimum Requirements for Antibiotic Products of Japan 1986*.

Seibutsugaku teki Seizai Kijun (Koseisho Yakumukyoku, Tokyo, Saikin Seizai Kyokai, 1985) covers the production and quality control of biological products comprising (1) vaccines and antitoxins where pathogenic micro-organisms are to be used, (2) blood products and (3) other biological products (e.g. interferon) where special techniques and facilities are required in addition to those commonly used in the production of other drugs. This book gives the minimum requirements for biological products covering, for example, handling, storage and so on. The list of products subject to the National Control Test is included. An English version was also published in 1986, entitled *Minimum Requirements for Biological Products 1986*.

Yakuji Handbook 1988 (Yakugyo Jiho Co., Ltd., Tokyo, Yakugyo Jiho Co., Ltd., 1988) is a year-book describing the activities of the pharmaceutical and related industries. There is an appendix of patient statistics, a list of representatives of foreign pharmaceutical companies and a list of pharmaceutical jargon with definitions. An English version is also published for foreigners entitled *Pharma Japan Yearbook*. But the English version has only the major items from the previous year's edition of *Yakuji Handbook*.

Nippon Yakkyokuho 11th ed. (The Society of Japanese Pharmacopoeia, Tokyo, Hirokawa Shoten, 1986) consists of two parts; part 1 includes monographs on basic medicines in frequent use and antibiotics, part 2 comprises mostly monographs on preparations of mixed active ingredients, crude drugs for home remedies, various kinds of vaccines, serum preparations and pharmaceutical aids. Also included are three indexes, one in Japanese, a second in English, and another in Latin. An English version was published in 1986 entitled *The Pharmacopoeia of Japan, Eleventh Edition*. It is distributed by Yakuji Nippo Ltd.

PRODUCTS AND FORMULATIONS

Information on products and formulations can be found in *Yakumei Kensaku Jiten* (*Japan Drug Index*) (Sogo Yakuji Kenkyujo, Tokyo,

Yakugyo Jiho Co., Ltd., 1984). The purpose of this book is to cross-reference drug names. It covers non-proprietary, official, trivial, chemical and brand names. All names given are completely cross-indexed. This edition has about 5,600 more drugs listed than the last edition published in 1981.

Iyakuhin, Iryo Eisei Yohin Kakakuhyo (Yakuji Nipposha, Tokyo, Yakuji Nipposha, 1988) is revised annually and provides retail and wholesale prices of over 40,000 prescription drugs, over-the-counter products, toiletries and other items sold in pharmacies. Products are listed in order according to the Japanese kana-syllabary system. At the end of the book there is a list of manufacturers with addresses and telephone numbers of their head office and branches.

Chiken Ishiyaku Joho (Taro Saito and Seiichi Omata, Tokyo, Iji Shuppansha, No. 18, 1987) is published as a supplement to *Shinryo to Shinyaku*. It is also published annually in book form. It covers products which are screened from the 131 Japanese medical journals. Products are listed by name (or number if the drug does not yet have a name) in kana and alphabetical order. Information provided includes actions, structural diagrams and bibliographic references.

Iyakuhin Yoran 4th ed. (Osaka-fu Byoin Yakuzaishi-kai, Tokyo, Yakugyo Jiho Co., Ltd., 1988) lists pharmaceutical specialities according to pharmacological/therapeutic group. It tabulates the following three items: (1) therapeutic group, non-proprietary name, composition, brand name, dosage, presentation and indication, (2) pharmacological activity, pharmacokinetics, warnings and toxicology and (3) indications and side-effects of drugs in the same therapeutic group. There is a generic and chemical name index.

Iryoyaku Nippon Iyakuhinshu (*Drugs in Japan, Ethical Drugs*) (11th edn Japan Pharmaceutical Information Center, Tokyo, Yakugyo Jiho Co., Ltd., 1987) covers about 15,000 ethical drugs arranged in order of therapeutic group. Each item provides therapeutic group, drug name, composition, indication, description, pharmacological action, warnings, precautions, and synonyms. The book contains three indexes: (1) kana-order index, (2) alphabetical index, (3) therapeutic index. It also has lists with identification codes of capsules and tablets and manufacturers' addresses. This book is revised annually.

Ippanyaku Nippon Iyakuhinshu 1988–1989 (*Drugs in Japan, OTC-Drugs 1988–1989*) (Japan Pharmaceutical Information Center, Tokyo, Yakugyo Jiho Co., Ltd., 1988) is a directory of over-the-counter drugs, revised biennially. It includes about 16,500 over-the-counter drugs arranged according to therapeutic group. In each therapeutic group products are listed in kana order, and composition, indications, dosages, retail prices and wholesale prices per packaging unit, are provided.

Saikin no Shinyaku (*New Drugs in Japan*) (39th edn Yakuji Nipposha, Tokyo, Yakuji Nipposha, 1988) is published annually and covers about 27,000 products. Products are arranged according to therapeutic group. At the beginning of each group there are reviews about trends in new drugs and government regulations. Each item provides drug name (official name), manufacturer, composition, action, dosage, indications, uses, warnings, description, price and bibliographic reference.

Jozai Kanbetsu Jiten (*Tablets Index*) (3rd edn, Nippon Byoin Yakuzaishikai, 1980) is a tablet and capsule identification guide that allows these drugs to be identified in several different ways, such as by markings, colour, manufacturers' code and size. It has two sections: (1) capsules and tablets without markings, (2) capsules and tablets with markings, 4,033 products manufactured by 197 companies are described. Packaging codes, manufacturers and products names are listed in the index.

Yakuzai Shikibetsu Handbook 1st edn. (Iyaku Joho Kenkyujo, Tokyo, Yakugyo Jiho Co., Ltd., 1986) covers 8,368 products. This book is used to identify capsules and tablets by the markings on packages as well as the capsules and tablets themselves. Its size is 10.5 x 15 cm and therefore it is easy to handle.

Iyakuhin Shikibetsu Code Ichiran (Nippon Byoin Yakuzaishikai, Tokyo, Yakuji Nipposha, 1986) contains 6,160 products manufactured by 207 companies. Manufacturers are listed alphabetically. Under each manufacturer, product identification codes are provided.

Yakuzai Shikibetsu Code Jiten 1987 Revised edn. (Masumitsu Takasugi, Osaka, Iyaku Journal Co., Ltd., 1987) covers products with markings such as tablets, capsules and so on. Products are arranged by manufacturer in kana order. There are five indexes: identification code (number order), identification code (alphabetically), ointments, product code or manufacturers' code and brand names.

10188 no Kagaku Shohin (Kagaku Kogyo Nipposha, Tokyo, Kagaku Kogyo Nipposha, 1987) is a typical databook of chemicals. Chemicals are arranged in order of their classification group. The chemical items usually provide description, packaging, composition, manufacturer, price, warnings and CAS registry number.

Joyo Iyakuhin Jiten (*Encyclopedia of Drugs*) (Ikuo Suzuki, Tokyo, Hirokawa Shoten, 1985) is intended primarily for physicians, pharmacists and other health sciences personnel. It provides chemical names, official names, brand names, dosages, indications, descriptions, analysis, stability, drug interactions, pharmacology, toxicity, metabolism, contra-indications and adverse effects. Drugs are arranged by therapeutic group. It is very similar to *Martindale: The Extra Pharmacopoeia*.

Iyakuhin Joho Index (Ikuo Suzuki, Tokyo, Hirokawa Shoten, 1987) is a typical drug compendium. All drugs are arranged in kana order. It provides official names, structure, therapeutic group, composition, pharmacology, indications, dosages, adverse effects and brand names. There is an English drug name index, a brand name index and a manufacturer list.

Nippon Yakkyokuho 11th edn. Kaisetsusho (Nippon Koteisho Kyokai, Tokyo, Hirokawa Shoten, 1986) is a complementary volume to *Nippon Yakkyokuho 11th edn. (The Pharmacopoeia of Japan 11th edn.)*. It provides supplementary data, including information on preparation, metabolism, pharmacological actions, adverse effects, toxicology, indications, dosage forms and bibliographic references. It contains historical notes on the *Pharmacopoeia of Japan* and Japanese, Latin and English indexes.

Nippon Yakkyokuho Gai Iyakuhin Seibun Kikaku 1986 (Koseisho Yakumukyoku Dai-2-ka, Tokyo, Yakugyo Jiho Co., Ltd., 1986) provides standards for 581 raw materials and 66 excipients and additives (*Pharmacopoeia of Japan* does not cover such ingredients of drugs). In 1987 a supplement was published, including 27 raw materials and 9 additives.

Tennen Yakubutsu Jiten (Takuo Okuda, Tokyo, Hirokawa Shoten Co., Ltd., 1986) is an encyclopedia of natural medicine. It contains over 5,000 items arranged in kana order. Each entry gives active ingredients, action and formulation. Entries are indexed by therapeutic group, pharmacological action, botanical and zoological name, English name, German name, Latin name and Chinese name.

Other books dealing with products and formulations are *Hoken yaku Jiten (1988 April edition)* and *Nippon Yakkyokuho 11th edn.*

DIRECTORIES

Yakugyo Kaisharoku (Yakugyo Jiho Co., Ltd., Tokyo, Yakugyo Jiho Co., Ltd., 1987) gives details of 365 manufacturers and 147 wholesalers including the date of establishment, head office, telephone number, cable and telex codes, branches, units, manufacturing facilities, research and development facilities, executives, business results, research and development to sales ratio, sales, products and number of employees. In 1988 an English version was published, *Pharmaceutical Manufacturers of Japan 1988–89*.

Iiku Kikan Meibo 1987–1988 (Yodosha, Tokyo, Yodosha, 1987) is a directory of medical facilities, medical colleges and medical laboratories. They are arranged according to local area. It provides names and addresses of the schools, professors, teaching staff and so on.

REGULATIONS

Yakuji Kanshi Shido Kankei Tsuchishu (Koseihso Yakumukyoku Shidoka, Tokyo, Yakugyo Jiho Co., Ltd., 1988) covers the official notices of the Pharmaceutical Affairs Bureau of the Ministry of Health and Welfare according to the Japan Pharmaceutical Affairs Law. Questions and answers about the notices are classified by subject. There is a list of the notices which is easy to search.

Seiyaku Kankei Tsuchishu 1987 (Nippon Koteisho Kyokai, Tokyo, Yakugyo Jiho Co., Ltd., 1987) covers the notices and questions and answers on procedures for approval to manufacture or import drugs up to August, 1987. It supplements the information provided in *Iyakuhin Seizo Shishin*.

Yakujiho, Yakuzaishiho no Tebiki 1987 (Koseisho, Tokyo, Yakugyo Jiho Co., Ltd., 1987) is a guidebook to the Pharmaceutical Affairs Law and related laws.

Iyakuhin Shonin Shinsa Jireishu 1986 (Nippon Koteisho Kyokai, Tokyo, Yakugyo Jiho Co., Ltd., 1986) is a guidebook to the procedures required for filing applications for approval and licensing of drugs in Japan. It also includes questions and answers, and examples of applications. It is a companion guidebook to *Iyakuhin Seizo Shishin* (*Drug Approval and Licensing Procedures in Japan*).

Shinyaku Shonin Shinsa Shitsugi Otoshu 1987 (Koseisho Yakumukyoku Dai-l-ka, Tokyo, Yakugyo Jiho Co., Ltd., 1987) provides comments on the regulation of application for approval and licensing of new drugs.

Seiyaku Kankei Tsuchishu 1987 (Nippon Koteisho Kyokai, Tokyo, Yakugyo Jiho Co., Ltd., 1987) collects notices related to the Pharmaceutical Affairs Law.

Other books containing information on regulations are *Saikin no Shinyaku* (*New Drugs in Japan*) 39th edn, *Iyakuhin Seizo Shishin 1987*, *Nippon Kosei Busshitsu Iyakuhin Kijun Kaisetsu 1986*, *Seibutsugaku teki Seizai Kijun*, *Nippon Yakkyokuho 11th edn*, *Nippon Yakkyokuho 11th edn. Kaisetsusho* and *Nippon Yakkyokuho Gai Iyakuhin Seibun Kikaku 1986*.

TERMINOLOGY

Iyaku Yogo Jiten (Nippon Iyakuhin Oroshigyo Rengokai, Tokyo, Yakugyo Jiho Co., Ltd., 1985) contains about 4,130 words which are most useful in pharmaceutical science. This dictionary is intended to assist wholesalers. It includes legal, economic words and the jargon of the pharmaceutical industry.

Yakugaku Dai Jiten (Masuo Akagi et al., Tokyo, Nippon Kogyo Gijutsu Renmei, 1982) is an encyclopedia of pharmaceutical sciences.

Hirokawa Yakukagaku Dai Jiten (Yakukagaku Dai Jiten Hensyu Iinkai, Tokyo, Hirokawa Shoten Co., Ltd., 1983), a comprehensive dictionary of pharmaceutical sciences, contains over 21,000 words with detailed descriptions. Each entry includes English, German, French and Latin words arranged in kana order.

Yakugaku Yogo Jiten (Nippon Yakugakkai Yakugaku Yogo Iinkai, Tokyo, Hirokawa Shoten Co., Ltd., 1971) is a dictionary of pharmaceutical sciences, containing 6,000 words in pharmaceutical science, organic chemistry, inorganic chemistry, physical chemistry, analytical chemistry, pharmacology and biochemistry. Each entry is given a definition by an expert.

Yakugaku Yogoshu (Nippon Yakugakkai, Tokyo, Maruzen Co., Ltd., 1985) lists Japanese pharmaceutical terms without definitions, compiled by the Pharmaceutical Society of Japan. It consists of two main parts; one is Japanese to English, the other is English to Japanese.

GENERAL

Yakuji Handbook 1988, *Yakugaku Dai Jiten* and *Hirokawa Yakukagku Dai Jiten* contain general information. The major clinical guidebook in use in Japan is *Konnichi no Chiryo Shishin (Today's Therapy) Vol.30, 1988* (Shigeaki Hinohara and Masakazu Abe, Tokyo, Igaku Shoin Ltd., 1988). It is revised annually. In this edition 891 professional medical doctors describe treatment information on 932 diseases. Every disease description contains details of drug therapy. It also includes acute poisoning cases and special considerations in drug therapy such as Chinese medicine, antineoplastics, antiinflammatory agents and antibiotics. There are both Japanese and English word indexes.

Library networks and document delivery centres

The pharmaceutical industry requires more information than ever before and even the largest companies do not have all the information they require. Therefore, pharmaceutical companies must use large, well equipped medical libraries in their vicinity (e.g. the University of Tokyo's Medical Library, Osaka University's Nakanoshima Library and Keio University's Medical Library and Information Center). But in recent years online document ordering systems have become well developed, so it is becoming easier for pharmaceutical companies to order the documents which they require. The most frequently used foreign source is the British Library Document Supply Centre (BLDSC), because it has large collections of medical and scientific literature, and provides a prompt photocopying service. The main Japanese information networks and document delivery systems are listed below:

The Japan Information Center of Science and Technology (JICST), 5–2, Nagatacho 2-Chome, Chiyoda-ku, Tokyo 100.

JICST was established in 1957 in order to promote scientific technology in Japan and is engaged in collecting, processing, and abstracting Japanese and foreign scientific and technical information. It creates and publishes the largest abstracts journal in Japan, *Kagaku Gijutsu Bunken Sokuho* (*Current Bibliography on Science and Technology*) and creates and maintains Japanese databases such as *JICST File on Science and Technology*. It supplies its online service through JOIS (JICST Online Information System) which was developed by JICST itself. It also provides copying, translation and other related services.

Japan Pharmaceutical Information Center (JAPIC), Kyodo Building 16, Chiyoda-ku, Tokyo 102.

The centre was established in 1972 with the goal of supplying all available drug information to medical facilities, administrative organizations and the pharmaceutical industry. It undertakes the collection, processing and abstracting of Japanese drug literature thereby creating and publishing the pharmaceutical abstracts journal *Nihon Iyaku Bunken Shorokushu* (*Abstracts Journal for Japanese Pharmaceutical Literature*) which is also available online (*JAPICDOC*).

Japan Patent Information Organization (JAPIO), 5–16, Toranomon 1-Chome, Minato-ku, Tokyo 105.

JAPIO was set up in 1985 in order to supply patent information. It does this by collecting, processing and indexing Japanese and foreign patent information and supplies its online service through the *Patent Online Information System* (*PATOLIS*). An English version; namely *Japanese Patent Abstracts in English* (*JAPIO*) is also available through ORBIT outside Japan. A photocopy service for Japanese patent documents is provided.

International Medical Information Center (IMIC), Daikyo-cho 26, Shinjuku, Tokyo 160.

The centre originated in 1972 to provide medical information to the medical and paramedical professions. It provides document delivery, information retrieval, translation and abstracting services.

Japan Association for International Chemical Information (JAICI), 4–16, Yayoi 2-chome, Bunkyo-ku, Tokyo 113.

The association has operated since 1974 with the aim of promoting scientific technology information activities, distributing chemical information and encouraging friendship and active interchange of personnel between foreign countries and Japan. As the Japanese agent of the *Chemical Abstracts Services*, it supplies the products and online services of CAS.

Medical Information System Developing Center (MEDIS-DC), 3–4, Akasaka 2-Chome, Minato-ku, Tokyo 107.

The centre offers an emergency medical information system and local medical information systems. It has also developed software for the automation of hospital information systems. At the present time it is developing a drug information system with JAPIC.

AMS Medical Information Service, 3–1, Marunouchi 2-Chome, Chiyoda-ku, Tokyo 100.

The Advanced Medical Information Service (AMS) plans to disseminate to physicians information on medical news, drugs, diagnosis support, workshops management and leisure via an online system. This medical information service was started in April 1986. It contains *Hoken Yaku Jiten* (*pdp Japanese Ethical Pharmaceutical Tariff*) and information on emergency management of poisoning.

National Diet Library, 10–1, Nagatacho 1-Chome, Chiyoda-ku, Tokyo 100.

The Library was founded in 1948. In addition to materials in other fields, it has one of the most comprehensive collections of Japanese and foreign journals, technical reports and books on science and technology. The National Diet Library is one of the major 'back-up' libraries in Japan. It has three lending methods, namely (1) mail, (2) facsimile and (3) visit and, in addition a reference service available by telephone.

Technomic Information Service, Inc., 8–7, Nihonbashi Tomizawacho, Chuo-ku, Tokyo 103.

The Service initiated and now maintains the *Pharmacasts* (*Asu no Shinyaku*) database which it supplies online through JIP. It also supplies foreign online drug services, such as *de Haen, Drugdex, and IOWA Drug Information Service*. Japan Medical Abstracts Society, 5–18, Takaido Higashi 2-Chome, Suginami-ku, Tokyo 168.

The Society edits and publishes *Japana Centra Revuo Medicina*, creates its database (*Japana Centra Revuo Medicina Title Guide*) and co-operates with JICST primarily by supplying clinical abstracts. It also provides a document delivery service of articles in journals covered by *Japana Centra Revuo Medicina*.

Information Science and Technology Association, 5–7, Koishikawa 2-Chome, Bunkyo-ku, Tokyo 112.

Formerly this Association was called the Japan Documentation Society. It organizes meetings and offers seminars on information science. It also publishes *The Journal of the Information Science and Technology Association* (formerly *The Journal of the Japan Documentation Society*).

The Japan Medical Library Association (JMLA), Gakkai Center Building, 4–16, Yayoi 2-Chome, Bunkyo-ku, Tokyo 113.

JMLA was founded in 1927 to promote the services of medical libraries, especially the establishment of inter-library loans. At the present time the membership (mainly Medical School Libraries) stands at over 100. Only three pharmaceutical companies are members, because the association limits membership to individual participating libraries which must meet specific requirements, such as minimum number of titles, facilities, employees and so on.

The Japan Pharmaceutical Library Association (JPLA) Library, Faculty of Pharmaceutical Sciences, University of Tokyo, 3–1, Hongo 7-Chome, Bunkyo-ku, Tokyo 113.

JPLA was founded in 1955 taking JMLA as a model. As of 1987 membership consisted of 44 pharmaceutical school libraries and 42 pharmaceutical company libraries. One of its main activities is the promotion of interlibrary loans.

Future information developments in the Japanese pharmaceutical industry

Japan is not an active producer of information, but a mass consumer of information from an international viewpoint. For instance, according to *Directory of Databases 1986*, the total number of databases available in Japan increased by about 15% to 1,483 whilst the Japanese databases listed increased by approximately 20.0% to 296, only 20% of the total. At the present time, Japanese databases are relatively few but their number is increasing because most Japanese secondary sources are adopting computer-aided editing; consequently they can be made available as databases. The electronically-prepared major Japanese newspapers are likewise available online as full text databases. Japan, one of the leading countries in the economic area, often does not publish Japanese information in English and many developed countries have criticized Japan for neglecting to do this and for being only a consumer of foreign information. To promote greater understanding of Japan, the Japanese government is encouraging the publication of Japanese work in English. As a result of this policy, the number of papers in English written by Japanese will grow year by year.

Advances in computer technology inevitably lead to changes in information storage forms. In the near future, most primary and secondary sources will be available in electronic form. For instance, from 1990, Japanese patent applications will be accepted in electronic form. The paperless society is becoming a reality just as F. W. Lancaster (1978), *Toward Paperless Information Systems* (Academic Press), predicted. Communication networks have also developed and most Japanese can use foreign materials without any time-lag. In the early 1980s communication speed was 300 bps; during the latter half of the 1980s it has reached 1200 bps; by the end of the decade 2400 bps will be commonplace. Microcomputers have come into wide use, and a lot of application software has been developed. The micro is already changing information handling techniques and environments. Artificial intelligence (AI) will soon be able to accurately translate Japanese to English automatically.

This will bring about a further increase in the number of English-language Japanese papers available. Most important information is being transferred from printed material to electronic form.

CD-ROM is in the spotlight as a revolutionary tool in information science. But although remarkable progress has been made in the development of CD-ROM hardware in Japan, the software has been somewhat neglected and the market is undeveloped. According to *Data Net*, the Japanese newsletter of CD-ROM and database services, the number of CD-ROM software programmes in Japan was 55 (26.3%) compared to 124 (59.3%) in the US. CD-ROM software produced in Japan is mainly used for newspapers, directories of books and journals, dictionaries, maps, telephone directories, who's who lists and so on. Half of the 58,000 units of CD-ROM software sold by 1988 were automobile navigation systems. In the area of science and technology there are very few Japanese CD-ROM software programmes. An example of one is a CD-ROM version of *Tokkyo Koho* (*Japanese Patent Documents*). This CD-ROM programme supplies original patent documents; the hit is searched through the *Patent Online Information System* (*PATOLIS*).

One reason that most organizations and libraries have not introduced it yet is that when CD-ROM software is introduced new hardware is usually required. This is because most CD-ROM software is for the IBM PC which is expensive and not in general use in Japan. If more CD-ROM software was available for Japanese personal computers, or if the software were standardized, CD-ROM systems would be more popular.

The Japanese pharmaceutical industry will expand not only in the domestic market but in the international market as well. Because of the difficulties in finding and developing new drugs, and the limited Japanese market, the large Japanese pharmaceutical companies are eager to develop into multinational corporations in order to expand their markets and increase their research and development capabilities. In Japanese society today the theme of 'internationalization' is being heavily stressed, and indeed, to be a multinational company entails a deep understanding of the culture and customs of other countries.

A significant problem in the Japanese pharmaceutical industry is the integration and establishment of internal documentation systems. Japanese pharmaceutical companies do not have large documentation staffs. Therefore, the techniques associated with and experience of information sources usage are most important.

Acknowledgement

I thank Hiroaki Mori (Tokyo Tanabe Co., Ltd.) and Ritsuo Kiyota (Kowa Co., Ltd.) for their helpful comments on the draft text. Special thanks are due to Douglas Lamont, without whose invaluable assistance this chapter would hardly have been possible.

References

Yakugyo Jiho Co., Ltd., (1988). *Yakuji Handbook '88*. (Yakugyo Jiho Co., Ltd.).
The Ministry of Health and Welfare. (1987). *Yakuhi Kogyo Seisan Dotai Tokei Nenpo*. (Yakugyo Keizai Kenkyujo).
Gaitz, C. M. (1985). *Aging 2000: Our Health Care Destiny, Vol.1: Biomedical Issues*. (Springer-Verlag).
Codling, M. D. (1987). *The Outlook for the U. S. Pharmaceutical Industry to 1991*. (Arthur D. Little Decision Resources).
James, B. G. (1977). *The Future of the Multinational Pharmaceutical Industry to 1990*. (Associated Business Programmes).
Mathieu, M. P. (1987). *New Drug Development: A Regulatory Overview*. (OMEC International Inc.).
Sasamoto, M. (1981). *Iyakuhin Joho*. (Hirokawa Shoten).
Ito, Y. (1974). *Yakugaku Joho Kagaku Gairon*. (Chijin Shokan).
Saito, T. (1980). *Drug Information (Rinsho Yakugaku Koza 12)*. (Chijin Shokan).
Tsuda, K. et al.(1970). *Iyakuhin Kaihatsu Gairon (Iyakuhin Kaihatsu Kiso Koza 1)*. (Chijin Shokan).
Farumashia Review Henshu Iinkai. (1983). *Kusuri to Joho no Dynamics (Farumashia Review 11)*. (Nippon Yakugakkai).
Matsunaga, T., Fujii, A., Oyama, F. and Sato, T. (1985). 'Igaku Chuo Zasshi Japana Centra Revuo Medicina History and Activities.' *Igaku Toshokan*, **32**(Supplement), 44–57.
Saito, K. (1985). 'JICST Activities with Particular Emphasis on online Medical Literature Information Service.' *Igaku Toshokan*, **32**(Supplement), 58–67.
The Ministry of International Trade and Industry. (1988). *Database White Paper 1988*. (Database Promotion Center).
Lancaster, F. W. (1978). *Toward Paperless Information Systems*. (Academic Press).
Ash, J. E. (1985). *Communication, Storage and Retrieval of Chemical Information*. (Ellis Horwood Ltd.).
Harada, M. (1987). *Mirai no Toshokan*. (Shoraisha).
The Health and Welfare Statistics Association. (1987). 'Kokumin Eisei no Doko.' *Kosei no Shihyo*, **34**(9), Supplement.
The Japan Information Center of Science and Technology. (1986). *Kagaku Gijutsu Joho Handbook*.
Takeuchi, H. (1989). 'CD-ROM; The enormous world beyond small disks.' *Science and Technology Information Service*, No.87, 1–5.
Nakamura, Y. et al. (1988). 'Internationalization of Japanese information,' *Joho no Kagaku to Gijutsu*, **38**(1), 1–31.

CHAPTER SIXTEEN

Information needs and availability in Australia

T. KOT AND N. PETTIT-YOUNG

Introduction

Australia is a democracy with a three-tier system of government comprising the Federal Government, a State Government in each of the five states and two territories, and Local Government. Most revenue is raised by the Federal Government and about one-third of federal budget outlays are for payments to the states and territories for education, health, housing, transport and assistance for local government. The Australian population is approximately 16 million with many people of British descent. Before World War 2 Australia depended largely on primary production but the demands of the war and a strong post-war immigration programme spurred economic growth. The manufacturing sector grew rapidly after the war but the main expansion in the past 20 years has been in the tertiary sector. During this time, considerable investment has also taken place in export-oriented mining and energy projects. Rural industries account for some 7% of production, mining contributes 6%, manufacturing 20% and the tertiary sector the balance. Australia enjoys a high standard of living and shares many cultural and social similarities with western nations.

Australian health-care system

The health service system in Australia is dominated by the legal structure which dates back to Federation in 1901 when some powers were handed over to the Commonwealth but the states retained powers for the provision of health services. Public health services are undertaken at federal, state and local government levels. The Commonwealth Government of Australia has increasingly become involved with developing and co-ordinating national policies, model legislation, guidelines and uniform standards for such areas as preventive medicine, communicable diseases, licit and illicit drugs, environmental health, nutrition and toxicology. The Commonwealth Government provides funding and professional expertise for public health services such as family planning, aboriginal health and arbovirus disease control.

Whilst the Commonwealth has powers to make laws, it has little responsibility for direct provision of health services. The State Governments exhibit some diversity in the delivery of health services but in the main they are responsible for hospital and community services, registration of health professionals, control over the scheduling of drugs and control of licensing of private hospitals and nursing homes. The guiding principle for the State authorities is that all people have assured access to an appropriate range of health services.

Australia's health services are based largely on a two-tier system. Private medical practitioners provide primary care and a comprehensive public hospital system provides institutional care. The high degree of urbanization in Australia has resulted in the capital cities having large, well-equipped hospitals, both public and private, providing a full range of general and specialist services. Country centres tend to have comprehensive base hospitals which serve areas of a wider radius. The Royal Flying Doctor Service provides medical care for people living in isolated regions by means of radio advice, regular outpost clinic visits and aerial evacuations in emergencies. The Royal Flying Doctor Service is financed by both Commonwealth and State governments and by private donors and benefactors.

The National Health and Medical Research Council (NHMRC) established in 1936, is recognized as the authoritative body in Australia on matters of general health, public health and medical research. A significant proportion of medical research expenditure is allocated to individuals and institutions by the NHMRC. In 1987 some Aus\$71 million was directed to medical research by the Commonwealth Government.

Health-care expenditure in Australia accounts for 7.6% of the Gross Domestic Product or approximately 10% of the Commonwealth Government budget outlay. (Commonwealth Government budget outlays for the various government functions are presented in the General Appendix.) Total health expenditure in Australia is shared by the Commonwealth, State and Local Governments and the private sector. Recurrent health expenditure by both government and private sectors is contained in the General Appendix.

Expenditure on pharmaceuticals in Australia, both private and government, represents approximately 10% of total health expenditure. The Commonwealth Government pays for the drugs prescribed by doctors and dentists minus the patient contribution for scheduled prescription items.

The Pharmaceutical Benefits Scheme was introduced in 1953 as part of the National Health Act. It provides subsidized medicines to all Australians under three classifications – general patients, concessional beneficiaries and pensioners. The Pharmaceutical Benefits Advisory Committee makes recommendations to the Minister for Health on the drugs which will be available as pharmaceutical benefits. At the present time, the scheduled list includes some 600 drugs, and 1200 items, if forms and strengths are included.

In 1983, the National Insurance Act was amended with the specific purpose of providing benefits for medical, optometrical, dental and pathology services for all Australians. The health insurance scheme, known as Medicare, replaced private medical and hospital insurance. The major features of Medicare include automatic entitlement to medical services at 85% of the scheduled fee which is set by the government in consultation with medical associations; access without direct charge to public hospital accommodation, and to inpatient and outpatient treatment by hospital doctors; reduced charges for private treatment in shared hospital wards; increased subsidy to private hospitals; and funding of community health programmes. However, it is still possible to take out private health insurance for private hospital cover, doctor of choice in public hospitals and ancillary health services. Medicare is currently financed in part by a 1.25% levy on taxable incomes, with low income cut-off points.

The Australian health profile

Some of the changes in the health status of Australians are dramatic such as the change in life expectancy, perinatal and infant mortality, ageing of the population and a reduction in death and illness from infectious diseases.

Life expectancy has risen steadily in Australia during the twentieth century. Average life expectancy in Australia at the turn of the century was 55.20 years for males and 58.84 years for females, for 1985 the figures are 72.6 for males and 79.3 for females. Since the turn of the century the young have survived to maturity in increasing numbers and female mortality has declined steadily. Between

1900 and 1945 the Australian male infant mortality declined from 101 deaths per 1000 births to 39 deaths per 1000.

Major contributions to the gains in life expectancy are a decline in deaths from infectious diseases (see the General Appendix for notifiable disease cases reported in Australia over the period 1980–1984); a fall in infant and child deaths; improved housing, better sanitation and food supply; and advances in medical research and technologies. Comparative data for 25 countries listing expectation of life, crude birth rates, crude death rates and infant mortality rates are in the General Appendix.

The 'modern epidemic', Acquired Immune Deficiency Syndrome (AIDS) appears to bear some relationship to aspects of lifestyle. At October 1988, 1024 cases of AIDS had been reported to the NHMRC Unit in Aids Epidemiology and Clinical Research.

Instead of infectious diseases, the most common causes of death in Australia are now diseases of the circulatory system, heart diseases, cancers, respiratory diseases, motor vehicle accidents and other types of accidents (see Table 16.1). Causes of death are classified according to the Ninth Revision of the International Classification of Diseases by the World Health Organization. Many of these diseases are referred to as 'lifestyle' diseases and have emerged in association with a distinct form of social structure known as industrial capitalism. Other health problems include epilepsy which affects approximately 100,000 Australians, two people in every hundred have diabetes and over one million people suffer from fifteen varieties of arthritis.

Changes in the population age structure have resulted in a higher proportion of older people. There is therefore a greater demand on health services as the disease complexity increases and the pattern changes from acute to chronic diseases.

For many years, Australians believed that they were a relatively classless, egalitarian society; however this view has changed in the last 30 years. Indicators of socioeconomic status are not clearly defined but a person's occupation can reflect many factors including health. Limited Australian studies on stress, work and ill-health associated with unemployment following factory closures have shown an increase in suicide, peptic ulcer disease, anorexia, weight loss, emotional problems, increased medical consultations, higher medication intake and disturbed family relationships.

The Australian Aboriginal population has a very different health profile from that of the general Australian population. The life expectancy of an Aboriginal infant at birth is about 20 years less than than of the general population. In the Northern Territory, where the Aboriginal population forms almost one-quarter (23.7%) of the total, the neonatal death rate in 1985 was 21.2 for

Table 16.1 Causes of death in each age group in Australia, 1984

Causes of death	Age groups (years)									Percentage (d)
	Under one	1–14	15–24	25–34	35–44	45–54	55–64	65–74	75 and over	Total (a)
Infectious and parasitic diseases	1.3	1.7	0.6	0.6	0.5	0.6	0.5	0.4	0.4	0.5
Neoplasms	0.6	17.7	7.1	15.6	30.8	37.6	36.8	29.3	16.6	23.8
Endocrine, nutritional and metabolic diseases and immunity disorders	0.8	3.1	1.5	1.4	1.5	1.7	2.1	2.4	2.2	2.1
Diseases of the nervous system and sense organs	1.6	8.5	2.4	2.5	2.6	1.5	1.2	1.2	1.2	1.4
Diseases of the circulatory system	0.7	2.5	4.0	10.3	24.6	34.6	42.8	50.7	60.9	49.4
Diseases of the respiratory system	2.5	5.4	2.6	2.8	3.7	4.1	5.4	8.0	8.3	7.1
Diseases of the digestive system	0.3	1.0	0.6	2.8	5.0	5.7	4.5	2.8	3.2	3.3
Congenital anomalies(b)	28.8	9.1	1.4	1.1	0.8	0.3	0.2	0.1	–	0.8
All other diseases(b)	37.9	1.7	4.6	6.5	2.5	2.4	2.3	2.8	4.7	4.3
Signs, symptoms and ill-defined conditions	23.6	2.5	0.8	0.8	0.5	0.2	0.1	0.1	0.3	0.7
Accidents, poisonings and violence	2.0	46.7	74.4	55.5	27.4	11.2	4.2	2.3	2.1	6.6
All causes	100.0	100.0	100.0	100.0	100.0	100.0	100.0	100.0	100.0	100.0

(a) Total includes 20 deaths where age is not known.
(b) Includes 823 deaths from conditions originating in perinatal period and 1,715 deaths from diseases of the genitourinary system.
(c) Rates are per 100,000 of population at risk except for children under one year of age which are per 100,000 live births registered.
(d) Percentage of all deaths within each age group.
Due to abnormal delays in the registration process in the NSW Registry of Births, Deaths and Marriages, the number of births and deaths recorded for 1984 is lower than could be otherwise expected. For further information see *Births, Australia 1984* (3301.0) and *Deaths, Australia 1984* (3302.0) (both published by ABS).
Source: Australian Bureau of Statistics. (1986). *Year Book Australia 1986.* (ABS, p. 219).

Aboriginals and 6.1 for non-Aboriginals (calculated for the number of deaths of infants under four weeks of age per 1,000 live births).

Whilst Aborigines are entitled to use health services provided to the general community, many do not use these services because of geographical, economic and cultural barriers. Despite the gradual disintegration of traditional tribal society, the majority of Aborigines live a separate existence where their economic and physical environments are so poor that their health status compares with that for developing countries. Health problems of Australian Aborigines include diseases of the respiratory system, hepatitis, liver diseases, diabetes mellitus and diseases of the circulatory and urinogenital systems.

The Australian pharmaceutical industry

The economic and industrial environment of the pharmaceutical industry is affected by the size of the market and the policies of the Australian Government. Almost all the leading pharmaceutical companies worldwide have operations in Australia and when these are added to local businesses, there are over 100 companies involved in the supply of pharmaceuticals. The value of the Australian ethical pharmaceutical market in March 1987 was US$1500 million and in comparison, the world market was estimated at some US$110,000 million.

In the *Annual Report* of 1987–1988, the Australian Pharmaceutical Manufacturers Association reveals that demand for products is strong and sales revenue has increased, although somewhat unevenly across the industry. One major problem that the industry faces is the continued low pricing of pharmaceuticals. In the 12 months to April 1988, the Consumer Price Index (a general rate of inflation) increased by 6.9% and ex-manufacturers prices of pharmaceuticals rose by some 5.4%. This is not a singular occurrence because over the past 13 years, prices paid to manufacturers for drugs listed on the Pharmaceutical Benefits Scheme have increased on an average by just over 4% per annum, yet the Consumer Price Index has increased by almost 10% per annum.

The Pharmaceutical Benefits Scheme is the most important feature of the Australian pharmaceutical industry's environment because it is the authoritative government body for setting product prices. When a product is approved for general marketing, the company concerned applies for listing with the Pharmaceutical Benefits Advisory Committee because it would be difficult to secure sales on the private prescription market. The Pharmaceutical Benefits Advisory Committee negotiates a price with the manufacturer and since the early 1970s the Committee has been securing generally low prices for the majority of products. In May 1983, generic drug pricing was introduced and currently the government has set a maximum price differential of 20 cents between the alternate brands of the same drug, or between generic and trade brands.

There is optimism that certain specific changes, such as extended patent life and the identification of factors justifying increases in recommended prices, introduced in September 1987 and January 1988 could have a fundamental and beneficial impact on the Australian pharmaceutical industry.

The pharmaceutical industry is heavily dependent upon imported products and the necessary active ingredients. Although three-quarters of the pharmaceuticals on the Australian market are manufactured locally, only 13% are drugs formulated from active ingredients manufacturerd and packaged locally, the remaining 65% of local drugs are formulated from imported active ingredients. 22% of products are imported fully finished and often only require packaging.

All imported drugs, both active ingredients and packaged drugs, are subject to import control. The Therapeutic Goods Act 1966 sets the standards for imported therapeutic substances and the Customs Regulations (Prohibited Imports) 1970 deals with the importation of drugs.

Imported therapeutic substances in Australia are subject to detailed regulations and controls as they are in all developed nations. These controls are contained in the Therapeutic Goods Regulation 1967 (amended 1987) and the Therapeutic Goods Act 1966. The legislation in 1987 established the Therapeutic Goods Committee which brings together expert medical advice and group representation pertaining to specific standards for drugs. In 1963, the Minister of Health established the Australian Drug Evaluation Committee to ensure that only safe drugs are imported into Australia. It has 6–8 expert members, appointed by the Minister for Health and their role is to advise the Minister on the safety and efficacy of therapeutic substances based on an assessment of the data which has been generated during the development of the drug. The Drug Evaluation Branch of the Department of Health (Therapeutic Division) investigates the pharmaceutical company's application for marketing approval of a new drug. There are other branches which provide support relating to areas of toxicology, medical services advisors, pharmaceutical chemistry evaluation, and drug information.

The pharmaceutical industry has continually expressed concern about the lengthy delays in approving general marketing applications for new drugs, applications for changes to approved indications and reformulations. A Public Service Board Review of Drug Evaluation Procedures in 1987 recommended changes to the organizational structure of the drug evaluation process and industry awaits the outcome of the recommendations. Major problems, experienced by the pharmaceutical industry with the regulatory process, include the erosion of effective patent life and the poor return on investment because of the delays involved in evaluating marketing approvals.

The overall market (see Table 16.2) for pharmaceuticals sold through community practice pharmacy is broadly comparable with the world market. The major areas are cardiovascular, respiratory, gastrointestinal and central nervous system reflecting in part the introduction of new 'breakthrough' products.

The pharmaceutical industry in Australia has opportunities for technology activities and involvement in research and development. Local research and development activities in 1987 amounted to AUS\$32.4 million, of which 46% was internal research and development carried out within firms, 30% was clinical trials in Australia, and 15% was research and development contracted out

Table 16.2. Main therapeutic areas for ethical pharmaceutical markets in the world and Australia

Therapeutic class	World market share (%) 1987	Australian market share (%)	
		1987	1977
Cardiovascular	18.1	19.5	14.5
Anti-infective	16.1	9.1	10.9
Alimentary tract	15.6	15.3	17.1
Central nervous system	11.1	11.1	13.2
Respiratory	7.1	12.4	12.2
Musculoskeletal	6.3	7.2	5.5
Blood products	6.1	1.6	1.1
Genitourinary	4.2	4.4	4.7
Dermatologicals	3.5	7.0	9.2
Cytostatics	3.0	0.6	0.2
Sensory organs	2.1	2.9	2.0
Hormonal products	1.7	0.7	0.7
Parasitology	0.2	1.0	1.0
Other	4.5	7.2	7.7
	99.6	100.0	100.0

Source: Parry, T. G. and Thwaites, R. M. (1988). *The pharmaceutical industry in Australia: a benchmark study.* (Australian Pharmaceutical Manufacturers Association Inc., p. 3, 7).

to private and public organizations. The value of total research and development expenditure (including overseas involvement) was 2.9% of the total sales of the pharmaceutical industry. There has been a considerable investment in the clinical stages of drug development, partly to meet the requirements of the Australian government regulatory authorities and partly because data from trials conducted by Australian clinical investigators has proven acceptable to regulatory authorities in many other countries.

There have been three major ventures by the pharmaceutical industry into basic pharmacological research and development in Australia: the Nicholas Institute for Medical and Veterinary Research, the Riker Laboratories Australia Research Division and the Roche Institute for Marine Pharmacology. Although all three were small by international standards, they achieved a number of scientific goals, their closure being due to financial decisions taken outside Australia. A number of other pharmaceutical firms have supported research units and work in basic pharmacology in University departments in Australia. For example, a Smith, Kline and French Research Institute operated within the Department of Pharmacology in the University of Sydney from 1962–1969.

In recent years, several medical and related research establishments and companies have developed expertise in selected areas, the best known being biotechnology. There has also been a strong focus on 'delivery system' technology and diagnostics.

The pharmaceutical industry in Australia is active in the sales and marketing area, and the Australian Pharmaceutical Manufacturers Association has promoted The Code of Conduct for companies involved in pharmaceutical advertising and promotion. The Australian Pharmaceutical Manufacturers Association is represented along with other outside medical and professional bodies on the Therapeutic Advertising Code Council.

In 1985, after petitioning by eight consumer groups and the Australian Federation of Consumer Organizations, the Consumer Health Forum was established to give consumers a voice in the policy decisions of the Commonwealth Department of Health. The issues that the Consumer Health Forum addresses and prepares policy advice on include a rational drug policy, patient package inserts, drug advertising and drug pricing. The Australian Pharmaceutical Manufacturers Association is a member organization of Consumer Health Forum and is involved in the issues and policies that are presented to the Minister. Australians have not been vocal consumerist activists to date; however experiences from overseas indicate that this area is likely to be of growing interest.

As in many other countries, medical representatives in Australia promote products directly to prescribers, either in general practice or in the hospital setting. In addition, they provide information and awareness of their company's products to pharmacists, nurses and other health professionals. The pharmaceutical industry ensures high standards of representative training and also accepts responsibility for providing educational meetings for prescribers and other allied health professionals.

Australian information and communications technology

The communications industry in Australia comprises the postal, domestic telecommunications, overseas telecommunications, commercial broadcasting, and national broadcasting and television services. The discussion which follows focuses on domestic and telecommunications services because to date technical barriers have been a hindrance to the internationalization of information services.

Telecommunications in Australia are provided by three public carriers: Telecom Australia, the national telecommunications carrier which operates a public switched telecommunications network as well as leased line facilities; Overseas Telecommunications Commission (OTC) which provides international services and Aussat Pty Ltd. which operates the domestic communications satellite system. All telecommunication services in Australia are under the control of the government with some public investment.

Telecom offers Datel and Integrated Digital Data Services (IDDS) for data transmission via modems or dedicated lines, over a range of speeds, and over short or long distances. Transmission sevices are based on international standards with equipment built to the same standard. The Integrated Services Digital Network, to be introduced soon, will integrate access to voice, data and text, by supporting voice services and office automation.

Telecom's Austpac is Australia's public 'packet switched' data service which switches data between terminals and computers. Many users operate on this public network which means costs are shared by all users who can access single or multiple destinations in Australia, as well as switching overseas via OTC's Data Access Service. Austpac's packet switching system is used mainly where access is required to databases within Australia and overseas.

OTC has access to a diverse world network through more than 240 international agreements giving access for direct-dial international telephone services to 183 countries.

Whilst there have been increased foreign earnings from telephone services, OTC has invested in transmission capacity and this remains OTC's largest cost element comprising Aus\$766.1 million for 1987–1988. Successful contracts for hubbing private services for transnational organizations in the Pacific and South East Asian regions will result in a large network of telecommunications services in the near future.

An optical fibre cable, Tasman 2, is underway to link Australia and New Zealand by the end of 1991. Tasman 2 is only the first stage of OTC's plans for a Pacific Ocean network of optical fibre submarine cables. An agreement has already been signed with major partners to plan for the provision of the Pac Rim East cable between New Zealand and Hawaii, and the Pac Rim West cable between Australia and Guam. From there, they will link with similar cables to Japan and North America, and between North America and Japan to form a digital loop around the Pacific. With demand increasing at 20% per annum, OTC expects to invest Aus\$2,000 million in new facilities and networks over the next decade to provide the capacity to meet its customers needs leading up to the turn of the century. This investment, coupled with initiatives for technology transfer to Australia through contract specifications, as in Tasman 2, is contributing to the strengthening of Australian technological expertise. OTC's packet switched data service has been upgraded over the last year, changes include a simplified log-on procedure, statistics following each user session, and faster (2400 bps) dial-up access. The OTC Data Access network provides access to the world's major databases in some 41 countries. As part of a campaign to promote Australian databases, OTC has produced a *Database Directory* listing more than 150 online information sources in Australia.

Integrated Services Digital Networking will provide additional capability between national telecommunication systems and with its greater speeds, compared with using modems, will significantly reduce delays for information transfer and will enable photo-videotex services to be offered in the future. New technology, including optical fibre cables and new generation of satellites and ground station facilities, will be in service in the 1990s to further increase transmission capacity.

Besides the international satellite network, the Satellite Communications Act 1984 commissioned AUSSAT to provide domestic satellite communications for Australia and neighbouring regions. Three satellites are currently in place, 36,000 km above the equator, acting as relay stations for transmitting all types of signals – voice, video and data. The satellite system is aimed to complement ground-based communications and has already found application in services to remote areas.

Technological developments are closely related to the performance of information transfer networks and services because of the great distances between cities and towns in Australia. There is only one obstacle towards continued expansion in telecommunications and that is the lack of investment of private enterprise capital, which could take research and development into a further stage. Nonetheless, Australian satellite technology is already sufficiently advanced that it competes with that of the Japanese, the field leaders.

Australian pharmaceutical information resources

The Australian scientific, academic and research communities have the same requirements for pharmaceuticals information as their colleagues in the Northern Hemisphere. Most of the usual major database vendors such as DIALOG, NLM, Orbit, BRS, STN and Data-Star are accessible through the developing networks but these are augmented by a number of national services.

Similarly, various international indexing and abstracting services that cover the pharmaceutical field, are held by many specialized libraries and research establishments. *Chemical Abstracts, International Pharmaceutical Abstracts, Inpharma, Clin-Alert* and *Reactions*, for example, would be regarded as standard reference sources by most researchers. Many of the core pharmaceutical journals are held by either academic, government and hospital libraries and access to lesser known journals is facilitated by well developed library networks and the existence of serial union lists.

Many of the published texts of interest in the pharmaceutical area are held by most institutions, although essentially only English language texts are acquired. Microfiche services such as Iowa Drug Information Service's *Drug Literature Microfilm* file and DeHaen *Drugs in Use* are housed in various libraries, hospital pharmacies and research centres.

Conversely, certain information services contribute data to international agencies, for example the Australian Adverse Drug Reactions Advisory Committee contributes data to the WHO Collaborating Center for International Drug Monitoring.

Computers and communications technology have largely overcome the problems of geographical isolation and scattered populations. There is, as a result, closer liaison between researchers which has facilitated exchange of information.

Some of the Australian resources relate to specialized information, whilst others have broader use. They are available in various forms including computerized databases, microfiche services, journals, newsletters and reference books. Lists of major publications and resources are in the General Appendix.

Australian databases

THE AUSTRALIAN MEDLINE NETWORK

The Australian *Medline* network was the first contact with online information for many health science professionals. The National Library of Australia first commenced courses for *Medline* users in late 1976 but, at that time, only a few people were trained, mostly from universities as only a few institutions could afford the high telecommunications costs. In 1979, OTC launched its packet switched network known as Midas, which allowed low cost access to Orbit and Lockheed databases. Wider use of the Australian *Medline* network was seen in 1980 when the Health Communications Network introduced a leased line carrying Commonwealth Department of Health traffic and the National Library of Australia began to offer training courses in several locations. In the early 1980s initial training and access to computerized databases such as *Medline, Biosis, RTECS* and *Health* were provided by the funding bodies with the assistance of Telecom (Department of Telecommunications).

The Medlars network is the shared responsibility of the National Library of Australia and the Commonwealth Department of Health. Medlars in Australia does not include the same files that are found in other countries. The Australian *Medline* network, for instance, does not routinely include *Biosis, Toxicology Data Bank, Registry of Toxic Effects of Chemical Substances* or *Toxnet* although access can be arranged for users who need to search these files regularly. At present, some 70 organizations have direct access to these other NLM databases; for those who require only occasional access, the National Library of Australia will undertake searches on a contract basis. It is anticipated that the *Toxline* file will be added to the Australian Medlars system in the near future. The Australian *Medline* network has remained relatively inexpensive and is the resource most frequently used by Australian researchers in the medical and pharmaceutical areas.

A Life Sciences Consultative Committee exists to provide feedback to the National Library of Australia to ensure that the system is responsive to the needs of its users. The committee comprises network representatives from various Australian states, the National Library of Australia and the Commonwealth Department of Community Services and Health.

At present, the Australian Medlars systems offers *Medline, Australian Medical Index, Serline, Catline* and the *Health Planning and Administration* database. Access to some of the backfiles of the *Medline* database is limited to certain time slots and there is some down-time each night to allow for system maintenance. *Hemloc* which is also on the system is an online catalogue of the holdings of the Commonwealth Department of Community Services and Health and various other state health department libraries. A feature of the Australian *Medline* network is an online ordering and electronic mail facility which is available for inter-library loans among network users.

AUSTRALASIAN MEDICAL INDEX

Perhaps the most interesting development in the Australian medical area has been the creation of a unique database covering Australasian publications.

In the early 1980s a group of medical librarians and health professionals established an Australian database to retrieve information from health, medical and biomedical material published in Australia and New Zealand. The database was named *Australasian Medical Index (AMI)* and consists primarily of data published from 1983 onwards though some major journals have been indexed back to 1980. The database first appeared in April 1985 and was designed to be used in conjunction with the *Medline* files. Material indexed includes journals, professional newsletters, conference proceedings, reports and books. Of particular interest is the inclusion of the reports of the National Health and Medical Research Council and records for the statistical series in the National Compendium of Health Statistics from the Australian Institute of Health.

Over 150 journals are indexed on *AMI*; those of interest in the pharmaceutical area include *Australian Pharmacist, Australian Journal of Hospital Pharmacy, Australian Adverse Drug Reactions Bulletin, Current Therapeutics* and *Patient Management*. Proceedings and conference reports of professional organizations are also indexed, e.g. *The Proceedings of the Australian Physiological and Pharmacological Society* since 1983. Several New Zealand health and medical journals are covered; journals of interest include *New Zealand Pharmacy and Therapeutics Notes*. Within *AMI* there are several subject specific files such as the *Alcohol and Drug File, Biology of Ageing*, and *Aboriginal Health*.

The *Australasian Medical Index* has been mounted on the Australian Medlars system for use with *Medline* search strategies. The *AMI* file is regularly updated and in January 1989 the file consisted of 27,268 citations.

NATIONAL DRUG INFORMATION SERVICE

In 1975 the Hospital and Allied Services Advisory Council identified the need for a computerized national drug information service, a recommendation accepted in 1977 by the Australian Health Ministers Conference. From 1979 to 1981 major Drug Information Centres were set up in each Australian State. Since that time, drug information centres have also appeared in teaching hospitals to meet local needs.

When the National Drug Information Service (NDIS) was established, the primary function was related to the need to promote the rational and consistent use of drugs in the treatment of patients. A second function was to co-ordinate drug information services and prevent duplication of resources and the wastefulness of undirected efforts. The major role of the National Centre was the production and publication of drug profiles in a standard format: approved name, trade name, manufacturer, pharmacology, therapeutic indications, dosage, preparations and strengths available, use in pregnancy and lactation, contra-indications, warnings, precautions, adverse reactions, significant interactions, pharmacokinetics, poisoning and overdosage, storage conditions and shelf-life, and product identification. The NDIS drug profiles prepared by specialist staff were the only independently produced Australian reviews of established and newly marketed therapeutic drugs and were promoted as a comprehensive and unbiased source of information for use amongst hospital and community pharmacists, general practitioners and community health centres.

However, in 1987, it was recommended that the National Drug Information Centre be dismantled because it had not lived up to the original expectations. Nevertheless, the major centres in the States still operate today, relying on the residual framework of the National Drug Information Service but lacking the central co-ordinating and information service functions of the National Centre. The production of drug profiles has been discontinued but those profiles already in existence (about 300) are still available online through the National Drug Information Database on the Commonwealth Department of Community Services and Health central computer.

ADVERSE DRUG REACTIONS ADVISORY COMMITTEE

The reporting of suspected adverse drug reactions in Australia and the utilization of the data are the responsibility of the Adverse Drug Reactions Advisory Committee (ADRAC) and the Congenital Abnormalities Subcommittee (CASC) which report regularly to the Australian Drug Evaluation Committee.

The Australian Drug Evaluation Committee has a very important role as the pharmaceutical regulatory affairs body. The Committee was established by the Minister of Health in 1963 in response to the thalidomide tragedy. It consists of eminent specialists in medicine and pharmacology, acting as an advisory body to the Minister, to report on the safety and efficacy of therapeutic substances.

The majority of adverse drug reaction reports are submitted voluntarily by doctors, dentists and pharmacists. All reports received are classified by ADRAC or CASC, reports are assessed without reference to previous reports or the published literature. Four broad classifications are employed for adverse drug reactions: certain, probable, possible sole drug suspected, and possible other drugs also suspected. Since the data is largely derived from a 'spontaneous' drug surveillance programme, it is subject to considerable bias and the information serves only as a guide to the sorts of problems that may be associated with any particular agent.

Information on adverse drug reactions is compiled and made available to enquirers in a number of ways. The monographic serial *Report of Suspected Adverse Drug Reactions* lists in tabulated form all suspected adverse drug reactions reported in Australia since 1964. The corresponding computerized database runs from 1972 and may be accessed by any health professional who has direct access to the Commonwealth Department of Community Services and Health central computer. The Committee also produces the *Adverse Drug Reaction Bulletin* which discusses topics of current interest and news reports regarding therapeutic substances in Australia.

NATIONAL REGISTER OF THERAPEUTIC GOODS

In 1982 the Therapeutic Goods Act was amended to allow the provision of a National Register of Therapeutic Goods comprising the National Register of Therapeutic Drugs and the National Register of Therapeutic Devices. The purpose of the computerized register is to provide a comprehensive record of all therapeutic drugs and devices on sale in Australia and it contains information regarding product identification number, date notified, proprietary name, dosage form, route(s) of administration, active substance, excipients, the Australian sponsor, whether the substance is manufactured overseas or locally, poisons schedules, therapeutic category and claimed indications or uses.

The database is not publicly available because it contains information provided by manufacturers and is, therefore, held in strict confidence. However, access may be permitted with the agreement of the Therapeutics Division of the Commonwealth Department of Community Services and Health.

AUSINET

Ausinet is an Australian vendor system operated by ACI Computer Services and accessible to overseas users via Austpac, the emphasis being Australian business information with few databases being directly relevant for pharmaceutical information. Of interest is the APAIS database, an Australian public affairs database that indexes information from newspapers, magazines and journals and is useful for retrieving items of current interest in the Australian public forum.

AUSTRALIS

The Australis information retrieval service is mounted on CSIRONET, CSIRO's communications system. Australis provides access to information concerning Australian scientific and technical developments, the databases which include *CSIRO Research in Progress, Australian Conferences*, and *Scientific and Technical Research Centres in Australia* present factual, bibliographic and research in progress information.

Microfiche services

NATIONAL POISONS REGISTER

The National Poisons Register is compiled by the National Poisons Information Service of the Commonwealth Department of Community Services and Health. In the mid 1970s the information was published in hardcopy form but in the 1980s it was updated and transferred to microfiche, at the same time, *Poisindex*, a poisoning management service was introduced. The National Poisons Register which is updated quarterly is intended for use in accident and emergency departments, in hospital pharmacies and poison information centres. It assists with the identification of household items, over-the-counter preparations and prescription products in the event of a poisoning where appropriate treatment of a patient is indicated.

In August 1988 the Register contained over 50,000 entries providing information on trade names, manufacturers and active ingredients. The product formulations are provided by manufacturers on the understanding that the information is confidential and only to be used by medical personnel who deal with actual cases of poisoning. A similar microfiche service The Pesticides Register is produced to assist with the identification of pesticide products.

Australian journals

Many of the professional associations and societies, government departments at State and Commonwealth levels, and statutory bodies produce publications that report on the work of the members or the agency. Access to this literature was previously haphazard but with the inception of the *Australasian Medical Index*, researchers and information specialists are now able to search and retrieve it.

The Commonwealth Department of Community Services and Health (Therapeutics Division, Drug Information Branch) has been producing and distributing the *Australian Prescriber* since 1975. This quarterly journal contains information on therapeutic drugs and its value is widely recognized. In 1982, the Government decided to discontinue the journal, but it was quite unprepared for the international outcry that followed and the publication was reinstated in 1983.

The previously mentioned *Australian Adverse Drug Reactions Bulletin* is another government publication that is published irregularly.

Government publications such as *Communicable Diseases Intelligence* provide statistics, analysis of disease trends and information relating to treatment.

Journals from professional societies such as the *Australian Journal of Hospital Pharmacy* contain valuable information relevant to clinical trials, laboratory studies, reports and guidelines from the speciality practice committees, and papers and abstracts of posters and oral communications presented at national and state conferences. Other professional society publications are *Australian Pharmacist* and *Australian Journal of Pharmacy*.

Several commercially-produced journals such as *Current Therapeutics, Patient Management* and *Modern Medicine in Australia* have therapeutics as a major theme with physicians and specialists providing information for doctors, pharmacists and other health professionals regarding current therapy and research.

Information emanating from research centres is often difficult to trace. However, the *Proceedings of the Australasian Society of Clinical and Experimental Pharmacologists* and the *Proceedings of the Australian Physiological and Pharmaceutical Society* are key publications providing invaluable information in the pharmaceutical area. A more comprehensive list of Australian journals relevant to the pharmaceutical area is provided in the General Appendix.

Australian reference texts

Researchers and health professionals in Australia have general access to world reference texts but there are some reference texts that are uniquely prepared for the Australian market. Important amongst these are the compiled lists of products and associated product information on prescription and over-the-counter drugs.

The *Prescription Products Guide* is an annual publication arranged alphabetically by trade name and cross-referenced to generic names providing information regarding manufacturer, product description, indications, precautions, drug interactions, adverse effects, use in pregnancy, lactation and other special patient groups, dosage and administration, and poison scheduling in each State. The product information is that which has been approved by the Department of Community Services and Health. However, products which have not been the subject of Department evaluation are presented in a general fashion only. The *Guide* is published by the Australasian Pharmaceutical Publishing Company.

MIMS Annual and *MIMS Bi-monthly* are also valuable guides to products on the Australian pharmaceutical market, the annual publication is of particular interest in providing illustrations and markings for tablet identification. The bi-monthly issue is a unique source for published information relating to new products and discontinued items. The *MIMS* publications are published by IMS Publishing a division of Intercontinental Medical Statistics (Australasia) Pty Ltd. and are readily available to prescribers and pharmacists.

Information for the lay person regarding drugs on the Australian Market is provided in the *A-Z of Australian Family Medicines*, now in its third edition. This book includes information on therapeutic uses, precautions to observe and common side effects of drugs.

The *Australian Pharmaceutical Formulary and Handbook* originally contained therapeutic and pharmaceutical formulae and some appendices. Recently, the formulary has been reduced in size, and the title now reflects more closely the book's content. Included are sections on poisons treatment, drugs and breastfeeding, dose tables with pharmacokinetic data, dosage tables for children and drug prescribing in renal impairment. The *Australian Pharmaceutical Formulary and Handbook* is now in its 14th edition, and is published by the Pharmaceutical Society of Australia.

Several texts specializing in pediatric drug therapy have been produced by various Australian children's hospitals, for example *Drug Doses for Children, Paediatric Pharmacopeia, Prescribing Medications for Children* and the *Paediatric Handbook*.

In recent years, there has been an emphasis on reference manuals or handbooks that provide therapeutic guidelines for the treatment of particular medical conditions. These books have found a receptive audience that is keen to learn from centres of excellence and teaching. In some cases, the State Departments of Health have adopted the guidelines to standardize treatment practices. A particular example is *Antibiotic Guidelines* now in its fifth edition which recommends antibiotics of choice for various disease states together with dosage and administration advice. Australian publications for information relating to drug use in pregnancy and lactation include the *Royal Women's Hospital Reference Guide on Drugs and Pregnancy*, and the Westmead Hospital publication *Drugs in Pregnancy and Lactation*. A more comprehensive list of Australian texts relevant to pharmaceuticals is provided in the General Appendix.

Australian published books and pamphlets are recorded in the publication *Australian National Bibliography* which is published by the National Library of Australia on a monthly basis with cumulative four-monthly and annual volumes. *Australian National Bibliography* is arranged alphabetically by author and title and by subject classification.

Australian pharmaceutical industry information

The pharmaceutical manufacturer remains the best source of information on all aspects of a drug. The manufacturer needs to amass considerable information for the government regulatory agency when applying for marketing approval for a new drug. With applications for new indications or reformulation of a drug, further information will be added for regulatory approval. This information is generally provided by the multinational parent company, or is researched by the local manufacturer.

The greater proportion of Australian pharmaceutical manufacturers have implemented information storage and retrieval systems to address their local needs and answer enquirers. Many companies have computerized access to Australian and overseas databases and to the parent pharmaceutical company databases via satellite link-up. The local manufacturer collects Australian information and experience or reports on any particular facet of a therapeutic drug and submits this data to the parent pharmaceutical company.

Pharmaceutical companies often are the only users of business and patent databases and may be able to assist other enquirers with information from *World Patents Index* and *Inpadoc*; these are the only databanks to which Australia contributes patent data. There is not a specifically created database of Australian patent information.

The pharmaceutical manufacturer may be required by the Drug Evaluation Branch of the Department of Community Service and Health to carry out special monitoring of a drug, and maintain a register of reports.

In the process of collecting information, and through the work of medical representatives, the pharmaceutical company is ideally placed to be aware of clinical trials being carried out with a drug. Not only that, the pharmaceutical company sponsors trials in various institutions and therefore is able to maintain a register of researchers. The pharmaceutical manufacturer's position and contact with specialists and researchers means that they are able to direct enquirers to other avenues for information on a particular drug.

Various associations exist in Australia to represent trade and professional sectors of the pharmaceutical environment, a listing is provided in the General Appendix.

Australian centres of research

Apart from texts, journals, professional newsletters and databases, current information regarding ongoing research in Australia is difficult to locate. There is always a time lag between research and publication. It is, therefore, considered worthwhile to briefly discuss current relevant Australian research in universities, research centres, hospitals, pharmaceutical companies and in government departments (see also the General Appendix).

Universities

Pharmacology departments in medical schools play an important role in the development of the discipline in Australia. Medical schools have been established in 10 of the 16 Australian universities and a school of medical research was established in 1948 in the Australian National University in Canberra.

The first Chairs of Pharmacology in Australia were founded in the University of Sydney (1949) and in the University of Melbourne (1958). The Chair of Clinical Pharmacology was founded at Melbourne in 1974, with the assistance of a grant from the pharmaceutical industry. At the same time, units of Clinical Pharmacology were established in all three of the general teaching hospitals of the University of Melbourne. Chairs in clinical pharmacology were established in the 1970s and 1980s in the Universities of Adelaide,

Queensland, Western Australia, New South Wales, Tasmania and Newcastle.

Most of the major areas of pharmacology are represented and actively pursued in Australia by researchers and pharmacologists. There is considerable strength in certain areas, such as cardiovascular pharmacology, neutrotransmitter mechanisms, endocrine pharmacology, pharmacokinetics and respiratory pharmacology. Clinical pharmacology is well established in Australia, current research in pharmacology can be seen in the topics presented at the plenary lectures, symposia and oral presentations at 10th International Congress of Pharmacology held in Sydney in August 1987.

The first Pharmacy School was founded in Melbourne in 1881 and it was modelled on the School of the Pharmaceutical Society of Great Britain. In New South Wales, pharmacy was established at the University of Sydney in 1900. In 1960, full-time degree courses in pharmacy were established at Melbourne, Sydney and Queensland Universities. Pharmacy Schools in the Institutes of Technology in South Australia and Western Australia and in the College of Advanced Education in Tasmania followed suit in the mid 1960s.

Research in the Department of Pharmacy at the University of Sydney covers a broad spectrum ranging from the design, synthesis and testing of new drugs, through to studies on methods of delivery and on the fate of drugs in man and animals.

At the College of Pharmacy in Victoria research is directed towards an examination of drug structure-activity relationships, resolution of the problems associated with drug formulation and reformulation and establishing the mechanism of drug action. The Pharmacy Practice Group conducts surveys and interviews on topics related to the practice of hospital and community pharmacy.

The Department of Pharmacy, at the University of Queensland continues research and studies into similar broad areas described for other Australian pharmacy departments. Studies into drug delivery systems particularly controlled release and bioadhesive formulations and delivery of drugs through the buccal mucosa are being investigated. Other topics include antibiotic effectiveness, radiopharmaceuticals and non-prescription medicines.

The School of Pharmacy at the University of Tasmania has undertaken a wide range of pharmacological projects, in a similar vein to the larger Pharmacy Departments in the other States but notably in the prostaglandin field. Epidemiological research in the School has focused on the community and the hospital setting with a view to investigating the patterns of drug use, attitudes and current prac-

Footnote: A more detailed review of the work of Australian institutions is given in the General Appendix.

tices. The aim is to evaluate future strategies for improving patient compliance and in turn, therapeutic response.

Research into pharmaceuticals in the School of Pharmacy at the South Australian Institute of Technology has focused on the broad areas of biopharmaceutics and pharmacokinetics, drug metabolism and toxicity.

Australian research centres

Medical research is also conducted in institutes devoted specifically to this purpose. These institutes have made important discoveries in biomedical sciences and through this work, have contributed to pharmaceutical knowledge in general.

In Australia, the National Health and Medical Research Council, has been associated in recent times with the formation of specific units devoted to full-time medical research. The first and most successful of the institutes is the Walter and Eliza Hall Institute for Medical Research. This inspired the establishment of similar institutions, the Baker Medical Research Institute, the Kanematsu Institute, the Kolling Institute and the Institute of Medical and Veterinary Science.

Another group of medical research institutes was established in the years following the Second World War. The Queensland Institute of Medical Research, the John Curtin School of Medical Research in Canberra, the Cancer Institute in Victoria and the Howard Florey Institute of Experimental Physiology and Medicine all undertake relevant research.

In the late 1950s and early 1960s came further activity with the establishment of the Children's Medical Research Foundation in Sydney, and the Royal Children's Hospital Research Foundation in Melbourne. The Garvan Institute of Medical Research in Sydney and the Mental Health Centre in Melbourne also came into being about this time.

More recently, the Ludwig Institute for Cancer Research in Melbourne was established along with the Murdoch Institute for Research into Birth Defects. The Anton Breinl Centre for Tropical Medicine in Queensland, the Heart Research Institute in Sydney and the Menzies School of Health Research have also been established in recent years.

The National Health and Medical Research Council provided funding for the establishment of a National Clinical Trials Centre in 1988. The primary goal of this Centre is to undertake clinical trials of drug efficacy and safety, community-based trials and non-pharmacological trials. Additionally, the National Clinical Trials Centre is providing expertise and resources for the design, implementation and analysis of clinical trials. The Centre will facilitate larger randomized multi-centre clinical trials. This is an important development for Australia as local trials can now be conducted and reliance on overseas studies is thereby reduced. Ultimately the National Clinical Trials Centre will cover all medical areas and will include studies related to organizational health and biomedical trials. Initially, during the establishment phase, the Centre will be concentrating on a few clinical areas including cancer and cardiovascular disease.

There are no plans in the immediate future to establish a register of clinical trials in Australia although plans may change if the concept becomes feasible. Currently all clinical trials in cancer therapy are registered with the Clinical Oncological Society of Australia and this information is entered into the world databases.

Cancer research is undertaken by a number of independent institutes and centres. However, there is also an overall organization which has a wider function to interact with the media, provide patient information and support, undertake educational campaigns and fund a cross-section of research projects.

Each State in Australia has a local cancer council or foundation which undertakes those functions mentioned above, including the NSW Cancer Council, ACT Cancer Society, Anti-Cancer Council of Victoria, Queensland Cancer Fund, Cancer Foundation of Western Australia, Anti-Cancer Foundation of South Australia, Northern Territory Anti-Cancer Foundation, and the Tasmanian Cancer Committee.

Each State cancer organization can be involved in compiling a cancer registry of statistical information on the incidence of the disease and deaths reported. As well, each State cancer organization recognizes the major cancer centres for clinical treatment of the disease. For example, in New South Wales, the major cancer centres are the hospitals affiliated with the universities, namely Prince of Wales, St. Vincents, Westmead, Royal Prince Alfred and Royal North Shore Hospitals. These major cancer centres are involved in experimental phase studies and clinical trials with new drugs and new protocols for the treatment of cancer.

The co-ordinating organization for the state cancer councils or foundations is the Australian Cancer Society. For researchers or clinicians seeking information on centres of cancer research, enquiries can be directed to the State cancer organizations. However the central position of the Australian Cancer Society will ensure that enquiries and enquirers are referred to the appropriate source or sources in Australia.

Several institutes undertake research into cancer in Australia, both the Ludwig Institute for Cancer Research and the Peter Mac-

Callum Cancer Institute are devoted exclusively to research in this speciality.

The Peter MacCallum Cancer Institute incorporates the Peter MacCallum Hospital and Peter MacCallum Clinics. The role of the Institute is research work, with a strong emphasis for chemotherapy research with numerous Phase II and Phase III clinical trials, and drug therapy trials.

The Queensland Institute of Medical Research today has a major focus for cancer research, especially malignant melanoma, and the identification of growth factors. At the clinical-laboratory interface, there is interest in the development of resistance of cancer cells to therapeutic drugs and regimens. Work with pediatric oncology has developed methods to predict chemotherapeutic responses of patients with acute myeloid leukemia.

The Institute's epidemiology studies are concentrated on cancer, in particular large bowel cancer and skin cancer, and infectious diseases. In 1987, data collection for the Queensland Cancer Registry and the Melanoma Project continued.

The Walter and Eliza Hall Institute of Medical Research was established in 1915, from income derived from a charitable trust. Since the 1950s, cellular and molecular immunology has remained the focus of the Institute's research work. Cancer research has taken on greater importance.

The Kanematsu Laboratories are a Department of Royal Prince Alfred Hospital in Sydney and their research primarily concerns cellular biology of human malignant cells, with emphasis on genotypic, phenotypic and proliferative abnormalities. Clinical activities predominantly involve the care of patients with leukemia, lymphoma, myeloma, neutropenia, and aplastic anemia.

In 1946 the John Curtin School of Medical Research was proclaimed, it is the largest and most diverse medical research institute in Australia. Current topics of study include cellular immunology, immunosuppressive agents and anti-viral vaccines.

There is considerable research being pursued into, what is generally called, diseases of the aged population. The research programmes of the Garvan Institute of Medical Research have such a focus and areas currently under study include osteoporosis, diabetes mellitus, cancer, and dementia. A research programme on cardiovascular disease commenced in 1988. The establishment in 1987 of a Clinical Investigation Unit and in 1988 of a Clinical Research Group will enhance research in clinical and pharmaceutical areas.

The Baker Medical Research Institute was founded in Melbourne in 1926, and since the beginning of 1975 has devoted its entire effort to cardiovascular research. The broad aims are to increase understanding of the basic causes of hypertension and atherosclerosis.

The Heart Research Institute located at Royal Prince Alfred Hospital is a new centre, still in the establishment phase. The focus of research is to be atherosclerosis.

The work of the Howard Florey Institute of Experimental Physiology and Medicine began in 1948 with a small group of medical researchers at the University of Melbourne. Since 1972, there has been considerable expansion of activities in the clinical fields of reproductive endocrinology, peptide physiology, biotechnology, and more recently hypertension and salt appetite, the control of hormone secretions and molecular biology. There is also interest at the Institute into how the various hormones released during stress affect the cardiovascular system.

The efforts of the Clinical Research Unit at the Walter and Eliza Hall Institute of Medical Research are directed mainly towards understanding the mechanisms of autoimmune diseases in order to improve the diagnosis, prevention and treatment of these diseases. The major disease targets are diabetes, thyroid autoimmunity, rheumatoid arthritis and other chronic, immuno-inflammatory soft tissue disorders. Since 1983, the clinical research team has undertaken immunological investigations of patients with AIDS.

The foundation stone for the Kolling Institute of Medical Research was laid in 1930. In 1948, the unit of Clinical Investigation was added. By 1974, the broad area of research was concentrating on immunology and the closely related fields of leukocyte biology, inflammation and cancer. The work in the Kolling Institute can be divided into six units: cell-mediated immunity, autoimmunity, cancer research, allergic diseases, immunosuppressive drugs and tropical diseases.

The University of Adelaide's Faculty of Medicine is one of Australia's earliest academic foundations. After World War 1 the need for medical laboratory services became apparent, and in 1938 the Institute of Medical and Veterinary Science was created. The areas of particular interest are clinical microbiology, human immunology and medical virology.

A research institute which has undergone change in research activities is the St. Vincent's Institute of Medical Research in Melbourne. The basic research being carried out is bone disease and treatment using hormones and pharmaceuticals.

Australia's geographic position close to the equator and the Tropic of Capricorn has led to considerable research into diseases associated with such tropical conditions.

An independent institute devoted to basic research was seen as the most effective way of tackling prevalent diseases such as Q fever, viral and tropical diseases and malaria research in North Queensland, and in 1945 the Queensland Institute of Medical Research was established. However, priorities have changed and today there are four broad areas of science research with malignant melanoma now a major focus of cancer research, especially the identification of growth factors. The Queensland Institute contributes research to the Australian Joint Venture Malaria Vaccine Programme for developing controls for this disease.

In 1984, the Menzies School of Health Research was established in the Northern Territory. The research priorities are tropical health and infectious diseases in northern Australia and especially the Northern Territory.

The Anton Breinl Centre for Tropical Medicine which is located at James Cook University of North Queensland, Townsville, has the broad aim to undertake tropical disease surveillance for the Northern Australian population covering the region bounded by the Tropic of Capricorn and the northern tip of Australia. Along with the research work of the Queensland Institute of Medical Research and the University of Queensland Medical School, a major programme in tropical health will be established in Queensland.

The Army Malaria Research Unit was in operation during World War 2 when Australian troops fought in Papua New Guinea. The Unit was re-formed in the 1960s, during the conflict in Vietnam because there was a very real problem with drug resistance in the treatment and prophylaxis of malaria.

At the Walter and Eliza Hall Institute for Medical Research immunoparasitology represents the largest single research effort to develop diagnostic aids and vaccines against parasitic diseases such as malaria, leishmaniasis and schistosomiasis.

The study of birth defects is a major theme for several institutes including the Children's Medical Research Foundation and Foundation 41 in Sydney; Melbourne has the Royal Children's Hospital Research Foundation, and the Murdoch Institute for Research into Birth Defects where a teratogenic sub-unit collects information on drug use and pregnancy outcomes.

The collection of data on birth defects is also the responsibility of the National Perinatal Statistics Unit which is located in Sydney.

There is research activity into the causes of mental illness and several institutes such as the Garvan Institute of Medical Research are investigating the role of brain chemical messengers and genetic markers.

The Mental Health Research Institute was established some 20 years ago and was originally an epidemiological unit. However, in recent times the Institute has entered into the neurological and biological research fields with particular interest in schizophrenia.

A National Research Institute in Gerontology and Geriatric Medicine was established in the late 1960s where research has three directions – biological, therapeutic and social.

University teaching hospitals

Hospital pharmacy in Australia has seen many changes as practice has moved from a purely supply function to a clinically-orientated service. The need for a ward-based pharmacy service has arisen because of the developments in pharmaceuticals. Firstly, the increased potency of drug products led to more effective treatment of many illnesses and, secondly, the explosion which has occurred in scientific knowledge about drugs and their role in health and disease.

Hospital pharmacists working in the hospital environment have adapted to these changing practices and have also recognized the need to undertake research. In the main the topics studied are frequently related to hospital pharmacy practice, pharmacokinetics, formulation and stability, and pharmacy service delivery. Most research is carried out in University teaching hospitals, where the medical specialities and access to resource personnel and laboratories, have provided a suitable environment for studies.

Australian information bureaux and networks

Information on pharmaceuticals in Australia is not collected or co-ordinated into a unified system but is distributed amongst research centres, pharmaceutical companies, government departments, private information services and a range of libraries. The library and information services industry in Australia has some 14,000 outlets and employs up to 50,000 people. Public libraries alone are used regularly by nearly 7 million people and many people use academic and special libraries. Each type of library from the National Library of Australia, State libraries, university libraries and so forth serves a a special clientele. The libraries rarely co-ordinate resources or services but there have been developments in recent years to establish networks in the medical and scientific areas to facilitate rationalization of resources. Major contact organizations and information bureaux are listed in the General Appendix.

The National Library of Australia

The National Library of Australia funded by the Commonwealth Government is a publicly available library that aims to meet the

issues which was historically carried out by the governments of Europe and the US has been more recently taken up by those agencies which compete successfully in terms of international commerce (e.g. the multinational companies). (See, for example, Gunnemann, 1975 and Bennett and Sharpe, 1985.) Hence, the modern equivalent of colonialism and direct political control is economic control. Most evident in the literature are the multinational companies, with economic resources far greater than many African states, which, it is argued (Bennett and Sharpe, 1985), have been responsible for dominating trade relations. Of course there are documented exceptions to this: 'In India . . . tariff barriers were raised to keep out external competition and the US slice of the market dropped from 80% to 40%. IBM left India in a huff in 1978 rather than submit to the condition that 40% of the Indian subsidiary be locally owned' (New Internationalist, August 1986).

The capacity of governments to control and monitor technology (e.g. pharmaceuticals) and the necessary information related to this is dependent upon, among other things, stability and consistent legal structures. There is evidence that governments are frequently unable to resist the pressure of the multinational: 'Very few Third World governments are in a position to control promotional activities, much less foot the bill for providing objective drug information' (Melrose, 1983).

Thus, in Melrose's view it seems that the pharmaceutical industry is as prone as any other to take advantage of slack central government control (and possibly the lack of effective competition). Furthermore, at issue is not just the matter of providing information on the effects and values of different drugs. There are cases where the types of drugs which would be of real value to developing countries are not available because the return from such a specialized product does not warrant the expense of research and development: 'Tropical disease research lags while the sales of pharmaceuticals developed for use against the special health problems of the industrial world continue to increase, both in volume and in profits. And woe betide the unwary traveller taken ill in a Third World country.' (Norris, 1982).

The disease profile of many African countries, unlike most of the North is concerned with children: some estimates indicate that 50% of all deaths are of children under 5 (Watt, 1983). Therefore the view has been expressed that the Northern pharmaceutical industries have in the past seemed not so concerned with the problems of impoverished African children suffering from diarrhoeal diseases and respiratory infection as the affluent Northern market for cosmetic drugs and dietary-related illness (Medawar, 1986). This can be explained in financial terms: 'The worldwide pharmaceutical market is running at around £75–80 billion per annum at manufacturers' price levels. Something like four-fifths of this market is in the developed industrial countries of the Northern hemisphere wherein live only about a quarter of the world's population', (Fry, 1987).

Yet African dependency upon this international business is immense, e.g. a recent survey in Ghana found that one-third of the Ministry of Health's recurrent budget was taken up in purchasing pharmaceutical products (Watt, 1983). One of the points at issue here is the capacity of Africa to exercise effective demand (and thus some form of consumer control) upon the industry.

Another current issue of pharmaceuticals use in the Developing Countries is that of dual standards in terms of drugs supplied. This may only be a very infrequent and minor issue in terms of the total industry effort in Africa; however, there are examples of pharmaceutical companies giving balanced information of their products in Europe and the US and not in Africa (Medawar, 1986).

Such examples only account for a small minority of the total effort made by international pharmaceutical companies in Africa but to some extent they overshadow the good work which is being carried out. However, in this chapter some of the problems on the ground which impede good information flow are reviewed. In this same vein Melrose points out that sometimes the local representative of the drug company will convince, by doubtful practices, the medical practitioners to adopt certain products. This may occur even though the representative is not trained to justify accurately his enthusiastic promotion. Melrose goes on to give an example from North Yemen of an individual who worked as a salesman for a drug company whilst at the same time being employed in the office of the Supreme Board of Drug Control.

It is well documented that the research and development costs of the pharmaceutical industry are very high and that the lifetime of any one product may be quite short. Melrose quotes the UN Centre for Transnational Corporations that '20% of all drugs sales at the manufacturer's level goes for promotion.'

The manufacturer of drugs is primarily interested in identifying and securing a market, selling pharmaceuticals and making profits. It is the business of national and international monitoring agencies such as the World Health Organization of the United Nations to make sure that this profit is made legally. However, we have already mentioned (Melrose) that occasionally companies may capitalize on the inability of Developing Countries to put up effective resistance to aggressive marketing. Therefore, in order to avoid unfair criticism, it is essential that the companies be seen to live by their own voluntary undertaking voiced through the International Federation of Pharmaceutical Manufacturers Associations that:

'Information on pharmaceutical products should be accurate, fair and objective', (IFPMA, 1981). Any information system which is to serve the region should follow this undertaking and, on the positive side, there are many examples of the major companies working co-operatively with development agencies (e.g. various American drugs companies co-operated with the agency AFRICARE to improve medical distribution systems in Gambia and Sierra Leone).

AIDS: a new cause for concern

Some of the range of issues prevalent in the African pharmaceutical situation have been sketched above. Recently into this arena has come the disease AIDS and its attached complications. It is at present projected that AIDS will spread at an alarming rate (Kreiss, 1986; *New Scientist*, 1987) and that although not a major problem for African economies at present it is almost certain to become one in the near future due to the way in which the disease affects the active, working population of the continent.

The regional, if not global, effects of AIDS makes the provision of unbiased, timely pharmaceutical information to and from the continent a life and death necessity.

Information systems and Africa

Modern information systems are still at an early stage of emergence and proliferation in much of Africa (for details concerning the development of online data processing centres see ECA 1988). The countries of Africa are generally described as 'developing' but this term is only of use to us if it describes a set of conditions prevalent in all these countries or quantifies some general relationship shared with the industrialized world of the North. Amin demonstrated an historical triangular relationship with the North in terms of trade. Turn (1979) showed a similar triangular relationship in terms of the flow of information (see *Figure 18.1*). Turn demonstrates that the Developing Countries are net exporters of raw data and importers of the finished products, refined in the North. The secondary relationship of dependency shown between Europe and the US is of less importance to us here. The important point is that the countries of the South as a whole are dependent upon the information and communication industry of the North and in a dependency pattern similar to that of the oil-exporting nations of the Gulf prior to the formation of OPEC. In the case of *Figure 18.1* the countries of the South find themselves exporting raw data only to pay for the expensive refined information and communication products exported to

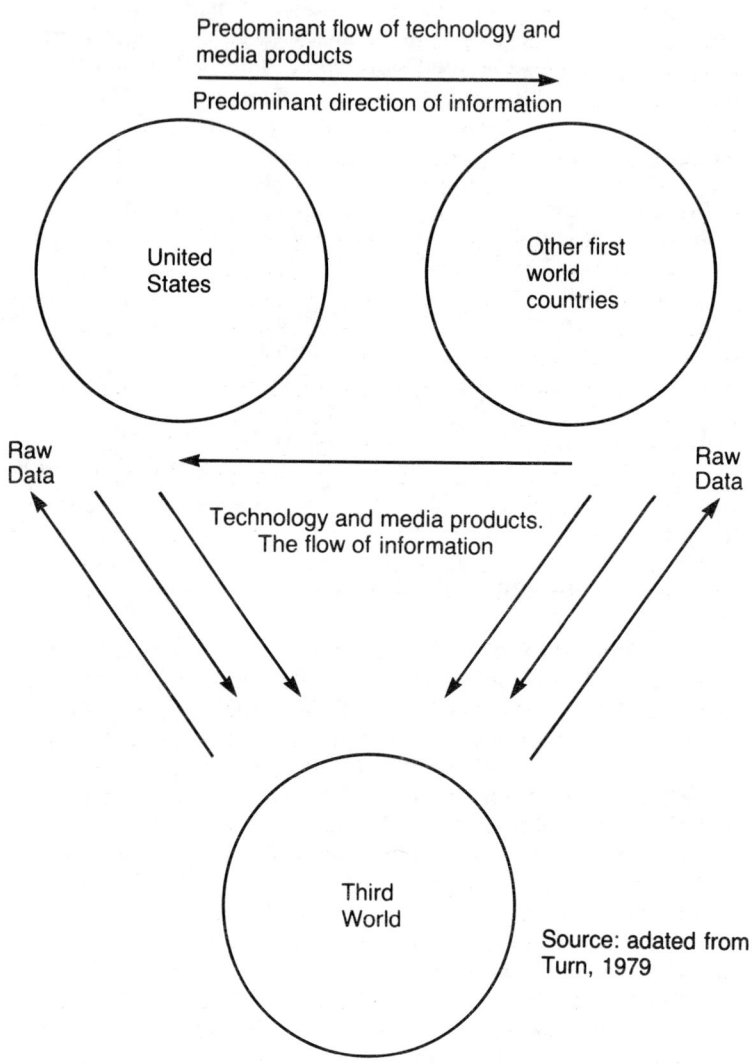

Figure 18.1   The nature of transborder data flows: Turns principles

them. The question of whether the South really requires information in this format and at this cost is rarely asked. The reasons for the present situation might be seen to be derived from the relationships and trading agreements originating during the colonial experience. Of course it is not just down to external factors. Some of the range of socioeconomic, political, cultural and environmental factors operating within most Developing Countries and which have such an inhibiting effect upon the development of a strong national information and communication superstructure, have been outlined by Eres (1981) and include such factors as the low availability of capital, the lack of available trained personnel, the poor quality of telephone services, etc. The disparity which inevitably arises between the capacity of decision makers in the North and the South to make good decisions based on sound, timely information is well known. This point is recognized by many professionals working in the field: 'Information is an international resource and publishable information should be freely available (i.e. without undue restriction) to all users in all countries' (Lohner, 1979).

But despite this recognition the information availability and access does not seem to be improving sufficiently rapidly. Some professionals question the overall need for the South to conform to the information system structures and protocols imposed by the North. Indeed, some argue that: 'A good information system for a developed country is not always adequate for a developing country' (Nakamura, 1983).

In developing this thinking one can look at the work of Durrani. In his various papers (1980a, 1980b, 1984) he criticizes the entire Western information concept of the library as the information focus of most value to Africa. The question for Durrani would seem to resolve itself into a debate on what information is intended for whom? If the information systems of Africa are intended for a minority, urban elite of decision-makers using the European languages, then the existing arrangements, though inadequately funded, have a methodology and structure consistent with this aim. But, if the systems are to bring information to rural communities so that local people can make informed decisions then a new and very different structure will be required. In the introduction it was stated that the problem is not so much that the North has little access to information on the health and pharmaceutical situation in Africa. Certainly there are difficulties for the information seeker and these become apparent in this book if one compares the situation between Europe and Africa (for example). But it is the problem of getting such vital information *within* Africa itself which causes concern. The type of alternative model that Durrani is indicating might be seen, in the health area, to be represented by a Mozambique example where: 'Village health workers were trained, not in the Institutes of Health Sciences but in special small provincial schools or health centres nearer to the communal villages they were to serve' (Watt, 1983).

The village level health worker, by definition working in rural areas where most of the population of Africa still lives, might be a much more appropriate channel for an information system on pharmaceutical products than anything that the major companies can produce on their own. The health worker is thus in a position to give information concerning the relative cost of various products, appropriate dosage and what the drug should be used for and by whom.

As already stated, information systems development is at an early stage in much of Africa. Quite often there is no indigenous pharmaceutical industry and present information is not tailored to the special needs of Africa. Often the only information sheet available to African Doctors is the *Monthly Index of Medical Specialities* (*MIMS*) which produces an equivalent African publication *MIMS Africa*. But distribution and access become problems for this as with all other forms of information. The above indicates that there is need for the resources of the Northern aid effort and the information requirements of the African people to be more closely and formally linked.

Sources

It should be recognized at the outset that there is no attempt here to differentiate between the various types of information in the field, e.g. (i) information important to the pharmaceutical industry and research, financial, research and technical development information, marketing and business information, (ii) issues of pharmaceuticals in health, clinical, preclinical, health and safety, and (iii) issues of pharmaceuticals in development, intellectual property, business conduct. All three subjects overlap and we have not separated them in the African context.

External to Africa

There is little need to indicate information sources on the pharmaceutical industry in the North. Existing services already mentioned in other chapters each have their 'African content' and 'African connection'. Some of the major movers in this field will be mentioned shortly. It is more appropriate and informative to focus on information access and diffusion 'first-hand' from the countries themselves rather than working through information system intermediaries. Therefore, we have carried out a small survey in the UK. This was intended to gain some insight into the capacity of African Embassies and High Commissions to provide information on their own national situation. Only 50% of those contacted replied and of these almost all passed on addresses in the home country to contact. At the time of writing very few of these secondary contacts have replied. This can be assessed as an indication of the state of the information process in Africa to the international community.

We will indicate some of the more specialized services available in the North which do cover African issues.

ONLINE

In general terms the Pan African Documentation Information System (PADIS) initiative with its focus on scientific and technical information looks hopeful for the future. Also ALDOC (The Arab League Documentation Centre) and ARISNET (Arab Information System Network) should be mentioned as being possible sources (Zahawi, 1986). In more general terms Dialog offers the *Dialindex* index of databases and in a relatively recent (November 1987) search for pharmaceutical company and health information on Tanzania, Kenya and Nigeria revealed almost 17,000 items drawn from such sources as *BIOSIS, Pharmaceutical News Index* and *EMBASE*. In a more specific search on the *Pharmaceutical News Index* concerning Algeria, Nigeria, Kenya, Tanzania, Uganda and Zimbabwe almost 1000 items were forthcoming. But these sources are more readily available in the North than in Africa itself.

More selectively for Africa we have the work of the databases which are more geared towards the development issue although not specifically the pharmaceutical industry. Of particular interest to us is the *Commonwealth Agricultural Bureau International (CABI)* which is available direct or on Dialog and also the database run by the Bureau of Hygiene and Tropical Diseases called *Public Health and Tropical Medicine (BHTD)* which is also available on the *CABI* direct connection.

SPECIALIST CENTRES AND OUTREACH ACTIVITIES

First we mention patents information, e.g. *INPADOC* (the *International Patent Documentation Centre*). Most of the multinational pharmaceutical companies working in Africa have their own information sources on performance and sales which should be obtainable from them. We do counsel the inquirer against haste however. As a small survey exercise we contacted the 20 largest multinational companies working in the field of pharmaceuticals (as given in Norris, 1982). We asked them for details of the information services they supply and arrange in the African context. Over a month after writing no replies had been forthcoming. Eventually some companies (e.g. Hoechst and Schering-Plough) sent details concerning (for example) their companies' sales and corporate details of subsidiaries (e.g. Nigerian Hoechst). The response was disappointing more so because it was known from anecdotal sources that many of the companies contacted had spent considerable efforts in streamlining their efforts in Africa. The fact that there seemed little willingness to represent this effort in terms of literature or direct communication is unfortunate. Other good corporate information sources in the UK include: The Association of the British Pharmaceutical Industry, 12 Whitehall, London SW1A 2DY; Mr. Tom Zomoyski, *Medicine Digest (Africa)*, HZI International Ltd., 28 Great Queen Street, London WC2 5BB; Mr. Kevin Morrissey, *Postgraduate Doctor (Africa Edition)*, Barker Publications Ltd., Barker House, 539 London Rd., Isleworth, Middlesex TW7 4DA.

The Commonwealth Pharmaceutical Association which is based at the headquarters of the Pharmaceutical Society of Great Britain can direct the inquirer to information on the activities of pharmacists within the Commonwealth nations in Africa.

Another source is the *Monthly Index of Medical Specialities* (*MIMS*) which produces an equivalent African publication, *MIMS Africa*. Of value to the academic inquirer might be the centres of excellence in terms of African studies. Much general information (although not always specific to pharmaceuticals) can be gained from the Africa Studies Centre at School Lane, Cambridge and the Library of the School of Oriental and African Studies (SOAS) in London.

In the international context the main information sources available are those provided by the United Nations and in particular the World Health Organization (WHO). WHO in particular provides information via Division of Diagnostic, Therapeutic and Rehabilitative Technology, Pharmaceuticals Unit, 20 Avenue Appia, CH-1211 Geneva 27, Switzerland which is a documentation unit and clearing house/referral centre. The UN gives the subject coverage of the unit as being drug addiction, medical care, toxicity and pharmaceuticals. The product of the unit is a *Quarterly Abstracts of Cumulated Data* which is also available as a computerized service. Approximately 20,000 items are added each year. Also produced by the unit is the *International Nonproprietary Names for Pharmaceutical Substances* (*INN*) which is produced as a cumulative list and is also computerized. At the same address is the Health and Biomedical Information Programme Health Legislation Unit. This unit covers such areas as health, legislation, environmental legislation, aging, human nutrition, food standards, pharmaceuticals, poisons, occupational hygiene, health statistics. It produces an *International Digest of Health Legislation*. The UN agency which covers pharmacology is: Vienna International Centre library, PO Box 100, Wagramserstrasse 5, A-1400 Vienna, Austria, which produces a range of current awareness products. Finally, for those interested in the wider, corporate issues relating to the pharmaceutical business, the agency that used to deal with the industrial nature of the pharmaceuticals business was: United Nations Industrial Development Organization, Chemicals, Pharmaceuticals and Building Materials Industries Section, Industrial Information Section, PO Box 707, 1011 Vienna, Austria. This group produced a number of reports including the *Directory of Industrial Information Services*. The agency was not listed in the 1985 edition of the *Directory of United Nations Databases and Information Systems* and the possible alternative is: Industrial and Technological Information Bank, Vienna International Centre, PO Box 300, Wagramerstrasse 5, A-1400 Vienna, Austria.

Internal to Africa

This is the most ambitious and problematic area. Other than by completing an in-country study of pharmaceutical information sources for all the countries in the continent it is impossible to indicate the true situation on the ground. The *Directory of Industrial Information Services* and *Systems in Developing Countries*, produced in 1985 by the Industrial Information Section of the United Nations Industrial Development Organization (UNIDO) listed no services at all in the pharmaceuticals area in Africa. There is an African publishing business growing up around pharmaceuticals (e.g. the *African Journal of Pharmacy and Pharmaceutical Sciences* published by the Nigerian Pharmaceutical and Medical Co.). It does seem that on an Africa-wide basis the actual industry is very limited although there are some specific examples of a developing local capacity (e.g. the Ministry of Health in Kenya quoted 21 pharmaceutical companies manufacturing locally, the vast majority being indigenous). A list of the centres of pharmaceutical knowledge for most of the African countries with an interest in the industry is presented in the General Appendix. This listing may well contain some information which is dated in part at least because of the ever changing situation in Africa. The listing builds on the previous work carried out by UNIDO (1976).

The future

If the overall effect of what has been said so far has been to point to an unbridgeable gulf between the North and South this has not been the intention. In order to see the situation in context the passage of the process of what has been called 'development' has been outlined. This included a brief review of what authors working in the continent have expressed as being certain elements of both the African experience of development and of the pharmaceutical industry. From this basis the chapter then scanned some of the problems which have been observed as regards the specific area of information provision and therefore has taken in such items as some of the more extreme actions of multinational companies in this context and what might be called an indigenous conception of the priorities (Durrani). There was then a brief exposition of the information systems available at the present time. This underlined the disparity between the Northern, 'technologist' approach of online systems which some have argued from an information technology perspective might result in the South 'leapfrogging' the industrial age into the information age (Chitty, 1983), with the stark facts of the basic needs of Africans today and what the resources they have can do for them.

Although there are many indications that the industry has been and is becoming more sensitive to the needs of Africa, one possible innovation for future development would be a co-ordinated pharmaceutical information policy within countries and in Africa as a whole. This would require the bringing together of the disparate groups (such as multinational companies, governments, national consortia, local groups, health workers, etc.) to co-operatively structure the information priorities with regard to health and development levels as a whole. Such an exercise might be expected to link information need with information provision and thus institute a process of planning, diffusion and access area by area. This would obviously be a macroplanning exercise and would need to take into account the motives and requirements of each of the primary actors. There are examples of this type of operation being attempted. In particular UNESCO has in the past attempted to co-ordinate information policy in specific geographic and subject areas. Within the NATIS, UNISIST and GPI programmes are examples which demonstrate the range of problems which can arise. In the pharmaceuticals field however there is little hope that the necessary resources could be drawn together for such an ambitious project. Possibly one sign of hope for the future is the work of agencies such as Information for International Development (IFID) and Appropriate Health Technology Action Group (AHTAG) which, among its many activities, acts as a clearing house for information, runs information projects in the Developing Countries and produces newsletters on key health issues. These agencies are at a much more modest level but perhaps in their structure we see some sign of the way information systems may work within Developing Countries in the South as well as between North and South in the future. In the case of IFID, Wills (1978) outlined a format for an information system which would offer a great deal to the South and yet work within the resource base available. *Figure 18.2* depicts the form of this organization and also demonstrates its operation. The inquirer has access to a development counsellor working in this locality. The counsellor can then, by a system of referral and voluntary help from subject specialists, either link the inquirer to the network of advisers (the key idea here being that South-South linkages and intergovernmental co-ordination on information issues would be developed) or to the information systems and pharmaceutical multinationals of the North. The system would work on the basis of voluntary assistance, self-interest and an almost organic growth. As more companies, information suppliers and end-users come to see the benefits of shared information so the list of co-operating agencies and institutions would grow. Another benefit of this system would be the involvement of local people in the creation, planning and running of their own system. An IFID type system is at present being planned to deal with AIDS information in East Africa by the African Aids Project Team of the School of Development Studies at the University of East Anglia. The Team is focusing on the impact of AIDS on rural labour supply and the ability of communities to cope with sudden demographic imbalance, but has a wider interest in the communication of information about the disease and the operation of information systems dealing with this communication. One of the aims of the information component of the research is to attempt to link up-to-date epidemiological information with the results of socioeconomic surveys on the rural labour situation. The resulting linked information set would be used to map disease intensity and areas of greatest need. This in turn would be matched against available donor assistance. Other functions of the information programme would be to develop South-South information linkages on the spread of disease and the sharing of ideas. Therefore, although the system would work in a North-South and South-North context, its real mission would be purpose designed to inform local people on local/regional issues by use of the whole range of local, national, regional and international facilities (for further details see *New Scientist*, 14 January, 1988).

All the above is at a very early stage of development but does offer the possibility of creating an information network which can

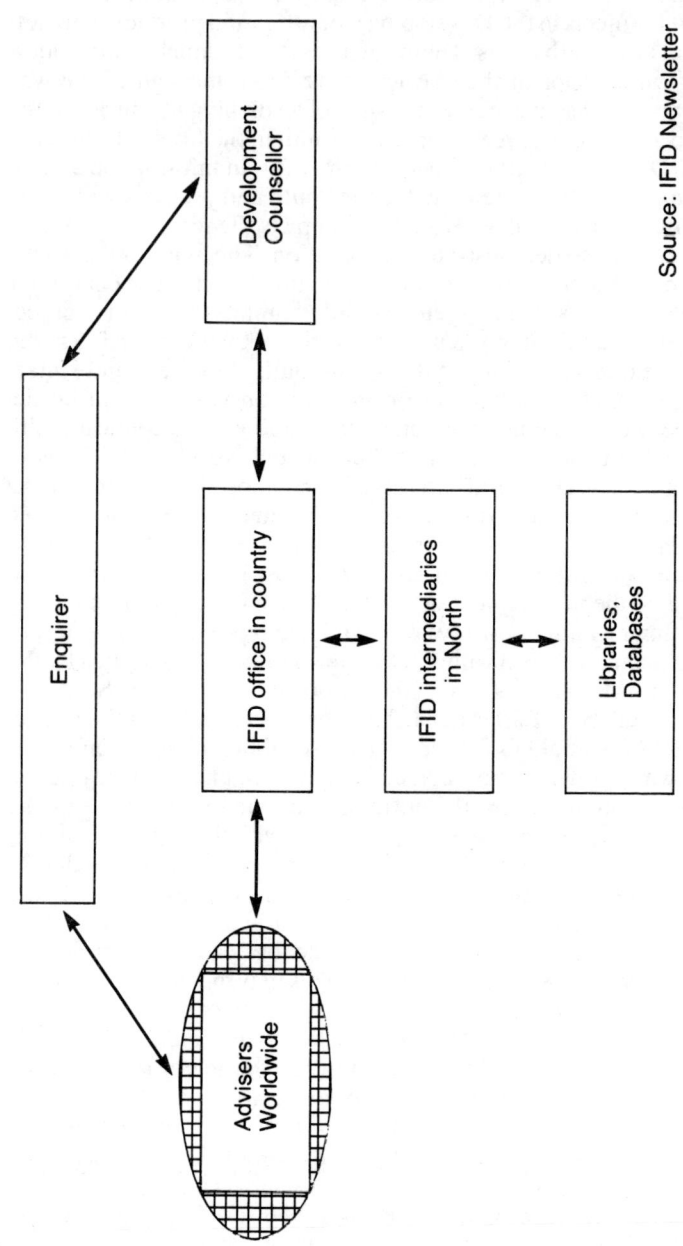

Source: IFID Newsletter

Figure 18.2   Information for International Development

serve both the interests of international organizations and indigenous populations.

Acknowledgements

I thank my colleagues G. Coleman, S. Cross, and N. Abel and also T. Eizenschitz (City University, London).

References

Amin, S. (1978). *Unequal development* (Monthly Review Press).
Amin, S. (1974). *Accumulation on a world scale* (Monthly Review Press).
  Bahaa El-Hadidg. (1983). 'Delayed online search: an alternative access mode for developing countries', *Journal of Information Science*, **5**, (5) 1983., 173–185.
Barnett, R. J. (1980). *The lean years: politics in the age of scarcity* (Abacus).
Bennett, D. C. and Sharpe, K. E. (1985). *Transnational corporations versus the state* (Princeton University Press).
Chitty, N. J. (1983). 'Chips and development', *Development Forum*, October 1983, p. 9.
Durrani, S. (1980a). *A relevant library service* (University of Nairobi Library).
Durrani, S. (1980b). *Will Njanjiru be silenced?* (University of Nairobi Library).
Durrani, S. (1984). 'Libraries and rural development', *Ideas and Action*.
ECA. (1988). *Directory of African electronic data processing centres and EDP experts in Africa*, (Economic Commission for Africa, United Nations).
El-Shishiny, N. et al. (1987). 'Public information services of ENSTINET', *Proceedings of the Third International Conference on Systems Research, Information and Cybernetics*.
Eres, B. K. (1981). 'Transfer of information technology to less developed countries: a systems approach', *Journal of the American Society for Information Science*, **32**(3), 97–102.
Ford, J. (1971). *The Role of the Trypanosomiases in African ecology: a study of the tsetse fly problem* (Clarendon).
Frank, G. F. (1978). *World accumulation 1492–1789*, (Macmillan).
Fry, M. (1987). 'Why generics? and how many medicines are really necessary?' *The Crown Agents Review*.
German Pharmaceutical Manufacturers' Association. (1982). *Drugs and the third world* (Frankfurt).
Gunnemann, J. P. (1975). *The nation-state and transnational corporations in conflict* (Praeger).
Harrison, P. (1984). *Inside the third world* (Penguin).
Industrial Information Section. (1985). *Directory of industrial information services and systems in developing countries* (UNIDO).
International Federation of Pharmaceutical Manufacturers Association (IFPMA). (1981). *The IFPMA code of pharmaceutical marketing practice*.
Kjekshus, H. (1977). *Ecology control and economic development in East African history* (Heinemann).
Kreiss, J. K. (1986). 'AIDS virus infection in Nairobi, prostitutes: spread of the epidemic in East Africa'. *New England Journal of Medicine*, no. 314., 414–18.
Lohner, W. (1979). 'Intergovernmental conference on scientific and technological information for development (UNISIST II): main issues and results', *International Cooperative Information Systems Seminar* (IDRC).

Medawar, C. (1986). *Insult or injury? an enquiry into the marketing and advertising of British food and drug products in the third world* (War on Want).

Melrose, D. (1982). *Bitter pills: medicines and the third world poor* (Oxfam).

Nakamura, Y. (1983). 'Information technology and strategy for developing countries, in Brown, K. (ed.) *The challenge of information technology* (North Holland Publishing).

New Internationalist. (1986). *The third dimension.*

New Scientist. (1987). *Uganda acts to stem epidemic.*

Norris, R. (1982). *Pills, pesticides and profits* (North River Press, Inc.)

Owen, R. and Sutcliffe, B. (1976). *Studies in the theory of imperialism* (Longman).

PADIS: Pan African documentation and information system, (UN Economic Commission for Africa, 1981).

Ransford, O. (1983). *Bid the sickness ceases: disease in the history of black Africa* (John Murray).

Slamecka, V. et al. (1985). 'A longitudinal profile of a national search service, *Information Processing and management*, **22**(3), 203–216.

Turn, R. (1979). '*Transborder data flows: concerns in privacy protection and free flow of information, vol. 1*, (US Federation for Information Processing Society, Inc.).

UNIDO. (1976). *Information sources on the pharmaceutical industry*, (UNIDO Guides to Information Sources, No. 20).

United Nations. (1985). *Directory of United Nations databases and information systems* (United Nations).

Walt, G. and Melamed A. (eds.). (1983). *Mozambique: towards a peoples health service*, (Zed Books Ltd.).

Wills, J. B. (1978). *Information for development: a system approach to assessing user needs*, (International Association of Agricultural Librarians and Documentalists) (Saur Publishing).

Zahawi, F. (1986). *The development, activities and plans of the Arab League Documentation Centre ALDOC*, Paper presented to Seminar on Databases and Networking in Development, (Brighton, 1986).

CHAPTER NINETEEN

Information needs and availability in the People's Republic of China

Y. GONG

History

Background

China has a long history full of scientific accomplishment. It is the originator of paper-making, printing, compasses and gunpowder, which have made great contributions to the development of mankind. The sense of antiquity prompts people's confidence, enthusiasm and pride. But this is the better side of the coin. On the other, China has somehow survived two centuries of foreign occupation and civil war to enjoy less than a decade of single-minded social and economic progress. People suffered from poverty, illiteracy and disease. Life expectancy was about 35 years. The terrible suffering undergone by the Chinese people is beyond description and, perhaps, one of the worst in the world.

Present situation

China is the largest nation in the world with a huge and diverse population of over 1000 million. It is at last on the road to becoming an industrial power matching in prosperity and creativity the most successful nations elsewhere. The primary aim, set by the 12th National Congress, is to quadruple the gross annual output by the end of this century and then proceed to raise, within the first 30–50 years of the next century, China's economy close to the level of the developed countries.

So at present, 'reform' is the slogan. The reopening of China to the outside world is one sign and the adjustment of the national economic system is another. For now, the economic consequences of reform are its sufficient justification. The door has been opened to the outside world to encourage foreign investment (in terms of the pharmaceutical industry, there have been five major joint venture pharmaceutical companies set up since 1980). Market forces, besides central planning, are to determine the pattern of industrial production. Peasants are allowed to farm the land they are responsible for and to sell their surplus produce in free markets wherever they have access to them.

This development is no surprise. The reforms are themselves a reflection of the huge changes in social and economic life that have taken place as a result of pragmatic policy.

The social and economic goals of reform will be accomplished by adapting science and technology to China's distinctive talents, of which the Chinese are self-conscious and proud. Dramatic changes are taking place in the organization of science and technology in China. In this respect, the essence of the reform is that 'science and technology must be oriented to economic reconstruction'. To that end, 'the defect of relying on purely administrative means in science and technology management, coupled with the State undertaking too much and exercising too rigid a control' is being remedied. So far, the main changes and undertaking are as follows:

- to remedy the drawback of research institutes being isolated from industrial enterprises, scientists are encouraged to undertake projects with industrial potential. From now on, whilst research projects with national priority will remain under the control of the State, other activities conducted by the scientific and technological institutes are to be managed by means of 'economic levers and market regulation'. This is a departure from the past system, under which projects were allocated from on high. This reflects the emphasis on research inside enterprises;
- a new appropriation system has been introduced in order to guarantee that the State's science policy is implemented. The budget will no longer be allocated as an average according to the number of staff in institutes; instead, projects in applied sciences will be carried out under contract, and there will be a system of competitive grants open to all scientists in basic research;
- to help with the effective management of scientific activities, the power of scientific directors is being restored. Directors will now be firmly responsible for the scientific direction of their institutes;
- many national open laboratories have been set up for visiting scientists both at home and abroad. This will certainly help to promote exchange personnel and increase international co-operation.

These changes may sound unsurprising but they do represent radical breaks with the past. In many ways, the reforms have changed people's life and thinking.

Hierarchical structure in pharmaceutical sciences

Although China is a centralized nation, as far as the pharmaceutical industry is concerned, the command chain is very complicated just because of its broad-ranging nature and the virtue of its roots in science and technology (see *Figure 19.1*).

The backbone of the pyramid of organizations involves four commissions/ministries: the State Science and Technology Commission, the State Economy Commission, the State Education Commission and the Ministry of Public Health. Higher up the hierarchy are the State Council (equivalent to the Cabinet in the UK), the Standing Committee and the National People's Congress (the highest representative body of the nation, equivalent to Parliament in the UK).

As far as the pharmaceutical industry is concerned, the Ministry of Public Health (MPH), the State Education Commission (SEC) as well as the State Science and Technology Commission (SSTC) are mainly involved in research and development. At present, 48 pharmaceutical colleges and universities are managed jointly by SEC and MPH. The Chinese Academy of Medical Sciences (CAMS) supervised by the MPH, has some research institutes under its direct control.

The Chinese Academy of Sciences (CAS), the highest research centre in the country and the best known to academic visitors from the West, has dominated and carried out the bulk of China's basic scientific research ever since the founding of the People's Republic of China in 1949. Because of its comprehensive coverage of subjects and the extensive geographic spread of its institutes throughout the country, CAS is certainly strong in the pharmaceutical sciences.

Immediately under the SEC come the National Pharmaceutical Administration Bureau (NPAB) and the National Administration Bureau of Traditional Chinese Medicine (NABTCM), which have characteristics of both an industrial and a commercial nature.

NPAB was founded in 1979, with responsibility for manufacturing and trading in Western medicine, Chinese patent medicine and medical apparatus and instruments. Before then all these functions had been conducted under the supervision of the MPH.

At present, NPAB has four companies and two academies under its direct control. The National Pharmaceutical Industry Company manufactures Western medicine, while the National Pharmaceutical Company markets the products; the National Medicinal Materials Company is responsible for the management of Chinese patent medicines and the National Medical Apparatus and Instrument Company wholesales medical devices. The two academies, the Graduate School of Pharmaceutical Industry and the Designing

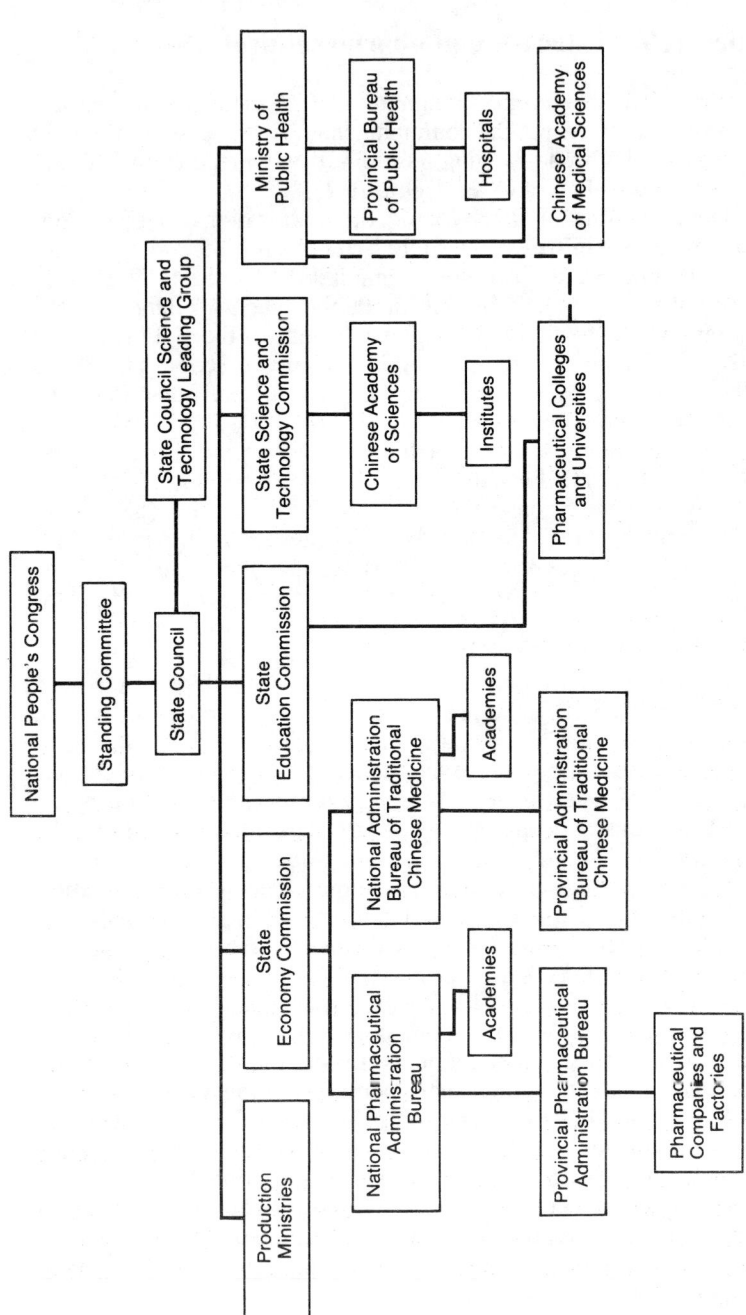

Figure 19.1 Hierarchical structure in the pharmaceutical industry in China.

Institute of Pharmaceutical Industry, are mainly involved in research and development.

At the provincial and municipal level of the government, much of the above pattern is repeated on a smaller scale. For example, in Shanghai, there is the Shanghai Pharmaceutical Administration Bureau and corresponding lower rank companies, which supervise 44 pharmaceutical factories, 14 traditional Chinese medicine factories and 34 medical apparatus and instrument factories.

NABTCM was set up in 1987. At present, the emphasis is on the side of traditional Chinese medical science and it is mainly involved in research and development.

Information systems in the pharmaceutical industry

Figure 19.2 shows the information systems in the pharmaceutical industry are diverse and are combined in a number of different systems, such as the National Library System, the College and University Library System and the Library and Information System of the Chinese Academy of Sciences, etc. in order to cover the wide-ranging nature of the field. There is no single information centre for the pharmaceutical industry in China.

Colleges and universities

In terms of tertiary education in pharmaceutical sciences, there are altogether 48 colleges and universities, amongst which the China Pharmaceutical University in Nanjing and the colleges of pharmacy in Shanghai, Shenyang, Beijing and Huaxi are the elite which attract the most funds and the best students.

The China Pharmaceutical University, China's first comprehensive university in pharmacy, was established in 1986 by the combination of the former Nanjing College of Pharmacy and the Nanjing College of Chinese Traditional Pharmacy. The University, offering courses in pharmacy, Chinese traditional pharmacy, pharmaceutical chemistry, biochemical pharmacy, pharmacology and pharmaceutical management, has been authorized by the State Council to confer degrees. It has 26 teaching departments, 1 research institute and 8 research laboratories, with a teaching and research staff of over 400. The number of students has reached 1500, including more than 100 postgraduates.

The newly built library, with a seating capacity of 800, has a collection of 235,000 items. Also attached to the University are a pharmaceutical factory and a medicinal botanical garden (The University, 1986).

From 1949 to the end of 1984, China trained more than 121,600 pharmacists. At present there are 11,000 students in colleges and universities, with about 3,000 graduates each year (Jin *et al.*, 1987).

Besides the above colleges and universities, there are another 28 colleges of Chinese traditional pharmacy, which is, perhaps, the unique feature of China's pharmaceutical sciences. These colleges usually offer the following specialities: Chinese traditional pharmacy, traditional Chinese medical science, acupuncture, massage, traumatology, etc. The Shanghai College of Chinese traditional pharmacy offers the first four specialities. Its library has a rich collection of 320,000 items, among which 110,000 are in the field of Chinese traditional pharmacy, the best collection in China in this respect.

Academies

Apart from the research projects in colleges and universities, China has parallel research organizations, academies, carrying out basic research. For the pharmaceutical sciences, the Chinese Academy of Medical Sciences (CAMS) is best known and dominant in the field. In terms of status, facilities, time available for research and size of budget other academies such as the Graduate School of Pharmaceutical Industry and the Chinese Graduate School of Traditional Chinese Medicine have been poor cousins of the CAMS.

In 1983, the Institute of Scientific and Technical Information affiliated to the CAMS provided batch-form computer retrieval services in China using *MEDLARS* files with the help of the Australian National Library. In 1986, the Institute imported the *MEDLARS-MEDLINE* tapes, mounted them on its own IBM computer and established the *MEDLARS* file and *SDILINE* file. Up to 1988 the system has more than 700,000 records from 1984 onwards, offering offline, online and SDI services to the users throughout the country.

Societies

The Chinese Pharmaceutical Association (CPA), one of the earliest mass organizations in China, was first established in Tokyo, Japan in 1907 by a group of overseas Chinese students who studied pharmacy there. When the Chinese Science and Technology Association was founded in 1958, the Association immediately became a corporate member.

In its constitution, the Association has the obligation and commitment of organizing pharmaceutical activities, to make contributions to the development of education, research and production in the pharmaceutical industry, to exchange views on concepts, theories and experience in order to raise the competence of its members in their work and to explore and promote means of co-operation and exchange of personnel both at home and abroad. With 24,695 full members, the Association has set up 8 special committees (medicament, pharmaceutical analysis, pharmaceutical chemistry, traditional Chinese medicine and natural drugs, antibiotics, pharmacology, biochemical drugs and medical history) and 6 working committees (academic, popularization and education, editing and publishing, organizing, inquiry and international academic exchange). At the provincial level, branches have been set up to co-ordinate activities and the basic structure is repeated at lower levels.

Besides the CPA, another influential and parallel association the Chinese Medical Association (CMA) is also much involved in the fields of public health, medicine and pharmacy. The CMA founded in 1915, now has a membership of 81,000, one of the largest mass organizations in China. At present, it has 36 special societies. Its library founded in 1931 became the present Library of Shanghai when the CMA moved its headquarters to Beijing from Shanghai in 1951. The Library has now built a collection of more than 115,750 items. In terms of the foreign medical and pharmaceutical periodicals, it is the second largest in China.

Clearing-houses of information

Apart from libraries in China, there are parallel institutions called institutes of information science (the principal one being the Institute of Scientific and Technical Information of China, ISTIC) which act as clearing-houses of information, focusing their interests on information analysis, information retrieval and database provision etc. From *Figure 19.2* one can see that nearly all governmental ministries and committees have their own information institutes. For example, the Shanghai Institute of Scientific and Technical Information is under the Shanghai Pharmaceutical Administration Bureau. It publishes three pharmaceutical journals and regularly holds news conferences to release information on new drugs, market trends and comments and criticism of established drugs.

Scientific communication

The rapid growth of scientific information is already posing serious problems for both scientists and information specialists. In today's fast-moving technological scene, especially in interdisciplinary fields like pharmaceuticals, the prompt and accurate transfer of information is of paramount importance.

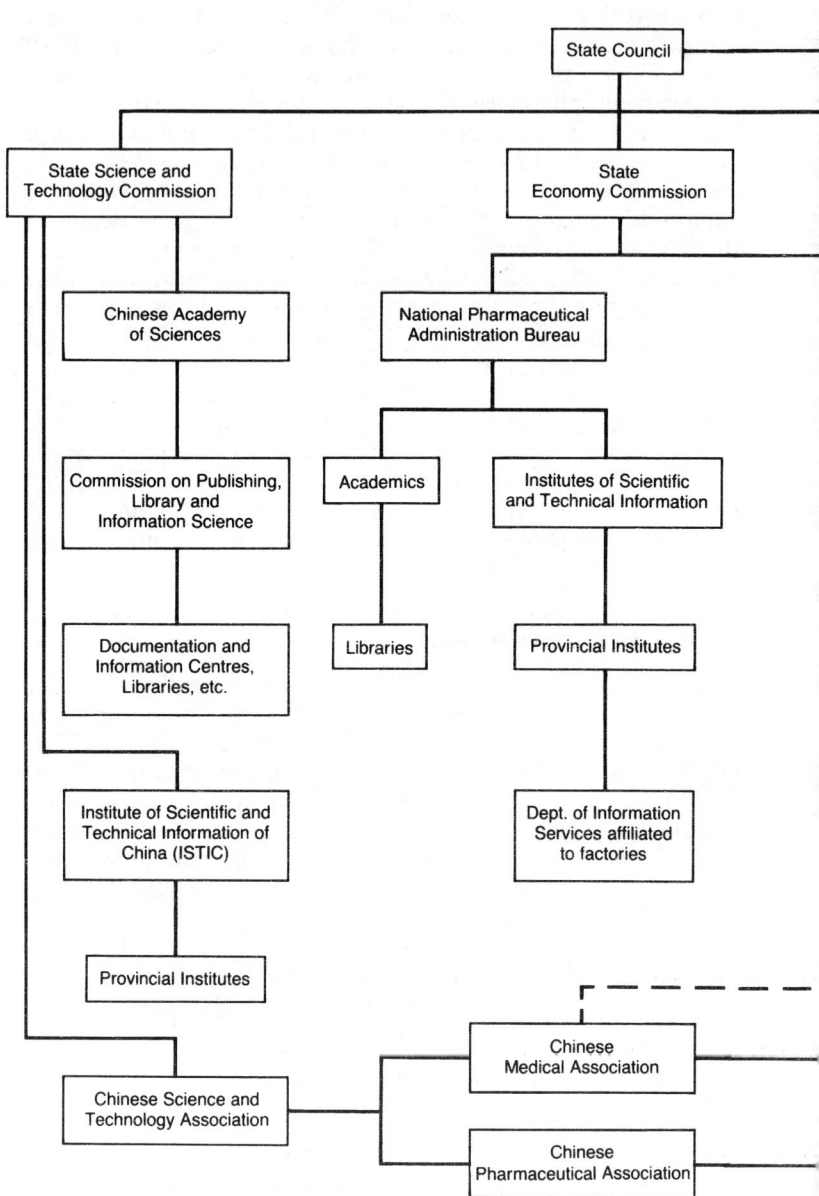

Figure 19.2   Information Systems in the Pharmaceutical Industry in China.

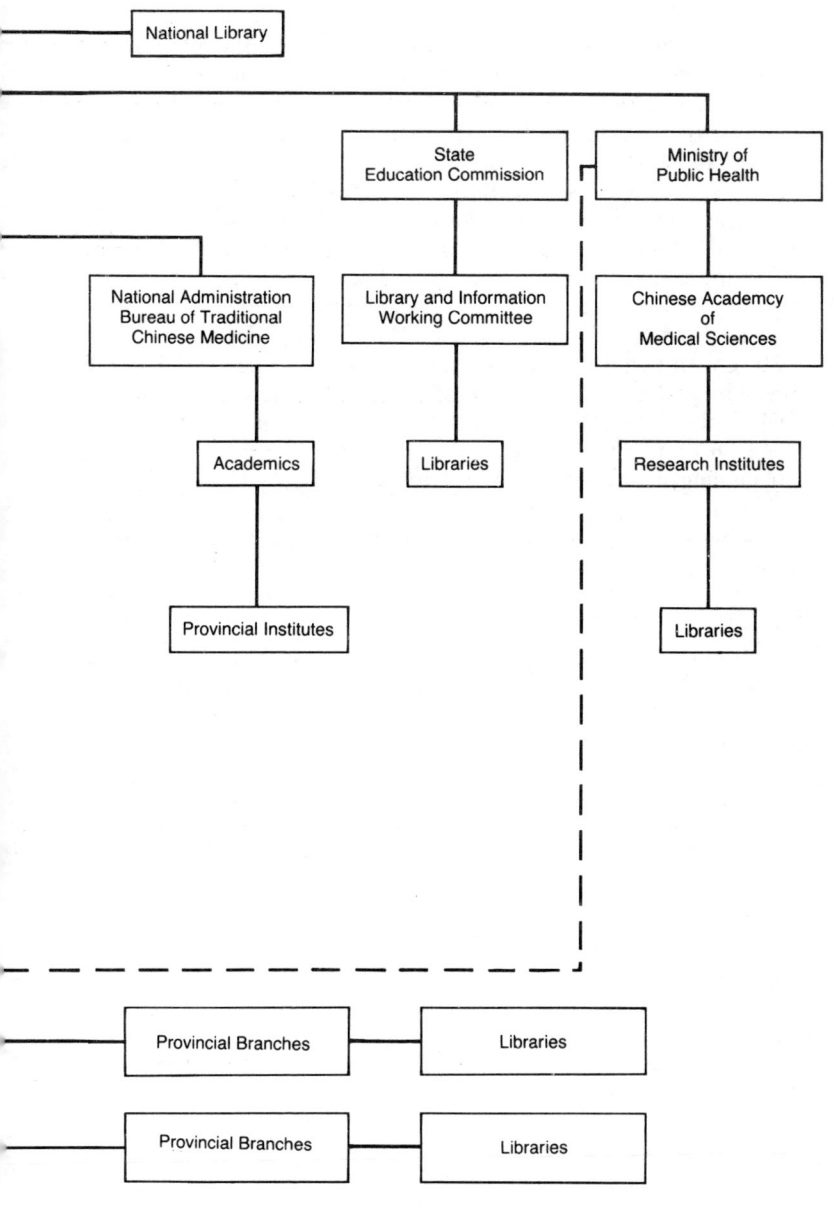

National Library

State
Education Commission

Ministry of
Public Health

National Administration
Bureau of Traditional
Chinese Medicine

Library and Information
Working Committee

Chinese Academcy
of
Medical Sciences

Academics

Libraries

Research Institutes

Provincial Institutes

Libraries

Provincial Branches

Libraries

Provincial Branches

Libraries

User pattern

In China, efficient though the above mentioned information systems are, the levels of service and automation are outdated. Whilst information retrieval in the West tends now to be concentrated on online services backed up by strategically placed conventional libraries, these services in China are mainly carried out manually. Online searching is still in its infancy and only a few privileged organizations which have their own mainframe computer have mounted either imported or 'home-grown' databases for Chinese literature to provide online and SDI services.

However, as a part of China's plans for modernization, computerised information retrieval is being given priority. In March, 1980, China began to rent an online terminal to access DIALOG from Hong Kong. Three years later, the Institute of Scientific and Technical Information of China (ISTIC) could search ESA-IBS, DIALOG and ORBIT systems using leased satellite lines, thus introducing a new era of computerised information retrieval in China.

37 bibliographic databases have now been imported on tape from America and Europe, such as *CA SEARCH*, *BIOSIS PREVIEWS*, *INSPEC*, *CAB*, *MEDLARS*, *NTIS*, *WPI* etc. (Chen, 1985) and, coincidentally, also 37 international terminals scattered throughout the country are available for direct access to ESA, DIALOG, ORBIT, INFOLINE, DMSS/DRI, and FIESTA systems (Mei, 1986).

Nevertheless, several obvious impediments have militated against progress in this respect. First, because of the tight budget, very few machines possessed by documentation centres have the capacity for a large number of information records, and the vast area of China adds telecommunication constraints to the bottleneck. Secondly, availability of information produced in foreign countries is another serious problem. Although lots of money has been spent subscribing to foreign materials, the majority have been placed in large cities such as Beijing and Shanghai; therefore, for those practising medicine or undertaking research projects outside the big cities, accessing information required can be extremely limited. The problem of information retrieval divorced from document delivery is obvious. Finally, secondary sources are poor; more energy and manpower needs to be directed toward secondary services. Even for Chinese literature, scientists often encounter the frustration of wasting plenty of time and obtaining nothing useful. In a word, unless overall planning is carefully scheduled, the effects of new, faster methods of dealing with the mass of literature will not be fully felt in the community in the near future. The need for a period of skilful planning and improvement is very strong.

Table 19.1. Chinese pharmacists' reading profile in terms of subjects

Subject	Pharmacy	Chemistry	General biology	Biochemistry
%	21.8	10.0	3.1	4.8

Botany	Microbiology	Pathology	Chemical medicine	General subjects	Others
1.0	2.1	1.8	21.5	23.4	10.5

In such an information environment, it seems that scientific communication in the pharmaceutical industry relies a great deal on individual initiative and personal contact.

Nevertheless, categories which, to a large extent, reflect the user pattern in the pharmaceutical industry can be classified into three (in terms of research, production and use of drugs) as follows.

1. 'Archival use' by those involved in research projects. Scientific research is, by definition, an exploration of the unknown. The aim of the pharmacist is essentially to find new drugs which can, no matter whether they come from biological or chemical agents, modify the body's biochemistry in such a way as to achieve the desired remedial or therapeutic effect. From this premise, scientists have to draw on others' experiences and findings because the development of science is regarded as a process of continuous renewal and cumulation (Brookes, 1981). Scientists have to keep abreast of the current trend and to place themselves in the forefront of their disciplines. *Table 19.1* shows the Chinese pharmacists' reading profile in terms of subjects. This means scientists lay heavy emphasis on archival materials (Mei, 1987).

Table 19.2 illustrates the format distributions of citations by the Chinese pharmacists. It clearly shows that periodical literature, rapid in publication and new in knowledge, is the main medium Chinese pharmacists prefer to use in written scientific communications.

Scientists doing research projects regard written records as the

Table 19.2. Distribution of citations by Chinese pharmacists.

	Periodical literature	Monographs	Other* literature	Total
Citations	4100	597	822	5519
%	74.3	10.8	14.9	100

* technical reports, government publications, standards, patents and conference papers.

main means of communication and, in turn, publications are necessary qualifications for promotion.

2. 'Personal contacts' by physicians and surgeons as well as those working in industrial research and development laboratories. Within the clinical environment prompt and accurate information is of vital importance. In industrial research and development laboratories, information seeking is very much problem-related. So communication in these groupings is of interest because present-day laboratory and clinical medicine have more affinities with technology (even experience and skill) than with science.

A recent study has shown that personal contact is an important channel for the diffusion of information since it is informal and casual. It is found that information discovered by chance, whether during casual conversation with colleagues or while attending a conference, may be valued more highly than information sought specifically or deliberately, because it seems to have been elicited spontaneously.

It then seems likely that scientists concerned with the use of drugs rate personal contacts higher as an information source than documentary sources such as original papers or reviews, while workers in other medical and pharmaceutical fields feel it is quite another story.

3. 'Intermediaries' for production personnel. A step further, in terms of information dissemination, there is a mechanism by which the process is heavily dependent on a few individuals, often prominent figures, who function as 'intermediaries' and direct the orientation of the information flow. This is a small, close community and a very stable one at that. On the one hand, they meet together regularly or irregularly to make comments on a certain topic or exchange views on an event. They inform each other of current events and share experiences and lessons and then pass these on to

their fellow workers. On behalf of their own institutes or at least their project teams, they internally monitor current literature because they have the time and privilege to read foreign materials. This is the role played by certain authoritative individuals and supports Price's dogma 'success breeds success'.

Extent and potential of information requirement

Thanks to the principles of giving priority to manufacturing medicinal materials and drugs which can cure the diseases most seriously affecting people's health, China (with an area of 9,600,000 km^2 and a population of 1,000,000,000) has made much progress to improve the health of the nation. Life expectancy has reached 72.5 years. The pharmaceutical industry has been developed to a stage where it has the capacity to satisfy the increasing demands of the nation. According to the statistical data in 1984 (Shen, 1986), the output was substantial (*Table 19.3*).

However, considering the vast area in the countryside and the terminal disease profile (in the order: heart attack, cranial vascular diseases, cancer, respiratory system diseases, digestive system diseases, wounds, poisoning, pulmonary tuberculosis, urogenital diseases and other newly emerging diseases), the pharmaceutical industry in China is far from perfect, and is very conscious that it:

- lags behind in comparison with the developed countries;
- requires a great deal of information produced outside China in order to place it in the forefront of the field;
- pays great attention to disseminating information in terms of cost-effectiveness because of the constraints of the present information retrieval services.

Table 19.3. Pharmaceutical industry in 1984.

Industries	Number of factories	Output (million yuan)
Chemical pharmaceutical industry	800	9,095
Medical apparatus equipment industry	291	786
Medical material industry	51	258
Pharmaceutical machinery industry	28	86
Chinese patent drugs industry	481	1,907
Others	48	226
Total	1,699	12,340

As far as the pharmaceutical industry is concerned, China is a large potential market. The market seen both from inside and outside is worthy of investment. Information services will play an important role in its development.

Information sources

Pharmacopoeia of the People's Republic of China

The *Chinese Pharmacopoeia* was first published in 1930. The first edition of the *Pharmacopoeia of the People's Republic of China* (*PPRC*) was published by the MPH in 1953. Then came the second and third editions in 1963, and 1977 respectively.

The latest edition, *PPRC 1985*, was published in two volumes in accordance with the terms of a Convention held in November, 1979 by the new Pharmacopoeia Commission.

In order to make preparations for the fourth edition, a new Pharmacopoeia Commission was appointed by the MPH on the approval of the State Council in 1979. The Pharmacopoeia Commission consists of 101 members who have acquired weighty scientific reputations in medical and pharmaceutical sciences, with Dr. Xinzhong Qian, former Minister of the MPH, as its chairman. In November, 1979, the Commission held a general session and laid down the provisions for the elaboration of *PPRC 1985*. The Commission believed that the role of the fourth edition of the *PPRC* in providing publicly available standards that apply to a product at any time during its shelf-life is of considerable value in developing the pharmaceutical industry and safeguarding people's health.

PPRC 1985 covers 1,489 specific drugs. Volume 1 contains 713 monographs describing Chinese traditional and herbal drugs and set prescription preparations, whilst Volume 2 deals with 776 monographs on medical and pharmaceutical substances, such as chemical drugs, antibiotics and biological products, etc. All substances and products included in the *Pharmacopoeia* are listed in the four indexes at the end of each volume: indexes arranged by Chinese generic name according to Chinese scripts, Chinese generic name based on the 'pingying' system and, therefore, arranged alphabetically; by Latin name and by its proposed name in the Latin form.

The opportunity has been taken to introduce a number of changes in style throughout the fourth *Pharmacopoeia*. Most notable among these has been the omission of certain monographs for materials that were present in the previous edition. *PPRC 1985* contains 455 Chinese traditional and herbal drugs, 281 less than

PPRC 1977. Furthermore, a considerable number of Latin names of drugs have been changed for the purpose of simplicity and accuracy which could most explicitly describe the characteristics of specific drugs. The nomenclature is used in such a way that a modifier has been used to precede the generic Latin form. For example, Rhizoma Zingiberis Recens is used instead of Rhizoma zingiberis; here the adjective 'Recens' is used to imply the meaning of 'fresh' which is certainly in accordance with the fact (Luo, 1987). Additionally the International System of Units (SI units) has been introduced wherever practicable. An approximate equivalent expressed in the traditional way is given in parentheses for the convenience of users. For example, atmospheric pressure is expressed in the form 101.3 kPa (760 mmHg) (Ren and Zhang, 1987).

Periodicals

Although there is a great amount of literature published in the form of confidential internal reports, because of the nature and characteristics of the pharmaceutical industry, the periodical literature is still the preferred way for scientific communication. The very rapid development of techniques, concepts and factual knowledge in the pharmaceutical industry means that any delay in publication would result in devastating effects – failure in competition with rivals.

Consequently, publication in scientific periodicals is of the highest importance. Furthermore, the pharmaceutical industry has deep roots in medicine, biochemistry and chemistry, etc. and this interdisciplinary nature also demands publication in the periodical literature. There is a large and increasing number of major journals covering the pharmaceutical industry in China. The most important of these (with the year of foundation and publisher) are:

Acta Pharmaceutica Sinica (1965), a journal of very high standards and the preferred channel of publication for most Chinese pharmacologists. It is published monthly by the Chinese Pharmaceutical Association.

Chinese Pharmaceutical Bulletin (1965), produced monthly by the Chinese Pharmaceutical Association. It carries research news, information on new drugs and reviews etc. besides original papers and therefore it is highly appreciated by the Chinese research workers.

Journal of China Pharmaceutical University (1956), one of the leading journals dedicated to the pharmaceutical sciences. Its original name was *Journal of the Nanjing College of Pharmacy*, which first appeared in 1956. Since the renaming of the College as the China Pharmaceutical University in 1987, the journal started to use the present title.

Chinese Journal of Pharmaceutical Analysis (1981), rapidly established itself as one of the most respected scientific journals in the pharmaceutical industry. It is a bimonthly publication of the Chinese Pharmaceutical Association. As its title suggests, it specializes in analytical techniques such as chromatographic and electrophoretic methods, mass spectrometry, and fluorescence assays.

Bulletin of Chinese Materia Medica, dedicated to research and development of the traditional Chinese medicine. It is published monthly by the Chinese Pharmaceutical Association.

In addition, there are an increasing number of journals devoted to the pharmaceutical industry, of which the most significant are:

West China Journal of Pharmaceutical Sciences (1986, quarterly), School of Pharmacy, West China Medical University;

New Drugs and Clinical Remedies (1982, bimonthly), Shanghai Pharmaceutical Administration Bureau Information Institute;

Journal of Marine Drugs (1981, quarterly), Chinese Pharmaceutical Association;

Chinese Journal of Clinical Pharmacology (1985, quarterly), by the Chinese Pharmaceutical Association;

Journal of Shenyang College of Pharmacy (1984, quarterly), Shanyang College of Pharmacy;

Chinese Journal of Antibiotics (1975, bimonthly), editorial board of the *Journal*;

Chinese Journal of Parasitology and Parasitic Diseases (1983, quarterly), Institute of Parasitic Diseases, Chinese Academy of Preventive Medicine;

Acta Pharmacologica Sinica (1980, bimonthly), Chinese Pharmacological Society;

Chinese Journal of Biomedical Engineering (1981, quarterly), Chinese Academy of Medical Sciences.

In addition to these, there are also several journals purely dedicated to traditional Chinese medicine, of which the best known is *Acta Medica Sinica* (1986, quarterly), All-China Association of Traditional Chinese Medicine. Others are:

Chinese Traditional and Herbal Drugs (1969, monthly), Information Centre of Chinese Traditional and Herbal Drugs, National Pharmaceutical Administration Bureau;

Journal of Traditional Chinese Medicine (1959, monthly), All-China Association of Traditional Chinese Medicine and China Academy of Traditional Chinese Medicine;

Acta Chinese Medicine and Pharmacology (1978, bimonthly), Hailongjiang College of Traditional Chinese Medicine;

Bulletin of Chinese Materia Medica (1975, monthly), Chinese Pharmaceutical Association;

Journal of Traditional Chinese Medicinal Herbs (1975, bimonthly), Information Centre of Traditional Chinese Medicinal Herbs, National Pharmaceutical Administration Bureau.

Finally, there are an increasing number of essentially 'national' journals, some of which are published irregularly, which contribute to the rapid growth and interdisciplinary nature of the field. These include *Chinese Journal of Hospital Pharmacy, Modern Applied Pharmacy, Chinese Journal of Pharmacology and Toxicology, Bulletin of Chinese Pharmacology, Pharmaceutical Industry, Pharmaceutical Design, Research in Chinese Patent Medicine, Pharmaceutical Information Communications, Journal of Zhejiang Pharmacy, Journal of Medical Apparatus and Instruments* and *New Drugs and Therapeutics.*

Abstracting and indexing services

Sources of information concerning guides and indexes to the literature in the pharmaceutical industry are principally in the relevant sections in *New Bibliography* published by the Chinese Book Import and Export Co-operation and *Index to National Periodicals and Newspapers* by the Shanghai Library. These services are of value both to scientists and librarians.

The only service exclusively devoted to the pharmaceutical industry is *Chinese Pharmaceutical Abstracts*. It is published bimonthly by the Institute of Pharmaceutical Technology Information, National Pharmaceutical Administration Bureau. The coverage of more than 220 Chinese journals which cover all aspects of the field speaks for its authoritative status in China. Furthermore, the *Abstracts* is one of the first to be published with the assistance of computer typesetting by the Institute of Scientific and Technical Information of China. Its contents are also held on magnetic tape and a card output form is available. It currently publishes approximately 1200–1500 abstracts per issue. The abstracts are generally well written and informative and, arranged in classified order with author, generic name and subject indexes, which are cumulated annually.

Chinese Medical Abstracts is, without doubt, a remarkably comprehensive source of information in medical sciences, but readers will also find it useful for a search of the pharmaceutical literature. It is published in 14 sections (*Table 19.4*). There is wide journal coverage in the series, and the abstracts are of high quality.

There are many other abstracting services providing a conspectus of the literature of particular parts of the pharmaceutical industry. It is worth emphasizing the value of comprehensive services such as the *Chinese Biological Abstracts* (Shanghai Library of the Chinese Academy of Sciences), which reviews the literature on pharmacology, botany and physiology and the *Chinese Chemical Database* (Library of the Institute of Organic Chemistry, Chinese Academy

Table 19.4. The 14 sections of *Chinese Medical Abstracts*

Title	Frequency	Year of foundation	Publisher	Abstracts per issue
Traditional Chinese Medicine	bimonthly	1960	Library and Information Centre, Graduate School of Traditional Chinese Medicine	350
Internal Medicine	monthly	1980	Guangxi Institute of Medical Information	350
Surgery	bimonthly	1981	Jiangsu Institute of Medical Information	650
Pediatrics	bimonthly	1982	Liaoning Institute of Medical Information	200
Birth Control and Gynaecology	quarterly	1982	Sichuan Institute of Medical Information	250
Basic Medicine	bimonthly	1984	Shanghai Medical University	350
Skin Science	quarterly	1984	Department of Skin Diseases, Xian Medical Universty	300
Otorhinolasyngology	quarterly	1984	Beijing Institute of Otorhinolasyngology	170
Hygiene	bimonthly	1984	Institute of Nutrition and Food Health, Chinese Academy of Preventive Medicine	300
Ophthalmology	quarterly	1986	Beijing Institute of Ophthalmology	150
Nursing	quarterly	1986	Wuhan Institute of Medical Sciences	250
Stomatology	quarterly	1986	Nanjing Medical College	250
Radiology and Diagnostics	quarterly	1987	Tongji Medical University (Hubei Province)	300
Laboratory and Clinical Medicine	bimonthly	1987	Tongji Medical University (Hubei Province)	250

numerous information needs of Australian users of all kinds. The National Library's collection aims to complement resources already held by other libraries. As a national depository it has the largest collection of Australian publications and also has a strong representation of South Pacific resources. Journal subscriptions include many fringe titles and often the National Library is the only Australian source.

State libraries

A State library exists in each State, funded by the State government and often complementing the resources held by other government libraries. Each holds vast reference collections of journals, books, newspapers and microform services as well as providing retrieval tools including abstracting and indexing services together with access to all the major international database networks. The State libraries are available to all but material cannot be borrowed and researchers must work in the library. Computer searches can be undertaken if required and a small fee is charged to cover overheads.

Public libraries

Public libraries operate at the local level with general collections covering a wide range of subject areas. Many of these libraries are computerized and are increasingly accessing computer databases to satisfy the information needs of their users. Public libraries generally serve as the first point of access for information for the general public and when resources are not immediately available for particular enquiries, public libraries tap into the national network of libraries and information services at various levels. A loan service and provision of photocopies operates within the Australian interlibrary system thus permitting access to all available resources.

Department of Health libraries

A Department of Health exists in each state of Australia with a reference library operational for departmental employees. All the major hospitals in Australia have their own medical libraries serving the hospital staff and, additionally, physicians and health personnel in the district.

The Commonwealth Department of Community Services and Health has several specialized information sections which are able to access all the database carriers of interest to pharmaceutical researchers, e.g. Dialog, Orbit, BRS and Data-Star. Dialog is regularly used, other vendors from the Northern Hemisphere have down times that mean Australia is without access for several hours during the working day. The major databases searched include *Medline, Biosis, Embase, International Pharmaceutical Abstracts, Diogenes, Pharmline* and *Pharmaceutical News Index*. The library also has dial-up facilities to the British database of the Department of Health and Social Services through Data-Star which has only recently been installed on the Australian market. Another network used by the Department of Health libraries but not by many others is the German system *DIMDI* (Deutsch Institute for Medical Documentation and Information). These libraries also have good collections of printed pharmaceutical resources.

The Commonwealth Department of Community Services and Health is also responsible for the computer hardware and telecommunications support for the Australian Medline system, the National Drug Information Service database and the Australian Adverse Drug Reactions Committee database.

Academic libraries

At campuses, where pharmacy and pharmacology courses are offered, the collections and computer facilities of academic biomedical libraries are available to students, academic staff and researchers, as well as specialized resources housed in pharmacy department libraries. Departmental libraries, however, do not usually access international databases as university libraries have comprehensive online services. Apart from serving their own campus personnel, these libraries will often undertake searches for the public at a fee covering costs.

Special libraries

Smaller specialized collections support the work of research centres, pharmaceutical companies and similar organizations. These libraries generally only serve their own organization's staff although most libraries participate in library networks and make their collections available through inter-library loan. Many special libraries are well equipped with extensive access to computerized systems.

Information services

Several private information consultancies have been established in Australia in recent years to provide research and literature searches in the medical and scientific area. A leading broker of information in the science and technology area is Scansearch Pty Ltd. which provides scientific, technical, business, patent and engineering information and accesses all major databases. The company offers, in addition, manual searching and document delivery services. Scansearch also currently acts as the contact point in Australia for Data-Star and is one of the few users of Questel-Télésystèmes in Australia, this network is particularly helpful in providing information on world trade marks compared with many other patent databases which are usually region specific. Scansearch have developed their own pool of scientific and technical experts covering a wide range of areas including chemistry, engineering, biology, physics and agriculture.

The CSIRO's Information Resources Unit provides an information service called Search Party offering information work for a fee. Admittedly most of their work covers scientific, technical and business areas, but all subject areas are covered. The team will design and compile databases, both for CSIRO applications or on a contract basis for non-CSIRO clients, which are mounted on the Australis network. Search Party represents STN and has access to numerous other international information systems including Dialog, ESA and Pergamon Orbit Infoline. The main Australian databases are also searched. Search Party provides users with extensive listings of information found on a particular topic, updates searches, provides information on how to locate documents listed and can provide document delivery if required. There are no upfront subscription charges and only costs incurred in retrieving information are passed on.

Another commercially-available information consultancy is Health Information Services which is based in Sydney and geared more to the medical information needs of the general public. The computerized systems used by Health Information Services are Australian *Medline* network, *Cancerlit*, *Dialog*, the *Australian National Drug Information Service* database and *The Australian Adverse Drug Reactions* database and *PDQ* (*Physicians Desk Query*), an online database designed by the National Cancer Institute in the United States to assist physicians in the treatment of cancer patients.

A different type of information service is provided by Intercontinental Medical Statistics (Australasia) Pty Ltd. In Australia, as in many other countries around the world, IMS gathers and publishes marketing information on the pharmaceutical industry. Although there are few competing companies involved in pharmaceutical market surveys, IMS in Australia is the leading company collecting and providing pharmaceutical business information.

Continuous studies and surveys are conducted on pharmaceutical consumption and patterns of disease and treatment. These studies and surveys are known as *Pharmaceutical Audits* and *Medical Indices*. The information is primarily available for the pharmaceutical industry, although access for government, professional and academic researchers and the like can be arranged for suitable enquirers on a commercial basis.

Information exchange in Australian libraries and information bureaux is facilitated by national union catalogues and various library networks.

The Australian Bibliographic Network is a national online bibliographic system based on co-operative participation with representation from every state of Australia and covering state, university, college, public and special libraries, as well as the National Library of Australia. The principal goals of the Australian Bibliographic Network are to assist libraries minimize the costs of developing and controlling their collections, to enhance the national document delivery system and to foster co-operative collection development. The main functions of this network are shared cataloguing and bibliographic verification providing locations for materials held by the various libraries. By showing where books, serials and other materials may be found in Australia, ABN enables libraries to establish co-operative acquisition schemes and to quickly identify other libraries from which specific items may be borrowed via the inter-library loan system.

The Australian Bibliographic Network became an operational system within the National Library of Australia in 1980 and became publicly available in late 1981. ABN is operated on computers located at the National Library of Australia and is made available online throughout Australia via the Telecom Datel communications network and Telecom's Austpac. ABN is a co-operative service through which Australian libraries share their resources on a voluntary basis. By mid-1988, more than 800 libraries were using the services offered by ABN and the database had grown to 5.6 million bibliographic records and 7.2 million holdings. The bibliographic holdings of the system are also produced on microfiche, the current ABN catalogue is called *Nucom 4* and excludes non-book materials. Predecessors of the ABN catalogue were known as *National Union Catalogue of Monographs*) with each stage representing a specific time period.

Locating journals in Australian libraries is somewhat easier than tracing books as several bibliographic guides have existed for many years. The earliest bibliographic guide for locating journals in Australia was the CSIRO's publication *Scientific Serials in Australian Libraries*. This was upgraded in the early 1980s with the production of the microfiche product, *NUCOS (National Union Catalogue of Serials)*. *Nucos* is produced semiannually and includes the journal holdings of national, state, university and college libraries.

Australian collaborative networks

The Australian Library and Information Association is the professional body representing librarians and information scientists in Australia. Each state of Australia has an active association with various interest groups representing specific areas, e.g. the Health Libraries Section which meets in each state on a regular basis, with national discussion usually at two yearly intervals. The Health Libraries section represents government and hospital libraries, pharmaceutical company librarians and other health centre librarians.

Apart from the professional bodies, collaborative networks have been established in most states of Australia in the health and medical areas. In late 1982 librarians from Sydney hospital and allied health libraries held a meeting with a view to establishing co-operative inter-library loan provision. The network of New South Wales health libraries was soon christened Gratis, thus embodying its central concept. It aims to maximize use of available resources within the group, to supply a free and efficient inter-library loan service between members, to ease the pressure on net lenders and to produce a union list of journal holdings of member libraries. New South Wales Gratis now has approximately 70 members representing hospital libraries, government department libraries, pharmaceutical company libraries and other health research organizations. A union list of journal holdings is regularly prepared which is arranged by priority ranking, it is produced in microfiche format with the assistance of the University Co-operative Bookshop Limited and is a valuable reference tool for readily locating journals.

Health library networks have been established in most other States of Australia along the same lines as Gratis and many are producing union lists of their own. Contact has been established between the health networks in the different States so that information exchange can be further extended. Reciprocal inter-library loans have been initiated between Queensland, South Australia, New South Wales, Victoria and the Australian Capital Territory, with each member library having a copy of each network's union list. A Joint Networks Co-ordinator has recently been appointed to ensure dissemination of information and the collection and distribution of statistics. It is hoped that a national union list of journals in the health networks may be produced in the future to facilitate rational use of Australia's resources in the medical and allied health fields.

Hemloc, a bibliographic file, available on the Australian medline network is useful for locating journals and monographs available in major health department libraries up until 1986. After this time a new facility called *Healthnet* was established as a computerized catalogue and acquisitions system for libraries participating in the Healthnet group. The Healthnet database has not been publicly mounted on the Australian *Medline* network and is only available to participating libraries at this time. Also available on *Healthnet* is the *Bibliography of Australian Medicine and Health Services to 1950* which contains over 11,000 entries of both monographs and serials concerning medicine and health in Australia covering the period 1790 to 1950.

Australia at this time is witnessing various attempts at national co-operative networks aimed at the rationalization of resources. A recent Australian Libraries Summit accepted an objective of ensuring an appropriate collection of resources, adequately recorded and accessible for the benefit of all Australians. It was resolved that such a national collection should be based on the principles of a distributed national collection which is the aggregation of all library collections in Australia in the public or private sector.

Future trends and investments

Australia is physically isolated from other English-speaking countries. While Australian science has a high reputation, the results of research originating in Australia can easily be overlooked and it is uncertain whether improved telecommunications will entirely resolve the problems of isolation because scientists and researchers essentially need the funding to attend international meetings and work more closely with fellow scientists in overseas countries.

Satellite technology has allowed Australia access to the information resources of 47 countries in the world to date. Australian researchers, clinicians, information specialists, and scientists have used this technology to gather information to support their own research work and for the general benefit of the local scientific community. However, the one drawback in this encouraging picture is the fact that venture capital which would further advance research and development activities is virtually non-existent. To the present time, government expenditure has provided the facilities for telecommunications and opened up access to information resources. With reliance on public funding, however, changing political circumstances could easily alter fundamental attitudes and priorities and constrain even the successful telecommunications programs.

Government grants for research in Australian universities and research institutions offer some advantages but the disadvantages are more highly publicized. Bequests, donations and private grants from industry help to make up the deficiency brought about by limited government funding. The Medical Research Council of the National Health and Medical Research Council has focused attention on the commercial development of research discoveries but there is a funding gap which hampers the transfer of research results to commercial development.

The developing links between research institutes and industry may facilitate commercial outlets. For instance, the Garvan Institute of Medical Research and California Biotechnology Incorporated, California and C. P. Ventures Limited, Australia participated in the establishment of Pacific Biotechnology Limited (Pac Bio) in June 1987. Pac Bio is a commercial biotechnology company with ambitions to lead the development of a new biopharmaceutical industry in Australia.

There is no comprehensive central register and index of Australian researchers, research projects or clinical research. It is possible to draw upon the material of annual and biennial conferences to gain an insight into current research but individual professional organizations and societies can only provide incomplete information. The pharmaceutical industry has, in fact, more intimate knowledge than most of relevant ongoing research through its contacts with academic, hospital and community environments. However, nowhere is this information collected and collated, and is usually only known by those people working in a particular field (the Invisible College). The Commonwealth Department of Industry, Technology and Commerce did initiate a directory of Australian research relevant to pharmaceuticals in 1984 entitled *Directory of Pharmaceutical, Veterinary and Agrochemical R and D in Australia* which provided details of researchers and their work. Unfortunately, the directory has not been updated since and the problem of locating current research in the pharmaceutical field remains.

There are attempts underway to establish information exchange with major research centres through a recently created Pharmaceutical Section of the Department of Industry, Technology and Commerce and the section is endeavouring to commence a register of Australian research in progress. Likewise, the Medical Council of the National Health and Medical Research Council is trying to redress some of the deficiencies identified and intends to compile a similar register of scientific, medical and technical research. The Australian Society for Medical Research can also be contacted for ascertaining current research in Australia.

There is still a need in Australia for a greatly improved and more closely co-ordinated system to collect scientific and technological information and to disseminate it in an accurate and timely manner. At the moment, Australia is without a national information policy and does not have an integrated information system. No Australian authority is empowered to give financial encouragement to the systematic development of scientific and technological information sources, as is the case in the UK and the US. Australia is also without a national scientific and technological information policy in contrast to such countries as Belgium, Sweden and the Netherlands.

Many of Australia's information services are of world standard but, for a variety of reasons, numerous potential users are inadequately served, or fail to take full advantage of resources and services at their disposal. Libraries are the largest and most highly co-ordinated providers of scientific and medical information services but, again, limited funding which is not commensurate with the continuing expansion in recorded knowledge places the system under strain.

Australia can occasionally experience difficulty and delays in obtaining reports or articles. However, a recently commenced trial of *Adonis*, a document delivery service on CD-ROM at the National Library of Australia may help reduce both costs and time delays. *Adonis* currently supplies 219 biomedical journals published in 1987 and 1988 on CD-ROM.

In these competitive times, scientific and technological knowledge must be accessible to researchers and others to avoid wasteful duplication of effort and to convert this effort into new products or processes in the market place. To some extent this depends upon policies and programmes in government. The efficiency and success of all of these processes is and will be influenced by the availability and effective use of information.

References

Grant, C. and Lapsley, H. M. (1988). *The Australian Health Care System, 1987* (School of Health Administration, University of New South Wales).

Commonwealth Department of Health. (1987). *Annual Report 1986-87* (Australian Government Publishing Service).

Bates, E. and Linder-Pelz, S. (1987). *Health Care Issues* (Allen and Unwin Australia Pty. Ltd.).

Commonwealth of Australia. (1986). *Year Book Australia 1986* (Australian Bureau of Statistics).

Commonwealth Department of Health. (1987). *Health Statistical Supplement 1986–87* (Australian Government Publishing Service).

Australian Bureau of Statistics. (1985). *Social Indicators No. 4 1984* (Australian Government Publishing Service).

Parry, T. G. and Thwaites, R. M. A. (1988). *The Pharmaceutical Industry in Australia: A Benchmark Study* (Australian Pharmaceutical Manufacturers Association Inc.).

CHAPTER SEVENTEEN

Information needs and availability in Southern America with special reference to Argentina

F. ROZANSKI

Introduction

'If you cannot keep on your toes with up-to-date information you are out of the game. A businessman can be ruined overnight if he does not read the newspapers and does not maintain a constant awareness of changes in the political and economic circumstances which may affect his business. The same applies to the scientist because his mission is to identify the relevant unknown questions and unveil the explanations. Both what is known and what is unknown must be documented in the information sources, which must be readily available and properly used if your job is to be accomplished'. These are the recent words of Andres O. M. Stoppani, a senior Argentine scientist, academician, and university professor specializing in research on tropical diseases. 'But, unfortunately in the last decades, the volume and quality of our available sources of information have been neglected and they have seriously deteriorated at a time when sources of information throughout the rest of the world are constantly being handled more efficiently. This is a vital factor that has led to the dwindling level of our contribution to scientific knowledge'.

The reasons for this are not purely economic but poor results from lagging economic activity have played a major part. Latin America began to borrow large sums of money in the latter part of the 1970s and its indebtedness rapidly reached unsupportable levels. Many of these loans were made to Latin American governments or were covered by government warranties. When in 1982, Mexico (whose total indebtedness had reached around $100 billion) announced its default on some of its overdue obligations, the rest of the countries in the region soon followed suit. Argentina's foreign debt is more than US$ 50 billion and Brazil's more than US$ 100 billion. Unfortunately, not even a small proportion of the money borrowed was used in an effort to satisfy information requirements and the Latin American foreign debt is now a severe constraint on access to information.

Research and development is mainly performed in the US, Europe and Japan and the information business is also centred in the Northern Hemisphere, so information on such research has to be paid for in foreign currency, which is in short supply in Latin American countries. In addition, the importance of sources of information for solving problems is a concept which is not well understood and the unfortunate outcome is that even less of the scarce available funds is assigned to information requirements.

Health profile

Some Latin Americans are very sensitive to generalizations about the region which ignore the differences existing among their countries and this feeling is well founded. In fact, although Latin American countries share a common heritage, there are significant variations as regards the extent of economic and social development.

The data in this chapter refer to the Argentine Republic. It is one of the three largest Latin American countries, the other two being Brazil and Mexico. Together with Chile and Uruguay, it is the southern most country in America, and in area, it is the eighth largest country in the world. The population is estimated at 30 million and is concentrated in cities located in the central area and on the coastline. Argentina is at an intermediate stage of economic and social development between the highly developed and very low-income countries. It has a long tradition in research, mainly in the field of medicine, with three Nobel Prizes: Bernardo A. Houssay (1947 – physiology), Luis F. Leloir (1970 – chemistry) and Cesar Milstein (1984 – immunology).

Although Argentina is a relatively advanced country within the region, many of the observations in this chapter are also applicable to other Latin American countries. Whenever possible, comparative data are provided.

It is a fact that the impact of modern medicine has benefited mankind in general and has contributed to an ever improving quality of life. However, when considering the region's health profile, it should be recognized that at least two main groups of problems coexist. On the one hand, there are health problems affecting the majority of the population which are similar to those

Table 17.1. Life expectancy at birth

	Years	Data
Argentina	69.7	1983
Bolivia	50.7	1981
Brazil	63.4	1980/1985
Chile	67.0	1984
Mexico	66.7	1983
US	74.0	1982
Uruguay	70.3	1977
Venezuela	67.8	1983

Source: reference (2).

prevailing in the developed countries, whilst on the other there are those affecting people lacking proper housing and adequate nutrition which are more characteristic of backward rural regions and urban peripheries.

As politician and health specialist Aldo Neri says: 'We live in two epidemiological times. We suffer from the sicknesses which affect high industrialized societies, but we are also affected by those which are prevalent in the low-income traditional societies.'[1]

A comparison between Argentina and a highly industrialized

Table 17.2. The five main causes of death

Argentina 1981	Proportion of all deaths (%)	US 1983	Proportion of all deaths (%)
1. Heart diseases	29	Diseases of the circulatory system	37
2. Cancers	17.3	Malignant neoplasms	22
3. Cerebrovascular diseases	9.5	Cerebrovascular diseases	7.7
4. Accidents	5.1	Accidents	4.5
5. Arteriosclerosis	4.8	Pneumonia	2.6

Source: reference (3).

society such as the US (*Tables 17.1* and *17.2*) reveals similarities in life expectancy and the main causes of death. But, in spite of these general similarities, there is concern about the quality and quantity of available health-care services. In addition, there are imbalances between urban and rural regions, and significant disparity between the different levels of society as a result of socioeconomic conditions.

Infant mortality

Although infant mortality has decreased, reflecting an improvement in the quality of life, it is still significantly high. Latest data available indicates a rate of 33.6 per million live births for 1981, against 62 per million in 1970. To illustrate the imbalances referred to above, in the city of Buenos Aires the rate is 17.7 per million, whereas in the northern province of Salta it reaches a peak of 51.4 per million.[4]

Argentina's figures compare favourably with other developing countries and with the least developed countries where the average infant mortality per 1000 live births is 141.2 but unfavourably with developed societies at 12.8 per million.[5] The most important death-causing diseases are influenza, pneumonia, intestinal infections and septicaemia. Diarrhoeal diseases are significant as are nutritional deficiencies.

The elderly

People are tending to live longer. Heikkinen (1987) when referring to Europe mentions that in ancient Rome people over 45 were called seniores. Nowadays, some people do not like to be referred to as 'elderly' even though they may be over 80. [6] The developing countries have a younger population due to demographic progression and a shorter life expectancy. However, for two Latin American countries, Argentina and Uruguay, Fustinoni (1985) states that birth rates are comparable with more developed societies. People over 65 years old now represent about 9% of the population, against 4% in the 'young' countries, and 12% in the 'old' countries. [7]

The health requirements of the ageing population represent a great challenge for society. The elderly, together with children, are those who demand most medical attention. They are also the sector of population that consumes most pharmaceutical products.

In order to provide health and social care, the Argentine government has the PAMI programme which covers about 2.6 million people.

Table 17.3 Evolution in the use of prescribed medicines in Argentina according to therapeutic groups 1973–1982

	December 1973 (%)	December 1982 (%)
Cardiovascular therapy	7.2	14.4
Psycholeptics (tranquillizers, hypnotics, sedatives and neuroleptics)	9.2	10.2
Antibiotics	9.4	7.6
Dermatological	4.0	5.5
Anti-inflamatory and anti-rheumatic	3.0	5.4
Vitamins	6.2	4.3
Against flu and coughs	3.1	3.8
Analgesic	3.9	3.3
Psychoanaleptics (anti-depressives, psychostimulants and neurotonics)	3.1	3.1
Antispasmodics	2.8	3.1
Subtotal	51.9	60.7
Other therapeutic groups	48.1	39.3
Total	100.0	100.0

Source: reference (8).

The working population

Information on fatal diseases is available (*Table 17.2*), though less is known about morbidity. Consumption of pharmaceuticals may be an indication and this is presented in *Table 17.3*. In analysing *Table 17.3*, it must be appreciated that usage follows the existing trends in more developed societies. Accidents, mental health and hearing problems are fields in which public health specialists feel there is a pressing need for improvement.

Infectious and parasitic diseases. Tropical diseases

In spite of the fact that Argentina is not located in tropical Latin America, infectious and parasitic diseases have a severe impact on health and the quality of life.

According to Garaguso (1988), it is believed that about 3 million people are infected with Chagas disease. The majority of Chagas infections are transmitted by triatomid bugs, but an increasing number of cases of infections result from blood transfusions or are acquired congenitally. The number of infected people in the South American continent is estimated to be in the range 16–18 million. As Garaguso states, parasitic diseases are related directly to poverty, malnutrition, poor housing conditions, deficient water supply and sanitary facilities, as well as illiteracy. [9] Other relevant endemic diseases in the region are hemorrahagic fever, uncinariasis, leishmaniasis, tuberculosis, malaria and hidatidosis (white cancer).

Factors affecting Argentine health policies

Some of the challenges facing the health sector in Argentina have been briefly described. It is widely accepted that the level of health in a society is closely related to its economic development.

But there are other factors, and besides the foreign debt crisis mentioned above, there are further considerations which influence the availability and dissemination of information.

International influences: WHO/PAHO (World Health Organization and the Pan-American Health Organization)

These two international organizations sponsor health policies which have a profound influence on the decisions taken by the authorities of developing countries including Argentina. Unfortunately, there is confusion on the practical measures which should be adopted to implement WHO's strategies. Their slogans are often used as a way of giving precise content to vague and abstract ideologies characteristic of domestic politics. In this context, it is worth noting that many health officials regard the egalitarian goal of 'Health for all by the year 2000' (to be attained mainly through primary health care and essential medicines) as the essence of the policy to be pursued by the governments in this area. Paradoxically, the content of the egalitarian goal itself is subject to different and often antagonistic interpretations. To illustrate this, the list of basic pharmaceuticals prepared by WHO, may be considered as a guideline to a minimal set of pharmaceuticals everyone should have access to, or alternatively, as an indication that these basic medicines are the only useful ones because they can allegedly treat most of the diseases. The latter is a biased interpretation which pays little attention to the potential of pharmaceutical research as a means of securing future health-care improvements and ignores or undervalues the crucial role of past research and industrial efforts in achieving today's medicines. Needless to say, the philosophical stance adopted has definite consequences for the type of information seen to be necessary.

WHO's Revised Drug Strategy approved in 1986 may represent some advancement. In this new strategy WHO acknowledges that clinical pharmacology is relevant to primary health-care. In addition, WHO's new strategy includes the promotion of collaborative research aimed at developing pharmaceuticals that are badly needed in major health areas such as tropical diseases, enteric disorders and human reproduction. It would have a very positive effect if WHO/PAHO emphasized even more the role of research in these spheres.

Political ideologies: government intervention

Due to the history of the country and to the impact of political ideologies, government intervention is likely to occur in all aspects of Argentine social and economic life, thereby reducing the scope for private initiative.

The perception of health as a fundamental human right is the reason given to increase the role of the government and as a consequence, all health-care services are heavily regulated. Industries and services supplying the health sector may be owned by private citizens but they work mostly for official or quasi-official institutions (i.e. social funds established by law and run by the labour unions which receive compulsory contributions from the employees and employers).

Under these circumstances, what the government believes and dictates is overwhelmingly important; the public official and not the consumer is the one to be satisfied. As a result of chronic political instability and the transition from military dictatorship to civilian rule, health policies and practices have fluctuated widely and the generation and availability of information has been affected. Therefore, the political factor is critical and may even be more persuasive than the scientific, academic or industrial requirements.

Nature of health services

In Argentina, health-care is provided by means of three main channels:

i)   the government institutions (hospitals, dispensaries and the like) which are administered at national, provincial or city levels;
ii) the social security or social fund agencies (Obras Sociales) which are administered by the government or labour unions. As they handle

significant economic resources, they exert a strong political influence on health-policy decisions;

iii) the private sector, formed by independent professionals and the private hospitals, health-care centres, clinics, and the like, which offer services directly to the public, in addition to rendering services to social funds.

Total health expenditure was estimated at US$ 5.4 billion in 1985, representing 8.2% of the Gross National Product (GNP).

By comparison, Ginés González García *et al.* (1987) state that Brazil spends 5.6% of GNP, the UK 6.2% and the US 10.7%. In Argentina expenditure per capita was US$ 178 in 1985, against US$ 96 in Brazil, US$ 531 in the UK, and US$ 1646 in the US.[10]

Of the total health expenditure, government disburses 23%, the social funds (with the money received from employees and employers) 39%, and the private citizen, by means of its direct payments, 38% (*Table 17.4*).

In developing countries, the incidence of pharmaceuticals in total

Table 17.4. Argentina in 1985

Total health expenditure	US$ 5.4 billion
Per capita health expenditure	US$ 178
Pharmaceutical sales	US$ 1.6 billion
Per capita expenditure in pharmaceuticals	US$ 53
Percentage of pharmaceuticals on total health expenditure	US$ 29.6%

Source: reference (11).

Table 17.5. Argentina in 1985. Percentage health and pharmaceutical expenditure by sector

	Total health expenditure	Total pharmaceutical expenditure
1 Government	22.7	6.3
2 Social funds	39.2	26.6
3 Private citizens	38.1	67.2
Total	100.0	100.0

Source: reference (12).

health expenditure is higher than in developed countries. There are several factors which influence this situation. For instance, the total amount spent in health is much lower in developing countries, and most raw materials, machinery, technology, information and some finished medicines must be imported. The private citizen bears almost 70% of the total expenditure on pharmaceuticals (*Table 17.5*).

The situation regarding pharmaceuticals in Argentina may be summarized as follows:

– medicines have an important role in the health sector, as they represent a significant portion of total health expenditure;
– notwithstanding the importance of medicines, the government and the social funds finance a relatively small part of the cost of medicines, compared with the portion of the total health bill they pay;
– pharmaceutical sales in Argentina represent only about 1.5% of the world pharmaceutical market. The whole of Latin America consumes approximately 7% of world sales. Thus, whilst the consumption of pharmaceuticals in Latin America may seem high from a domestic standpoint they are relatively low compared to sales in the US, Europe and Japan.

Sources and users of information

The pharmaceutical industry

The more innovative pharmaceutical companies have not only been able to generate and develop inventions, but have also transferred know-how to developing countries. Many of the modern medicines used in industrialized societies are also available in the developing Latin American countries.

The large pharmaceutical research-oriented laboratories from the US, Europe and Japan play an active role in the South American market. They manufacture medicines or license their technology and they submit a wealth of scientific and technical information regarding their latest achievements to the government control institutions, who must grant official authorization before the medicines can be manufactured or sold.

Because they are highly research-oriented, these manufacturers engage in local sophisticated clinical research in order to determine the efficacy and toxicity of their new products and provide abundant medical literature for the promotion of their products in the market. In summary, these laboratories are the main and most dynamic source of information, both for health authorities and for medical doctors.

Local companies, large, medium and small firms, are also very active. In Argentina, there are approximately 200 laboratories who compete amongst themselves and with the international subsidiaries. However, local companies (unlike the internationals) are not a good source of information on new developments because they concentrate on promoting and supporting the products they market. International affiliates receive the information they need from their headquarters, and also generate information from the results of the local research undertaken, mostly clinical research.

In the case of local laboratories it is different. They may work under licences from innovators, in which case they are also provided with the information they require, which in turn is submitted to the authorities and circulated to suppliers and consumers. However, as there is no effective patent protection to cover inventions, few licence contracts are awarded and many local firms prefer to engage in copying the successful pharmaceutical formulae of the innovators.

In order to be able to engage in imitating original medicines, the local laboratories have extraordinary information requirements. They must study new compounds, their efficacy and safety problems, their market potential and, where these new drugs can be obtained (generally from unauthorized dealers) they must copy the preclinical and clinical information for registration purposes, the manufacturing techniques and advertisement brochures.

Professional organizations help locate all the required information, and the large local laboratories have their own people abroad surveying useful information. Also, there are manufacturers whose business is to produce compounds for copiers all over the world. Industrial spying and free-riding (taking advantage of the researchers' efforts) is facilitated by weak intellectual property protection.

Therefore, there are two parallel worlds. On one hand, there is the regular network of pharmaceutical information generated by the pharmaceutical producers and the scientific community. But, on the other, there is the irregular or parallel information network which is very active as a source of information for copiers of the original research. Research documentation, patents, publications, the wealth of data on new compounds presented at scientific meetings and congresses together with the data submitted to the government control institutions (the list is not complete) are used for both purposes. Technology piracy has become big business and an issue of increasing concern for those who work and who invest in research.

The authorities help the local firms by lowering their requirements for information which must be submitted for the granting of authorization for production and marketing, in order that the copiers may obtain the official approval to manufacture their imitations. Hundreds of these copies can be found in the market. The serious consequences of this irregular behaviour to both public health and research activities requires no further elaboration.

When well known laboratories, such as Squibb or Eli Lilly, close their affiliates because of political and economic reasons, the community is widely affected, as they are not only manufacturers and promotion centres, but play an important role in providing information and training.

In summary, information on pharmaceuticals originates in the US, Western Europe and Japan and reaches South America through the regular and irregular networks.

However, there is some information which is generated locally. A few local companies have made limited efforts to design new chemical and biological agents, and they have filed patents to protect their inventions in the industrialized world. Likewise, some independent university researchers also engage in the design of new drugs. Preclinical research in animals, although not very frequent, is mainly carried out to satisfy regulatory requirements. Clinical research is more widespread and some researchers participate in worldwide multicentre studies under the supervision of the headquarters of international pharmaceutical companies to confirm with registration procedures and for marketing purposes. Finally, technical development for local production is required in order to adapt information received from abroad. As most active ingredients are not manufactured locally but have to be imported, most technical problems are in the area of formulation. Production and storage activities originate technical problems which require creative solutions. There is exchange of technical information between suppliers and producers.

These activities create information which may be published locally by the professional bodies, the trade associations, and in specialized technical newspapers (see General Appendix).

The scientific community

Argentina's research in medical science is significant and much research is financed by government funds. Of the units engaged in research, 85% belong to the government and an additional 6% are quasi-governmental.[13]

In spite of its achievements, it is generally recognized that research in the field of health is at a critical point and this view is supported by the lack of continuity in research lines, the obsolescence of equipment and the shortage of adequately skilled manpower. The brain-drain of researchers is probably a direct reflection of this situation.

Unreliable data circulate as to how much is invested in research.

Nature has estimated that it amounts to US$ 350 million per year.[14] CONICET (the National Board of Science and Technology) employs about 250 researchers in medical sciences. There are about 200 scholarships granted in the field yearly and 30% of all CONICET's subsidies go into medical research.[15]

The National University of Buenos Aires, the largest in the country, employs about 500 researchers in medical science, working on 300 projects. Several other national universities located in the provinces also undertake projects. The special Programme for Research and Training in Tropical Diseases is financed by UNDP/World Bank/WHO, mainly in the field of Chagas disease.[16]

The Secretariat of Science and Technology published (1985) a list of 128 government or quasi-governmental units or institutes which offer services in pharmacological research. *Guía de Servicios en Farmacológica y Farmacoquímica* (*Directory of Services in Pharmacology and Pharma-Chemistry*) is a useful source of information, and identifies the name and address of the units, the names of the directors, has very brief references on the research projects undertaken, the techniques applied, and states whether they have a library and which services they offer pharmaceutical companies.

These research centres are generally small and lack resources. They are fiercely independent and resist attempts at integration into larger and more efficient centres. The most specialized in pharmacology are ININFA (Institute of Pharmacological Research), with premises at the School of Pharmacy of the University of Buenos Aires, and CEFAPRIN (Center of Studies in Pharmacology and Natural Substances), located in the premises of a private foundation. These are but two of the 128 listed. For their information, the research units depend on their own libraries or on central libraries.

Researchers find many barriers in the pursuit of information which they need not only for their projects but also for teaching purposes. These obstacles adversely affect their work and productivity. The picture is very different from the one described when referring to the pharmaceutical laboratories. Amongst the most frequent difficulties are the following:

- long delays by libraries in making available the material that is requested. Most of the delays are due to lack of funds to buy publications, to management difficulties, to shortage of manpower, to space problems, and the like. The researchers' desire is to be able to buy or subscribe to the publications they need most, which is seldom possible because of the costs involved;
- interruption of subscriptions to important journals due to lack of resources;
- non-availability of specialized books and publications that cannot be found in any library;

- dispersion of libraries scattered in distant premises. Co-operation and exchange among them exists, but is very slow and inefficient;
- libraries receiving little attention from decision-makers and they do not have adequate budgets to cover their needs.

Private foundations and pharmaceutical companies bring some relief to the situation. International organizations also help but the requirements are enormous and to date remain unfulfilled.

Regulatory authorities

Every pharmaceutical product must be approved by the Secretariat of Public Health before it may be manufactured or sold in the country. The laboratories interested in manufacturing the medicines must submit information which, according to existing regulations, is necessary to prove the efficiency and safety of the preparations. The government agency in charge of analysing this information is the Instituto Nacional de Farmacología y Bromatología (National Institute of Pharmacology and Bromatology).

The Institute's main source of information is the data presented by the laboratories. They also receive reports from the US Food and Drug Administration and from WHO/PAHO on new drugs approved, toxicity, adverse rections, as well as publications such as *Medicines and Therapeutics* and the *Bulletin on Essential Medicines*. There is an *Argentine Pharmacopeia* which is in its sixth edition, published since 1978 by Editorial Codex.

In theory, the Institute should be the leading institution in promoting the availability of up-to-date information, as it is staffed with well trained personnel. But this is not the case, as politics and budgetary restrictions intervene. The Institute does publish an interesting bulletin called *Revista del INFyB* on technical matters. In addition, there is a Latin American network comprising all the official pharmaceutical quality control laboratories which exchange information and publish a bulletin on their activities prepared by the Institute.

The medical profession

The medical profession receives information on medicines mainly from the manufacturing laboratories. Medical representatives who call on doctors to promote their products must be qualified, and as such require adequate training as well as access to information. Their presentations are generally oral, and supported with certain printed material in some cases.

While studying, doctors take a course on pharmacology, and generally use recommended text books such as (i) *Las Bases Farm-*

acológicas de la Terapéutica (*The Pharmacological Basis of Therapeutics*), (Goodman and Gilman, Editorial Médica Panamericana); (ii) *Farmacología Experimental y Clínica* (*Experimental and Clinical Pharmacology*) (M. Litter, Editorial El Ateneo); (iii) *Farmacología: Bases Bioquímicas y Patológicas* (*Pharmacology: Pathological and Biochemistry Bases*), (Bowman and Rand, Editorial Interamericana); (iv) *Interacciones de las Drogas* (*Drug Interactions*), (P. D. Hansen, Editorial Médica Panamericana).

Members of the medical profession generally attend a wide variety of scientific and professional congresses. FECIC, a private foundation promoting science and culture, publishes a yearly guide with the list of scientific and technical meetings. In 1987, there were 268 listed, many of which were on medical and pharmacological subjects.

The professional associations, and in particular the Argentine Medical Association, have prestigious libraries. Nevertheless, there are many complaints about the quality of information. It is felt that little is provided on subjects such as the rational use of pharmaceuticals, adverse drug reactions and drug interactions, while too much emphasis is placed on promotion. There are several lists of the medicines on the market which are given in the General Appendix.

Published literature and libraries

The professional research and technical staff of the business firms subscribe to key international periodicals: about 20 good foreign magazines are received at each of the main laboratories. Publications such as *Scrip*, *Inpharma*, *Drugs of Today*, *Drugs of the Future*, *Lancet*, *British Medical Journal*, *New England Medical Journal*, *Presse Medicale*, *Hypertension*, *Currents Contents* and *Chemical Abstracts* are often quoted.

Several important laboratories have libraries for their own use and some of these are open to the health professionals. Some have access to databases in the US and Europe. International affiliates are also connected to the databases of the parent company and receive ample information on their products. Several important laboratories print publications of their own, dealing not only with their products but also with subjects of interest to health professionals. They also help research centres and institutions of learning (such as the medical schools) to publish their journals.

BIREME, the Latin American and Caribbean Centre of Information in Health Services, was created in 1967 by PAHO and operates the Regional Library of Medicine at Rua Botucatu 862, 04023 Sao Paulo, Brazil. BIREME uses the *MEDLINE* database to service requests originating from the Latin American region. It has created its own database *LILACS* and consolidates Latin American literature on health sciences. BIREME issues two publications: *IMLA* (*Index Medicus Latinoamericano*) printed quarterly with abstracts taken from 300 Latin American periodicals, and *LILACS – SP*, including documents which have been incorporated into the *LILACS* database. In addition, BIREME's *Information Bulletin* deals with the Centre's activities. The General Appendix lists the libraries in Latin American countries that co-operate with BIREME as the National Co-ordination Centre. The General Appendix lists more than 300 Latin American publications included in *IMLA*.

The Association of Argentine Biomedical Libraries (ABBA) is an association formed by about 80 libraries throughout the country specialized in the medical field (see General Appendix). Pharmaceutical laboratory libraries are also included in this list.

In the field of medicine the most prestigious libraries are those belonging to the National Academy of Medicine, which participates in the BIREME Regional Centre, and the Argentine Medical Association. In the pharmaceutical sector, the most renowned is the library of the School of Pharmacy and Biochemistry of the University of Buenos Aires which has 14,000 books and lists the names of 1000 periodicals. However, very few are complete collections and most are not up-to-date. In 1983, 315 periodicals were purchased and 100 were received as donations but during 1984 and 1985, none could be purchased although it is hoped that in the future 250 titles will be renewed. The library would need about US$ 100,000 a year to keep its best collections up-to-date, but would be satisfied if at least 30% of its budgetary requirements could be met.

Nowinsky, Ripa and Villar (1983) state that approximately 8000 medical publications are published in the world, of which about 600 are from Latin American countries[17]; 300 are covered by BIREME, and 29 are from Argentina.

ABBA issues a *Catalogue of Periodical Publications* in the medical field, but the last edition was published in 1983. It lists more than 3000 titles of periodical publications and the libraries that receive them. Most of the publications come from developed countries. A list of the main local publications, which are usually irregular, is included in the General Appendix.

One of the most valuable publications in the medical field is *MEDICINA* because it includes information on selected research projects but it should be noted that most researchers prefer to publish in journals with an international circulation. As for the medical profession, there is preference for publications such as *La Prensa Médica Argentina*, *Orientación Médica* and *La Semana*

Médica. The most important issues produced by pharmaceutical laboratories are received by the libraries and are listed in ABBA's *Catalogue of Periodical Publications.*

PAHO's publications in Spanish have wider acceptance than WHO's publications in English as they eliminate the language barrier problem. *Medicamentos* is issued on a quarterly basis and translates into Spanish articles from *Drug and Therapeutics, Bulletin,* (London Consumers' Association), *The Medical Letter on Drug Reaction Bulletin* (D. M. Davies, Shotley Bridge General Hospital, UK). *Medicamentos* is available through professional bodies. *Boletín de Medicamentos Esenciales* is a PAHO publication promoting the implementation of the list of essential drugs. In addition, PAHO's *Bulletin* also includes pharmacological information.

In the famous *Ulrich's International Periodicals Directory* (Bowker), more than 500 periodicals from all parts of the world are listed under the headings Pharmacy and Pharmacology. Only two of these originated in Argentina: *Revista Farmacéutica* and *Manual Farmacéutico.* The former is published by the Argentine Academy of Pharmacy and Biochemistry and is the oldest specialized journal in this field. It is indexed by *Chemical Abstracts.* The *Manual Farmacéutico* provides general information on medicines and the list of those currently marketed.

For the businessmen and marketing people, information about what is being sold is essential. In this field, PMA and Close-Up audit the market and publish data for subscribers. There are also market surveys by specialized professional firms.

The impact of databases and the benefits of telecommunication improvements

Databases are designed to make information more easily searchable and they are a blessing at a time when the amount of healthcare information is expanding so rapidly. Thanks to technical improvements in informatics and telecommunications, access to information on a worldwide basis is gradually becoming possible. However, for technical and economic reasons, the Latin American countries can only insert themselves in the international networks and use modern information technology at a slower pace than the more industrialized countries.

The Secretariat of Science and Technology published (1988) a guide to the Argentine databases in the fields of science and technology under the title, *Directorio de Bases de Datos en Ciencia y Tecnología en la Argentina,* which lists 57 local databases, out of which 8 cover the medical sciences. This guide states that 34 new databases are in the process of development, 6 of which will be in the health field. Over recent years there is a growing tendency to develop and improve databases. UNESCO's recommended software for databases, CDS/ISIS, is being increasingly used by the Argentine databases. One of the major difficulties of these databases is that most of them are not available online because they are not linked to ARPAC, which is the national network for data delivery. ARPAC is the network offered by the State enterprise, ENTEL (the Argentine telephone company), to link computers and terminals both in the country and abroad through foreign nets such as TELENET, TRANSPAC, IBORPAC, EURONET and others. The reasons argued for not making online service available is that the volume of business is small and therefore it does not justify the costs of sophisticated telecommunications.

There are two databases on medicines. *DAME* files essential information on 3000 medicines on the market. The Colegio de Farmacéuticos de la Provincia de Buenos Aires runs a database on 400 essential medicines mostly recommended by WHO. A list of databases in the health services is provided in the General Appendix.

Several institutions offer access to international databases. For example, the Argentine Centre of Scientific and Technological Information CAICYT subscribes to *Dialog Information Retrieval Service* (US), and offers professional assistance with enquiries. A more modest, but similar service is provided by the library of the School of Pharmacy of the University of Buenos Aires. Frequent users have their own terminals for direct communication with databases not only in the US but also in Europe.

There are limiting factors to the use of databases. In the first place, training is required to take full advantage of the benefits offered by these information systems and there is a shortage of expert assistance. Furthermore, the cost of inquiries cannot be easily met by the impoverished research units. In addition the indigenous libraries, as already explained, are not in a position to provide all the supporting full text of the references retrieved and this detracts from potential benefits.

Finally CD-ROM, such as that offered by Cambridge Scientific Abstracts (*COMPACT CAMBRIDGE MEDLINE*), is gaining acceptance among users because of its lower costs by comparison with online enquiries.

The professional bodies

The professional organizations afford good opportunities to their members to obtain up-to-date information and to participate in discussion groups. Qualified professionals and scientists are members of both foreign and domestic professional organizations and receive their publications. The main local professional pharmaceutical bodies are: Academia Argentina de Farmacia y Bioquímica, Asociación Bioquímica Argentina; Confederación Farmacéutica Argentina (grouping all the associations representing pharmacists); Asociación Química Argentina; AMAIFA, Asociasión de Médicos de la Industria Farmacéutica (medical doctors working in the pharmaceutical industry); CEDIQUIFA, Centro de Estudios para el Desarrollo de la Industria Químico-Farmacéutica Argentina (multidisciplinary study centre for the development of the pharmaceutical industry); SAFYBI, Sociedad Argentina de Farmacia y Bioquímica Industrial, SAFE, Sociedad de Farmacología Clínica y Experimental; Sociedad Argentina de Farmacéuticos de Hospital; SAFyT, Sociedad Argentina de Farmacología y Terapéutica and Sociedad Argentina de Marketing Farmacéutico.

The Colegio de Farmacéuticos de la Provincia de Buenos Aires (professional association of pharmacists) is active in the information field. In addition to its database, already mentioned, it has a co-operative agreement with the University of Iowa that provides approximately 1000 articles per annum on microfilm, indexed according to drug and disease. There is a connection to the Dialog and Delphi lists and this service is offered to interested parties. The Association produces two publications, *Acta Farmacéutica Bonaerense* on scientific and technical matters and *BIFASE*, with educational abstracts. In addition, there is a monthly *Bulletin* containing practical information for pharmacists.

CEDIQUIFA, the Study Centre for the Development of the Pharmaceutical Industry in Argentina, is mostly concerned with stressing the role pharmaceutical research plays in health-care improvements. It produces three publications: a *Newsletter*, a *Bulletin* and a *Technical Magazine*. These contain general information and comments, and stress the importance of introducing patent protection for inventions. They deal with registration issues and subjects such as original drugs versus copy-cat medicines and bio-equivalence.

The trade associations

Several business associations represent the firms in the pharmaceutical sector. One of the oldest is CAEME, Cámara Argentina de Especialidades Medicinales, which was founded in 1925. The members are mainly international pharmaceutical manufacturers. CAEME forms part, at the national level, of FAIS, the federation grouping the health industries. They also form part at the inter-

American level of FIFARMA and at the international level of IFPMA, International Federation of Pharmaceutical Manufacturers' Associations, located in Geneva. As CAEME is a member of IFPMA, they abide by the regulations of the Code of Ethics approved by IFPMA.

The domestic firms are mostly represented by CILFA and CO-OPERALA. At the Latin American level, CILFA is associated with ALIFAR, Asociación Latinoamericana de Industrias Farmacéuticas. Those firms which manufacture active ingredients for pharmaceuticals also participate in PROQUIFARMA, or its counterpart CAPDROFAR, depending if they are international or domestic companies. Importers of bulk active ingredients for pharmaceuticals are also members of the Cámara Argentina de Productos Químicos.

All these trade associations provide ample information on economic and industrial matters.

Conclusions

The highly research-oriented international pharmaceutical laboratories with local subsidiaries are the most fruitful sources of up-to-date information on therapeutic agents. Unfortunately, the survival of some of these sources is threatened. During the 1980s, several local subsidiaries, mostly of US origin, closed down their operations in Argentina due to legal and economic reasons. Other companies may take a similar decision in the 1990s. One of the main factors affecting these firms is the inadequacy of Argentine legislation which does not protect intellectual property rights effectively coupled with the implementation of political and administrative policies which ignore the distinction between original and pirate medicines. While the former medicine is backed by a wealth of scientific information which is the result of long and sound research, the pirate medicine is based on whatever information is available regardless of its quality.

The possibilities of obtaining information for the scientific community are restricted by the lack of an appropriate financial support and because of inefficiency in the public sector. The former problem is not easily rectified in view of the economic difficulties the country faces. In addition to the many constraints already discussed, the lack of resources slows down the incorporation of microcomputers in the research centres which could speed up the flow of vital information. As to management in the public sector, there are increasing signs of changing attitudes. The research community is pressing to participate more actively in research projects, both locally and in co-operation with foreign institutions. To this end, there is a need to have extensive and up-to-date information. These pressures, as well as the general public dislike of the so-called 'brain-drain', may eventually succeed in bringing about marked improvements in the relevant policies.

The medical profession and the consumers are demanding more copious and unbiased information on the rational use of drugs and on adverse reactions. Unfortunately, first, the local publications are not as numerous as necessary nor is their circulation sufficient in view of the importance of pharmaceutical products in the health area, and, secondly, the coexistence of original and pirate medicines creates confusion and uncertainty for prescribers. For instance, there is little information on the safety and efficiency of pharmaceutical imitations.

In Argentina there is an increasing concern about the causes of the country's problems and their solutions. Isolation has already been acknowledged as a deterrent to the country's development. It seems likely that democratic and stable government, being established towards the end of the 1980s, will wish to promote international co-operation. Deregulation of the public sector has also become a common theme. Economic advisors are recommending the establishment of the principles of competition in the health-care services in order to improve quality.[18] All this is encouraging.

It is reasonable to think that the information flow would be accelerated if:

- WHO/PAHO launched a more positive campaign on the value of pharmaceutical research;
- the strategy of promoting essential drugs was clearly to acknowledge that the development of new and original medicines is also a goal to be actively pursued and promoted;
- political decisions to suppress pirate medicines were taken soon, thus encouraging investments in research activities by more laboratories and at the same time the implementation of stricter quality requirements by the national control authorities.

This would encourage investment in and wider use of new information technology which would help close the gap with the more advanced nations.

It must always be remembered that the aim should be to help those who are suffering and to preserve the health and longevity of all.

References

(1) Neri, A. (1982). *Salud y Política Social*, p. 83 (Hachette).
(2) World Health Organization. (1986). *World Health Statistics*, WHO pp. 47–48 (WHO).
(3) Ministry of Health and Social Action, PAHO/WHO. (1985). *Argentina: Descripción de su situación de salud*, p. 401 (Ministry of Health and Social Action); World Health Organization. (1986). *World Health Statistics* pp. 254–261 (WHO).
(4) Ministry of Health and Social Action PAHO/WHO. (1985). *Argentina: Descripción de su situación de salud*, p. 97 (Ministry of Health and Social Action).
(5) World Health Organization. (1986). *World Health Statistics*, p. 45 (WHO).
(6) Heikkinen E. (1987). 'Health implications of population aging in Europe', *World Health Statistics*, **40**(1), 22(WHO).
(7) Fustinoni, O. (1985). 'Geriatría '85', *Hoja de Noticias*, **46**, 2 (CEDIQUIFA).
(8) *Statistics prepared for the Business Association* (Cámara Argentina de Especialidades Medininales (CAEME), unpublished data.
(9) Garaguzo P. (1988). 'Casi un venenon cotidiano', *Revista la Nación*, 24 January 1988, p. 22. (La Nación).
(10); (11); (12). González García, G., Abadie, P., Llovet, J. J. and Ramón, S. (1987). *El gasto en salud y en medicamentos Argentina 1985* pp. 96–97; p. 118; pp.103 and 116 (CEDES, Centro de Estudios de Estado y Sociedad).
(13) Secretaría de Ciencia y Técnica. (1983). *Boletín 4*, 18 (SUBCYT).
(14); (15); (16). As quoted by Ministry of Health and Social Action, PAHO/WHO (1985). *Argentina: Descripción de su situación de salud*, pp. 322 and 325 (Ministry of Health and Social Action).
(17) Novinski, A., Ripa, J. C. and Villar H. (1983). *Biblioteca Hospitalaria, Medicina Sanitaria y Administración de la Salud. Vol. II Atención de la Salud*. p. 730 (ed. A. Sonis, El Ateneo).
(18) Panadeiros, M. (1988). 'Sistema de atención médica en la Argentina. Propuesta para su reforma, pages 1 and 2, FIEL' *Documento de Trabajo no. 17* (Fundación de Investigaciones Económicas Latinoamericanas).

CHAPTER EIGHTEEN

Information needs and availability in Central African countries

S. BELL

The African context

The purpose of this chapter is not only to supply details concerning the sources of information for pharmaceuticals in Africa but also to look at some of the issues which surround information availability in the continent. To attempt to review the present condition of 'information sources and flow for pharmaceuticals in Africa', as if Africa was one homogeneous entity, is ambitious. This is so because of the physical enormity of the region and the immense diversity in historical, socioeconomic, climatic and ecological conditions. Therefore, recognizing that this study must be superficial, the context in which any modern information system, including international pharmaceuticals research, access and distribution must work is briefly underlined.

This chapter will scan the:

- historical context of development, specifically as it relates to health, in Africa (chiefly sub-Saharan);
- development of the information industry in Africa;
- information sources for the pharmaceutical industry at present both within and outside the region;
- possibilities for an integrated and co-ordinated approach to the distribution of information for the future.

At the outset it should be stated that information access, supply and delivery are *not* such great problems to individuals working in what has been termed the North (the US, Japan, Europe and the Eastern bloc) who wish to get information on the African situation and who have access via large institutions to a whole range of specialized information sources. Informing, and indeed misinforming, is more of a problem to those who do not have sophisticated access facilities, i.e. the indigenous populations of the countries which make up the region. This population is often hard pressed to evaluate the objectivity and appropriateness of information which comes before it and is therefore vulnerable to what some authors have argued to be the considerable commercial forces which the modern international pharmaceutical industry can bring to bear (Norris, 1982). It is not the intention here to criticize unfairly those companies and individuals which have made a great contribution to the health of the continent. However, one of the points which will be made is that the pharmaceutical industry does have reason to state with regard to the third world: 'The roots of the problem lie in the sheer poverty of these countries and their people; the international suppliers of drug products cannot be made responsible for the situation' (German Pharmaceutical Manufacturers' Association, 1982).

Nevertheless it can be argued that the suppliers of drugs share a common history with the economic and political forces which have contributed to the poverty prevalent in developing countries today.

Historical background

To understand the present situation as regards the pharmaceutical information industry in Africa a brief review of the historical relationships experienced between the region and the more wealthy countries of the world is presented. The experience can be summed up by what many see as an ambiguous word, development. Various schools of thought exist which attempt to explain the nature of the development process (e.g. Marxism, Stages Theory, Dependency Theory). Here the dependency view of development is elaborated whilst recognizing that other interpretations can be made. Briefly, authors such as Frank argue that the African experience of development has been dominated by the European colonialism of the eighteenth and nineteenth century. It was during this period that trade and power relations between the industrializing countries of what is now called the 'North' and the largely rural economies of the 'South' were set in place. With the arrival of the slave trade: '. . . Africa lost its independence. It began to be shaped according to external requirements, namely, those of mercantilism' (Amin, 1976).

Frank (1978) and Amin (1974, 1976) have indicated the nature of the mercantile relationship between Africa on the one hand and Europe and America on the other. Of particular interest when looking at the internationalization of pharmaceutical information is the work of Amin who indicated (1974) the nature of the triangular pattern of exploitative trade that has accompanied the relationships between the North and South. It can be argued that this triangular form is maintained in the modern information business. Overall, the result of the imperial and colonial experience in Africa was to set in place relationships which have often been characterized by dependency and exploitation.

Historical ramifications have far-reaching implications for the health profile of the region. Africa had its fair share of indigenous diseases, but there is evidence that in the past ecological control mechanisms were used by local populations in order to reduce the severity and incidence of outbreaks. The depopulation of the African interior which occurred with the acceleration of the slave trade and the introduction and spread of new diseases in the eighteenth century is estimated by Kjekshus (1977) to have resulted in millions of East Africans being removed from their homes and being sold into slavery. Kjekshus argues that one of the long-term results of this process was a collapse in the system of ecological control contrived by the native population over many years. This in turn encouraged the spread of tsetse fly and thus sleeping sickness (Ransford, 1983). The important fact to note is that economic and social actions can have widespread effects upon (1) the ecology of a region and (2) the related health profile of that region, and thus (3) the economic and political independence of the region. It is thus possible to argue that Africa's reputation as a centre of sickness and disease is in part a reputation historically perpetuated by the North because understanding of the African health profile was only begun in earnest after the system of ecological control had already broken down (Kjekshus, 1977).

The health profile of the African continent arises from a large number of inter-relating factors but, typically, it is seen as comprising high birth and infant mortality rates, a wide incidence of diet-related illness, low life expectancy and more recently in East and Central Africa particularly with the arrival of the AIDS pandemic, growing mortality among the working population.

More recent developments

Concerning the relationship between much of Africa and the North, although it is no longer openly colonial, as with much of the third world, it is often argued to be largely exploitative (for example Barnet, 1980; Harrison, 1984 to name two examples taken at random). From a brief review of the African situation in the 1970s and 1980s it has been argued that the function in controlling regional

of Sciences) with its coverage of analytical, toxic and spectral papers.

Monographs

It is somewhat arbitrary, in a subject so diverse, to make a selection of significant books and monographs. Moreover, it is more true in the pharmaceutical industry than in most subjects that a book has an inherent tendency to date rather quickly. There are several good textbooks published by the professors of the former Nanjing College of Pharmacy (now the China Pharmaceutical University). *Pharmacy* (Guojie Liu, 2nd edn, 1985, People's Health Publishing House) is a current work which provides a bridge between basic pharmaceutical theory and clinical usage of drugs. *Pharmaceutical Chemistry* (Nanjing College of Pharmacy, People's Health Publishing House, 1978) is published in collaboration by members from the Shenyang College of Pharmacy and the Sichuan Medical College. This acknowledged textbook deals with inorganic, synthetic and natural drugs and their quality control. Similar is *Pharmaceutical Analysis* (Department of Pharmaceutical Analysis, Nanjing College of Pharmacy, Jiangsu Scientific and Technical Publishing House, 1981) which is divided into two parts. *Part 1* describes in detail analytical techniques such as thin-layer chromatography, gel filtration, gas-phase chromatography and infrared spectroscopy, etc., whilst *Part 2* provides information on chemical structure and physical and chemical properties of drugs. The textbook is of value particularly because the authors have drawn on a large number of sources (altogether 1400 references to the work of others) for the information used, relying heavily upon official publications.

Adverse drug reactions and interactions are discussed in *Drug Interactions* (Jingbo Tang, Henan Scientific and Technical Publishing House, 1981) and *Drug Interactions and Warnings* (Shanghai Branch, Chinese Pharmaceutical Association, Shanghai Scientific and Technical Publishing House, 1980). Apart from these, the most important handbook in the area is probably the updated *Handbook of Adverse Drug Reactions* (edited by Haixue Kuang and Lixia Zhang, Heilongjiang Scientific and Technical Publishing House, 1985), which covers the properties of 654 drugs. The format of the handbook has been planned in tabular form for ease in grasping all the pertinent facts at one glance and it is divided into 34 chapters based on pharmaceutical and clinical effects.

A number of large multi-volume treatises covering part of the subject in great depth exist to fulfil the need for background materials or specific items of reference. Foremost is *Collection of Drugs* (Shanghai Graduate School of Pharmaceutical Industry) which was

Table 19.5. 11 volumes of *Collection of Drugs* by the Shanghai Graduate School of Pharmaceutical Industry

Volume	Title	Editor(s)	Number of drugs contained	Date of publication
1	*Drugs Acting on the Cardiovascular System*	Jiaqi Dong and Yinghe Yao	420	1976
2	*Antibiotics and Anti-infectives*	Daizhong Chen and Jichen Zhang	840	1978
3	*Drugs for Relieving Coughs, Eliminating Phlegm and Calming Dyspnea*	Jichen Zhang	238	1979
4	*Anti-allergic Drugs and Immunopotentiators*	Yinghe Yao and Qinglin Liu	165	1980
5	*Antineoplastic Drugs*	Huilian Cui	302	1983
6	*Drugs affecting the Digestive System*	Yongming Xu and Jingwen He	400	1983
7	*Biochemical Drugs*	Qihao Feng	500	1984
8	*Antiparasitic Drugs*	Ziyou Gao and Baozhen Chen	170	1985
9	*Drugs acting on the Nervous System*	Jingmei Sha, Yuegang Cai, et al.	748	1985
10	*Diagnostic Drugs*		257	in press
11	*Drugs for Blood Diseases*			in press

originally a multi-volume treatise, but after revision in 1976, has been converted into a serial publication. It is probably the most authoritative and comprehensive collection, amounting to 4500 drugs manufactured both at home and abroad. 11 volumes have been published (*Table 19.5*). Under each entry, the following is given: Chinese and foreign generic name, chemical name, structure, source, property, effects and usage, side effects and warnings, dosages, etc. At the end of each volume, there are indexes for drugs both in Chinese and English.

Other major reference monographs include:

Pharmacology (Changshao Zhang and Yi Zhang, People's Health Publishing House, 1965);

Methodology of Pharmacologic Experiments (Shuyun Xu, People's Health Publishing House, 1985);

Chemistry of Drugs (Qijie Xiao, People's Health Publishing House, 1964);

Chemistry of Drug Preparations (Guang Tang and Shikun Zhou, People's Health Publishing House, 1981);

Drug Analysis (Department of Drug Analysis, Nanjing College of Pharmacy, Jiangsu Scientific and Technical Publishing House, 1981);

Kinetics of Drug Metabolism (Changxiao Liu, Hunan Scientific and Technical Publishing House, 1980);

Organic Theory and Drug Analysis (Liqun Liu, People's Health Publishing House, 1984);

New Analytic Chemistry of Toxins (Naizhao Hu, Mass Publishing House, 1986);

Clinical Evaluation of Drugs (Shisong Xu, Shanghai Scientific and Technical Publishing House, 1982).

Finally, there are numerous good handbooks and manuals. The acknowledged standard in China has long been *New Pharmacology* (edited by Xinqian Chen and Youyu Jing, People's Health Publishing House). It first appeared in 1951. The 12th edition was published in 1985 and 2 million copies have been sold, making it a bestseller. This revision, as in the past, presents specific information on approximately 1500 selected drugs in order to provide a comprehensive source for quick reference. In this new edition, the nomenclature of all drugs is based on *PPRC 1985*. Those which do not have established names are named on the basis of WHO's *Non-Proprietory Names for Pharmaceutical Substances*. Drugs included are organized into categories of functions in the treatment of diseases and, thereafter are divided into 24 major groups.

Other leading handbooks are:

Handbook of Drugs (Zhen Yie et al., People's Health Publishing House, 1965);

Handbook of Clinical Drugs (Nanjing Medical College et al., 2nd edn, Shanghai Scientific and Technical Publishing House, 1986);

Handbook of Kinetic Parameters of Drugs (Wenyi Tang' Hunan Scientific and Technical Publishing House, 1987);

Manuals of Pharmaceutical Preparations (Department of Pharmacology, Shantong Medical College, Shantong People's Publishing House, 1959);

Note on Pharmaceutical Preparations (Xueqiu Gu, People's Health Publishing House, 1965);

Data Handbook of Pharmaceutical Preparations (Honghai Ge' People's Health Publishing House);

Handbook of Drugs and Formularies (Guozong Sun, People's Health Publishing House, 1965);

Practical Handbook of Gas-Phase Chromatography (Jilin Graduate School of Chemical Industry, Chemical Industry Publishing House, 1980);

Manuals of Drug Preparations (Shenghe Zhu and Anren Wang, Henan Scientific and Technical Publishing House, 1984);

Catalogue of National Reagents Products (the Editorial Board of the Catalogue, Chemical Industry Publishing House, 1979);

Chemical Manuals of Organic Preparations (Guangdian Han, Petroleum and Chemical Industry Publishing House, 1977).

Reference books for Chinese traditional and herbal drugs

Traditional Chinese medicine, based on its long clinical experience and sophisticated theory of balance, stresses the diagnosis of the symptoms of each individual patient and uses a milder method with natural herbs to restore the natural equilibrium and resistance of the patient. Today the treatment of diseases utilizing both Chinese and Western medicines is widespread practice in many hospitals throughout China and has achieved many encouraging results, particularly in treating certain chronic and degenerating diseases. It has reportedly less serious side effects and is as attractive as alternative medicine.

Traditional Chinese medicine is primarily the experience accumulated during the course of more than 3000 years by the Chinese people. Because of this, there is a rich collection of recorded material in this unique field. The following list is not exhaustive.

Compendium of Materia Medica is a famous and immortal masterpiece, first published in 1578 by Shizheng Li who spent 26 years (from 1552 to 1578) on it. The *Compendium* contains 1892 drugs with 11,096 formularies and more than 1000 illustrations. It covers all aspects of Chinese traditional and herbal drugs such as clinical usage, processing methods, identification and culture of drugs.

The *Compendium* was first translated into Japanese in 1606, and then Latin, French, German, English and Russian, etc., thus making it available in most of the world.

A standard reference work for many years, *Digest of Research Papers of Traditional Chinese Medicine*, edited by Shoushan Liu, is published periodically and reviews current developments for traditional Chinese medicine. Designed principally for professional research workers, the reference work provides a comprehensive coverage of scholarly and authoritative literature both at home and abroad. Volume 1, published in 1967, provides a selection of 4000 summaries or monographs published from 1820 to 1961 from 390 journals. All the literature contained is classified on the basis of the described drugs in terms of their biochemical, pharmacologic, toxic and clinical aspects. Volume 2 appeared in 1979 and contains 3300 monographs published from 1962 to 1974 from 711 journals, while the latest volume published in 1986, provides coverage of 4000 citations (1975–1979) from 330 journals. This remarkable compilation has inevitably entailed great expense, but has been rewarded with high renown from research workers.

The best detailed book (in 8 volumes) for the crude drug-oriented reader is *The Chinese Pharmaceutical Flora*, by Jian Pei and Taiyan Zhou. Each volume contains 50 pharmaceutical plants and gives good accounts of their geographical location, ecology, composition, pharmacology and effects etc. Other important works on the subject are: *A Collection of National Chinese Traditional and Herbal Drugs, Dictionary of Traditional Chinese Medicine, A Compilation of Preparations of Traditional Chinese Medicine, Pharmacology of Traditional Chinese Medicine, Identification of Traditional Chinese Medicine, Clinical Handbook of Traditional Chinese Medicine, Handbook of Preparations of Traditional Chinese Medicine* and *New Compilation of Processing Methods of Traditional Chinese Medicine*.

References

Brookes, B. C. (1981). 'The foundation of information science. Part IV. Information science: the changing paradigm.' *Journal of Information Science*, 3, 3–12.

Chen, G. (1985). 'Online information retrieval in China.' *Proceedings of the 9th International Online Information*, 111–119.

Jin, H. W. (1987). 'Prediction of the demand for pharmacists in 2000. *Chinese Pharmaceutical Bulletin*, 22, 3–6.

Luo, Q. H. (1987). 'The comparison of Latin names of drugs in PPRC 1977 and PPRC 1965.' *Chinese Pharmaceutical Bulletin*, 22, 307–309.

Mei, W. Q. (1986). 'The role of computers in information retrieval and their current trend.' *Journal of Zhejiang Graduate School of Medicine*, 37, 272–275.

Mei, X. C. (1987). 'A preliminary analysis of pharmaceutical information.' *Chinese Pharmaceutical Bulletin*, 22, 101–105.

Ren, Z. Y. and Zhang, Q. X. (1987). 'The measure units in the Pharmacopoeia of the People's Republic of China.' *Chinese Pharmaceutical Bulletin*, 22, 47–53.

Shen, J. X. (1986). *A chance to vigorously develop the pharmaceutical industry in China* (Nanjing College of Pharmacy).

The University. (1986). *An introduction to China Pharmaceutical University* (China Pharmaceutical University).

CHAPTER TWENTY

Trends and perspectives

W. R. PICKERING

Introduction

Throughout the preceding chapters, a number of themes emerge which reflect both the current status of information science and the impact of other factors affecting the availability of information essential to the manufacture and use of pharmaceuticals. These factors can be grouped into three categories: first, socio-economic conditions and trends which influence the supply of information; secondly, technically orientated developments contributing to the accessibility of information in general; and thirdly, an increasing awareness of the need for national and international policies and strategies to cope with the burgeoning demand for soundly based information.

Relevant socio-economic conditions and trends

Health care has advanced rapidly in the developed world but in many of these communities social security systems have reached a state of financial crisis. National health-care systems are financed in different ways because of their varying political and economic backgrounds but there are increasing pressures generally to hold or reduce the financial burden of expensive medicines.

The measures adopted are usually, on the one hand, budgetary constraints and price controls, and on the other, the easing of restrictions on cheaper alternatives (generics): the knock-on effects are complex and the subject of much contentious debate. In very poor regions, the adequacy of health care is further jeopardized not only by a shortage of suitable medicines but also by inadequate administration and a very high incidence of disease.

Improved health care and better living standards raise life expectancy but, at the same time, alter disease profiles significantly. In most developed countries, nowadays, the major causes of death are cardiovascular conditions, cancer and respiratory complaints; the emphasis is on 'lifestyle' complaints rather than infectious diseases as previously. This has brought about a corresponding change in the pattern of research and development in pharmaceuticals since the companies have to follow the more profitable market sectors.

The companies are conscious not only of the cost of generating research information but also of its potential value to competitors. Piracy erodes profitability and is a major threat which is difficult and expensive to counter. Add to this the normal risks of information leakage prior to patent application and it is hardly surprising that companies are very careful about what information they release and to whom; there is an understandable constraint on the free flow of commercially sensitive data.

Information is passed to regulatory authorities on a non-disclosure basis but otherwise information is released to the general public primarily to promote products and to ensure their safe and efficient use in the community. When information is not considered sensitive by virtue of its nature or age, it is permitted to pass into the public domain in order to satisfy scientists' desire to publish or otherwise to reflect the quality of their work and, in fact, the pharmaceutical industry contributes substantially to the general advancement of science.

Several contributors draw attention to political factors which have or will have impact on pharmaceutical manufacturers and on the movement of information. For instance, South America has a history of political instability and economic difficulty both of which hinder efficient information transfer since other problems are more pressing. Conversely, the moves in Europe arising from the Single European Act (1987) and leading to a single market in 1992, will have profound effects on the environment in which the parmaceutical industry operates. Harmonization of standards and relevant legislation is a necessary pre-requisite to a single economy and, even where unanimity is not achievable, some form of mutual recognition is essential. There are, therefore, opportunities to address issues (such as patent advantage, delays in granting product licences and the problems arising from parallel importing) which are key to the industry. In more general terms, the gradual unification of the Community must facilitate the declared policy of free and easy exchange of scientific and technical information.

Information accessibility

The expression 'information explosion' occurs throughout the foregoing chapters with almost monotonous regularity; there is no doubting the reality of the phenomenon and its impact. Nor is there any doubt that hard copy is still very popular and the so-called 'paperless society' is rapidly becoming a myth. Prices charged by publishers continue to rise sharply and more rigorous enforcement of copyright legislation is likely to add significantly to consumer costs. Libraries customarily work to closely constrained budgets even in the comparatively wealthy industrial sector, and it requires great skill to satisfy user requirements (or even most of them) under these circumstances. A major part of information work is concerned with obtaining and presenting published information in response to genuine user needs when funds are not unlimited and the costs of delivery are escalating. The differences between developed and developing countries are apparent; if the US is agonizing over literature costs what is the situation for the remote, poorer parts of the world? Simon Bell concludes that, because the 'countries of the South' as a whole are dependent on the information and communication industries of the North, the situation is critical. Not only are the net importers of information paying a high tariff for other peoples' data, but they are also paying for expensive refined products containing the data they have exported freely. Furthermore, the format in which such information is presented is generally more appropriate for developed countries than developing communities who require information in a form more in keeping with their social and educational status.

Online services have grown into a multi-million dollar industry and this is another area in which the consumer is faced with rising costs only partially because of the extended range of services offered. For the most part, online searching only identifies the existence of information and its retrieval has then to be completed. It is at this point that the consumer now anticipates the benefits of technology: it is already possible electronically to locate information, scan it and decide its value – it would seem to be a short step to add full text delivery, all of this from the user's desk. However, it remains to persuade suppliers that altered methodology does not deny them revenues. The application of technologies such as fibre optics, compact discs and artificial intelligence promises much and, together with extended downloading opportunities, may lead to new relationships between consumer organizations and suppliers. In this relationship, the universally desired free flow of scientific and technical information is fostered with much of the expensive processing being done efficiently and economically within the consumer organization whilst, at the same time, essential revenues are guaranteed for suppliers. Some go as far as to maintain that advanced technologies offer developing countries (and others) opportunities to leapfrog to improved information accessibility but others are sceptical, pointing out that an information system designed for the most developed countries is not always suitable for those that are less developed.

Those working in development and production sectors of the pharmaceutical industry have always had a requirement for data as opposed to textual information. The trend towards automation of data collections in other scientific and technical fields is important because it broadens the possibilities of shared knowledge bases related to expert centres and directly available to scientists. Such developments anticipate, perhaps, information systems that link members of the scientific community more directly. In this context, it will be interesting to observe how the use of communications networks such as JANET (Joint Academic Network) and EARN (European Academic Research Network) develops.

The role of technology in information retrieval and delivery

Information technology is a major theme throughout the entire book. Online services are accessible more or less universally via the available communications facilities and database hosts. However, use of these services may be restricted by inadequate local communications, the absence of suitable hardware and by general lack of finances to support running costs. Developments involving advanced technology such as satellites and optical fibre, continue to improve the potential for communication and most contributors have high expectations of CD-ROM although the eventual impact of optical storage technology is difficult to assess at present.

At its most sophisticated, information technology promises desk-top access to a wide range of information in flexible format. Continued enhancements in computer technology should ensure further reductions in the cost of data processing and an ever increasing emphasis on the role of personal computers as work stations. Such advances, however, require a corresponding adjustment in attitudes in matters of downloading, presentation of information in electronic or optical formats and realistic pricing policies; progress in information science is dependent not only on technical developments but equally on the relaxing of operational constraints. The information industry responds to the same economic laws as the pharmaceutical industry and needs to sustain and develop its position in an identical manner but there is a perception that information suppliers are sometimes too conservative in their accommodation of new technology, being content to rely upon proven conventional products and an established market sector. If this is so, then the realization of the potential of information technology would be hampered.

Computer graphics have allowed great strides to be made in areas such as chemical structure representation and there have been corresponding advances in the techniques of structure handling. Parallel developments in relational technology have afforded a very flexible basis for database management and all these technologies in combination create a powerful tool for those engaged in pharmaceutical research as well as those primarily concerned with information storage and retrieval.

Because of the power of the work station and the rising level of software sophistication, there is a trend towards progressive decentralization of information handling, placing it nearer to the end user. The knock-on effects are numerous and varied but the issues raised centre on user-friendliness, cost containment, information management and the changing role of the information professional.

To enable the effective use of information sources by end users, organizations either have to arrange suitable training or apply techniques to simplify access to sources in variable format. The latter course is to be preferred and research continues into potential user aids such as intelligent front-ends, expert system interfaces and natural language interrogation.

The costs of installing workstations, telecommunications and sophisticated software are awesome and, when coupled with inadequately controlled access to external services, threaten to become prohibitive. The insertion of information technology to handle information from both external and internal sources requires management skills of a high order if the investment of capital and labour is to be balanced by measurable benefit in a reasonable timescale which ensures an acceptable payback period. Furthermore, organisations that enter wholeheartedly into information technology would be ill-advised to ignore the information management problems that lurk for the unwary; it is not sufficient to invest in hardware and software and assume that the movement of information will automatically fall into place in a controlled manner. There are, therefore, correspondingly large organizational adjustments to be made to accompany the application of technology and it is, perhaps, to the realm of information management that the information professional should turn his attention.

The need for the traditional skills persists, particularly in smaller units, but by and large, the trend towards decentralization must alter the demands placed upon information staff. Ideally, the accommodation of sophisticated information handling based on advanced technology is best accomplished by a tripartite, equal partners working relationship between end-users, information professionals and technology experts such as computer personnel.

Where does all this lead? Is information technology the prerogative of the wealthy or will it enable the less privileged improved access to the information so desperately needed? The answer seems to rest, at least in part, with national authorities and international organizations but also, and one suspects mainly, with the suppliers of technology and information, respectively. The technical, economic, political and financial problems are enormous and far beyond the scope of this volume but the rewards in terms of extended markets, wealth creation in depressed areas, political stability and improved quality of life for many millions of people would more than justify the effort. It seems that the current imbalance in the distribution of wealth and the inadequacy of administrations to deal with the limited funds available are major obstacles to be overcome. And, even if technology could be dropped in place, would it be appropriate without the sponsorship of corresponding educational programmes? Possibly a compromise solution is achievable along the lines suggested by Simon Bell. Information agencies built into the fabric of national or regional information policies might provide the focus that the multinationals, the information suppliers and the technologists require to assure themselves that the investment of time, effort and money would not be wasted and could lead to both improved standards for the local community and to a return on the investments made.

In summary, the potential of information technology is almost limitless but the technology presents its own problems in the process of solving others.

The nature and extent of information flow

An inescapable fact, perhaps not so surprising to information professionals but possibly to the lay person, is that, despite the current emphasis on technology, most scientists and technicians (and one suspects others) still talk to one another and read hard copy by choice. Possibly the reason is simply that electronic systems at this time do not deliver all the information that users require but an alternative explanation is that most people do not relate so readily to machines as to more personal forms of communication.

Several authors spend time discussing the 'invisible college' and the value of personal contacts. Provisions for conferences and seminars, the role of professional bodies and their meetings, interchange of staff, centres of expertise or excellence, and directories of research are all viewed as vital elements in effective information transfer.

The continued preference for and dependence on the written word ensures that libraries retain their pivotal position in information transfer throughout the developed world. They remain the principal means by which users obtain full text and the inter-library loan system is the bulwark on which much information flow rests. The automation of this process is, therefore, of particular interest and the several references to online catalogues, computer networks, electronic document delivery and so on are encouraging even if some are at the planning stage only. However, there is a long way to go in this sphere and also in the transformation of libraries into more active information centres offering different kinds of services. The holdings of the major libraries, such as the British Library, remain critical to the functioning of library networks even as they become automated. Despite the undoubted reliance of most nations on the library system, Simon Bell questions the value to Africa of the western concept of the library as major information focus speculating that if rural communities are to be served then a new and very different structure will be required.

The idea of focusing information retrieval and presentation on centres of excellence or information centres has a great deal of merit but still has patchy application. UNESCO is concerned with a long term programme of UNISIST action in the promotion and co-ordination of Information Analysis Centres at international level [Saracevic T. and Woods J. B. (1981), *Consolidation of information: handbook for evaluation, restructuring and re-packaging of scientific and technical information* (PG1/81/WS/16, UNESCO)]. The European Community also endorses the principle of specialized information centres in science and technology [*Spinks Report*, (1980) (HMSO, UK)] and has initiated independent action. Some socialist countries have been attracted to the formalized information network approach as instanced by the Soviet Scientific and Technical Information System which in 1977 comprised 10 major services with 86 central branches and 15 republican information institutes.

The principle can be extended less formally and into more specific areas of information. At its most sophisticated level, information work involves the extraction of ideas or data from published and other sources, and the validation of these leading to their presentation in a convenient form. At its simplest, however, it can be the accumulation and indexing of primary source material allowing a moderately effective means of entry to information. The meaning of the term 'centre of excellence' can thus vary fundamentally but the more valuable ones tend to be associated with areas of subject specialization. As long ago as 1963, Weinberg [*Science government and information – a report on the President's Science Advisory Committee* (The White House, Washington DC)] suggested the overall solution to the information explosion lay in the development of specialized information centres or information analysis centres to offer centres of expertise based on subject specialists. Those that exist are often the result of local initiatives and are not necessarily well co-ordinated, leading to the conclusion that there is scope for improvement.

One would have expected developments in information handling techniques and communications to have enhanced information networks and referral systems in a manner which encouraged the application of the information centre approach to a greater extent than has been the case. In fact, there are many hundreds of isolated examples and the literature relating to these is extensive. At a national level, both Japan and West Germany have promoted the concept using government funds.

Much of this book addresses issues of drug information; indeed, it would be surprising if it did not. There is mounting consumer interest in the developed countries in being better informed about the benefits and risks associated with marketed pharmaceuticals, presumably because consumers are not content to rely solely on the judgment of the regulatory authorities and because of an increasing desire to participate in medical care decisions. In fact, as Hugh Tilson points out, the accumulating database of post-marketing experience provides the ultimate basis for best and safest use of drugs but it takes time for evidence to be gathered and interpreted.

Since the pharmaceutical companies are the foci for information about their products, the manner in which they fulfil their responsibilities is crucial to the rational use of drugs and to the well being of consumers.

In developing countries, consumer interest is perhaps directed primarily onto other matters but there is still an overall requirement to obtain and prescribe the right drugs in the recommended manner. John Dunne's account of the problems that face both the industrialized and the developing countries is enlightening. He stresses the need for independent multi-disciplinary bodies to judge the acceptability of new products and to subject them to systematic review but points out that regulatory authorities are not constituted to develop as primary sources of drug information. He also emphasizes the need to ensure that information on pharmaceuticals is placed in the context of the overall social, economic and educational development of each community.

Policies and strategies

The efforts of WHO to promote and monitor the flow of drug information are based extensively on formal resolutions adopted by the World Health Assembly. There is a concomitant onus on member states to develop consistent policies of their own. However, this is only one aspect of a more generalized need to coordinate approaches to the exchange of scientific and technical information and, in particular, the application of appropriate technology.

Early developments in information retrieval benefited from Federal aid in the US and the general availability of the major online databases owes much to this fact. Japan in the immediate post-war years established policies aimed at recovery which included provisions for scientific information (NIST plan). Japan's neighbour China with a completely different political set up is also conscious of the need to plan not only in science but also in information retrieval. It is particularly interesting to note the mention of information analysis centres in this respect.

Simon Bell refers to the UNESCO's long term programme based on information analysis before proceeding to discuss information agencies and the ideas surrounding IFID (Information for International Development). In Africa, there is a paucity of national or regional policies but the aim, in his view, should be to collaborate on a wide basis not only within the continent but outside as well. Elsewhere, there is reference to problems arising from lack of Information and Information Technology policies and from inadequate funding and investment capital. Western Europe (or more strictly the European Community) has a set of problems all of its own arising from the nature of the alliance and differences in culture and language. There are current community initiatives and projects such as ESPRIT, RACE and ADONIS which receive political and financial support: some earlier ones such as DIANE (EURONET) have not worked out too well, but there is commitment to the principle of scientific and technical information exchange.

The overall impression, as before, is of a wide discrepancy

between the developed and developing regions. The more developed countries not only have the funds but also the administration necessary to attempt policy definition and the implementation of suitable strategies. Elsewhere, policies are either non-existent or incomplete and even when funds become available there is frequently a lack of an appropriate mechanism for making best use of them.

Conclusions

On reflection, this final section might be better entitled 'food for thought' rather than 'conclusions' since conclusions imply a predetermined prospectus and an act of judgement which would certainly be inappropriate. Anyone who has completed the challenge of reading the entire text will perhaps have been led to thinking about what will happen next or even, hopefully, about what should happen next. What has emerged is that one or two areas of endeavour stand out amongst many others as being vital to further improvements in information flow.

The increasing incidence of collaborative databanks, often associated with centres of excellence, might be viewed optimistically as the prelude to systems and networks enabling research workers a greater degree of direct data exchange. Similarly, the promotion of information analysis centres within national and regional settings could become the means of focusing information flow in inadequately served geographical areas.

The impact of information technology threatens to become overwhelming in many respects but sensibly applied and managed, promises great benefits which are only partially realized at this time. The majority look forward eagerly to exciting new developments but others are less convinced of information technology's relevance to their immediate problems.

Will the relationship between information suppliers and the consumer enter a new era? The full exploitation of the likely benefits of information technology may well depend in part on changing attitudes and collaborative efforts between numerous interested parties including those who purvey information commercially or in support of other business activities.

The ever-widening gap between the developed world and the rest remains the subject of political summit meetings and of public concern and debate. Perhaps too much reliance has been placed on political solutions and too little on alternative initiatives. Although the supply of drugs and information relevant to their development, manufacture and use is only an aspect of the total disparity, nothing is more fundamental than health-care and survival. Could it be that an adjustment of business motives is the way forward? This would require subtle market investigation and development, careful risk analysis, collaborative arrangements to share and reduce risks but, in addition, guarantees that ensure progress and security of investment, with commitment to long-term planning. These ideas are not necessarily immediately attractive in simple business terms and would require exceptional vision and courage to implement. However, the scale of the problem is such that only those who wield power can shape the framework within which the rest of us can contribute: it is the policy makers – the Inner Cabinets and the Boardrooms – that we should seek to inform and influence.

GENERAL APPENDIX TO PART III

Editorial Note

Tables included in the General Appendix have been extracted from the authors' original contributions in order to maintain the flow of the text without sacrificing potentially valuable information. As a consequence, the Appendix itself displays a degree of imbalance because of the varying perspectives of the contributors, and appropriate adjustment will be considered in any future edition.

Literature sources

Pharmaceutical Information Sources used in the US

Guides to the Literature

Andrews, T. *Literature of Pharmacy and the Pharmaceutical Sciences* (Libraries Unlimited, 1986).
Brunn, A. L. *How to Find Out in Pharmacy* (Pergamon, 1969).
Fenichel, C. H. ed. *Pharmaceutical Information* (Drexel University, 1982) (*Drexel Library Quarterly*, **18**, no. 2).
Hall, V. B., and S. W. Schwerzel. *Index of Sources of Data and Statistics in Pharmacy and the Health Care Field* (American Pharmaceutical Association, 1981).
Pasztor, M., and J. Hopkins. *Bibliography of Pharmaceutical Reference Literature* (Pharmaceutical Press, 1968).
Piermatti, P. A., and B. M. Hill. *A Basic Booklist and Core Journals for Pharmaceutical Education* (2nd edn, American Association of Colleges of Pharmacy, 1986).
Revill, J. P., ed. *Drug Information Sources: A World-Wide Annotated Survey* (2nd edn., Gothard House Publications, 1984).
Sewell, W. *Guide to Drug Information* (Drug Intelligence Publications, 1976).

Indexing and Abstracting Publications

Biological Abstracts (BioSciences Information Service).
Chemical Abstracts (American Chemical Society).
Excerpta Medica Section 38, Adverse Reactions Titles (Excerpta Medica Foundation).

Excerpta Medica Section 37, Drug Literature Index (Excerpta Medica Foundation).
Excerpta Medica Section 30, Pharmacology (Excerpta Medica Foundation).
Government Reports Announcement Abstracts and Index (U.S. Technical Information Service (NTIS)).
Index Guide to Drug Information and Retrieval (H. Fukushima, et al., Elsevier/North-Holland, 1979).
Index Guide to Rational Drug Therapy (H. Fukushima, et al., Elsevier, North-Holland, 1982).
Index Medicus (U.S. National Library of Medicine).
International Pharmaceutical Abstracts (American Society of Hospital Pharmacists).
International Pharmaceutical Technology and Product Manufacture Abstracts (Childwall University Press).
Medicinal and Aromatic Plants Abstracts (New Delhi, India, Publications and Information Directorate).
Pharmaceutical News Index (Data Courier, Inc.).
Science Citation Index (Institute for Scientific Information).
Toxicology Abstracts (Cambridge Scientific Abstracts).

Reviews and Surveys

Advances in Biochemical Psychopharmacology (Raven Press).
Advances in Drug Research (Academic Press).
Advances in Pharmaceutical Sciences (Academic Press).
Advances in Pharmacology and Chemotherapy (Academic Press).
Analytical Profiles of Drug Substances (Academic Press).
Annual Reports in Medicinal Chemistry (Academic Press).
Annual Review of Chronopharmacology (Pergamon Press).
Annual Review of Pharmacology and Toxicology (Annual Reviews).
Antibiotics and Chemotherapy (S. Karger).
Cancer Chemotherapy (Excerpta Medica/Elsevier).
Forschritte der Chemie organischer Naturstoffe, Progress in the Chemistry of Organic Natural Products (Springer-Verlag).
Modern Methods in Pharmacology (Alan R. Liss).
Progress in Biochemical Pharmacology (S. Karger).
Progress in Drug Research: Fortschritte der Arzneimittelforschung: Progres der recherches pharmaceutiques (Birkhauser Verlag).
Progress in Medicinal Chemistry (Elsevier Science Publishers).
Radiopharmacy and Radiopharmacology Yearbook (Gordon and Breach Science Publishers).
Recent Progress in Hormone Research (Academic Press).
Reviews in Biochemical Toxicology (Elsevier Biomedical).
Reviews of Physiology, Biochemistry, and Pharmacology (Springer-Verlag).
Year Book of Drug Therapy (Yearbook Medical Publishers).

Pharmaceutical Information Sources used in Japan

List of Japanese Sources in English

PRIMARY SOURCES (NEWSLETTERS)

Japan Pharmaceutical Letter Tokyo, Media Services, 1974–, weekly
Pharma Japan Tokyo, Yakugyo Jiho Co., Ltd., 1984–, weekly

PRIMARY SOURCES (JOURNALS)

Acta Paediatrica Japonica (overseas edition) Tokyo, Japan Pediatric Society, 1958–, 4 times per year, ISSN 0001–6632
Analytical Sciences Tokyo, Japan Society for Analytical Chemistry, 1985–, bimonthly, ISSN 0910–6340
Biomedical Research Tokyo, Biomedical Research Foundation, 1980–, bimonthly, ISSN 0388–6107
Bulletin of the Chemical Society of Japan Tokyo, The Chemical Society of Japan, 1926–, monthly, ISSN 0009–2673
Cell Structure and Function Kyoto, Japan Society for Cell Biology, 1975–, quarterly, ISSN 0386–7196
Congenital Anomalies Nagoya, Nippon Senten Ijo Gakkai, 1963–, quarterly, ISSN 0037–2285
Chemical and Pharmaceutical Bulletin Tokyo, The Pharmaceutical Society of Japan, 1953–, monthly, ISSN 0009–2362
Chemistry Letters Tokyo, The Chemical Society of Japan, 1972, monthly, ISSN 0366–7022
Endocrinologica Japonica Tokyo, Japan Endocrine Society, 1954–, bimonthly, ISSN 0013–7219
GANN Monographs on Cancer Research Tokyo, Japanese Cancer Association, 1966–, twice a year, ISSN 0072–0151
Gastroenterologia Japonica Tokyo, Japanese Society of Gastroenterology, 1966–, bimonthly, ISSN 0435–1339
Japanese Circulation Journal Kyoto, Japanese Circulation Society, 1935–, monthly, ISSN 0047–1828
Japanese Heart Journal Tokyo, Japanese Heart Journal Association, 1960–, bimonthly, ISSN 0021–4868
Japanese Journal of Cancer Research (GANN) Tokyo, Japanese Cancer Association, 1941–1, monthly, ISSN 0016–450X
Japanese Journal of Medicine Tokyo, The Japanese Society of Internal Medicine, 1962–, quarterly, ISSN 0021–5120
Japanese Journal of Ophthalmology Tokyo, University of Tokyo, 1957–, quarterly. ISSN 0021–5155
Japanese Journal of Pharmacology Kyoto, The Japanese Pharmacological Society, 1951–, monthly, ISSN 0021–5198
Japanese Journal of Physiology Nagoya, Physiological Society of Japan, 1951–, bimonthly, ISSN 0021–521X
Japanese Journal of Surgery Tokyo, Japan Surgical Society, 1970–, bimonthly, ISSN 0047–1909
Journal of Antibiotics Tokyo, Japan Antibiotics Research Association, 1953–, monthly, ISSN 0368–1173

Journal of Biochemistry Tokyo, Japanese Biochemical Society, 1922–, monthly, ISSN 0021–924X

Journal of Cardiography Tokyo, Cardiograph Society, 1971–, quarterly, ISSN 0386–2887

Journal of Japanese Society of Hospital Pharmacist Tokyo, Japanese Society of Hospital Pharmacist, 1965–, monthly, ISSN 0914–0697

Journal of Pharmacobiodynamics Tokyo, The Pharmaceutical Society of Japan, 1978–, monthly, ISSN 0386–846X

The Journal of Toxicological Sciences Tokyo, The Japanese Society of Toxicological Sciences, 1976–, quarterly, ISSN 0388–1350

PRIMARY SOURCES (REFERENCE BOOKS)

pdp Japanese Ethical Pharmaceutical Tariff 1988, 4 Yakugyo Kenkyukai, Tokyo, Yakugyo Jiho Co., Ltd., 320pp, 1988

Pharmaceutical Manufactures of Japan 1988–89, Yakugyo Jiho Co., Ltd., Tokyo, Yakugyo Jiho Co., Ltd., 309pp, 1988

Drug Approval and Licensing Procedures in Japan 1987, Nippon Koteisho Kyokai, Tokyo, Yakugyo Jiho Co., Ltd., 924pp, 1987

Minimum Requirements for Antibiotics Products of Japan '86 Japan Antibiotics Research Association, Tokyo, Yakugyo Jiho Co., Ltd., 1088pp, 1986

Minimum Requirements for Biological Products 1986 The Ministry of Health and Welfare, Tokyo, Association of Biologicals Manufacturers of Japan, 1986

Pharma Japan Yearbook 1988–89 Yakugyo Jiho Co., Ltd., Tokyo, Yakugyo Jiho Co., Ltd., 124pp, 1988

The Pharmacopoeia of Japan 11th Edition Nippon Koteisho Kyokai, Tokyo, Yakuji Nippo Co., Ltd.,1586pp, 1987

Guidelines for Clinical Evaluation of New Drugs 1986 The Ministry of Health and Welfare, Tokyo, Yakuji Nippo Co., Ltd., 190pp, 1986

Standards and Certification Systems Concerning Drugs in Japan The Ministry of Health and Welfare, Tokyo, Yakugyo Jiho Co., Ltd., 237pp, 1988

Pharmaceutical Administration in Japan 3rd Edn. The Ministry of Health and Welfare, Tokyo, Yakuji Nippo Co., Ltd., 253pp, 1986

Requirements for the Registration of Drugs in Japan 2nd Edn. The Ministry of Health and Welfare, Tokyo, Yakuji Nippo Co., Ltd., 290pp, 1986

JAPTA List: Japanese Drug Directory 3rd Edn. Nippon Yakugyo Boeki Kyokai, Tokyo, Yakuji Nippo Co., Ltd., 680pp, 1987

Guide to Medical Device Registration in Japan 2nd Edn. The Ministry of Health and Welfare, Tokyo, Yakuji Nippo Co., Ltd., 230pp, 1987

GMP Regulations of Japan 3rd Edn. The Ministry of Health and Welfare, Tokyo, Yakuji Nippo Co., Ltd., 280pp, 1988

SECONDARY SOURCES (PRINTED MATERIALS)

Pharmacast Tokyo, Technomic, 1982–, 4 times per year

Japan Pharmaceutical Abstracts Tokyo, Drug Business Research, 1970–, monthly

Journals published outside Japan which are most often used by the Japanese pharmaceutical industry

Agents and Actions (ISSN 0065–4299)
American Journal of Medicine (ISSN 0002–9343)
Antimicrobial Agents and Chemotherapy (ISSN 0066–4804)
Archives Internationales de Pharmacodynamie et de Therapie (ISSN 0003–9780)
Archives of Biochemistry and Biophysics (ISSN 0003–9861)
Arzneimittel-Forschung (ISSN 0004–4172)
Biochemical and Biophysical Research Communications (ISSN 0006–291X)
Biochemical Pharmacology (ISSN 0006–2952)
British Journal of Clinical Pharmacology (ISSN 0306–5251)
British Journal of Pharmacology (ISSN 0007–1188)
British Medical Journal (ISSN 0007–1447)
Cancer Chemotherapy and Pharmacology (ISSN 0344–5704)
Chemotherapy (ISSN 0009–3157)
Clinical Pharmacology and Therapeutics (ISSN 0009–9236)
Current Therapeutic Research (ISSN 0011–393X)
Drugs (ISSN 0012–6667)
European Journal of Clinical Pharmacology (ISSN 0031–6970)
European Journal of Pharmacology (ISSN 0014–2999)
FASEB Journal (Formerly *Federation Proceedings*) (ISSN 0892–6638)
Journal of Antimicrobial Chemotherapy (ISSN 0305–7453)
Journal of the American Medical Association (ISSN 0002–9955)
Journal of Cardiovascular Pharmacology (ISSN 0160–2446)
Journal of Clinical Investigation (ISSN 0021–9738)
Journal of Medicinal Chemistry (ISSN 0022–2623)
Journal of Pharmaceutical Sciences (ISSN 0022–3549)
Journal of Pharmacology and Experimental Therapeutics (ISSN 0022–3565)
Journal of Pharmacy and Pharmacology (ISSN 0022–3573)
Lancet (ISSN 0023–7507)
Life Science (ISSN 0024–3205)
Molecular Pharmacology (ISSN 0026–895X)
Nature (ISSN 0028–0836)
Naunyn-Schmiedeberg's Archives of Pharmacology (ISSN 0028–1298)
New England Journal of Medicine (ISSN 0028–4793)
Proceedings of the Society for Experimental Biology and Medicine (ISSN 0037–9727)
Science (ISSN 0036–8075)
Toxicology and Applied Pharmacology (ISSN 0041–008X)

Reference books published outside Japan which are most often used by the Japanese pharmaceutical industry

AMA Drug Evaluations 6th edn. American Medical Association. Saunders, 1986

American Drug Index 32nd edn. Norman F. Billups, Lippincott, 1988
British Pharmacopoeia 1988 British Pharmacopoeia Commission, Her
   Majesty's Stationery Office, 1988
Index Nominum 1987 Scientific Documentation Center of the Swiss
   Pharmaceutical Society, 1987
*Merck Index: an Encyclopedia of Chemicals, Drugs and Biologicals 10th
   edn.* Marth Windholz, Merck, 1983
Physicians' Desk Reference 42nd edn. Barbara B. Huff, Medical
   Economics, 1988
Physicians' Desk Reference for Nonprescription Drugs 9th edn. Barbara
   B. Huff, Medical Economics, 1988
Rote Liste 1988 Bundesverband der Pharmazeutischen Industrie Editio
   Cantor 1988
USAN and the USP Dictionary of Drug Names 1988 Mary C. Griffiths,
   U.S. Pharmacopeial Convention Inc., 1987
Martindale: The Extra Pharmacopoeia 28th edn. James E. F. Reynolds,
   Pharmaceutical Press, 1982
Remington's Pharmaceutical Sciences 17th edn. Arthur Osol, Mack, 1985
*Goodman and Gilman The Pharmacological Basis of Therapeutics 7th
   edn.* Louis S. Goodman and Alfred Gilman, Macmillan, 1985
Dictionnaire Vidal 63th edn. Jacques Medan, O.V.P., 1987
*United States Pharmacopeia 21st revision and National Formulary 16th
   edn.* U.S. Pharmacopeial Convention Inc., 1984
AHFS Drug Information 1988 Gerald K. McEvoy, American Society of
   Hospital Pharmacists, 1988
Handbook of Nonprescription Drugs 8th edn. American Pharmaceutical
   Society, 1986
USP DI: United States Pharmacopeia Dispensing Information William M.
   Hetler, U.S. Pharmacopeial Convention, 1988
Meyler's Side Effects of Drugs 10th edn. M. N. G. Dukes, Elsevier, 1984
Side Effects of Drugs Annual 11 M. N. G. Dukes, Elsevier, 1987
Registry of Toxic Effects of Chemical Substances R. J. Lewis, NIOSH,
   1987
Pharmacological and Chemical Synonyms 8th edn. E. E. J. Marler,
   Elsevier, 1985
British National Formulary, No. 12, 1987 British Medical Association and
   the Pharmaceutical Society of Great Britain, 1987
The Pharmaceutical Codex 11th edn. The Pharmaceutical Society of Great
   Britain, The Pharmaceutical Press, 1979
Organic-Chemical Drugs and their Synonyms 6th edn. Martin Negwer,
   Akademie-Verlag, 1987

Pharmaceutical Information Sources used in Australia

Australian Journals

Australian Adverse Drug Reactions Bulletin
Australian Journal of Biological Sciences

Australian Journal of Chemistry
Australian Journal of Hospital Pharmacy
Australian Journal of Pharmacy
Australian Pharmacist
Australian Prescriber
Cancer Forum
Cancer Review
Chemistry in Australia
Communicable Diseases Intelligence
Concepts in Clinical Pharmacy
Congenital Abnormalities Monitoring Report
Current Drug Information
Current Therapeutics
Hospital Therapeutics
Laboratory Information Bulletin, National Biological Standards Laboratory
Medical Journal of Australia
Medical Research Bulletin
Modern Medicine of Australia
NSW Pharmacist
Patient Management
Perinatal Newsletter
Pharmabulletin
Pharmacy Trade
Proceedings of the Australian Physiological and Pharmaceutical Society
Report of the National Health and Medical Research Council
Your Pharmacy

Australian Reference Books

Aboriginal Communities of the Northern Territory of Australia. (1988). *Traditional Bush Medicines: an Aboriginal Pharmacopeia* (Greenhouse Publications).

Adelaide Children's Hospital Inc. (1984). *Drug Doses for Children* (Adelaide Children's Hospital Inc.).

Analgesic Guidelines Sub-Committee, Victorian Drug Usage Advisory Committee. (1988). *Analgesic Guidelines* (Victorian Medical Postgraduate Foundation).

Antibiotic Guidelines Sub-Committee, Standing Committee on Infection Control, Health Department of Victoria. (1987). *Antibiotic Guidelines*, (5th edn., Health Department of Victoria).

Australian Medical Association. (1986). *Optimal use of analgesic drugs in chronic cancer pain control* (South Australian Health Commission).

Batagol, R. Compiler. (1983). *The Royal Women's Hospital Reference Guide on Drugs and Pregnancy* (Royal Women's Hospital).

Buchanan, N. and Paird-Lambert, J. (1988). *Prescribing Medications for Children* (Williams and Wilkins).

Commonwealth Department of Health. (1986). *Australian Approved Names and Other Names for Therapeutic Substances* (4th edn., Australian Government Publishing Service).

Commonwealth Department of Health. (1986). *Index of Australian Approved Names and Synonyms for Therapeutic Substances* (Australian Government Publishing Service).

Health Link, Westmead Hospital. (1988). *A Guide to Consumer Health Information* (Westmead Hospital).

Hunyor, S. (1987). *Cardiovascular Drug Therapy* (Williams and Wilkins).

Kucers, A. and Bennett, N. (1987). *The Use of Antibiotics*, 4th edn., Heinemann).

Mims Annual 1988 (1988). (IMS Publishing).

Paediatric Pharmacology Unit, Westmead Hospital. (1985). *Drugs in Pregnancy and Lactation*, (2nd edn., Westmead Hospital).

Pharmaceutical Society of Australia. (1988). *Australian Pharmaceutical Formulary and Handbook*, (14th edn.).

Psychotropic Guidelines Sub-Committee, Victorian Drug Usage Advisory Committee. (1989). Psychotropic Drug Guidelines (Victorian Medical Postgraduate Foundation).

Queen Elizabeth Hospital, Drugs Committee. (1988). *The Queen Elizabeth Hospital Therapeutic Handbook/Formulary 1988* (Adelaide, Queen Elizabeth Hospital).

Royal Alexandra Hospital for Children. (1988). *Paediatric Handbook* (Royal Alexandra Hospital for Children).

Royal Children's Hospital, Melbourne. (1985). *Paediatric Pharmacopeia* (Royal Children's Hospital).

Spencer, R. (1986). *The A-Z of Australian Family Medicines*, (3rd edn., Butterworths).

Thomas, J., ed. (1988). *Prescription Products Guide*, (17th edn., Australian Pharmaceutical Publishing Company Ltd.).

Wilson, J. (1986), *Drug Use in Respiratory Disease* (Williams and Wilkins).

Pharmaceutical Information Sources used in South America

South American Journals (*Index Medicus Latinoamericano*, 1987)

ARGENTINA

Acta Bioquimica Clinica Latinoamericana
Acta Gastroenterológica Latinoamericana
Acta Phisiologica et Pharmacologica Latinoamericana
Acta Psiquiatrica y Psicológica de América Latina
Archivos Argentinos de Dermatología
Archivos Argentinos de Pediatría
Archivos de Oftalmologia de Buenos Aires
Boletin de la Academia Nacional de Medicina de Buenos Aires
Cuadernos Medico Sociales
Medicina
Obstetracia y Ginecología Latino-Americanos

Prensa Médica Argentina
Quiron
Revista Argentina de Anestesiología
Revista Argentina de Cirugía
Revista Argentina de Dermatología
Revista Argentina de Mastología
Revista Argentina de Microbiología
Revista Argentina de Tuberculosis, Enfermedades Pulmonares y Salud Publica
Revista de la Asociación Médica Argentina
Revista de la Facultad de Ciencias Médicas de Cordoba
Revista de la Facultad de Ciencias Médicas de la Plata
Revista de la Facultad de Ciencias Médicas de la Universidad Nacional de Cuyo
Revista del Hospital de Clínicas
Revista del Hospital Materno Infantil Ramon Sarda
Revista del Hospital de Niños
Revista Neurologica Argentina
Revista de Oxigenoterapia Hiperbarica
Revista de la Sanidad Militar Argentina

BRAZIL

ABCD: Arquivós Brasileiros de Cirurgia Digestiva
ACM: Arquivos Catarinenses de Medicina
Acta AWHO
Acta Cirúrgica Brasileira
Acta Oncológica Brasileira
AMB: Revista de Associação Médica Brasileira
Anais de Academia Brasileira de Ciéncias
Anais Brasileiros de Dermatologia
Anais de Faculdade de Medicina da Universidade Federal de Minas Gerais
Anais do hospital da Siderurgica Nacional
Anais Paulistas de Medicina e Cirurgia
Arquivos de Biologia e Tecnologia
Arquivos Brasileiros de Cardiologia
Arquivos Brasileiros de Endocrinologia e Metabologia
Arquivos Brasileiros de Medicina
Arquivos Brasileiros de Neurocirurgia
Arquivos Brasileiros de Oftalmologia
Arquivos de Gastroenterologia
Arquivos Médicos do ABC
Arquivos de Neuro-Psiquiatria
APS Curandi
APS Curandi em Cardiologia
APS Curandi em Gastroenterologia
Boletim do C.B.R.: Centro de Biologia da Reprodução
Boletim Epidemiológico: Ministério da Saude (Rio de Janeiro)
Boletim da Sociedade Brasileira de Hematologia e Hemoterapia
Boletin Informativo de la Union

Brazilian Journal of Medical and Biological Research
CCS: Centro de Ciências da Saude – UFPB
CCS: Ciência, Cultura, Saude
Ciencia e Cultura
Cirurgia Vascular & Angiologia
Estomatologia e Cultura
Femina
Folha Médica
GED: Gastroenterologia Endoscopia Digestiva
Ginecologia e Obstetricia Brasileiras
HFA – Publicação Técnico Cientifica
Hileia Médica
HMK - Cancer
HU Revista
Jornal Brasileiro de Ginecologia
Jornal Brasileiro de Nefrologia
Jornal Brasileiro de Psiquiatria
Jornal Brasileiro de Urologia
Jornal de Pediatria
Jornal de Pneumologia
Klinikos
Medicina HUPE–UERJ
Medicina: Revista do Centro Acadêmico Rocha Lima
Memorias do Instituto Oswaldo Cruz
Neurobiologia
Pediatria (São Paulo)
Pediatria Moderna
RBM: Revista Brasileira de Medicina
RBM: Revista Brasileira de Medicina (Cardiologia)
Reprodução
Residencia Médica (Rio de Janeiro)
Revista ABP–APAL
Revista da AMRIGS
Revista da Associação Paulista de Cirurgiões Dentistas
Revista Baiana de Saude Publica
Revista Brasileira de Analises Clínicas
Revista Brasileira de Anestesiologia
Revista Brasileira de Biologica
Revista Brasileira de Cancerologia
Revista Brasileira de Ciências do Esporte
Revista Brasileira de Saude Ocupacional
Revista Brasileira de Ciências Morfologicas
Revista Brasileira de Cirurgia
Revista Brasileira de Clínica e Terapeutica
Revista Brasileira de Colo-Proctologia
Revista Brasileira de Educação Médica
Revista Brasileira de Génetica – Brazilian Journal of Genetics
Revista Brasileira de Ginecologia e Obstetricia
Revista Brasileira de Malariologia e Doenças Tropicais

Revista Brasileira de Neurologia
Reviste Brasileira de Odntologia
Revista Brasileira de Oftalnologia
Revista Brasileira de Ortopedia
Revista Brasileira de Otorrinolaringologia
Revista Brasileira de Patologia Clínica
Revista Brasileira de Psicanalise
Revista Brasileira de Reumatologia
Revista Brasileira de Saude Ocupacional
Revista de Ciéncias Biomédicas
Revista Cientifica Casl
Revista Cientifice: Maternidade, Infancia e Ginecologia
Revista do Colégio Brasileiro de Cirurgiões
Revista da Escola de Enfermagem da USP
Revista da Faculdade de Odontologia de Ribeirão Preto
Revista de Farmacia e Bioquímica da Universidade de São Paulo
Revista da Fundação SESP
Revista do HCPA e Faculdade de Medicina da Universidade Federal do Rio Grande do Sul
Revista do Hospital das Clínicas: Faculdade de Medicina da Universidade de São Paulo
Revista IATROS: Faculdade de Medicina de Santo Amaro
Revista da Imagen
Revista do Instituto de Medicina Tropical de São Paulo
Revista Médica do HGF
Revista Médica do IAMSPE
Revista Médica do Parana
Revista de Medicina
Revista Médica
Revista de Microbiologia
Revista Paulista de Enfermagem
Revista Paulista de Hospitaís
Revista Paulista de Medicina
Revista Paulista de Pediatria
Revista de Psiquiatria Clinica
Revista de Psiquiatria do Rio Grande do Sul
Revista de Saude Publica
Revista da Sociedade Brasileira de Cirurgia Plástica
Revista da Sociedade Brasileira de Medicina Tropical
Seara Médica Neurocirurgica
Temas: Teoria e Pratica do Psiquiatra

CHILE

Archivos Chilenos de Oftalmología
Boletín de Cardiología
Boletín del Hospital de San Juan de Dios
Boletín del Hospital de Viña del Mar
Boletín Micológico
Cuadernos Médico-Sociales

Parasitologia al Dia
Revista Chilena de Neuro-Psiquiatría
Revista Chilena de Nutrición
Revista Chilena de Obstetricia y Ginecología
Revista Chilena de Pediatría
Revista Médica de Chile
Revista Médica del Maule
Revista de Otorrinolaringología y Cirugía de Cabeza y Cuello

COLOMBIA

Acta Médica Colombiana
Archivos de la Sociedad Americana de Oftalmología y Optometria
Colombia Médica
Medicina
Revista Colombiana de Anestesiología
Revista Colombiana de Pediatría y Puericultura
Revista Colombiana de Psiquiatría
Revista de la Federacion Odontológica Colombiana
Revista Hospital de Colombia: Hospitales y Sociología
Revista del Hospital Mental de Antioquia
Revista Latinoamericana de Psicología
Revista de la Universidad Industrial de Santander: Medicina
Salud Uninorte

COSTA RICA

Acta Médica Costarricense
Neuroeje
Notas de Población: Revista Latinoamericana de Demografia
Revista Costarricense de Ciencias Médicas
Revista Médica de Costa Rica
Revista Médica del Hospital Nacional de Niños Dr. Carlos Saenz Herrera

CUBA

Archivos de Medicina Interna
Revista Cubana de Administración de Salud
Revista Cubana de Cirugía
Revista Cubana de Estomatologia
Revista Cubana de Medicina
Revista Cubana de Pediatria

ECUADOR

Revista Ecuatoriana de Medicina y Ciencias Biológicas
Revista de la Universidad de Guayaquil

EL SALVADOR

Revista del Instituto de Investigaciones Médicas

ESTADOS UNIDOS

Boletín de la Oficina Sanitaria Panamericana
Educación Médica y Salud

GUATEMALA

Archivos Latinoamericanos de Nutrición

JAMAICA

Cajanus
West Indian Medical Journal

MEXICO

Anales de la Sociedad Mexicana de Oftalmologia
Anales de la Sociedad Mexicana de Otorinolaringología
Archivos del Instituto de Cardiología de México
Archivos de Investigación Médica
Boletín de Estudios Médicos y Biológicos
Boletín Médico del Hospital Infantil de México
Gaceta Médica de México
Infectología
Revista de Gastroenterología de México
Revista Latinoamericana de Microbiología
Revista de Medicina
Revista Médica
Revista Médica del IMSS
Revista Mexicana de Anestesiología
Revista Mexicana de Micología
Revista Mexicana de Pediatría
Revista Mexicana de Radiología
Salud Mental

PANAMA

Revista del Hospital de Niño
Revista Médica de la Caja de Seguro Social
Revista Médica de Panama
Revista Panameña de Obstetricia y Ginecología

PUERTO RICO

Boletín de la Asociación Médica de Puerto Rico

REPUBLICA DOMINICANA

Acta de Odontologia Pediatrica
Revista Médica Dominicana

URUGUAY

Anestesia – Analgesia – Reanimación
Archivos de Medicana Interna
Cirugía del Uruguay
Revista Médica del Uruguay

VENEZUELA

Acta Científica Venezolana
Acta Odontológica Venezolana
Archivos del Hospital Vargas
Archivos Médicos de Guayana
Archivos Venezolanos de Psiquiatría y Neurología
Boletín de la Dirección de Malariologia y Saniamiento Ambiental
Boletín Médico de Postgrado
Centro Médico
Gaceta Médica de Caracas
GEN
Investyigación Clínica
Isabelica Médica
Medicina Crítica Venezolana
Medicina Interna
Revista de la Facultad de Medicina
Revista de la Fundación Jose Maria Vargas
Revista de Obstetricia y Ginecología de Venezuela
Revista Oftalmológica Venezolana
Revista de la Sociedad Médica Hospital San Juan de Dios
Revista de la Sociedad Médico-Quirurugica del Hospital de Emergencia
  'Perez de Leon'
Revista Venezolana de Cirugía
Técnica Hospitalaria

Journals in Spanish Published in Argentina

Edited by WHO/PAHO
  Boletín de Medicamentos Esenciales
  Medicamentos y Terapéutica. (Buenos Aires through *Agremiación Médica Platense*)
  Boletín Oficina Sanitaria Panamericana (Washington)
Edited by Instituto Nacional de Farmacología y Bromatología (Argentine Institute of Pharmacology and Bromatology)
  Revista INF y B
  Red Latinoamericana de laboratorios oficiales de control de calidad de medicamentos del sector salud (Latin American net of official quality control laboratories of medicines)
Edited by Academia Argentina de Farmacia y Bioquímica
  Revista Farmacéutica
Edited by Departamento de Organización de la Investigación Científica y Técnica de la Academia de Ciencias de Buenos Aires

Resúmenes de Estudios Científicos sobre Fármacos en uso en la Argentina
Edited by Colegio Farmacéutico de la Provincia de Buenos Aires
Acta Farmacéutica Bonaerense
Boletín
BIFASE (Bibliografía farmacéutica seleccionada)
Edited by CEDIQUIFA, Centro de Estudios para el Desarrollo de la Industria Químico-Farmacéutica Argentina
Hoja de Noticias
Boletín CEDIQUIFA
Investigación y Técnica
Edited by SAFYBI, Asociación Argentina de Farmacia y Bioquímica Industria
Revista SAFYBI
Intended mainly for the medical profession
El Día Médico
La Semana Médica
Orientación Médica
Medicine
Intended mainly for industry
Laboratorios (Emma Fiorentino)
Industria Farmacéutica (Edimarket)
Lists of Medicines
Agenda Farmacéutica Kairo
DPF Indice Actualizado de Especialidades Medicinales
Manual Farmacéutico
Vademecum Therapia
Vademecum VDB, Vallory
Directories
Catálogo Colectivo de Publicaciones Periódicas (a list of periodicals issued by Asociación de Bibliotecas Biomédicas Argentinas, ABBA including both foreign and domestic publications which can be found in Argentine libraries)
FECIC, Guía de reuniones científicas y técnicas en la Argentina (a yearly list of scientific and professional meetings which take place in Argentina)
Guía de servicios en farmacología y farmacoquímica (a directory of research units in the pharmacology field which offer services to private laboratories, published in 1985 by the Secretariat of Science and Technology)
Directorio de Bases de Datos en Ciencia y Tecnología en la Argentina (a list of databases in science and technology in Argentina published by the Undersecretary of Informatics and Development, SID, 2nd. edn., September 1988)
Except where otherwise stated the publications are produced in Buenos Aires.

Online Databases

Databases Used in the US

Pharmaceutical Sciences

Consumer Drug Information Fulltext (American Society of Hospital Pharmacists). Current edition. Complete text of *Consumer Drug Digest*, containing detailed information for patients about the generic drugs which comprise the majority of prescription drugs.

De Haen Drug Data (Paul de Haen International, Inc.). 1980–. Information from international biomedical journals on investigational and marketed drugs, clinical studies, adverse effects and drug interactions. Includes four subfiles: *De Haen's Drugs in Use, De Haen's Drugs in Research, De Haen's Drugs in Prospect,* and *De Haen's ADRID* (*Adverse Drug Reactions and Interactions*).

Drug Information Fulltext (American Society of Hospital Pharmacists). Current editions. Complete text of current editions of *The American Formulary Service* (commercially available drugs) and *The Handbook on Injectable Drugs* (commercially available and in use in the US).

International Pharmaceutical Abstracts (IPA) (American Society of Hospital Pharmacists). 1970–. Indexes and abstracts literature on drug development, drug use, and pharmacy practice as reported in pharmaceutical, medical, and related journals. Corresponds to printed *International Pharmaceutical Abstracts*.

Martindale Online (The Pharmaceutical Society of Great Britain). Current edition. Full-text database composed of evaluated information on drugs in clinical use worldwide. Nomenclature, properties, adverse effects, administration, etc. Contains more data and is updated more frequently than printed counterpart, *Martindale. The Extra Pharmacopoeia*.

The Merck Index Online (Merck and Co.). Late 1800s–. Expanded and updated version of latest edition of print counterpart, *The Merck Index.* Each record treats a specific chemical entity or group of closely-related chemicals and gives information on structure, uses, toxicity, chemical and brand names, CAS (Chemical Abstracts Service) registry numbers, and company codes. Bibliographic references provided.

Biomedicine

Biosis Previews (BIOSIS). 1969–. Covers all aspects of biosciences and medicine, including pharmacognosy. Abstracts included for those records which appear in printed *Biological Abstracts*; searchers may retrieve these or use *BIOSIS* numbers to lead to entries in printed source. Also includes *Biological Abstracts/RRM*, formerly entitled *Bioresearch Index*.

CANCERLIT (US National Cancer Institute and US National Library of Medicine). 1963–. Bibliographic references, with abstracts, on all aspects of cancer.

CCRIS (US National Cancer Institute). Current information. Chemical

Carcinogenesis Research Information System. Organized by chemical record; contains carcinogenicity, tumour promotion, and mutagenicity test results.

Combined Health Information Database (Produced by group of federal health information agencies). 1973–. References to and abstracts of publications and programs for patients and the general public, plus the full text of public information pamphlets on selected health topics. Subfiles on arthritis, diabetes, digestive diseases, high blood pressure, health education.

EMBASE (Elsevier). 1974–. Computerized version of *Excerpta Medica*, specialty abstract journals. International in scope and enriched by approximately 100,000 records annually which are not included in printed journals.

HSDB (Hazardous Substances Data Bank) (US National Library of Medicine). Non-bibliographic, factual 'databank' which contains peer-reviewed information on toxicology of potentially hazardous chemicals. Includes references to sources from which information was taken.

MEDLINE (US National Library of Medicine). 1966–. Covers clinical and basic sciences literature; international in scope. Includes abstracts for over 40% of records added since 1975. Printed counterparts: *Index Medicus, Index to Dental Literature*, and *International Nursing Index.*

PDQ Databases (US National Cancer Institute and US National Library of Medicine). Current information. Information on cancer prognoses and treatments; directory information for physicians and organizations which specialize in caring for cancer patients; and descriptions of active treatment protocols, including details of drug dosages and regimens.

RTECS (US National Library of Medicine). Current information. Based on most current printed edition of *Registry of Toxic Effects of Chemical Substances. RTECS* record for a chemical includes toxicological/carcinogenicity review information, threshold limit values, air standards, and National Institute for Occupational Safety and Health (NIOSH) Criteria Document availability, plus references to sources from which information was taken.

SCISEARCH (Institute for Scientific Information). 1974–. Like its printed counterpart, *Science Citation Index*, provides not only author and title word access to references, but also citation indexing. This permits retrieval of newly published articles through their authors' citing of earlier documents. Includes extensive coverage of medicine and pharmaceutical sciences. Reduced online rates for subscribers to the printed index.

Smoking and Health Database (US Department of Health and Human Services, Office of Smoking and Health). 1960–. Corresponds to printed *Smoking and Health Bulletin*. References and abstracts to the literature on the effects of smoking on health.

TOXLINE (US National Library of Medicine). 1965–. Bibliographic database with records drawn from *MEDLINE, BIOSIS*, and unique databases such as *Environmental Mutagenicty Information Center File*. Toxicology, poisoning, adverse effects, environmental and industrial toxicology, etc.

Biotechnology

Biotechnology (Derwent Publications, Inc.). 1982–. Online counterpart to printed *Biotechnology Abstracts*. Covers journal articles, conference proceedings, international patent literature on all technical aspects of biotechnology.

Life Sciences Collection (Cambridge Scientific Abstracts). 1978–. Contains entries, with abstracts, corresponding to those in a series of 17 abstracting journals in animal behaviour, ecology, genetics, immunology, toxicology, and other areas.

Chemistry

CA Search (Chemical Abstracts Service). 1967–. Computerized equivalent to *Chemical Abstracts* from *8th Collective Index* (1967–1971) to present. As mounted on BRS and Dialog online service, contains bibliographic citations only; on STN International, abstracts are included. Use is facilitated by a group of chemical name dictionaries (*CHEMNAME, CHEMSIS* and *CHEMZERO*) which contain CAS Registry Numbers, molecular formulae, synonymous names, etc.

RINGDOC Databases (Derwent Publications, Ltd.). 1964–. Include information on the chemical properties, preparation and toxicology of drugs. Designed for the pharmaceutical industry.

Pharmacy Industry, Regulations and Patents

Chemical Industry Notes (Chemical Abstract Service/American Chemical Society). 1974–. Indexes business-oriented periodicals and newspapers which cover worldwide chemical industry.

Claims Databases (IFI/Plenum Data Corporation). Several 1950; others, current. References to US patents from 1950 to date; patents which cite other patents; compounds listed in patents; and patents which have been reassigned, re-examined, or which have expired.

DIOGENES (FOI Services, Inc., and Washington Business Information, Inc.). 1976–. Helps locate US Food and Drug Administration regulatory information. Contains news stories, unpublished documents, and complete text of some FDA publications. Listings of approved products, experience reports for medical devices, and documentation of regulatory actions and recalls.

Federal Register Abstracts (National Standards Association). 1977–. Like its printed counterpart, *Federal Register*, covers actions of US federal regulatory agencies.

Pharmaceutical News Index (UMI/Data Courier). 1974–. References to articles about drugs, devices, and cosmetics from a selected group of publications from US, UK, and Japan. Covers legislation, regulations, and court actions; corporate sales, mergers, etc.; research grant applications; press releases, and other news items.

PHARMPAT (Chemical Abstracts Service). 1987–. Information on patents in pharmaceuticals, biotechnology, medicinal chemistry, and related fields. Patents from 27 national patent offices and 2 inter-

national organizations. Abstracts include chemical structure diagrams.
SUPERTECH (EIC/Intelligence, Inc.). 1973–. Resource for patent information in biotechnology. Corresponds to five monthly abstract journals, e.g., *Telegen*, and *Artificial Intelligence Abstracts*.
World Patents Index (Derwent, Inc.). 1963–. Includes coverage of pharmaceutical patents from 1963 to date. Abstracts, informative titles, International Patent Classification Codes, Derwent subject codes, information on equivalent patents.

Databases Used in Japan

Japanese Patent Abstracts in English (JAPIO)

Producer:	Japan Patent Information Technologies Corporation
Vendor:	ORBIT Information Technologies
Coverage:	1977 to the present
Updates:	monthly

JICST File on Science, Technology and Medicine in Japan

Producer:	The Japan Information Center of Science and Technology
Vendor:	The Japan Information Center of Science and Technology (JICST)
Coverage:	1985 to the present
Updates:	monthly

Foreign Secondary Sources Most Frequently Used in Japan

Title	Database name	Producer	Vendor
Index Medicus	*MEDLINE*	NLM	JOIS
			DIALOG
			BRS
Chemical Abstracts	*CAS-ONLINE*	CAS	STN
	CA-Search	CAS	JOIS
			DIALOG
			BRS
Ringdoc	*Ringdoc, Ringdoc UDB*	Derwent	ORBIT
			DIALOG
Biological Abstracts	*BIOSIS*	Bioscience	JOIS
Biological Abstracts/ RRM			DIALOG
			BRS
International Pharmaceutical Abstracts	*IPA*	Am. Hosp. Assoc.	DIALOG
			BRS
Pharmaprojects	*Pharmaprojects Online*	V & O	Data-Star
Excerpta Medica	*EMBASE*	Elsevier	JOIS
			DIALOG
			BRS

Science Citation Index	SCISEARCH	ISI	DIALOG BRS
	CANCERLIT	NLM	JOIS BRS
	TOXLINE	NLM	JOIS
de Haen	de Haen	de Haen Drug Inc.	DIALOG JIP
Journal of Synthetic Methods	CRDS	Derwent	ORBIT
Biotechnology Abstracts	Biotechnology Abstracts	Derwent	ORBIT
Farmdoc	WPI	Derwent	DIALOG ORBIT
CLAIMS	CLAIMS	IFI/Plenum	DIALOG
Pharmaceutical News Index	PNI	Data Courier	DIALOG
CAB Abstracts	CAB	CAB International	JOIS DIALOG BRS
Predicast	PTS	Predicast	DIALOG BRS Data-Star
Dissertation Abstracts	Dissertation Abstracts	UMI	DIALOG BRS
Current Contents	CCON	ISI	BRS

Japanese Secondary Sources

Printed title	Online file name	Producer	Online vendor
Igaku Chuo Zasshi	Ichushi Title Guide (English title: Japana Centra Revuo Medicina Title Guide)	Igaku Chuo Zasshi Kankokai	Maruzen, Kinokuniya
	JMEDICINE (Full title: JICST File on Medical Science)	JICST	JOIS
Kuguku-Gijutsu Bunken Sokuho	JICST Rikogaku Bunken (English title: JICST File on Science and Technology)	JICST	JOIS
	JICST Kokyo Shiryo (English title: JICST File on	JICST	JOIS

	Government Reports)		
	JCLEARING	JICST	JOIS
Nihon Iyaku Bunken Shorokushu	*JAPICDOC*	JAPIC	JIP, TIS
Asu no Shinyaku	*Asu no Shinyaku* (English Title: *Pharmacast*)	Technomic	Infostream (JIP)
Nihon Tokkyo	*PATOLIS*	JAPIO	JAPIO
Japan Pharmaceutical Market	*JPM-DB*	IMS Japan	TIS
	MDI-DB	IMS Japan	TIS

Databases used in Argentina

LILACS consolidates Latin American literature on health services and is operated by the Latin American and Caribbean Center of Information in Health Services (Rua Botucatu 862, 04023 São Paulo, Brazil).

CIMF, Centro de Información de Medicamentos consolidates information on 450 active substances, mostly the essential medicines, and is operated by the Colegio de Farmacéuticos de la Provincia de Buenos Aires (Calle 5, No 966, (1900) La Plata, Argentina).

DAME, Banco de Datos de Medicamentos consolidates information on 3000 medicines in the Argentine market and 1500 active substances, and is operated by EDyABE (Hidalgo 462, P.B. 'A', (1405) Buenos Aires, Argentina).

RECURSOS CYT consolidates information on scientific research projects and is operated by the Secretariat of Science and Technology (Rivadavia 1906, 2° piso, (1033) Buenos Aires, Argentina).

INCAR consolidates national Latin American and world information on cardiology, and is operated by Sociedad Iberoamericana de Investigación Científica (Solís 453, (1078) Buenos Aires, Argentina).

INEC consolidates information on Chagas disease and is operated by Sociedad Iberoamericana de Investigación Científica (Solís 453, (1078) Buenos Aires, Argentina).

INPED consolidates information on pediatrics and is operated by Sociedad Iberoamericana de Pediatría (Solís 453, (1078) Buenos Aires, Argentina).

INOG consolidates information on obstetrics and is operated by Sociedad Iberoamericana de Obstetricia y Ginecología (Solís 453, (1078) Buenos Aires, Argentina).

BRED consolidates information on biomedicine and is operated by the Universidad Nacional de Rosario (Moreno 750, (2000), Rosario, Santa Fe, Argentina).

MAMA consolidates information on breast pathology and is operated by the Universidad Nacional de Rosario (Moreno 750, (2000) Rosario, Santa Fe, Argentina).

Centres of Excellence
Australia

Australian Centres of Research
Departments of Pharmacology in Australian Universities

Institute	Main topics
University of Melbourne Department of Pharmacology	Platelet-activating factor, guinea pig isolated tracheal chains, histamine response. B-adrenoceptors in human mammary artery and coronary artery. Inotropic agents and reflex activation. Calcium antagonists – diltiazem, and reflex vagal control
University of Melbourne Departments of Medicine, Pathology and Physiology	Calcium antagonists, binding properties anipamil. Smooth muscle, hydralazine inhibits inositol phosphate production. ACE inhibitor, lisinopril
Monash University Department of Pharmacology	Postjunctional sensitivity to purines and blood pressure. Drug abuse, tolerance and dependence – receptor is dominant.

Departments and Schools of Pharmacy

University of Sydney Department of Pharmacology	Theophylline inhibit effects of platelet-activating factor on human lung. Bronchodilation and bronchoconstriction – prostaglandin F. Medication compliance with aerosol steroids in children with moderately severe asthma.
University of New South Wales Department of Pharmacology	N-acetylcysteine protects lungs of mice to bleomycin plus hyperbaric oxygen. Allopurinol dose and inhibition of xanthine oxidase in man. New cardiac stimulatory polypeptides from sea anemone Actinia tenebrosa.
University of Adelaide Department of Pharmacology	Study cimetidine inhibition of tubular secretion of amiloride. Comparison of low dose enteric coated, soluble and intravenous aspirin in man.
Flinders University Department of Clinical Pharmacology	Ibuprofen enantiomers and rat hepatocyte intermediary metabolism. Oxotremorine-induced changes in temperature. Localization of NADPH-cytochrome P450 reductase.

University of Queensland
Department of Physiology
and Pharmacology
University of Western
Australia
Department of Pharmacology

Diltiazem effect age-dependent and
spasmogen-dependent in rat.
PAF-antagonist drugs specificity.
Verapamil interacts with nicotinic
receptor.
Pharmacology of inflammation and
intra-articular steroids.

Institution	Main fields of research	Major topics.
University of Sydney. Department of Pharmacy	Membrane transport of drugs	Transnasal drug delivery system, microparticulate drug carriers.
	Drug-macromolecule interactions	Trimethoprim and methotrexate binding to Lactobacillus casei dihydrofolate reductase.
	Drug metabolism	Induction by DDT, gomphoside, hydroxychloroquine.
	Photochemical activation of drugs	Misonidazole, metronidazole and azathioprine.
	Development of new drugs and diagnostic agents	Methotrexate, digitalis and related glycosides, radiopharmaceuticals.
	Drug delivery and dose form design	Aerosols, behaviour of aggregates, regional aerosol deposition.
	Pharmacokinetics	Antirheumatic, anticoagulant, immunosuppressive agents; chloroquine and prenalterol.
College of Pharmacy in Victoria	Computer-assisted drug design group	CNS drug action, a structure-activity relationship, receptor mapping and drug-receptor interaction.
	Synthesis and characterization group	Inhibitors of pyridoxal phosphate-dependent enzymes, phosphopeptides, peptide analogues and nucleoside analogues.
	Nuclear magnetic resonance spectroscopy	Molecular conformations of drugs, peptides and proteins;

		and analysis of drug-receptor interactions.
	Drug formulation and delivery	Implanted solid delivery systems, dithranol and topical depigmenting agents.
	Drug assay and pharmacokinetics	Assays for reserpine, verapamil and prazosin; pharmacokinetics studies of propofol, alfentanil; and interaction study for changes in plasma protein binding.
	Pharmaceutical microbiology	Rapid identification of bacteria, fungi and yeasts contaminating pharmaceutical products.
	Drug action and toxicity	Drug screening, drugs in addiction, receptor classification and drug-receptor kinetics.
University of Queensland. Department of Pharmacy	Pharmacokinetics study	Amiloride and hydrochlorothiazide in young and elderly subjects, gallamine in dogs, cefaronide, ceftriaxone and cefoperazone in sheep.
	Drug delivery systems	Controlled release and bioadhesive formulation, and delivery through the buccal mucosa.
	Pharmacy services	Diabetic services and computer drug information systems; non-prescription medication and hypertension. Drug formulation particle adhesion and application to drug processing.
	Phytochemical studies	Genus Duboisia.
University of Tasmania. School of Pharmacy	Drug design	Uptake of solutes from their solutions by plastics used in infusion systems.

	Drug metabolism	Use of molecular graphics for elucidating mechanisms of drug metabolism.
	Pharmacokinetics	Morphine kinetics and active metabolic accumulation.
	Delivery systems	Stability of all-in-one total parenteral nutrition admixtures. Stability of various antibiotics in peritoneal dialysis fluids.
	Mechanisms of action	Skin transport mechanism and iontophoresis.
	Epidemiological research	Drugs in road accidents. Treatment response in rheumatoid arthritis patients. Antibiotic prescribing policy in hospitals.
South Australian Institute of Technology. School of Pharmacy	Biopharmaceutics	New dosage forms and drug interactions. Sustained-release pellet-filled theophylline capsules.
	Pharmacokinetics	Morphine and its conjugates, ketoconazole.
	Drug metabolism	Stereoselective drug disposition at molecular level. Cimetidine effect on disposition of ibuprofen enantiomers.
	Toxicity	Metabolism and toxicity of xenobiotics.

Research Institutes

Institution	Main fields of research	Major topics.
Ludwig Institute for Cancer Research	Molecular biology laboratory. Epithelial biology laboratory.	New therapeutic protocols using hemopoietic growth factors.
Tumour Biology Branch	Clinical research programme	Phase I study of recombinant human

		G-CSF in patient receiving melphalan. Immunoassays to measure levels of human GM-CSF in serum. In small cell lung cancer (SCLC), bombesin and two antagonists. Hematoporphyrin derivative of anthracyclines. Pharmacokinetics of unchanged carboplatin in patients with SCLC.
Garvan Institute of Medical Research	Disease of the aged population, osteoporosis, diabetes mellitus, cancer, dementia, and cardiovascular disease	Agents known to modulate the growth of human breast cancer cells. Identify genes involved in the etiology of osteoporosis and diabetes. Development of a low dose insulin infusion method, and insulin nasal spray. Synthesis of analogues of calcitriol. Quantitation of growth hormone binding protein. Structure of vitamin D receptors. Drug effects on oentral dopamine metabolism.
Howard Florey Institute of Experimental Physiology and Medicine	Clinical reproductive endocrinology. Biotechnology. Hypertension. Molecular biology	Radioimmunoassay for measuring relaxin in the serum of pregnant women. Vasopressin and control of secretion of pituitary hormone ACTH. Angiotension II. Atrial natriuretic factor. Ethanol intake inhibited by naloxone in rats.

		Angiotensin binding sites. Hormones being studied include corticotrophin releasing factor, adrenocorticotrophic hormone, vasopressin, somatostatin, and steroids.
John Curtin School of Medical Research	Biochemistry. Physical biochemistry. Experimental pathology. Medical molecular biology. Experimental pathology. Medicine and clinical science. Microbiology. Immunology. Human biology. Pharmacology. Physiology. Experimental neurology unit. Transplantation biology unit	Chemical properties of proteins and enzymes. Experimental allergic encephalomyelitis suppressed in rats with desferrioxamine and initiated Phase-I clinical trial. Amino acids and peptides as central transmitters and their role in pain and analgesia. Study of surface antigens of influenza virus, hemagglutin and neuraminidase.
Baker Medical Research Institute	Pharmacology group. Cardiovascular	Endothelium derived relaxing factor, biological half-life. Cardiovascular responses to clonidine in rats influenced by the adrenal cortex and medulla. Orally active inotropic drugs.
Kolling Institute of Medical Research	Cell-mediated immunity. Autoimmunity. Cancer research. Allergic diseases. Immunosuppressives. Tropical diseases	Anaphylactic reactions in anaesthesia. Immunoassays for detecting sensitivities for penicillin, cephalosporins and morphine developed. Production of standardized allergen preparations.

		Cyclosporin A delayed-type hypersensitivity and production of lymphokines. Rheumatoid arthritis, auranofin and normalized interleukin-l production.
Institute of Medical and Veterinary Science	Tissue pathology. Neuropathology. Clinical microbiology. Human immunology. Clinical chemistry. Virology. Medical virology	Effects of xylitol intravenously and metabolic acidosis. Detection of bismuth poisoning in colostomy patients. Fever in abbattoir workers. Aluminium in repeated hemodialysis and osteomalacia. Hepatitis B virus infection.
The Walter and Eliza Hall Institute of Medical Research	Genetic, molecular and cellular basis of immunity. Cancers of the white blood cells. Autoimmune diseases. New vaccines for major tropical diseases	Plasmodium falciparum blood-stage antigens characterized. Leishmaniasis and assessment of vaccine efficacy. Schistosomiasis enzymes and membrane proteins studied.
Peter MacCallum Cancer Institute	Radiotherapy. Clinical immunology and immunogenetics. Hematology. Biological and clinical research. Experimental chemotherapy.	
Queensland Institute of Medical Research	Viral and tropical diseases. Malaria research. Cancer research. Molecular biology. Biochemistry. Epidemiology studies. Entomology. Public health	Malignant melanoma surface merozoite and rhoptry antigens, parasitophorus vacuole antigens. Purification antigen of Giardia intestinalis. Drug therapy of filariasis. Epstein-Barr virus. Ross River virus.

		Monoclonal antibodies to Murray Valley encephalitis virus. Chlamydia vaccine
Menzies School of Health Research	Public health. Epidemiology. Infectious diseases	
National Biological Standards Laboratory	Monitoring. Infectious diseases	Japanese encephalitis virus vaccine. Assay procedure oleandomycin phosphate, neomycin sulphate, oxytetracycline hydrochloride and prednisolone
The Children's Medical Research Foundation	Gene expression. Oncogenes. Central nervous system disorders. Congenital malformations	Viral infections and drugs. Alcohol and other addictive drugs.
Reproductive Medicine Unit, Sir Charles Gardiner Hospital		Papaverine, syntolamine L-prostadil. Delivery of Lutenising Hormone Release Factor and agonists.
Army Malaria Research Unit	Drugs and pharmacology research in prophylaxis of malaria	Chlorproguanil pharmacokinetics in man.

Address Details for Australian Research Centres

Australian Capital Territory

Australian Institute of Health, GPO Box 570, Canberra, ACT 2601, Telephone 062–435000.

John Curtin School of Medical Research, The Australian National University, GPO Box 4, Canberra City, ACT 2601, Telephone 062–492550.

National Biological Standards Laboratories, Currong Street, Braddon, ACT 2601, Telephone 062-496187.

New South Wales

Australian Cancer Society, 500 George Street, Sydney, NSW 2000, Telephone 02–2671944.

Army Malaria Research Unit, Milpo, Ingleburn, NSW 2174, Telephone 02–6184111.

Children's Medical Research Foundation, c/o Royal Alexandra Hospital for Children, PO Box 61, Camperdown, NSW 2050, Telephone 02–5198761.

Department of Physiology and Pharmacology, University of New South Wales, NSW 2033, Telephone 02–6972222.

Department of Pharmacology, University of Newcastle, NSW 2308, Telephone 049–680401.

Department of Pharmacology, Department of Pharmacy, University of Sydney, NSW 2006, Telephone 02–6922222.

Foundation 41, Crown Street, Sydney, NSW 2010, Telephone 02–3315274.

Garvan Institute of Medical Research, St Vincents Hospital, Darlinghurst, NSW 2010, Telephone 02–336418

Heart Research Institute, Royal Prince Alfred Hospital, NSW 2050, Telephone 02–5166111.

Kanematsu Laboratories, Royal Prince Alfred Hospital, NSW 2050, Telephone 02–5166111.

Kolling Institute of Medical Research, Royal North Shore Hospital, Pacific Highway, St Leonards, NSW 2065, Telephone 02–4387111.

National Clinical Trials Centre, Dr J Simes, Director, School of Public Health, University of Sydney, NSW 2006, Telephone 02–6609222.

Northern Territory

Menzies School of Health Research, Building 4, Royal Darwin Hospital, Rocklands Drive, Casuarina, NT 5792, Telephone 089–208612.

Queensland

Anton Breinl Centre for Tropical Medicine, The Tropical Health Surveillance Unit, James Cook University, Townsville, QLD 4811, Telephone 077–212281.

Department of Physiology and Pharmacology School of Pharmacy, University of Queensland, QLD 4072, Telephone 07–3773191.

Queensland Institute of Medical Research, Bramston Terrace, Herston, QLD 4006, Telephone 07–2536222.

South Australia

Department of Clinical Pharmacology, Flinders University, Sturt Road, Bedford Park, SA 5042, Telephone 08–2753911.

Department of Clinical and Experimental Pharmacology, University of Adelaide, North Terrace, Adelaide, SA 5000, Telephone 08–2285333.

Institute of Medical and Veterinary Science, Frome Road, Adelaide, SA 5000, Telephone 08–2287222.

School of Pharmacy, South Australian Institute of Technology, North Terrace, Adelaide, SA 5000, Telephone 08 2362211.

Tasmania

School of Pharmacy, University of Tasmania, Churchill Avenue, Sandy Bay, TAS 7005, Telephone 002–202190.

Victoria

Department of Pharmacology, University of Melbourne, VIC 3052, Telephone 03–3877222.

Baker Medical Research Institute, Commercial Road, Prahran, VIC 3181, Telephone 03–5224333.

Department of Pharmacology, Monash University, VIC 3168.

Howard Florey Institute of Experimental Physiology and Medicine, University of Melbourne, Gratton Street, Parkville, VIC 3052, Telephone 03–3447286.

Ludwig Institute for Cancer Research, PO Royal Melbourne Hospital, Parkville, VIC 3050.

Mental Health Research Institute, Park Street, Private Bag No. 3, Parkville, VIC 3052,Telephone 03–3892260.

Murdoch Institute for Research into Birth Defects, Royal Children's Hospital, Flemington Road, Parkville, VIC 3052, Telephone 03–3455050.

National Research Institute of Gerontology and Geriatric Medicine, Poplar Road, Parkville, VIC 3052, Telephone 03–3872211.

Peter MacCallum Cancer Institute, 481 Lonsdale Street, Melbourne, VIC 3000, Telephone 03–6021333.

Royal Children's Hospital Medical Research Foundation, Flemington Road, Parkville, VIC 3052, Telephone 03–3455044.

St Vincent's Institute of Medical Research, 41 Victoria Parade, Fitzroy, VIC 3065, Telephone 03–4182373.

Victorian College of Pharmacy Ltd., 381 Royal Parade, Parkville, VIC 3052, Telephone 03–3877222.

Walter and Eliza Hall Institute of Medical Research, Royal Parade, Parkville, VIC 3052, Telephone 03–3452555 (within Australia), Telephone 61–3–3452555 (international).

Western Australia

Department of Pharmacology, University of Western Australia, Stirling Highway, Nedlands, WA 6009, Telephone 09–3802452.

School of Pharmacy, Curtin University of Technology, Kent Street, Bentley, WA 6102, Telephone 09–3507369.

Reproductive Medicine Unit, Sir Charles Gairdner Hospital, Verdun Street, Nedlands, WA 6009, Telephone 09–3893333.

Australian Information Bureaux

Australian Institute of Health, GPO Box 570, Canberra, ACT 2601, Telephone 062–435000.

Australian Library and Information Association, PO Box E441, Queen Victoria Terrace, ACT 2600, Telephone 062–851877, Fax 062–822249.

CSIRO, Information Resources Unit, 314 Albert Street, East Melbourne, VIC 3002, Telephone 03–4187253, Fax 03–4190459.

Health Information Services, PO Box 1608, Hornsby Northgate, NSW 2077, Telephone 02–4763955.

Intercontinental Medical Statistics (Australasia) Pty Ltd., 12 Holterman Street, Crows Nest, NSW 2065, Telephone 02–4396255, Fax 02–4381024.

National Health and Medical Research Council, PO Box 100, Woden, ACT 2606, Telephone 062–898826.

National Library of Australia, Canberra, ACT 2600, Telephone 062–621111.

Scansearch, GPO Box 973, Adelaide, SA 5001, Telephone 08–3722735 (within Australia), Telephone 61–8–3722735 (international).

Australian Pharmaceutical Organizations

Trade associations

Association of Medical Directors of the Australian Pharmaceutical Industry, c/o Roche Products, 4 Inman Road, Dee Why, NSW 2099, Phone: 02–9820222.

Australian Chemical Industry Council, GPO Box 1610m, Melbourne, VIC 3001, Phone: 03–6996299, Telex: 37754.

Australian Pharmaceutical Manufacturers Association Inc., Level 2, 77 Berry Street, North Sydney, NSW 2060, Phone: 02–9222699, Telex: APMAS 26953.

National Council of Chemical and Pharmaceutical Industries, 12th Floor, 65 Berry Street, North Sydney, NSW 2060, Phone: 02–9225111, Telex: 177214.

National Pharmaceutical Distributors Association, 102 Briens Road, Northmead, NSW 2152, Phone: 02–6836193, Fax: 02–6830639.

The Agricultural and Veterinary Chemicals Association of Australia Ltd., Private Mail Bag 938, North Sydney, NSW 2059, Phone: 02–9637690, Telex: 177214.

The Proprietary Association of Australia Inc., Private Mail Bag 938, North Sydney, NSW 2060, Phone: 02–9225111, Telex: 177214.

Professional associations

Association of Regulatory and Clinical Scientists, PO Box 87, North Sydney, NSW 2060.

Australian Society of Clinical and Experimental Pharmacologists, Department of Clinical Pharmacology, Royal North Shore Hospital, Pacific Highway, St Leonards, NSW 2065, Phone: 02–4387632.

Australian Pharmaceutical Science Association, c/o Victorian College of Pharmacy Ltd., 381 Royal Parade, Parkville, VIC 3052, Phone 03–3877222.

Australian Society for Medical Research, 145 Macquarie Street, Sydney, NSW 2000, Phone: 02–274461, Fax: 02–2313120.

Australian Society for Microbiology, Clunies Ross House, 191 Royal Parade, Parkville, VIC 3052, Phone: 03–3481441.

Pharmacy Guild of Australia, PO Box 36, Deakin, ACT 2600, Phone: 062–810911, Telex: PGOAC 62582.

Pharmacy Practice Foundation, c/o University of Sydney, Department of Pharmacy, Sydney, NSW 2006, Phone: 02–6922668.

Pharmaceutical Society of Australia, PO Box 21, Curtin, ACT 2605, Phone: 062–811366.

Society of Hospital Pharmacists of Australia, 4th Floor, 450 St Kilda Road, Melbourne, VIC 3004, Phone: 03–2673581.

South America

Argentinian Libraries

Association of Argentine Biomedical Libraries (ABBA). Participating libraries in 1984

Hospital Aleman, Asociacion Medica, Pueyrredón 1640, 1118 Buenos Aires, Speciality: German medicine.

Hospital Britanico de Buenos Aires, Perdriel 74, 1280 Buenos Aires, Speciality: Medicine.

Hospital Frances, Biblioteca Medica Francesa, La Rioja 951, 1221 Buenos Aires, Speciality: French medicine.

Hospital Italiano, Gascón 450, 2nd piso, 1181 Buenos Aires, Speciality: Medicine.

Hospital Militar Central, 'Cir. My. Dr Cosme Argerich', Av. Luis María Campos 726, 1426 Buenos Aires, Speciality: Medicine.

Hospital Nacional 'Prof. Alejandro Posadas', Martínez de Hoz y Marconi, 1704 Ramos Mejía, Speciality: Medicine.

Academia Nacional de Medicina, Andrés Pacheco de Melo 3081, 1st piso, 1425 Buenos Aires, Speciality: Medicine and related sciences.

Agremiacion Medica Platense, Calle 6 No 1137, 1900 La Plata, Speciality: Hospital administration, sport.

Asociacion Argentina de Dermatologia, General Urquiza 609, 1221 Buenos Aires, Speciality: Dermatology.

Asociation Argentine del Cancer (ASARCA), Tucumán 731, 3° F, 1049 Buenos Aires, Speciality: Cancer.

Asociacion de Bioquimicos de la Ciudad de Cordoba, Paso de los Andes 35, 5000 Córdoba, Speciality: Biochemistry.

Asociacion Ondontologica Argentina, Junín 959, 1113 Buenos Aires, Speciality: Dentistry.

Centro Interdisciplinario de Investigacion en Psicologia, Matematica y Experimental (C.I.I.P.M.E.), Cangallo 2158, 1040 Buenos Aires, Speciality: Experimental psychology, psychometry and statistics.

Centro Medico de Mar del Plata, San Luis 1978, 3°, 7600 Mar del Plata, Speciality: Medicine.

Centro Nacional de Reeducacion Social (CE.NA.RE.SO.), Combate de los Pozos 2133, 1245 Buenos Aires, Speciality: Pharmacodependence.

Colegio de Medicos (1° Circunscripcion), 9 de Julio 2464, 3000 Santa Fe, Speciality: General medicine.

Colegio Medico de Tucuman, Catamarca 15, 1st piso, 4000 San Miguel de Tucumán, Speciality: Medicine and related sciences.

Comision Nacional de Energia Atomica, Av. del Libertador 8250, 1429 Buenos Aires, Speciality: Nuclear energy and related topics.

Confederacion Medica de la Republica Argentina (COMRA), Hipólito Yrigoyen 2038, 1089 Buenos Aires, Speciality: Medical administration.

Departamento de Zoonosis Rurales, España 770, 7300 Azul, Speciality: Veterinary medicine and public health.

Fundacion Genetica Humana, Salta 661/3, 1074 Buenos Aires, Speciality: Human genetics.

Instituto de Biologia y Medicina Experimental, Obligado 2490, 1428 Buenos Aires, Speciality: Physiology, biology, endocrinology, experimental medicine.

Instituto de Investigaciones Bioquimicas 'Fundacion Campomar', Antonio Machado 151, 1405 Buenos Aires, Speciality: Biochemistry, cellular biology, plants, soil.

Instituto de Neurobiologia, Serrano 665, 1414 Buenos Aires, Speciality:

Neuroendocrinology, biology, animal virology, pharmacological principles, botanical ecophysiology, botany.

Sociedad Argentina de Cardiologia, Azcuénaga 980, 1115 Buenos Aires, Speciality: Cardiology.

Bago, Bernardo de Irigoyen 248, 1072 Buenos Aires, Speciality: Medicine.

Ciba-Geigy, Arias 1851, 1429 Buenos Aires, Speciality: Medicine.

Elea S.A.C.I.F. y A, Saladillo 2450/68, 1440 Buenos Aires, Speciality: Therapeutic endocrinology.

Farmerit, Juana Azurduy 1534, 1429 Buenos Aires, Speciality: Medicine, chemistry, pharmacology.

Gerardo Ramon y Cia, Intendente Amaro Avalos 4208, 1605 Munro, Ruta Panamericana y Paraná, Speciality: Medicine, pharmacology, chemistry.

Lazar y Cia. S.A., Av. Velez Sarsfield 5855, 1605 Munro, Speciality: Medicine, pharmacy.

Pfizer S.A.C.I., Miñones 2177, 1428 Buenos Aires, Speciality: Medicine.

Quimica Hoechst, 25 de Mayo 460, 3°, 1002 Buenos Aires, Speciality: Medical clinic and pharmacology.

Roux-Ocefa, Medina 138, 1407 Buenos Aires, Speciality: Pharmacy, pharmaceutical technology, chemistry, medicine.

Schering Argentina, Monroe 1378, 1428 Buenos Aires, Speciality: Gynaecology.

Facultad de Ciencias Exactas y Naturales, Universidad de Buenos Aires, Pabellón 2, Ciudad Universitaria, 1428 Nuñez.

Facultad de Farmacia y Bioquimica, Universidad de Buenos Aires, Junín 956 – 6°, 1112 Buenos Aires, Speciality: Pharmacy and biochemistry.

Facultad de Medicina, Biblioteca Central, Universidad de Buenos Aires, Paraguay 2155, 4°, 1121 Buenos Aires, Speciality: Medicine and related sciences.

Instituto de Oncologia 'Angel H. Roffo', Universidad de Buenos Aires, Av. San Martín 5481, 1417 Buenos Aires, Speciality: Cancer.

Facultad de Ciencias Medicas, Universidad Nacional de la Plata, Calle 60 y 120, 1900 La Plata, Speciality: Medicine.

Centro de Informacion y Documentacion en Ciencias de la Salud del Nordeste Argentino (CIDCSANA), Universidad Nacional del Nordeste, Hospital Julio C. Perrando, Av. 9 de Julio 1099, 3500 Resistencia, Chaco, Speciality: Health sciences, Chagas, infectious diseases.

Facultad de Ciencias Quimicas, Universidad Nacional de Cordoba, Pabellón Argentina, Estafeta 32, Ciudad Universitaria, 5000 Córdoba, Speciality: Chemical sciences and related topics.

Facultad de Odontologia, Universidad de Buenos Aires, Marcelo T. de Alvear 2142, 1122 Buenos Aires, Speciality: Dentistry.

Ministerio de Bienestar Social, Biblioteca Central, Av. Italia esq. Independencia, 4600 San Salvador de Jujuy, Speciality: Medical sciences.

Sociedad Argentina de Reumatologia, Austria 2469, 1° 'B', 1425 Buenos Aires, Speciality: Rheumatology.

Instituto de Perfeccionamiento Medico-Quirurgico 'Prof. Dr. Jose M. Jorge, IV Catedra de Cirugia (Hospital de Clinicas), Córdoba 2351, 7°, 1120 Buenos Aires, Speciality: Surgery, gastroenterology, bacteriology, immunology, radiology.

Facultad de Odontologia, Universidad Nacional de la Plata, Calle 51 e/l y 115, 1900 La Plata, Speciality: Dentistry.

Hospital Privado, Centro Medico de Córdoba, Biblioteca 'Pedro Martinez Est', Avda. Vélez Sarsfield 2350, 5000 Córdoba, Speciality: Medical sciences.

Asociacion Medica Argentina, Santa Fe 1171, 3°, 1059 Buenos Aires, Speciality: Medicine.

Servicio de Informatica Neuroendocrinologica, Sineu, Serrano 665, 1°, 1414 Buenos Aires, Speciality: Neuroendocrinology.

Ministerio de Salud de la Provincia de Buenos Aires, Biblioteca Central, Calle 51, No 1120, 1900 La Plata, Speciality: Public health, medical attention, social welfare.

Carrera de Ingenieria en Alimentos de la Universidad de Buenos Aires, Rutas 5 y 7, 6700 Luján, Speciality: Science and technology in food.

Instituto 'Sidus S.A.', Avda. Calchaquí 137, 1876 Bernal, Speciality: Biological sciences.

Asociacion Medica de Avellaneda, Laprida 193, 1°, 1870 Avellaneda, Buenos Aires.

Hospital General de Niños 'Ricardo Gutierrez', Gallo 1330, 1425 Buenos Aires, Speciality: Pediatrics.

WHO PAHO BIREME – Latin American and Caribbean Centres of Information in Health Sciences. List of Libraries participating as the national co-ordination centres

Argentina, Biblioteca, Academia Nacional de Medicina, Andre Pacheco de Melo 3081 – 1o, piso, 1425 Buenos Aires, Tel 83–24–15.

Bolivia, Biblioteca, Facultad de Ciencias de la Salud, Universidad Mayor de San Andres, Av. Saavedra, 2246 (Mira Flores), La Paz, Tel 36–90–22.

Brazil, Bireme – Centro Latino Americano e do Caribe de Informação em Ciencias, Organização Pan-Americana da Saude, Rua Botucatu 862, Vila Clementino, Caixa Postal 20381, 04023 São Paulo, Tel 549–26–11, Telex 1122143.

Chile, Biblioteca Central, Facultad de Medicina, Universidad de Chile, Avenida Independencia 1027, Casilla 70001, Santiago, Tel 776560 Anexo 5343, Telex 341264 BICMED CX.

Colombia, Centro de Documentacion del Sistema Nacional de Salud, Ministerio de Salud, Calle 16 N.7–39 of 901, Bogota, Tel 282–2531.

Costa Rica, Biblioteca Nacional de Salud y Seguridad Social, Caja Costarricense de Seguridad Social, Apartado de Correos 10105, San Jose, Tel 21–61–93.

Cuba, Sistema Nacional de Informacion de Ciencias Medicas y Centro Nacional de Informacion de Ciencias Medicas, Ministerio de Salud Publica, Calle 23 No. 177, Vedado, Habana.

Ecuador, Banco de Informacion Cientifico-Medica, Facultad de Ciencias

Costa Rica, Biblioteca Nacional de Salud y Seguridad Social, Caja Costarricense de Seguridad Social, Apartado de Correos 10105, San Jose, Tel 21–61–93.

Cuba, Sistema Nacional de Informacion de Ciencias Medicas y Centro Nacional de Informacion de Ciencias Medicas, Ministerio de Salud Publica, Calle 23 No. 177, Vedado, Habana.

Ecuador, Banco de Informacion Cientifico-Medica, Facultad de Ciencias Medicas, Universidad Central del Ecuador, Iquique y Sodiro, Apartado 6120, Quito, Tel 528–690 ext 166 and 148, Telex 22140.

El Salvador, Biblioteca, Facultad de Medicina, Universidad de El Salvador, Ciudad Universitaria Final, 25 Ave. Nte., San Salvador.

Guatemala, Centro de Informacion de la Direccion General de Servicios de Salud, Ministerio de Salud Publica y Asistencia Social, Guatemala, Tel 21801 ext 39.

Honduras, Biblioteca Medica Nacional, Edificio de Ciencias Medicas, Detras del Hospital Escuela, Tegucigalpa, Tel (504) 32–5804.

Mexico, Centro Nacional de Informacion y Documentacion en Salud, Rio Mixcoac 36, 9° piso, Apartado Postal 19–471, 03100 Mexico, Tel 534–4820.

Nicaragua, Centro Nacional de Informacion y Documentacion en Salud, Ministerio de Salud, Complejo Civico 'Camilo Ortega' 22, Managua, Tel 500–4549 ext 2408.

Panama, Centro de Informacion y Documentacion Cientifica y Tecnologica, Vicerectoria de Investigacion y Postgrado, Universidad de Panama, Ciudad Universitaria Octavio Mendes Pereira, Estafeta Universitaria, Panama.

Paraguay, Departamento de Recursos Humanos, Centro Medico Nacional, Ministerio de Salud Publica y Bienestar Social, Calle Brasil y Petirossi, Asuncion, Tel 91699.

Peru, Biblioteca, Facultad de Medicina, Universidad Peruana Cayetano Heredia, Calle Honorio Delgado 430, Apartado 5045, Lima, Tel 815772.

Republica Dominicana, Biblioteca, Instituto Tecnologico de Santo Domingo, Avenida de Los Proceres Gala, Apartado Postal 249–2 Zona 2. Santo Domingo, Tel 566–8187 and 566–6711.

Uruguay, Biblioteca Nacional de Medicina, Centro Nacional de Informacion en Medicina y Ciencias de la Salud, Facultad de Medicina, Universidad de la Republica, General Flores 2125, Montevideo. Tel 290113, Telex 6351.

Venezuela, Sistema Nacional de Documentacion e Informacion Biomedica, Instituto de Medicina Experimental, Universidad Central de Venezuela, Apartado de Correos 50587, Sabana Grande, Caracas, Tel 61–98–11 ext 2332, 662–65–40 (direct), Telex 27495 SINADIB.

Africa

African Centres Relevant to Pharmaceuticals

Algeria
Institut D'Hygiene et de Médecine D'Outre-Mer, Université D'Alger, 2 rue Didouche Mourad, Algiers.

Institut Pasteur D'Algérie, Rue du Dr Laveran, Algiers.
Université D'Alger, Faculté de Médecine et de Pharmacie, 2 rue Edouard Cot, Algiers.
Angola
Repartição Farmaceutica, Serviços de Saude Assitancia, Luanda.
Benin
Institut de Recherches pour les Huiles et Oleagineux (IRHO), Pobe.
Burundi
Laboratoire Medicale, Bujumbura.
Central African Republic
Institut de Recherches pour les Huiles et Oleagineux, BP 53, Grimari.
Institut Pasteur, BP 923, Bangui.
Chad
Ministère de la Santé Publique et des Affairs Sociales, Office de L'Inspection des Pharmacies, Fort Lamy.
Congo
Institut Pasteur, BP 120, Brazzaville.
Egypt
Cairo University, Faculty of Pharmacy, Sharia Kasr El Aini, Cairo.
Drug Research and Control Centre, Al-Ahram Street, Auberge Station.
Drug Research Institute, University of Alexandria, Alexandria.
The Egyptian General Organization for the Pharmaceutical Industry and Medical Equipment, 9 Emad Eldin Street, Cairo.
Institute of Research for Tropical Medicine, 10 Sharia Kasr El Aini, Cairo.
Pharmaceutical Sciences Laboratory, National Research Centre, Al-Tahrir Street, Sokki, Cairo.
Pharmaceutical Society of Egypt, Dar El-Hikma, 42 Sharia Kasr El-Aini, Cairo.
University of Assiut, Faculty of Pharmacy, Assiut.
Vaccine and Serume Laboratory, Agouza, Cairo.
Ghana
Central Clinical Laboratory, PO Box 300, Accra.
Institute of Health and Medical Research, Accra.
Pharmaceutical Society of Ghana, PO Box 2133, Accra.
Pharmacy Board, Ministry of Health, PO Box m-44, Accra.
Kenya
Federation of Kenya Pharmaceutical Manufacturers, PO Box 40411, Nairobi.
Kenya Association of the Pharmaceutical Industry, PO Box 41578, Nairobi.
Institute of Medical Research and Training, c/o Medical Research Laboratory, PO Box 30141, Nairobi.
Medical Research Centre, Narcotics Section, PO Box 9370, Nairobi.
Ministry of Health, PO Box 30016, Nairobi.
East African Industrial Research Organization, PO Box 30650, Nairobi.
Madagascar
Centre Orstom de Tananarive, Office de la Recherche Scientifique Outre-Mer, BP 434, Tananarive.
Institut Pasteur, BP 1274, Tananarive.

Mauritius
Mauritius Pharmaceuticals Co. Ltd., Bell Village, Port Louis.
Malawi
Pharmacy and Poisons Board, Ministry of Health, PO Box 351, Blantyre.
Morocco
Ministère de la Santé Publique, Quartier des Ministères, Rabat.
Conseil National de l'ordre des Pharmaciens, 21 bis, Av. Allal Ben Abdellah Passage Karratchou, Rabat.
Chambre des Pharmaciens du Maroc, 26 rue du Parc, Casablanca.
Chambre Syndicale des Fabricants, Representants et Groissistes au Maroc en Produits Pharmaceutiques, 16 rue Dumont d'Urville, Casablanca.
Institut Pasteur, BP 414, Tangier.
Banitex S A, Tanger, Km 2, Ancienne route de Rabat, Rabat.
Cooper Maroc, Société de Cooperation Pharmaceutique, 41 rue Mohamed Diouri, Casablanca.
Laboratories Nicholas, 65 Boulevard Moulay Ismail, Casablanca.
Laboservice, 7 rue Oqba, Rabat.
Lafrabiol, 12 rue Rachid Reda, Casablanca.
Laprophan, Laboratoires de Produits Pharmaceutiques d'Afrique du Nord, 21 rue de Oudayas, Casablanca.
Lepetit Pharmaghreb, 280 Boulevard Yacoub el Mansour, Casablanca.
Maphar S A, 8 rue Saint-Quentin, Casablanca.
Sopharma, Société Pharmaceutique Marocaine, 21 rue de L'Eglise, Casablanca.
Mozambique
Inspecção do Exercico Farmaceutica, Lourenço Marques.
Instituto de Pesquisas Medicas de Moçambique, Caixo Postal 1572, Avenida Fernandes Tomaz 179, Lourenço Marques.
Niger
Office National de Produits Pharmaceutiques et Chemiques (ONPPC), Ministère de la Santé, Niamey.
Nigeria
Government Chemist, Federal Ministry of Health, PMB 12525, Lagos, Lagos State
Pharmaceutical Society of Nigeria, PO Box 181, 4 Tinubu Square, Lagos.
Nigerian Pharmaceutical and Medical Co., Box 399, 21 Wharf Road, Apapa.
University of Ibadan, Department of Pharmacology, Ibadan, Western State.
University of Ife, Faculty of Pharmacy, Ile-Ife, Western State.
Nigerian Hoechst Ltd., Plot 144, Oba Akran Avenue, PO Box 261, Ikeja.
Senegal
Institut de Médecine Tropicale Appliquée, Université de Dakar, Dakar.
Institut de Recherches pour les Huiles et Oleagineux, Tivaouane M'Bamby.
Institut Pasteur, BP 220, Dakar.

Sudan
Sudan Medical Research Laboratories, PO Box 287, Khartoum.
University of Khartoum, Faculty of Pharmacy, PO Box 1996, Khartoum.
Tanzania
Government Chemical Laboratory, PO Box 164, Dar es Salaam.
Pharmacy and Poisons Board, Ministry of Health and Social Welfare, PO Box 9183, Dar es Salaam.
East African Institute for Medical Research, PO Box 1462, Mwanza.
Tunisia
Pharmacie Centrale de Tunisie, 51 rue Charles Nicole, El Menzah, Tunis.
Institut Pasteur de Tunis, 13 Place Pasteur, Tunis.
Uganda
Makerere University, Department of Pharmacology and Therapeutics, PO Box 7072, Kampala.
United Republic of Cameroon
Institut de Recherches pour les Huiles et Oleagineux, BP 243, Doula.
Institut Pasteur BP 888, Yaounde.
Upper Volta
Institut de Recherches pour les Huiles et Oleagineux, Niangoloko par Bobo, Dioulasso.
Pharmacy Bureau, Ministry of Public Health, Ouagadougou.
Zaire
Departement de la Santé Publique et Affaires Sociales, BP 3088, Kinshasa.
Direction des Dispensaires D'Etat, BP 3333, Kinshasa 1.
Fond Medical de Coordination, BP 169, Kinshasa 1.
Institut de Médecine Tropicale, Reine Astrid, BP 1697, Kinshasa.
Université Louvanium, Faculté de Médecine, BP 834, Kinshasa.
Zambia
The General Manager, General Pharmaceutical Ltd., PO Box 81131, Kabwe.
The General Manager, National Drugs Company Ltd., PO Box 31343, Lusaka.
The General Manager, Medical Stores Ltd., PO Box 30207, Lusaka.
Pneumoconiosis Medical and Research Bureau, Cornwall Avenue, PO Box 205, Kitwe.
University of Zambia, Department of Pharmacology, PO Box 2379, Lusaka.
The Director of Pharmaceutical Services, Ministry of Health, PO Box 30205, Lusaka.
Zimbabwe
Ministry of Health, Box 8204, Causeway, Harare.

Health Statistics

Comparative Expectation of Life, Crude Birth, Crude Death, and Infant Mortality Rates – Selected Countries

Country	Year	Expectation of life (years)				Year	Crude birth rate(a)	Crude death rate(b)	Infant mortality rate(c)
		At birth		At 65 years					
		Males	Females	Males	Females				
Australia	1984	72.6	79.3	14.5	18.7	1984	15.0	7.1	9.2
Austria	1985	70.4	77.4	13.6	17.0	1985p	11.5	11.8	11.0
Canada	1984	73.0	80.1	15.0	19.6	1985p	14.9	7.0	8.1
Denmark	1984	71.8	77.8	14.0	18.0	1985p	10.6	11.4	7.7(d)
England and Wales	1984	71.9	77.9	13.5	17.6	1984	12.8	11.4	10.2
France	1984	71.7	80.1	14.9	19.4	1985p	13.9	10.1	8.0
F.R. Germany	1985	71.6	78.3	13.7	17.7	1985p	9.6	11.5	9.6
Greece	1984	73.8	78.6	15.6	17.7	1985p	11.7	9.3	14.0
Hong Kong	1984	75.1	81.4	17.2	21.0	1984	14.4	4.8	7.6(e)
Hungary	1985	65.1	73.2	11.8	15.1	1985p	12.2	13.9	20.4
Ireland	1983	70.3	76.0	12.6	15.9	1985p	17.5	9.0	10.1(d)
Italy	1981	71.5	78.2	14.1	17.7	1985p	10.1	9.5	10.9
Japan	1985	75.0	81.0	15.8	19.5	1984	12.5	6.2	5.5(e)
Mauritius	1985	64.4	71.7	10.7	14.7	1985p	19.0	6.8	25.1
Netherlands	1984	73.0	79.9	14.1	19.0	1985p	12.1	8.5	7.9
New Zealand	1984	71.2	77.8	13.9	18.0	1985p	15.9	8.4	10.8
Northern Ireland	1985	70.3	76.5	12.7	16.5	1984	17.5	9.9	10.5
Norway	1984	73.0	79.8	14.5	18.8	1985p	12.4	10.6	8.3(d)

Poland	1984	66.8	75.0	12.5	15.9	1985p	18.2	10.3	18.5
Romania	1984	67.1	72.7	12.8	14.7	1984	15.5	10.3	23.4
Scotland	1985	70.1	75.9	12.5	16.3	1984	12.7	12.1	10.3
Singapore	1985	70.2	75.6	12.9	15.9	1985p	16.6	5.2	9.3
Sweden	1984	73.9	80.1	14.9	18.9	1985p	11.8	11.2	6.7
United States	1983	71.0	78.3	14.5	18.8	1985p	15.7	8.7	10.5
Yugoslavia	1982	67.8	73.7	12.9	15.4	1984p	15.9	9.1	28.8

(a) The number of live births per 1,000 mean population.
(b) Number of deaths per 1,000 mean population.
(c) The number of deaths of live born children within one year at birth per 1,000 live births.
(d) 1984 figures.
(e) 1985 figures.
From: Lapsley, H. and Grant, C. (1988), *The Australian Health Care System, 1987* (School of Health Administration, University of New South Wales).

Notifiable Diseases: Number of Cases Notified in Australia, 1980–1984

	1980	1981	1982	1983	1984
Amoebiasis	53	62	33	57	46
Ankylostomiasis	219	136	110	88	75
Anthrax	2	–	–	–	–
Arbovirus infection	18	17	221	33	1,577
Brucellosis	49	36	28	16	15
Cholera	3	2	1	4	–
Diphtheria	1	18	2	1	–
Gonorrhoea	11,487	11,197	12,805	10,646	8,894
Hepatitis A					
(infectious)	1,385	1,453	1,046	991	674
Hepatitis B (serum)	646	500	725	943	1,559
Hydatid disease	41	24	12	10	9
Leprosy	35	38	46	62	28
Leptospirosis	64	95	135	242	227
Malaria	541	408	548	570	640
Ornithosis	17	13	14	19	42
Poliomyelitis	1	–	–	–	–
Salmonella					
infections	2,292	2,269	1,866	2,989	2,092
Shigella infections	545	424	437	567	420
Syphilis	2,902	2,916	3,211	3,556	3,323
Tetanus	9	12	12	10	7
Tuberculosis (all					
forms)	1,554	1,460	1,363	1,218	1,299
Typhoid fever	19	26	15	22	50
Typhus (all forms)	–	–	11	21	8

Source: Australian Bureau of Statistics (1986). *Year Book Australia 1986*.
(ABS, p. 216).

Commonwealth of Australia Budget Outlays by Function

	1986–87 Actual	Estimate	1987–88 Change		Proportion of total outlays(a)
	$ m	$ m	$ m	%	%
1. Defence	7,208.8	7,404.0	195.2	2.7	9.4
2. Education	5,215.7	5,685.2	469.6	9.0	7.2
3. Health	7,499.3	8,212.9	713.6	9.5	10.4
4. Social security and welfare	20,533.5	22,599.0	2,065.5	10.1	28.6
5. Housing and community amenities	1,672.9	1,376.0	– 296.9	–17.7	1.7
6. Culture and recreation	948.6	1,032.5	83.9	8.8	1.3
7. Economic services					
A. Transport and communication	1,669.3	1,661.6	– 7.7	– 0.5	2.1
B. Industry assistance and development	1,180.3	1,224.4	44.1	3.7	1.5
C. Labour and employment	1,030.6	1,040.2	9.6	0.9	1.3
D. Other economic services	266.6	247.5	– 19.1	– 7.2	0.3
Total economic services	4,146.8	4,173.7	26.9	0.6	5.3
8. General public services					
A. Legislative services	474.3	573.8	99.5	21.0	0.7
B. Law, order and public safety	520.0	579.3	59.3	11.4	0.7
C. Foreign affairs and overseas aid	1,301.8	1,391.9	90.1	6.9	1.8
D. General and scientific research	521.3	525.2	3.9	0.8	0.7
E. Administrative services	2,496.1	2,723.0	226.8	9.1	3.4
Total general public service	5,313.6	5,793.2	479.6	9.0	7.3
9. Not allocated to function					
A. Assistance to other governments	14,436.8	14,868.8	432.1	3.0	18.8
B. Public debt interest	7,923.3	7,850.9	– 72.4	– 0.9	9.9
C. Contingency reserve	–	150.0	150.0	n.a	0.2
D. Asset sales	–	– 1,000.0	–1,000.0	n.a	n.a
Total not allocated to function	22,360.1	21,869.8	– 490.3	– 2.2	28.9
Total outlays	74,899.2	78,146.2	3,247.0	4.3	n.a

(a) Excluding 90 asset sales.
From: Lapsley, H. and Grant, C. (1988), *The Australian Health Care System, 1987* (School of Health Administration, University of New South Wales, p. 93).

Recurrent health expenditure: area of expenditure by source of funds in Australia 1984–1985

Area of expenditure	Government			Private			Total	Proportion of total recurrent expenditure
	Common-wealth	State and local	Total government	Health insurance	Other private	Total private		
	($m)			($m)			($ m)	(%)
1. Institutional services								
(a) Hospitals								
Public (i)	1,043(ii)	3,529	4,572	295	187	482	5,054	33.3
Private	166	..	166	533	169	703	869	5.7
Repat. and mental	274	538	812	4	51	55	867	5.7
Total hospitals	1,483	4,067	5,550	832	408	1,240	6,789	44.7
(b) Nursing homes	1,005	79	1,084	..	273	273	1,357	9.0
(c) Other institutional	64	111	175	6	116	122	297	2.0
Total institutional	2,551	4,257	6,808	838	797	1,635	8,443	55.9
2. Non-institutional services								
(a) Medical services	2,306	..	2,306	1	578	379	2,685	17.7
(b) Dental services	21	64	85	232	408	640	725	4.8
(c) Other professional services	63	..	63	109	360	469	532	3.5

								%
(d) Community health services	120	290	410	1	6	7	417	2.8
(e) Pharmaceuticals	637	..	637	25	520	545	1,182	7.8
(f) Aids and appliances	41	..	41	18	203	221	262	1.7
(g) Other non-institutional	9	57	66	7	..	7	72	0.5
Total non-institutional	3,197	411	3,608	392	1,876	2,268	5,876	38.7
3. Health promotion and illness prevention	9	112	121	..	..	..	121	0.8
4. Administration	220	114	334	204	..	204	538	3.5
5. Research	138	13	152	..	43	43	194	1.3
Total recurrent expenditure	6,115	4,907	11,022	1,434	2,715	4,150	15,172	100.0
Capital consumption	32	302	334			169	334	
Capital expenditure	42	391	433				602	
Total health expenditure	6,189	5,600	11,789			4,319	16,108	

.. Less than $ 1 million or no data available. (i) Public hospitals included here are those recognized under the Medicare agreement. (ii) Identified health grants (IHGs) of $ 1,374 million are not included as part of Commonwealth outlays on health, as they are part of general revenue grants from the Commonwealth to the States. IHGs effectively become part of State and Local Government outlays on health. (p. 28, *Payments to or for the States, the N.T. and Local Government Authorities, 1985–1986, 1985–86 Budget Paper No. 7*).

From: Lapsley, H and Grant, C (1988), *The Australian Health Care System, 1987* (School of Health Administration, University of New South Wales, p. 92).

Index